Selected Titles in This Series

272 **Eva Bayer-Fluckiger, David Lewis, and Andrew Ranicki, Ed.** forms and their applications, 2000

271 **J.P.C. Greenlees, Robert R. Bruner, and Nicholas Kuhn, Editors,** Homotopy methods in algebraic topology, 2000

270 **Jan Denef, Leonard Lipschitz, Thanases Pheidas, and Jan Van Geel, Editors,** Hilbert's tenth problem: Relations with arithmetic and algebraic geometry, 2000

269 **Mikhail Lyubich, John W. Milnor, and Yair N. Minsky, Editors,** Laminations and foliations in geometry, topology and dynamics, 2001

268 **Robert Gulliver, Walter Littman, and Roberto Triggiani, Editors,** Differential geometric methods in the control of partial differential equations, 2000

267 **Nicolás Andruskiewitsch, Walter Ricardo Ferrer Santos, and Hans-Jürgen Schneider, Editors,** New trends in Hopf algebra theory, 2000

266 **Caroline Grant Melles and Ruth I. Michler, Editors,** Singularities in algebraic and analytic geometry, 2000

265 **Dominique Arlettaz and Kathryn Hess, Editors,** Une dégustation topologique: Homotopy theory in the Swiss Alps, 2000

264 **Kai Yuen Chan, Alexander A. Mikhalev, Man-Keung Siu, Jie-Tai Yu, and Efim I. Zelmanov, Editors,** Combinatorial and computational algebra, 2000

263 **Yan Guo, Editor,** Nonlinear wave equations, 2000

262 **Paul Igodt, Herbert Abels, Yves Félix, and Fritz Grunewald, Editors,** Crystallographic groups and their generalizations, 2000

261 **Gregory Budzban, Philip Feinsilver, and Arun Mukherjea, Editors,** Probability on algebraic structures, 2000

260 **Salvador Pérez-Esteva and Carlos Villegas-Blas, Editors,** First summer school in analysis and mathematical physics: Quantization, the Segal-Bargmann transform and semiclassical analysis, 2000

259 **D. V. Huynh, S. K. Jain, and S. R. López-Permouth, Editors,** Algebra and its applications, 2000

258 **Karsten Grove, Ib Henning Madsen, and Erik Kjær Pedersen, Editors,** Geometry and topology: Aarhus, 2000

257 **Peter A. Cholak, Steffen Lempp, Manuel Lerman, and Richard A. Shore, Editors,** Computability theory and its applications: Current trends and open problems, 2000

256 **Irwin Kra and Bernard Maskit, Editors,** In the tradition of Ahlfors and Bers: Proceedings of the first Ahlfors-Bers colloquium, 2000

255 **Jerry Bona, Katarzyna Saxton, and Ralph Saxton, Editors,** Nonlinear PDE's, dynamics and continuum physics, 2000

254 **Mourad E. H. Ismail and Dennis W. Stanton, Editors,** q-series from a contemporary perspective, 2000

253 **Charles N. Delzell and James J. Madden, Editors,** Real algebraic geometry and ordered structures, 2000

252 **Nathaniel Dean, Cassandra M. McZeal, and Pamela J. Williams, Editors,** African Americans in Mathematics II, 1999

251 **Eric L. Grinberg, Shiferaw Berhanu, Marvin I. Knopp, Gerardo A. Mendoza, and Eric Todd Quinto, Editors,** Analysis, geometry, number theory: The Mathematics of Leon Ehrenpreis, 2000

250 **Robert H. Gilman, Editor,** Groups, languages and geometry, 1999

249 **Myung-Hwan Kim, John S. Hsia, Yoshiyuki Kitaoka, and Rainer Schulze-Pillot, Editors,** Integral quadratic forms and lattices, 1999

(*Continued in the back of this publication*)

Quadratic Forms and Their Applications

CONTEMPORARY MATHEMATICS

272

Quadratic Forms and Their Applications

QF 99

Proceedings of the Conference on
Quadratic Forms and Their Applications
July 5–9, 1999
University College Dublin

Eva Bayer-Fluckiger, David Lewis,
Andrew Ranicki, Editors

American Mathematical Society
Providence, Rhode Island

This volume contains the proceedings of the conference on "Quadratic Forms and Their Applications", which was held at University College Dublin July 5–9, 1999. Some of the papers are expanded versions of the one-hour survey lectures delivered at the conference. Others are devoted to current research and are based on the half-hour lectures, while others are concerned with the history of quadratic forms.

The conference was supported by the European Community under the auspices of the TMR network FMRX CT-97-0107 "Algebraic K-Theory, Linear Algebraic Groups and Related Structures".

2000 *Mathematics Subject Classification.* Primary 11Exx, 01Axx.

Library of Congress Cataloging-in-Publication Data

Conference on Quadratic Forms and Their Applications (1999 : University College Dublin)
Quadratic forms and their applications : proceedings of the Conference on Quadratic Forms and Their Applications, July 5–9, 1999, University College Dublin / Eva Bayer-Fluckiger, David Lewis, Andrew Ranicki, editors.
p. cm. — (Contemporary mathematics, ISSN 0271-4132 ; 272)
Includes bibliographical references.
ISBN 0-8218-2779-0 (alk. paper)
1. Forms, Quadratic—Congresses. I. Bayer-Fluckiger, Eva, 1951– II. Lewis, David. III. Ranicki, Andrew, 1948– IV. Title. V. Contemporary mathematics (American Mathematical Society) ; v. 272.
QA243 .C753 1999
512′.74—dc21 00-053584

∞ The paper used in this book is acid-free and falls within the guidelines
established to ensure permanence and durability.
Visit the AMS home page at URL: http://www.ams.org/

10 9 8 7 6 5 4 3 2 1 05 04 03 02 01 00

Contents

Preface ix

Conference lectures x

Conference participants xii

Conference photo xiv

Galois cohomology of the classical groups
EVA BAYER-FLUCKIGER 1

Symplectic lattices
ANNE-MARIE BERGÉ 9

Universal quadratic forms and the Fifteen Theorem
J. H. CONWAY 23

On the Conway-Schneeberger Fifteen Theorem
MANJUL BHARGAVA 27

On trace forms and the Burnside ring
MARTIN EPKENHANS 39

Equivariant Brauer groups
A. FRÖHLICH AND C.T.C. WALL 57

Isotropy of quadratic forms and field invariants
DETLEV W. HOFFMANN 73

Quadratic forms with absolutely maximal splitting
OLEG IZHBOLDIN AND ALEXANDER VISHIK 103

2-regularity and reversibility of quadratic mappings
ALEXEY F. IZMAILOV 127

Quadratic forms in knot theory
C. KEARTON 135

Biography of Ernst Witt (1911–1991)
INA KERSTEN 155

Generic splitting towers and generic splitting preparation of quadratic forms
MANFRED KNEBUSCH AND ULF REHMANN 173

Local densities of hermitian forms
MAURICE MISCHLER 201

Notes towards a constructive proof of Hilbert's theorem on ternary quartics
VICTORIA POWERS AND BRUCE REZNICK 209

On the history of the algebraic theory of quadratic forms
WINFRIED SCHARLAU 229

Local fundamental classes derived from higher K-groups: III
VICTOR P. SNAITH 261

Hilbert's theorem on positive ternary quartics
RICHARD G. SWAN 287

Quadratic forms and normal surface singularities
C.T.C. WALL 293

Preface

These are the proceedings of the conference on "Quadratic Forms and Their Applications" which was held at University College Dublin from 5th to 9th July, 1999. The meeting was attended by 82 participants from Europe and elsewhere. There were 13 one-hour lectures surveying various applications of quadratic forms in algebra, number theory, algebraic geometry, topology and information theory. In addition, there were 22 half-hour lectures on more specialized topics.

The papers collected together in these proceedings are of various types. Some are expanded versions of the one-hour survey lectures delivered at the conference. Others are devoted to current research and are based on the half-hour lectures. Yet others are concerned with the history of quadratic forms. All papers were refereed, and we are grateful to the referees for their work.

This volume includes one of the last papers of Oleg Izhboldin, who died unexpectedly on 17th April 2000 at the age of 37. His untimely death is a great loss to mathematics and in particular to quadratic form theory. We shall miss his brilliant and original ideas, his clarity of exposition, and his friendly and good-humoured presence.

The conference was supported by the European Community under the auspices of the TMR network FMRX CT-97-0107 "Algebraic K-Theory, Linear Algebraic Groups and Related Structures". We are grateful to the Mathematics Department of University College Dublin for hosting the conference, and in particular to Thomas Unger for all his work on the TEX and web-related aspects of the conference.

Eva Bayer-Fluckiger, Besançon
David Lewis, Dublin
Andrew Ranicki, Edinburgh

October, 2000

Conference lectures

60 minutes.

A.-M. BERGÉ, *Symplectic lattices.*
J.J. BOUTROS, *Quadratic forms in information theory.*
J.H. CONWAY, *The Fifteen Theorem.*
D. HOFFMANN, *Zeros of quadratic forms.*
C. KEARTON, *Quadratic forms in knot theory.*
M. KRECK, *Manifolds and quadratic forms.*
R. PARIMALA, *Algebras with involution.*
A. PFISTER, *The history of the Milnor conjectures.*
M. ROST, *On characteristic numbers and norm varieties.*
W. SCHARLAU, *The history of the algebraic theory of quadratic forms.*
J.-P. SERRE, *Abelian varieties and hermitian modules.*
M. TAYLOR, *Galois modules and hermitian Euler characteristics.*
C.T.C. WALL, *Quadratic forms in singularity theory.*

30 minutes.

A. ARUTYUNOV, *Quadratic forms and abnormal extremal problems: some results and unsolved problems.*
P. BALMER, *The Witt groups of triangulated categories, with some applications.*
G. BERHUY, *Hermitian scaled trace forms of field extensions.*
P. CALAME, *Integral forms without symmetry.*
P. CHUARD-KOULMANN, *Elements of given minimal polynomial in a central simple algebra.*
M. EPKENHANS, *On trace forms and the Burnside ring.*
L. FAINSILBER, *Quadratic forms and gas dynamics: sums of squares in a discrete velocity model for the Boltzmann equation.*
C. FRINGS, *Second trace form and T_2-standard normal bases.*
J. HURRELBRINK, *Quadratic forms of height* 2 *and differences of two Pfister forms.*
M. IFTIME, *On spacetime distributions.*
A. IZMAILOV, 2*-regularity and reversibility of quadratic mappings.*
S. JOUKHOVITSKI, *K-theory of the Weil transfer functor.*
V. MAUDUIT, *Towards a Drinfeldian analogue of quadratic forms for polynomials.*
M. MISCHLER, *Local densities and Jordan decomposition.*
V. POWERS, *Computational approaches to Hilbert's theorem on ternary quartics.*
S. PUMPLÜN, *The Witt ring of a Brauer-Severi variety.*
A. QUÉGUINER, *Discriminant and Clifford algebras of an algebra with involution.*
U. REHMANN, *A surprising fact about the generic splitting tower of a quadratic form.*

C. Riehm, *Orthogonal representations of finite groups.*
D. Sheiham, *Signatures of Seifert forms and cobordism of boundary links.*
V. Snaith, *Local fundamental classes constructed from higher dimensional K-groups.*
K. Zainoulline, *On Grothendieck's conjecture about principal homogeneous spaces.*

Conference participants

A.V. Arutyunov, Moscow	arutunov@sa640.cs.msu.su
P. Balmer, Lausanne	balmer@math.rutgers.edu
E. Bayer-Fluckiger, Besançon	bayer@vega.univ-fcomte.fr
K.J. Becher, Besançon	becher@math.univ-fcomte.fr
A.-M. Bergé, Talence	berge@math.u-bordeaux.fr
G. Berhuy, Besançon	berhuy@math.univ-fcomte.fr
J.J. Boutros, Paris	boutros@com.enst.fr
L. Broecker, Muenster	broe@math.uni-muenster.de
Ph. Calame, Lausanne	philippe.calame@ima.unil.ch
P. Chuard-Koulmann, Louvain-la-Neuve	chuard@agel.ucl.ac.be
J.H. Conway, Princeton	conway@math.princeton.edu
J.-Y. Degos, Talence	degos@math.u-bordeaux.fr
S. Dineen, Dublin	sean.dineen@ucd.ie
Ph. Du Bois, Angers	pdubois@univ-angers.fr
G. Elencwajg, Nice	elenc@math.unice.fr
M. Elhamdadi, Trieste	melha@ictp.trieste.it
M. Elomary, Louvain-la-Neuve	elomary@agel.ucl.ac.be
M. Epkenhans, Paderborn	martine@uni-paderborn.de
L. Fainsilber, Goteborg	laura@math.chalmers.se
D.A. Flannery, Cork	dflannery@cit.ie
C. Frings, Besançon	frings@math.univ-fcomte.fr
R.S. Garibaldi, Zurich	skip@math.ethz.ch
S. Gille, Muenster	gilles@math.uni-muenster.de
M. Gindraux, Neuchatel	marc.gindraux@maths.unine.ch
R. Gow, Dublin	rod.gow@ucd.ie
F. Gradl, Duisburg	gradl@math.uni-duisburg.de
Y. Hellegouarch, Caen	hellegou@math.unicaen.fr
D. Hoffmann, Besançon	detlev@vega.univ-fcomte.fr
J. Hurrelbrink, Baton Rouge	jurgen@julia.math.lsu.edu
K. Hutchinson, Dublin	kevin.hutchinson@ucd.ie
M. Iftime, Suceava	mtime@warpnet.ro
A.F. Izmailov, Moscow	izmaf@ccas.ru
S. Joukhovitski, Bonn	seva@math.nwu.edu
M. Karoubi, Paris	karoubi@math.jussieu.fr
C. Kearton, Durham	cherry.kearton@durham.ac.uk
M. Kreck, Heidelberg	kreck@mathi.uni-heidelberg.de
T.J. Laffey, Dublin	thomas.laffey@ucd.ie

C. Lamy, Paris	lamyc@enst.fr
A. Leibak, Tallinn	aleibak@ioc.ee
D. Lewis, Dublin	david.lewis@ucd.ie
J.-L. Loday, Strasbourg	loday@math.u-strasbg.fr
M. Mackey, Dublin	michael.mackey@ucd.ie
M. Marjoram, Dublin	martin.marjoram@ucd.ie
V. Mauduit, Dublin	veronique.mauduit@ucd.ie
A. Mazzoleni, Lausanne	amedeo.mazzoleni@ima.unil.ch
S. McGarraghy, Dublin	john.mcgarraghy@ucd.ie
G. McGuire, Maynooth	gmg@maths.may.ie
M. Mischler, Lausanne	maurice.mischler@ima.unil.ch
M. Monsurro, Besançon	monsurro@math.univ-fcomte.fr
J. Morales, Baton Rouge	morales@math.lsu.edu
C. Mulcahy, Atlanta	colm@spelman.edu
H. Munkholm, Odense	hjm@imada.sdu.dk
R. Parimala, Besançon	parimala@vega.univ-fcomte.fr
O. Patashnick, Chicago	owen@math.uchicago.edu
S. Perret, Neuchatel	stephane.perret@maths.unine.ch
A. Pfister, Mainz	pfister@mathematik.uni-mainz.de
V. Powers, Atlanta	vicki@mathcs.emory.edu
S. Pumplün, Regensburg	pumplun@degiorgi.science.unitn.it
A. Quéguiner, Paris	queguin@math.univ-paris13.fr
A. Ranicki, Edinburgh	aar@maths.ed.ac.uk
U. Rehmann, Bielefeld	rehmann@mathematik.uni-bielefeld.de
H. Reich, Muenster	reichh@escher.uni-muenster.de
C. Riehm, Hamilton, Ontario	riehm@mcmail.cis.mcmaster.ca
M. Rost, Regensburg	markus.rost@mathematik. uni-regensburg.de
D. Ryan, Dublin	dermot.w.ryan@ucd.ie
W. Scharlau, Muenster	scharlau@escher.uni-muenster.de
C. Scheiderer, Duisburg	claus@math.uni-duisburg.de
J.-P. Serre, Paris	jean-pierre.serre@ens.fr
D. Sheiham, Edinburgh	des@maths.ed.ac.uk
F. Sigrist, Neuchatel	francois.sigrist@maths.unine.ch
V. Snaith, Southampton	vps@maths.soton.ac.uk
M. Taylor, Manchester	mcbtayr@mail1.mcc.ac.uk
J.-P. Tignol, Louvain-la-Neuve	tignol@agel.ucl.ac.be
U. Tipp, Ghent	utipp@cage.rug.ac.be
D.A. Tipple, Dublin	david.tipple@ucd.ie
M. Tuite, Galway	michael.tuite@nuigalway.ie
T. Unger, Dublin	thomas.unger@ucd.ie
C.T.C. Wall, Liverpool	ctcw@liverpool.ac.uk
S. Yagunov, London Ontario	yagunov@jardine.math.uwo.ca
K. Zahidi, Ghent	karim.zahidi@rug.ac.be
K. Zainoulline, St. Petersburg	kirill@pdmi.ras.ru

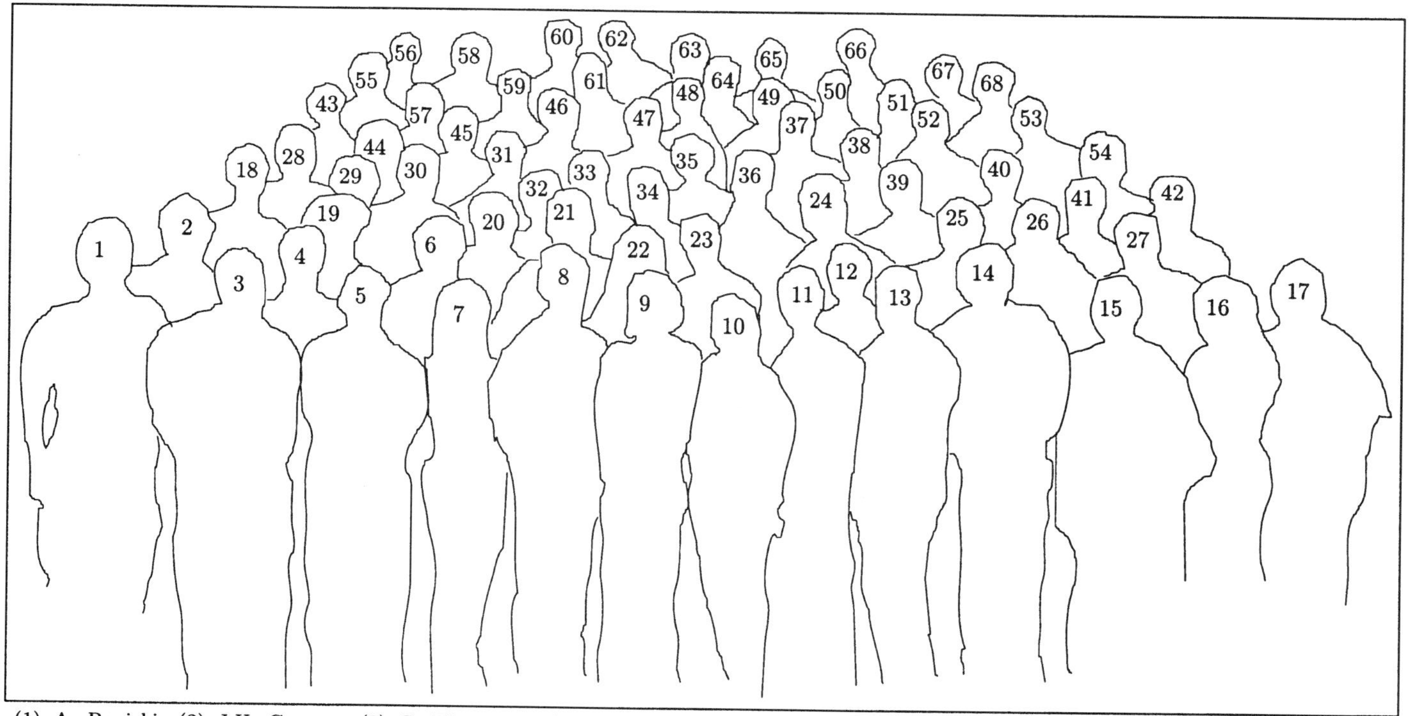

(1) A. Ranicki, (2) J.H. Conway, (3) G. Elencwajg, (4) K. Zainoulline, (5) M. Elomary, (6) V. Snaith, (7) S. Pumplün, (8) G. Berhuy, (9) A. Quéquiner, (10) M. Du Bois, (11) M. Monsurro, (12) V. Mauduit, (13) E. Bayer-Fluckiger, (14) D. Lewis, (15) R. Parimala, (16) V. Powers, (17) F. Sigrist, (18) H. Reich, (19) K. Zahidi, (20) C.T.C. Wall, (21) H. Munkholm, (22) M. Iftime, (23) A. Leibak, (24) C. Frings, (25) C. Riehm, (26) M. Mischler, (27) J.-P. Serre, (28) S. Perret, (29) A.-M. Bergé, (30) W. Scharlau, (31) M. Kreck, (32) S. Gille, (33) J.-Y. Degos, (34) R.S. Garibaldi, (35) S. McGarraghy, (36) O. Patashnick, (37) Ph. Du Bois, (38) G. McGuire, (39) C. Mulcahy, (40) D. Sheiham, (41) Ph. Calame, (42) L. Fainsilber, (43) J.-P. Tignol, (44) J.-L. Loday, (45) M. Taylor, (46) J. Morales, (47) A. Mazzoleni, (48) C. Scheiderer, (49) A.F. Izmailov, (50) K. Hutchinson, (51) A.V. Arutyunov, (52) J. Hurrelbrink, (53) A. Pfister, (54) M. Rost, (55) Y. Hellegouarch, (56) T. Unger, (57) P. Chuard-Koulmann, (58) D. Hoffmann, (59) M. Gindraux, (60) U. Tipp, (61) M. Epkenhans, (62) T.J. Laffey, (63) K.J. Becher, (64) M. Elhamdadi, (65) F. Gradl, (66) P. Balmer, (67) D. Flannery, (68) M. Tuite.

Contemporary Mathematics
Volume **272**, 2000

GALOIS COHOMOLOGY OF THE CLASSICAL GROUPS

Eva Bayer–Fluckiger

Introduction

Galois cohomology sets of linear algebraic groups were first studied in the late 50's – early 60's. As pointed out in [18], for *classical groups*, these sets have classical interpretations. In particular, Springer's theorem [22] can be reformulated as an injectivity statement for Galois cohomology sets of orthogonal groups; well–known classification results for quadratic forms over certain fields (such as finite fields, p–adic fields, ...) correspond to vanishing of such sets. The language of Galois cohomology makes it possible to formulate analogous statements for other linear algebraic groups. In [18] and [20], Serre raises questions and conjectures in this spirit. The aim of this paper is to survey the results obtained in the case of the classical groups.

1. Definitions and notation

Let k be a field of characteristic $\neq 2$, let k_s be a separable closure of k and let $\Gamma_k = \mathrm{Gal}(k_s/k)$.

1.1. Algebras with involution and norm–one–groups (cf. [9], [15]). Let A be a finite dimensional k–algebra. An *involution* $\sigma : A \to A$ is a k–linear antiautomorphism of A such that $\sigma^2 = id$.

Let (A, σ) be an algebra with involution. The associated *norm–one–group* U_A is the linear algebraic group over k defined by

$$U_A(E) = \{a \in A \otimes E \mid a\sigma(a) = 1 \}$$

for every commutative k–algebra E.

1.2. Galois cohomology (cf. [20]). For any linear algebraic group U defined over k, set $H^1(k, U) = H^1(\Gamma_k, U(k_s))$. Recall that $H^1(k, U)$ is also the set of isomorphism classes of U–torsors (principal homogeneous spaces over U).

1.3. Cohomological dimension. Let k be a perfect field. We say that the *cohomological dimension of k is $\leq n$*, denoted by $\mathrm{cd}(k) \leq n$, if $H^i(\Gamma_k, C) = 0$ for every $i > n$ and for every finite Γ_k–module C.

We say that the *virtual cohomological dimension of* k is $\leq n$, denoted by $\mathrm{vcd}(k) \leq n$, if there exists a finite extension k' of k such that $\mathrm{cd}(k') \leq n$. It is known that this holds if and only if $cd(k(\sqrt{-1})) \leq n$, see for instance [4], 1.2.

1.4. Galois cohomology mod 2. Set $H^i(k) = H^i(\Gamma_k, \mathbf{Z}/2\mathbf{Z})$. Recall that we have $H^1(k) \simeq k^*/k^{*2}$ and $H^2(k) \simeq \mathrm{Br}_2(k)$.

If $q \simeq < a_1, \dots, a_n >$ is a non–degenerate quadratic form, we define the discriminant of q by $\mathrm{disc}(q) = (-1)^{\frac{n(n-1)}{2}} a_1 \dots a_n \in k^*/k^{*2}$, and the Hasse–Witt invariant by $w_2(q) = \Sigma_{i<j}(a_i, a_j) \in \mathrm{Br}_2(k)$.

1.5. Galois cohomology and quadratic forms. Let q be a non–degenerate quadratic form over k and let O_q be its orthogonal group. Then $H^1(k, \mathrm{O}_q)$ is in bijection with the set of isomorphism classes of non–degenerate quadratic forms over k that become isomorphic to q over k_s (equivalently, those which have the same dimension as q) (cf. [20], III, 1.2. prop. 4).

2. Injectivity results

Some classical theorems of the theory of quadratic forms can be formulated in terms of injectivity of maps between Galois cohomology sets $H^1(k, \mathrm{O})$, where O is an orthogonal group. This reformulation suggests generalisations to other linear algebraic groups, as pointed out in [18] and [21]. The aim of this § is to give a survey of the results obtained in this direction, especially in the case of the classical groups.

2.1. Springer's theorem. Let q and q' be two non–degenerate quadratic forms defined over k. Springer's theorem [22] states that if q and q' become isomorphic over an odd degree extension, then they are already isomorphic over k. This can be reformulated in terms of Galois cohomology as follows. Let O_q be the orthogonal group of q. If L is an odd degree extension of k, then the canonical map

$$H^1(k, \mathrm{O}_q) \to H^1(L, \mathrm{O}_q)$$

is injective.

Serre makes this observation in [18], 5.3., and asks for generalisations of this result to other linear algebraic groups. One has the following

THEOREM 2.1.1. *Let U be the norm-one-group of a finite dimensional k–algebra with involution. If L is an odd degree extension of k, then the canonical map*

$$H^1(k, U) \to H^1(L, U)$$

is injective.

PROOF. See [1], Theorem 2.1.

The above results concern injectivity after a base change. As noted in [21], some well–known results about quadratic forms can be reformulated as injectivity statements of maps between Galois cohomology sets $H^1(k, U) \to H^1(k, U')$, where U is a subgroup of U'. This is for instance the case of Pfister's theorem :

2.2. Pfister's theorem. Let q, q' and ϕ be non–degenerate quadratic forms over k. Suppose that the dimension of ϕ is *odd*. A classical result of Pfister says that if $q \otimes \phi \simeq q' \otimes \phi$, then $q \simeq q'$ (see [15], 2.6.5.). This can be reformulated as follows. Denote by O_q the orthogonal group of q, and by $O_{q\otimes\phi}$ the orthogonal group of the tensor product $q \otimes \phi$. Then the canonical map $H^1(k, O_q) \to H^1(k, O_{q\otimes\phi})$ is injective. One can extend this result to algebras with involution as follows :

THEOREM 2.2.1. *Let (A, σ) and (B, τ) be finite dimensional k–algebras with involution. Let us denote by U_A the norm–one–group of (A, σ), and by $U_{A\otimes B}$ the norm–one–group of the tensor product of algebras with involution $(A, \sigma) \otimes (B, \tau)$. Suppose that $dim_k(B)$ is odd. Then the canonical map*

$$H^1(k, U_A) \to H^1(k, U_{A\otimes B})$$

is injective.

Note that Theorem 2.2.1 implies Pfister's theorem quoted above, and also a result of Lewis [10], Theorem 1.

For the proof of 2.2.1, we need the following consequence of Theorem 2.1.1.

COROLLARY 2.2.2. *Let U and U' be norm–one–groups of finite dimensional k–algebras. Suppose that there exists an odd degree extension L of k such that $H^1(L, U) \to H^1(L, U')$ is injective. Then $H^1(k, U) \to H^1(k, U')$ is also injective.*

PROOF OF 2.2.1. By a "dévissage" as in [1], we reduce to the case where A and B are central simple algebras with involution, that is either central over k with an involution of the first kind, or central over a quadratic extension k' of k with a k'/k-involution of the second kind. Using 2.2.2, we may assume that B is split and that the involution is given by a symmetric or hermitian form. We conclude the proof by the argument of [2], proof of Theorem 4.2.

It is easy to see that Theorem 2.2.1 does not extend to the case where both algebras have even degree.

2.3. Witt's theorem. In 1937, Witt proved the "cancellation theorem" for quadratic forms [26] : if q_1, q_2 and q are quadratic forms such that $q_1 \oplus q \simeq q_2 \oplus q$, then $q_1 \simeq q_2$. The analog of this result for hermitian forms over skew fields also holds, see for instance [8] or [15].

These results can also be deduced from a statement on linear algebraic groups due to Borel and Tits :

THEOREM 2.3.1. *([20], III.2.1., Exercice 1) Let G be a connected reductive group, and P a parabolic subgroup of G. Then the map $H^1(k, P) \to H^1(k, G)$ is injective.*

3. Classification of quadratic forms and Galois cohomology

Recall (cf. 1.5.) that if O_q is the orthogonal group of a non–degenerate, n–dimensional quadratic form q over k, then $H^1(k, \mathrm{O}_q)$ is the set of isomorphism classes of non–degenerate quadratic forms over k of dimension n. Hence determining this set is equivalent to classifying quadratic forms over k up to isomorphism. The cohomological description makes it possible to use various exact sequences related to subgroups or coverings, and to formulate classification results in cohomological terms. This is explained in [20], III.3.2., as follows :

Let SO_q be the special orthogonal group. We have the exact sequence

$$1 \to \mathrm{SO}_q \to \mathrm{O}_q \to \mu_2 \to 1.$$

This exact sequence induces an exact sequence in cohomology

$$\mathrm{SO}_q(k) \to \mathrm{O}_q(k) \overset{\det}{\to} \mu_2 \to H^1(k, \mathrm{SO}_q) \to H^1(k, \mathrm{O}_q) \overset{\mathrm{disc}}{\to} k^*/k^{*2}.$$

The map $H^1(k, \mathrm{O}_q) \overset{\mathrm{disc}}{\to} k^*/k^{*2}$ is given by the discriminant. More precisely, the class of a quadratic form q' is sent to the class of $\mathrm{disc}(q)\mathrm{disc}(q') \in k^*/k^{*2}$.

Note that the map $\mathrm{O}_q(k) \overset{\det}{\to} \mu_2$ is onto (reflections have determinant -1). Hence we see that

PROPOSITION 3.1. *In order that $H^1(k, \mathrm{SO}_q) = 0$ it is necessary and sufficient that every quadratic form over k which has the same dimension and the same discriminant as q is isomorphic to q.*

EXAMPLE. Suppose that k is a finite field. It is well–known that non–degenerate quadratic forms over k are determined by their dimension and discriminant. Hence by 3.1. we have $H^1(k, \mathrm{SO}_q) = 0$ for all q.

We can go one step further, and consider an H^2–invariant (the Hasse–Witt invariant) that will suffice, together with dimension and discriminant, to classify non–degenerate quadratic forms over certain fields.

Let Spin_q be the spin group of q. Suppose that $\dim(q) \geq 3$. We have the exact sequence

$$1 \to \mu_2 \to \mathrm{Spin}_q \to \mathrm{SO}_q \to 1.$$

This exact sequence induces the cohomology exact sequence

$$\mathrm{SO}_q(k) \overset{\delta}{\to} k^*/k^{*2} \to H^1(k, \mathrm{Spin}_q) \to H^1(k, \mathrm{SO}_q) \overset{\Delta}{\to} \mathrm{Br}_2(k),$$

where $\mathrm{SO}_q(k) \overset{\delta}{\to} k^*/k^{*2}$ is the spinor norm, and $H^1(k, \mathrm{SO}_q) \overset{\Delta}{\to} \mathrm{Br}_2(k)$ sends the class of a quadratic form q' with $\dim(q) = \dim(q')$, $\mathrm{disc}(q) = \mathrm{disc}(q')$ to the sum of the Hasse–Witt invariants of q and q', $w_2(q') + w_2(q) \in \mathrm{Br}_2(k)$ (cf. [23]).

Hence we obtain the following :

PROPOSITION 3.2. *(cf. [20], III, 3.2.) In order that $H^1(k, \mathrm{Spin}_q) = 0$, it is necessary and sufficient that the following two conditions be satisfied :*

(i) *The spinor norm $\mathrm{SO}_q(k) \to k^*/k^{*2}$ is surjective ;*

(ii) *Every quadratic form which has the same dimension, the same discriminant and the same Hasse–Witt invariant as q is isomorphic to q.*

EXAMPLE. Let k be a p–adic field. Then it is well–known that the spinor norm is surjective, and that non–degenerate quadratic forms are classified by dimension, discriminant and Hasse–Witt invariant. Hence by 3.2. we have $H^1(k, \mathrm{Spin}_q) = 0$ for all q.

4. Conjectures I and II

In the preceding §, we have seen that if k is a finite field then $H^1(k, \mathrm{SO}_q) = 0$; if k is a p–adic field and $\dim(q) \geq 3$, then $H^1(k, \mathrm{Spin}_q) = 0$. Note that SO_q is connected, and that Spin_q is semi–simple, simply connected. These examples are special cases of Serre's conjectures I and II, made in 1962 (cf. [18]; [20], chap. III) :

THEOREM 4.1. *(ex–Conjecture I) Let k be a perfect field of cohomological dimension ≤ 1. Let G be a connected linear algebraic group over k. Then $H^1(k, G) = 0$.*

This was proved by Steinberg in 1965, cf. [24]. See also [20], III.2.

CONJECTURE II. *Let k be a perfect field of cohomological dimension ≤ 2. Let G be a semi-simple, simply connected linear algebraic group over k. Then $H^1(k,G) = 0$.*

This conjecture is still open in general, though it has been proved in many special cases (cf. [20], III.3). The main breakthrough was made by Merkurjev and Suslin [13], [25], who proved the conjecture for special linear groups over division algebras. More generally, the conjecture is now known for the *classical groups.*

THEOREM 4.2. *Let k be a perfect field of cohomological dimension ≤ 2, and let G be a semi-simple, simply connected group of classical type (with the possible exception of groups of trialitarian type D_4) or of type G_2, F_4. Then $H^1(k,G) = 0$.*

See [3]. The proof uses the theorem of Merkurjev and Suslin [13], [25] as well as results of Merkurjev [12] and Yanchevskii [28], [29], and the injectivity result Theorem 2.1.1. More recently, Gille proved Conjecture II for some groups of type E_6, E_7, and of trialitarian type D_4 (cf. [7]).

5. Hasse Principle Conjectures I and II

Colliot–Thélène [5] and Scheiderer [16] have formulated real analogues of Conjectures I and II, which we will call *Hasse Principle Conjectures* I and II. Let us denote by Ω the set of orderings of k. If $v \in \Omega$, let us denote by k_v the real closure of k at v.

HASSE PRINCIPLE CONJECTURE I. *Let k be a perfect field of virtual cohomological dimension ≤ 1. Let G be a connected linear algebraic group. Then the canonical map*

$$H^1(k,G) \to \prod_{v\in\Omega} H^1(k_v,G)$$

is injective.

This has been proved by Scheiderer, cf. [16].

HASSE PRINCIPLE CONJECTURE II. *Let k be a perfect field of virtual cohomological dimension ≤ 2. Let G be a semi-simple, simply connected linear algebraic group. Then the canonical map*

$$H^1(k,G) \to \prod_{v\in\Omega} H^1(k_v,G)$$

is injective.

This conjecture is proved in [4] for groups of classical type (with the possible exception of groups of trialitarian type D_4), as well as for groups of type G_2 and F_4.

If k is an algebraic number field, then we recover the usual Hasse Principle. This was first conjectured by Kneser in the early 60's, and is now known for arbitrary simply connected groups (see for instance [13] for a survey).

In the case of classical groups, these results can be expressed as classification results for various kinds of forms, in the spirit outlined in §3. This is done in [17] in the case of fields of virtual cohomological dimension ≤ 1 and in [4] for fields of cohomological dimension ≤ 2.

References

[1] E. Bayer–Fluckiger, H.W. Lenstra, Jr., *Forms in odd degree extensions and self–dual normal bases*, Amer. J. Math. **112** (1990), 359–373.

[2] E. Bayer–Fluckiger, D. Shapiro, J.–P. Tignol, *Hyperbolic involutions*, Math. Z. **214** (1993), 461–476.

[3] E. Bayer–Fluckiger, R. Parimala, *Galois cohomology of the classical groups over fields of cohomological dimension* ≤ 2, Invent. Math. **122** (1995), 195–229.

[4] E. Bayer–Fluckiger, R. Parimala, *Classical groups and the Hasse principle*, Ann. Math. **147** (1998), 651–693.

[5] J.–L. Colliot–Thélène, *Groupes linéaires sur les corps de fonctions de courbes réelles*, J. reine angew. Math. **474** (1996), 139–167.

[6] Ph. Gille, *La R–équivalence sur les groupes algébriques réductifs définis sur un corps global*, Publ. IHES **86** (1997), 199–235.

[7] Ph. Gille, *Cohomologie galoisienne des groupes quasi–déployés sur des corps de dimension cohomologique* ≤ 2, preprint (1999).

[8] M. Knus, *Quadratic and Hermitian Forms over Rings*, Grundlehren Math. Wiss., vol. 294, Springer–Verlag, Heidelberg, 1991.

[9] M. Knus, A. Merkurjev, M. Rost, J.–P. Tignol, *The Book of Involutions*, AMS Coll. Pub., vol. 44, Providence, 1998.

[10] D. Lewis, *Hermitian forms over odd dimensional algebras*, Proc. Edinburgh Math. Soc. **32** (1989), 139–145.

[11] A. Merkurjev, *On the norm residue symbol of degree 2*, Dokl. Akad. Nauk. SSSR, English translation : Soviet Math. Dokl. **24** (1981), 546-551.

[12] A. Merkurjev, *Norm principle for algebraic groups*, St. Petersburg J. Math. **7** (1996), 243–264.

[13] A. Merkurjev, A. Suslin, *K–cohomology of Severi–Brauer varieties and the norm–residue homomorphis*, Izvestia Akad. Nauk. SSSR, English translation : Math. USSR Izvestia **21** (1983), 307-340.

[14] V. Platonov, A. Rapinchuk, *Algebraic Groups and Number Theory*, Academic Press, 1994.

[15] W. Scharlau, *Quadratic and hermitian forms*, Grundlehren der Math. Wiss., vol. 270, Springer–Verlag, Heidelberg, 1985.

[16] C. Scheiderer, *Hasse principles and approximation theorems for homogeneous spaces over fields of virtual cohomological dimension one*, Invent. Math. **125** (1996), 307–365.

[17] C. Scheiderer, *Classification of hermitian forms and semisimple groups over fields of virtual cohomological dimension one*, Manuscr. Math. **89** (1996), 373–394.

[18] J–P. Serre, *Cohomologie galoisienne des groupes algébriques linéaires*, Colloque sur la théorie des groupes algébriques, Bruxelles (1962), 53–68.

[19] J–P. Serre, *Corps locaux*, Hermann, Paris, 1962.

[20] J–P. Serre, *Cohomologie Galoisienne*, Lecture Notes in Mathematics, vol. 5, Springer–Verlag, 1964 and 1994.

[21] J–P. Serre, *Cohomologie galoisienne : progrès et problèmes*, Sém. Bourbaki, exposé **783** (1993–1994).

[22] T. Springer, *Sur les formes quadratiques d'indice zéro*, C. R. Acad. Sci. Paris **234** (1952), 1517–1519.

[23] T. Springer, *On the equivalence of quadratic forms*, Indag. Math. **21** (1959), 241–253.

[24] R. Steinberg, *Regular elements of semisimple algebraic groups*, Publ. Math. IHES **25** (1965), 49–80.

[25] A. Suslin, *Algebraic K–theory and norm residue homomorphism*, Journal of Soviet mathematics **30** (1985), 2556–2611.

[26] A. Weil, *Algebras with involutions and the classical groups*, J. Ind. Math. Soc. **24** (1960), 589–623.

[27] E. Witt, *Theorie der quadratischen Formen in beliebigen Körpern*, J. reine angew. Math. **176** (1937), 31–44.

[28] V. Yanchevskii, *Simple algebras with involution and unitary groups*, Math. Sbornik, English translation : Math. USSR Sbornik **22** (1974), 372-384.

[29] V. Yanchevskii, *The commutator subgroups of simple algebras with surjective reduced norms*, Dokl. Akad. Nauk SSSR, English translation : Soviet Math. Dokl. **16** (1975), 492-495.

Laboratoire de Mathématiques de Besançon
UMR 6623 du CNRS
16, route de Gray
25030 Besançon
France

E-mail address: bayer@Math.univ-fcomte.fr

Contemporary Mathematics
Volume **272**, 2000

SYMPLECTIC LATTICES

ANNE-MARIE BERGÉ

Introduction

The title refers to lattices arising from principally polarized Abelian varieties, which are naturally endowed with a structure of *symplectic* $\mathbb{Z}$-modules. The density of sphere packings associated to these lattices was used by Buser and Sarnak [B-S] to locate the Jacobians in the space of Abelian varieties. During the last five years, this paper stimulated further investigations on density of symplectic lattices, or more generally of *isodual lattices* (lattices that are isometric to their duals, [C-S2]).

Isoduality also occurs in the setting of modular forms: Quebbemann introduced in [Q1] the *modular lattices*, which are integral and similar to their duals, and thus can be rescaled so as to become isodual. The search for modular lattices with the highest Hermite invariant permitted by the theory of modular forms is now a very active area in geometry of numbers, which led to the discovery of some symplectic lattices of high density.

In this survey, we shall focus on isoduality, pointing out its different aspects in connection with various domains of mathematics such as Riemann surfaces, modular forms and algebraic number theory.

1. Basic definitions

1.1 Invariants. Let E be an n-dimensional real Euclidean vector space, equipped with scalar product $x.y$, and let Λ be a lattice in E (discrete subgroup of rank n). We denote by $m(\Lambda)$ its *minimum* $m(\Lambda) = \min_{x\neq 0\in\Lambda} x.x$, and by $\det\Lambda$ the determinant of the Gram matrix $(e_i.e_j)$ of any $\mathbb{Z}$-basis $(e_1, e_2, \cdots e_n)$ of Λ. The density of the sphere packing associated to Λ is measured by the *Hermite invariant* of Λ

$$\gamma(\Lambda) = \frac{m(\Lambda)}{\det\Lambda^{1/n}}.$$

The *Hermite constant* $\gamma_n = \sup_{\Lambda\subset E}\gamma(\Lambda)$ is known for $n \leq 8$. For large n, Minkowski gave linear estimations for γ_n, see [C-S1], I,1.

2000 *Mathematics Subject Classification.* Primary 11H55; Secondary 11G10,11R04,11R52.
Key words and phrases. Lattices, Abelian varieties, duality.

Another classical invariant attached to the sphere packing of Λ is its kissing number $2s = |S(\Lambda)|$ where

$$S(\Lambda) = \{x \in \Lambda \mid x.x = m(\Lambda)\}$$

is the set of *minimal vectors* of Λ.

1.2 Isodualities. The *dual lattice of* Λ is

$$\Lambda^* = \{y \in E \mid x.y \in \mathbb{Z} \text{ for all } x \in \Lambda\}.$$

An *isoduality* of Λ is an isometry σ of Λ onto its dual; actually, σ exchanges Λ and Λ^* (since ${}^t\sigma = \sigma^{-1}$), and σ^2 is an automorphism of Λ. We can express this property by introducing the group $\mathrm{Aut}^{\#}\,\Lambda$ of the isometries of E mapping Λ onto Λ or Λ^*. When Λ is isodual, the index $[\mathrm{Aut}^{\#}\,\Lambda : \mathrm{Aut}\,\Lambda]$ is equal to 2 except in the *unimodular* case, i.e. when $\Lambda = \Lambda^*$, and the isodualities of Λ are in one-to-one correspondence with its automorphisms.

We attach to any isoduality σ of Λ the bilinear form

$$B_\sigma : (x, y) \mapsto x.\sigma(y),$$

which is integral on $\Lambda \times \Lambda$ and has discriminant $\pm 1 = \det \sigma$.

Two cases are of special interest:

(i) *The form B_σ is symmetric*, or equivalently $\sigma^2 = 1$. Such an isoduality is called *orthogonal*. For a prescribed signature (p, q), $p + q = n$, it is easily checked that the set of isometry classes of σ-isodual lattices of E is of dimension pq. We recover, when $\sigma = \pm 1$, the finiteness of the set of unimodular n-dimensional lattices.

(ii) *The form B_σ is alternating*, i.e $\sigma^2 = -1$. Such an isoduality, which only occurs in even dimension, is called *symplectic*. Up to isometry, the family of symplectic $2g$-dimensional lattices has dimension $g(g+1)$ (see the next section); for instance, every two-dimensional lattice of determinant 1 is symplectic (take for σ a planar rotation of order 4). Note that an isodual lattice can be both symplectic and orthogonal. For example, it occurs for any 2-dimensional lattice with $s \geq 2$. The densest 4-dimensional lattice $\mathbb{D}_4$, suitably rescaled, has, together with symplectic isodualities (see below), orthogonal isodualities of every indefinite signature.

2. Symplectic lattices and Abelian varieties

2.1 Let us recall how symplectic lattices arise naturally from the theory of complex tori. Let V be a complex vector space of dimension g, and let Λ be a full lattice of V. The complex torus V/Λ is an *Abelian variety* if and only if there exists a *polarization on* Λ, i. e. a positive definite Hermitian form H for which the alternating form $\mathrm{Im}\, H$ is integral on $\Lambda \times \Lambda$. In the $2g$-dimensional real space V equipped with the scalar product $x.y = \mathrm{Re}\, H(x, y) = \mathrm{Im}\, H(ix, y)$, multiplication by i is an isometry of square -1 that maps the lattice Λ onto a sublattice of Λ^* of index $\det(\mathrm{Im}\, H)$ $(= \det \Lambda)$. This is an isoduality for Λ if and only if $\det(\mathrm{Im}\, H) = 1$. The polarization H is then said *principal*.

Conversely, let $(E, .)$ be again a real Euclidean vector space, Λ a lattice of E with a symplectic isoduality σ as defined in subsection 1.2. Then E can be made into a complex vector space by letting $ix = \sigma(x)$. Now the real alternating form $B_\sigma(x, y) = x.\sigma(y)$ attached to σ in 1.2(*ii*) satisfies $B_\sigma(ix, iy) = B_\sigma(x, y)$ (since σ is an isometry) and thus gives rise to the definite positive Hermitian form

$H(x,y) = B_\sigma(ix,y) + iB_\sigma(x,y) = x.y + ix.\sigma(y)$, which is a principal polarization for Λ (by 1.2 (ii)).

So, *there is a one-to-one correspondence between symplectic lattices and principally polarized complex Abelian varieties.*

REMARK. In general, if $(V/\Lambda, H)$ is any polarized abelian variety, one can find in V a lattice Λ' containing Λ such that $(V/\Lambda', H)$ is a principally polarized abelian variety. For example, let us consider the Coxeter description of the densest six-dimensional lattice $\mathbb{E}_6$. Let $\mathcal{E} = \{a + \omega b \mid a, b \in \mathbb{Z}\} \subset \mathbb{C}$, with $\omega = \frac{-1+i\sqrt{3}}{2}$ be the Eisenstein ring. In the space $V = \mathbb{C}^3$ equipped with the Hermitian inner product $H((\lambda_i),(\mu_i)) = 2\sum \lambda_i\overline{\mu_i}$, the lattice $\mathcal{E}^3 \cup (\mathcal{E}^3 + \frac{1}{1-\omega}(1,1,1))$ is isometric to $\mathbb{E}_6$, and the lattice $\frac{1}{\omega-\overline{\omega}}\{(\lambda_1,\lambda_2,\lambda_3) \in \mathcal{E}^3 \mid \lambda_1+\lambda_2+\lambda_3 \equiv 0 \quad (1-\omega)\}$ to its dual $\mathbb{E}_6^*$ (see [M]). The rescaled lattice $\Lambda = 3^{\frac{1}{4}}\mathbb{E}_6^*$ satisfies $i\Lambda \subset 3^{-\frac{1}{4}}\mathcal{E}^3 \subset \Lambda^*$: while the polarization H is not principal for Λ, it is principal on $\Lambda' = 3^{-\frac{1}{4}}\mathcal{E}^3$, and the principally polarized abelian variety $(\mathbb{C}^3/\Lambda', H)$ is isomorphic to the direct product of three copies of the curve $y^2 = x^3 - 1$.

2.2 We now make explicit (from the point of view of geometry of numbers) the standard parametrization of symplectic lattices by the *Siegel upper half-space*

$$\mathfrak{H}_g = \{X + iY, \quad X \text{ and } Y \text{ real symmetric } g \times g \text{ matrices}, Y > 0\}.$$

Let $\Lambda \subset E$ be a $2g$-dimensional lattice with a symplectic isoduality σ. It possesses a *symplectic basis* $\mathcal{B} = (e_1, e_2, \cdots, e_{2g})$, i.e. such that the matrix $(e_i.\sigma(e_j))$ has the form

$$J = \begin{pmatrix} O & I_g \\ -I_g & O \end{pmatrix},$$

(see for instance [M-H], p. 7). This amounts to saying that the Gram matrix $A := (e_i.e_j)$ is symplectic. More generally, a $2g \times 2g$ real matrix M is *symplectic* if ${}^tMJM = J$.

We give E the complex structure defined by $ix = -\sigma(x)$, and we write $\mathcal{B} = \mathcal{B}_1 \cup \mathcal{B}_2$, with $\mathcal{B}_1 = (e_1, \cdots, e_g)$. With respect to the $\mathbb{C}$-basis $\mathcal{B}_1$ of E, the generator matrix of the basis $\mathcal{B}$ of Λ has the form $(\,I_g \; Z\,)$, where $Z = X + iY$ is a $g \times g$ complex matrix. The isometry $-\sigma$ maps the real span F of $\mathcal{B}_1$ onto its orthogonal complement $F^\perp$, and the $\mathbb{R}$-basis $\mathcal{B}_1$ onto the dual-basis of the orthogonal projection $p(\mathcal{B}_2)$ of $\mathcal{B}_2$ onto $F^\perp$. Since $Y = \operatorname{Re} Z$ is the generator matrix of $p(\mathcal{B}_2)$ with respect to the basis $(-\sigma)(\mathcal{B}_1) = (p(\mathcal{B}_2))^*$, we have $Y = \operatorname{Gram}(p(\mathcal{B}_2)) = (\operatorname{Gram}(\mathcal{B}_1))^{-1}$; the matrix Y is then symmetric, and moreover Y^{-1} represents the polarization H in the $\mathbb{C}$-basis $\mathcal{B}_1$ of E (since $H(e_h, e_j) = e_h.e_j + ie_h.\sigma(e_j) = e_h.e_j$ for $1 \le h, j \le g$). Now, the Gram matrix of the basis $\mathcal{B}_0 = \mathcal{B}_1 \perp p(\mathcal{B}_2)$ of E is $\operatorname{Gram}(\mathcal{B}_0) = \begin{pmatrix} Y^{-1} & O \\ O & Y \end{pmatrix}$. Since the (real) generator matrix of the basis $\mathcal{B}$ with respect to $\mathcal{B}_0$ is $P = \begin{pmatrix} I_g & X \\ O & I_g \end{pmatrix}$, we have $A = \operatorname{Gram}(\mathcal{B}) = {}^tP\operatorname{Gram}(\mathcal{B}_0)P$, and it follows from the condition "A symplectic" that the matrix X also is symmetric, so we conclude

$$A = \begin{pmatrix} I_g & O \\ X & I_g \end{pmatrix}\begin{pmatrix} Y^{-1} & O \\ O & Y \end{pmatrix}\begin{pmatrix} I_g & X \\ O & I_g \end{pmatrix}, \text{ with } X + iY \in \mathfrak{H}_g.$$

On the other hand, such a matrix A is obviously positive definite, symmetric and symplectic.

Changing the symplectic basis means replacing A by ${}^tPAP$, with P in the symplectic modular group

$$\mathrm{Sp}_{2g}(\mathbb{Z}) = \{P \in \mathrm{SL}_{2g}(\mathbb{Z}) \mid {}^tPJP = J\}.$$

One can check that the corresponding action of $P = \left(\begin{smallmatrix}\alpha & \beta \\ \gamma & \delta\end{smallmatrix}\right)$ on $\mathfrak{H}_g$ is the homography $Z \mapsto Z' = (\delta Z + \gamma)(\alpha Z + \beta)^{-1}$.

Most of the well known lattices in low even dimension are proportional to symplectic lattices, with the noticeable exception of the above-mentioned $\mathbb{E}_6$: the roots lattices $\mathbb{A}_2$, $\mathbb{D}_4$ and $\mathbb{E}_8$, the Barnes lattice P_6, the Coxeter-Todd lattice K_{12}, the Barnes-Wall lattice BW_{16}, the Leech lattice Λ_{24}, ... In Appendix 2 to [B-S], Conway and Sloane give some explicit representations $X + iY \in \mathfrak{H}_g$ of them. A more systematic use of such a parametrization is dealt with in section 6.

3. Jacobians

The Jacobian $\mathrm{Jac}\, C$ of a curve C of genus g is a complex torus of dimension g which carries a canonical principal polarization, and then the corresponding *period lattice* is symplectic. Investigating the special properties of the Jacobians among the general principally polarized Abelian varieties, Buser and Sarnak proved that, while the linear Minkowski lower-bound for the Hermite constant γ_{2g} still applies to the general symplectic lattices, the general linear upper bound is to be replaced, for period lattices, by a logarithmic one (for explicit values, see [B-S], p. 29), and thus one does not expect large-dimensional symplectic lattices of high density to be Jacobians. The first example of this obstruction being effective is the Leech lattice. A more conclusive argument in low dimension involves the centralizer $\mathrm{Aut}_\sigma(\Lambda)$ of the isoduality σ in the automorphism group of the σ-symplectic lattice Λ: if Λ corresponds to a curve C of genus g, we must have, from Torelli's and Hurwitz's theorems, $|\mathrm{Aut}_\sigma(\Lambda)| = |\mathrm{Aut}(\mathrm{Jac}\, C)| \le 2|\mathrm{Aut}\, C| \le 2 \times 84(g-1)$. Calculations by Conway and Sloane (in [B-S], Appendix 2) showed that $|\mathrm{Aut}_\sigma(\Lambda)|$ is one hundred times over this bound in the case of the lattice E_8, and one million in the case of the Leech lattice!

However, up to genus 3, almost all principally polarized abelian varieties are Jacobians, so it is no wonder if the known symplectic lattices of dimension $2g \le 6$ correspond to Jacobians of curves: the lattices $\mathbb{A}_2$, $\mathbb{D}_4$ and the Barnes lattice P_6 are the respective period lattices for the curves $y^2 = x^3 - 1$, $y^2 = x^5 - x$ and the Klein curve $xy^3 + yz^3 + zx^3 = 0$ (see [B-S], Appendix 1). The Fermat quartic $x^4 + y^4 + z^4 = 0$ gives rise to the lattice $\mathbb{D}_6^+$ (the family $\mathbb{D}_{2g}^+$ is discussed in section 7), slightly less dense, with $\gamma = 1.5$, than the Barnes lattice P_6 ($\gamma = 1.512\ldots$) but with a lot of symmetries ($\mathrm{Aut}_\sigma(\Lambda)$ has index 120 in the full group of automorphisms.

The present record for six dimensions ($\gamma = 1.577\ldots$) was established in [C-S2] by the Conway-Sloane lattice $M(\mathbb{E}_6)$ (see section 7) defined over $\mathbb{Q}(\sqrt{3})$. This lattice was shown in [Bav1], and independently in [Qi], to be associated to the exceptional Wiman curve $y^3 = x^4 - 1$ (the unique non-hyperelliptic curve with an automorphism of order $4g$, viewed in [Qi] as the most symmetric Picard curve).

In the recent paper [Be-S], Bernstein and Sloane discussed the period lattice associated to the hyperelliptic curve $y^2 = x^{2g+2} - 1$, and proved it to have the form $L_{2g} = M_g \perp M'_g$, where M_g is a g-dimensional isodual lattice, and M'_g a copy of its dual. Here the interesting lattice is the summand M_g (its density is that of L_{2g}, and its group has only index 2): it turns out to be, for $g \leq 3$, the densest isodual packing in g dimensions.

REMARK. The Hermite problem is part of a more general systole problem (see [Bav1]). So far, although a compact Riemann surface is determined by its polarized Jacobian, no connection between its systole and the Hermite invariant of the period lattice seems to be known.

4. Modular lattices

4.1 Definition. Let Λ be an n-dimensional *integral* lattice (i.e. $\Lambda \subset \Lambda^*$), which is similar to its dual. If σ is a similarity such that $\sigma(\Lambda^*) = \Lambda$, its norm ℓ (σ multiplies squared lengths by ℓ) is an integer which does not depend on the choice of σ. Following Quebbemann, we call Λ a *modular lattice of level.* Note that level one corresponds to unimodular lattices.

For a given pair (n, ℓ), the (hypothetical) modular lattices have a prescribed determinant $\ell^{n/2}$, thus, up to isometry, there are only finitely many of them; as usual we are looking for the largest possible minimum m (the Hermite invariant $\gamma = \frac{m}{\sqrt{\ell}}$ depends only on it). In the following, we restrict to even dimensions and *even lattices.*

Then, the modular properties of the theta series of such lattices yield constraints for the dimension and the density analogous to Hecke's results for $\ell = 1$ ([C-S1], chapter 7). Still, for some aspects of these questions, the unimodular case remains somewhat special. For example, given a prime ℓ, there exists even ℓ-modular lattices of dimension n if and only if $\ell \equiv 3 \mod 4$ or $n \equiv 0 \mod 4$ (see [Q1]).

4.2 Connection with modular forms. Let Λ be an even lattice of minimum m, and let Θ_Λ be its theta series

$$\Theta_\Lambda(z) = \sum_{x \in \Lambda} q^{(x.x)/2} = 1 + 2sq^{m/2} + \cdots \qquad (\text{ where } q = e^{2\pi i z}).$$

Now, when Λ is ℓ-modular ($\ell > 1$), Θ_Λ must be a modular form of weight $n/2$ with respect to the so-called *Fricke group of level* ℓ, a subgroup of $\mathrm{SL}_2(\mathbb{R})$ which contains $\Gamma_0(\ell)$ with index 2 (here again, the unimodular case is exceptional).

From the algebraic structure of the corresponding space $\mathcal{M}$ of modular forms, Quebbemann derives the notion of extremal modular lattices extending that of [C-S1], chapter 7. Let $d = \dim \mathcal{M}$ be the dimension of $\mathcal{M}$. If a form $f \in \mathcal{M}$ is uniquely determined by the first d coefficients $a_0, a_1, \cdots, a_{d-1}$ of its q-expansion $f = \sum_{k \geq 0} a_k q^k$, the unique form $F_\mathcal{M} = 1 + \sum_{k \geq d} a_k q^k$ is called ℓ. *extremal,* and an even ℓ-modular lattice with this theta series is called an *extremal lattice.* Such a (hypothetical) extremal lattice has the highest possible minimum, equal to $2d$ unless the coefficient a_d of $F_\mathcal{M}$ vanishes. No general results about the coefficients of the extremal modular form and more generally of its eligibility as a theta series seem to be known.

4.3 Special levels. Quebbemann proved that the above method is valid in particular for prime levels ℓ such that $\ell + 1$ divides 24, namely $2, 3, 5, 7, 11$ and 23. (For a more general setup, we refer the reader to [Q1], [Q2] and [S-SP].) The dimension of the space of modular forms is then $d = 1 + \lfloor \frac{n(1+\ell)}{48} \rfloor$ (which reduces to Hecke's result for $\ell = 1$). The proof of the upper bound

$$m \leq 2 + 2\left\lfloor \frac{n(1+\ell)}{48} \right\rfloor$$

was completed in [S-SP] by R. Scharlau and R. Schulze-Pillot, by investigating the coefficients a_k, $k > 0$ of the extremal modular form: all of them are even integers, the leading one a_d is positive, but a_{d+1} is negative for n large enough. So, for a given level in the above list, there are (at most) only finitely many extremal lattices. Other kinds of obstructions may exist.

4.4 Examples.

• $\ell = 7$, at *jump dimensions* (where the minimum may increase) $n \equiv 0 \mod 6$. While a_{d+1} first goes negative at $n = 30$, Scharlau and Hemkemeier proved that no 7-extremal lattice exists in dimension 12: their method consists in classifying for given pairs (n, ℓ) the even lattices Λ of level ℓ (i.e. $\sqrt{\ell}\Lambda^*$ is also even) with $\det \Lambda = \ell^{n/2}$; for $(n, \ell) = (12, 7)$, they found 395 isometry classes, and among them no extremal modular lattice.

If an extremal lattice were to exist for $(n, \ell) = (18, 7)$, it would set new records of density. Bachoc and Venkov proved recently in [B-V2] that no such lattice exists: their proof involves spherical designs.

• Extremal lattices of jump dimensions are specially wanted, since they often achieve the best known density, like in the following examples:

Minimum 2. $\mathbb{D}_4$ $((n, \ell) = (4, 2))$; $\mathbb{E}_8$ $((n, \ell) = (8, 1))$.

Minimum 4. K_{12} $((n, \ell) = (12, 3))$; BW_{16} $((n, \ell) = (16, 2))$; the Leech lattice $((n, \ell) = (24, 1))$.

Minimum 6. $(n, \ell) = (32, 2)$: 4 known lattices, Quebbemann discovered the first one (denoted Q_{32} in [C-S1]) in 1984; $(n, \ell) = (48, 1)$: 3 known lattices P_{48p}, P_{48q} from coding theory, and a "cyclo-quaternionic" lattice by Nebe.

• Extremal even unimodular lattices are known for any dimension $n \equiv 0 \mod 8$, $n \leq 80$, except for $n = 72$, which would set a new record of density. The case $n = 80$ was recently solved by Bachoc and Nebe. The corresponding Hermite invariant $\gamma = 8$ (largely over the upper bound for period lattices) does not hold the present record for dimension 80, established at 8,0194 independently by Elkies and Shioda. The same phenomenon appeared at dimension 56.

We give in section 7 Hermitian constructions for most of the above extremal lattices, making obvious their symplectic nature.

5. Voronoi's theory

5.1 Local theory. In section 4, we looked for extremal lattices, which (if any) maximize the Hermite invariant in the (finite) set of modular lattices for a given pair (n, ℓ). In the present section, we go back to the classical notion of an *extreme lattice*, where the Hermite invariant γ achieves a local maximum. Here, the existence of such lattices stems from Mahler's compactness theorem. The same argument applies when we study the local maxima of density in some natural families of lattices such as isodual lattices, lattices with prescribed automorphisms etc. These families share a common structure: *their connected components are orbits of one lattice under the action of a closed subgroup $\mathcal{G}$ of* $\mathrm{GL}(E)$ *invariant under transpose.*

For such a family $\mathcal{F}$, we can give a unified characterization of the strict local maxima of density. In order to point out the connection with Voronoi's classical theorem

a lattice is extreme if and only if it is perfect and eutactic,

we mostly adopt in the following the point of view of Gram matrices. We denote by $\mathrm{Sym}_n(\mathbb{R})$ the space of $n \times n$ symmetric matrices equipped with the scalar product $< M, N >= \mathrm{Trace}(MN)$. The value at $v \in \mathbb{R}^n$ of a quadratic form A is then ${}^t vAv =< A, v^t v >$.

5.2 Perfection, eutaxy and extremality. Let $\mathcal{G}$ be a closed subgroup of $\mathrm{SL}_n(\mathbb{R})$ stable under transpose, and let $\mathcal{F} = \{{}^tPAP, P \in \mathcal{G}\}$ be the orbit of a positive definite matrix $A \in \mathrm{Sym}_n(\mathbb{R})$. We denote by $\mathcal{T}_A$ the *tangent space to the manifold* $\mathcal{F}$ *at* A, and we recall that $S(A)$ stands for the set of the minimal vectors of A.

• Let $v \in \mathbb{R}^n$. The gradient at A (with respect to $<,>$) of the function $\mathcal{F} \to \mathbb{R}^+$ $A \mapsto < A, v^t v > \det A^{-1/n}$ is the orthogonal projection $\nabla_v = \mathrm{proj}_{\mathcal{T}_A}(v^t v)$ of $v^t v$ onto the tangent space at A.

The $\mathcal{F}$-*Voronoi domain* of A is

$$\mathcal{D}_A = \text{ convex hull } \{\nabla_v,\ v \in S(A)\}.$$

We say that A is $\mathcal{F}$-*perfect* if the affine dimension of $\mathcal{D}_A$ is maximum ($= \dim \mathcal{T}_A$), and *eutactic* if the projection of the matrix A^{-1} lies in the interior of $\mathcal{D}_A$.

These definitions reduce to the traditional ones when we take for $\mathcal{F}$ the whole set of positive $n \times n$ matrices (and $\mathcal{T}_A = \mathrm{Sym}_n(\mathbb{R})$). But in this survey we focus on families $\mathcal{F}$ naturally normalized to determinant 1: the tangent space at A to such a family is orthogonal to the line $\mathbb{R}A^{-1}$, and the eutaxy condition reduces to "$0 \in \overset{\circ}{\mathcal{D}}_A$".

• The matrix A is called $\mathcal{F}$-*extreme* if γ achieves a local maximum at A among all matrices in $\mathcal{F}$. We say that A is *strictly* $\mathcal{F}$-*extreme* if there is a neighbourhood V of A in $\mathcal{F}$ such that the strict inequality $\gamma(A') < \gamma(A)$ holds for every $A' \in V$, $A' \neq A$.

• The above concepts are connected by the following result.

THEOREM ([B-M]). *The matrix A is strictly $\mathcal{F}$-extreme if and only if it is $\mathcal{F}$-perfect and $\mathcal{F}$-eutactic.*

The crucial step in studying the Hermite invariant in an individual family $\mathcal{F}$ is then to check the strictness of any local maximum. A sufficient condition is that any $\mathcal{F}$-extreme matrix should be *well rounded*, i.e. that its minimal vectors should span the space $\mathbb{R}^n$. It was proved by Voronoi in the classical case.

5.3 Isodual lattices. Let σ be an isometry of E with a given integral representation S. Then we can parametrize the family of σ-isodual lattices by the Lie group and symmetrized tangent space at identity

$$\mathcal{G} = \{P \in \mathrm{GL}_n(\mathbb{R}) \mid {}^tP^{-1} = SPS^{-1}\}, \qquad \mathcal{T}_I = \{X \in \mathrm{Sym}_n(\mathbb{R}) \mid SX = -XS\}.$$

The answer to the question

does σ-extremality imply strict σ-extremality?

depends on the representation afforded by $\sigma \in O(E)$. It is positive for symplectic or orthogonal lattices. A minimal counter-example is given by a three-dimensional rotation σ of order 4: the corresponding isodual lattices are decomposable (see [C-S2], th. 1), and the Hermite invariant for this family attains its maximum 1 on a subvariety of dimension 2 (up to isometry).

In [Qi-Z], Voronoi's condition for symplectic lattices was given a suitable complex form. It holds for the Conway and Sloane lattice $M(\mathbb{E}_6)$ (and of course for the Barnes lattice P_6 which is extreme in the classical sense) but not for the lattice $\mathbb{D}_6^+$. (An alternative proof involving differential geometry was given in [Bav1].)

In dimension 5 et 7, the most likely candidates for densest isodual lattices were also discovered by Conway and Sloane; they were successfully tested for isodualities σ of orthogonal type (of respective signatures $(4,1)$ and $(4,3)$). In dimension 3, Conway and Sloane proved by classification and direct calculation that the so called m.c.c. isodual lattice is the densest one (actually, there are only 2 well rounded isodual lattices, m.c.c. and the cubic lattice).

5.4 Extreme modular lattices. The classical theory of extreme lattices was recently revisited by B. Venkov [Ve] in the setting of spherical designs. That the set of minimal vectors of a lattice be a spherical 2- or 4-design is a strong form of the conditions of eutaxy (equal coefficients) or extremality.

An extremal ℓ-modular lattice is not necessarily extreme: the even unimodular lattice $\mathbb{E}_8 \perp \mathbb{E}_8$ has minimum 2, hence is extremal, but as a decomposable lattice, it could not be perfect. By use of the modular properties of some theta series with spherical coefficients, Bachoc and Venkov proved ([B-V2]) that this phenomenon could not appear near the "jump dimensions": in particular, any extremal ℓ-modular lattice of dimension n such that $(\ell = 1, and\ n \equiv 0, 8 \mod 24)$, or $(\ell = 2, and\ n \equiv 0, 4 \mod 16)$, or $(\ell = 3, and\ n \equiv 0, 2 \mod 12)$, is extreme.

This applies to the famous lattices quoted in section 4. [For some of them, alternative proofs of the Voronoi conditions could be done, using the automorphism groups (for eutaxy), testing perfection modulo small primes, or inductively in the case of laminated lattices.]

5.5 Classification of extreme lattices. Voronoi established that there are only finitely many equivalence classes of perfect matrices, and he gave an algorithm for their enumeration.

Let A be a perfect matrix, and $\mathcal{D}_A$ its traditional Voronoi domain. It is a polyhedron of maximal dimension $N = n(n+1)/2$, with a finite number of hyperplane faces. Such a face $\mathcal{H}$ of $\mathcal{D}_A$ is simultaneously a face for the domain of exactly one other perfect matrix, called *the neighbour of A across the face $\mathcal{H}$*.

We get, in taking the dual polyhedron, a graph whose edges describe the neighbouring relations; this graph has finitely many inequivalent vertices. Voronoi proved that this graph is connected, and he used it up to dimension 5 to confirm the classification by Korkine and Zolotarev. His attempt for dimension 6 was completed in 1957 by Barnes. Complete classification for dimension 7 was done by Jaquet in 1991 using this method. Recently implemented by Batut in dimension 8, Voronoi's algorithm produced, by neighbouring only matrices with $s = N, N+1$ and $N+2$, exactly 10916 inequivalent perfect lattices. There may exist some more.

This algorithm was extended in [B-M-S] to matrices invariant under a given finite group $\Gamma \subset \mathrm{GL}_n(\mathbb{Z})$): it works in the centralizer of Γ in $\mathrm{Sym}_n(\mathbb{R})$.

That there are only finitely many isodual-extreme lattices of type symplectic or orthogonal stems from their well roundness. But the present extensions of Voronoi's algorithm are very partial (see section 6).

5.6 Voronoi's paths and isodual lattices. The densest known isodual lattices discovered by Conway and Sloane up to seven dimensions were found on paths connecting, in the lattice space, the densest lattice Λ to its dual Λ^*: such a path turns out to be stable under a fixed duality, and the isodual lattice $M(\Lambda)$ is the fixed point for this involution. In [C-S2], these paths were constructed by gluing theory.

Actually, the Voronoi algorithm for perfect lattices provides another interpretation of them. In dimensions 6 and 7, the densest lattices $\mathbb{E}_6$ and $\mathbb{E}_7$ and their respective duals are Voronoi neighbours of each other. The above mentioned paths $\Lambda - \Lambda^*$ are precisely the corresponding neighbouring paths. For dimensions 3 and 5, we need a group action: for dimension 5, we use the regular representation Γ of the cyclic group of order 5, and the path $\mathbb{D}_5 - \mathbb{D}_5^*$ contains the Γ-neighbouring path leading from $\mathbb{D}_5$ to the perfect lattice $\mathbb{A}_5^3$; for dimension 3, we use the augmentation representation of the cyclic group of order 4, and the Conway and Sloane path $\Lambda - \Lambda^*$ is part of the Γ-neighbouring path leading from $\Lambda = \mathbb{A}_3$ to the Γ-perfect lattice called "axial centered cuboidal" in [C-S2].

5.7 Eutaxy. The first proof of the finiteness of the set of eutactic lattices (for a given dimension and up to similarity) was given by Ash ([A]) by means of Morse theory: the Hermite invariant γ is a topological Morse function, and the eutactic lattices are exactly its non-degenerate critical points. Bavard proved in [Bav1] that γ is no more a Morse function on the space of symplectic lattices of dimension $2g \geq 4$; in particular, one can construct continuous arcs of critical points, such as the following set of symplectic-eutactic 4×4 matrices

$$\{\left(\begin{smallmatrix} I & O \\ O & A \end{smallmatrix}\right), \qquad A \in \mathrm{SL}_2(\mathbb{R}) \text{ s.t. } m(A) > 1\}.$$

6. Hyperbolic families of symplectic lattices

This section surveys a recent work by Bavard: in [Bav2] he constructs families of $2g$-dimensional symplectic lattices for which he his able to recover the local and global Voronoi theory, as well as Morse's theory. The convenient frame for these constructions is the Siegel space $\mathfrak{h}_g = \{X + iY \in \mathrm{Sym}_g(\mathbb{C}) \mid Y > 0\}$, modulo homographic action by the symplectic group $\mathrm{Sp}_{2g}(\mathbb{Z})$.

In these families most of the important lattices ($\mathbb{E}_8, K_{12}, BW_{16}$, Leech ...) and many others appear with fine Siegel's representations $Z = X + iY$.

6.1 Definition.

In the following we fix an integral positive symmetric $g \times g$ matrix M.
To any complex number $z = x + iy, y > 0$ in the Poincaré upper half plane $\mathfrak{h}$, we attach the complex matrix $zM = xM + iyM \in \mathfrak{h}_g$, and we consider the family

$$\mathcal{F} = \{zM, z \in \mathfrak{h}\} \subset \mathfrak{h}_g.$$

On can check that the homographic action of $\left(\begin{smallmatrix} \alpha & \beta \\ \gamma & \delta \end{smallmatrix}\right) \in \mathbb{P}SL_2(\mathbb{R})$ on $\mathfrak{h}$ corresponds to the homographic action of $\left(\begin{smallmatrix} \alpha I & \beta M \\ \gamma M^{-1} & \delta I \end{smallmatrix}\right) \in \mathrm{Sp}_{2g}(\mathbb{R})$ on $\mathcal{F}$. This last matrix is integral when $\left(\begin{smallmatrix} \alpha & \beta \\ \gamma & \delta \end{smallmatrix}\right)$ lies in a convenient congruence subgroup $\Gamma_0(d)$ of $\mathrm{SL}_2(\mathbb{Z})$ (one may take $d = \det M$), thus up to symplectic isometries of lattices, one can restrict the parameter z to a fundamental domain for $\Gamma_0(d)$ in $\mathfrak{h}$.

The symplectic Gram matrix A_z associated to $z = x + iy, y > 0 \in \mathfrak{h}$, as defined in 2.2, is given by

$$A_z = \frac{1}{y} \begin{pmatrix} M^{-1} & xI \\ xI & |z|^2 M \end{pmatrix}.$$

For $g = 1$ and M a positive integer, this is the general 2×2 positive matrix of determinant 1, and we recover the usual representation in $\mathfrak{h}/\mathbb{P}\,\mathrm{SL}_2(\mathbb{Z})$ of the 2-dimensional lattices.

6.2 Voronoi's theory. The Voronoi conditions of eutaxy and perfection for the family $\mathcal{F}$, as defined in 5.2, can be translated in the space $\mathfrak{h}$ of the parameters, equipped with its *Poincaré metric* $ds = \frac{|dz|}{y}$.

• Fix $z \in \mathfrak{h}$, and for any $v \in \mathbb{R}^{2g}$ denote by ∇_v the (hyperbolic) gradient at z of the function $z \mapsto {}^t v A_z v$; we can represent the Voronoi domain of A_z by the convex hull $\mathcal{D}_z$ in $\mathbb{C}$ of the $\nabla_v, v \in S(A_z)$; it has affine dimension $0, 1$ or 2, this maximal value means "perfection" for z. As defined in 5.2, z is eutactic if there exist strictly positive coefficients c_v such that $\sum_{v \in S(A_z)} c_v \nabla_v = 0$.

Bavard showed that Voronoi's and Ash's theories hold for the family $\mathcal{F}$:

Strict extremality $\Leftrightarrow$ *extremality* $\Leftrightarrow$ *perfection and eutaxy,*

Hermite's function is a Morse function, its critical points are the eutactic ones.

Actually, these results are connected to the strict convexity of the Hermite function on the family $\mathcal{F}$ (see [Bav1] for a more general setting).

REMARK. In the above theory, the only gradients that matter are the extremal points of the convex $\mathcal{D}_z$. Following Bavard, we call *principal* the corresponding minimal vectors. In the classical theory, all minimal vectors are principal; this is no more true in its various extensions.

• The next step towards a global study of γ in $\mathcal{F}$ was to get an hyperbolic interpretation of the values ${}^t v A_z v$. To this purpose, Bavard represents any vector $v \in \mathbb{R}^{2g}$ by a point $p \in \mathfrak{h} \cup \partial\mathfrak{h}$ in such a way that for all $z \in \mathfrak{h}$, ${}^t v A_z v$ is an exponential function of the hyperbolic distance $d(p, z)$ (suitably extended to the boundary $\partial\mathfrak{h}$). In particular, there is a discrete set $\mathcal{P}$ corresponding to the principal minimal vectors.

• There is now a simple description of the Voronoi theory for the family $\mathcal{F}$. We consider the Dirichlet-Voronoi tiling of the metric space $(\mathfrak{h}, d)$ attached to the set $\mathcal{P}$: the cell around p is $C_p = \{z \in \mathfrak{h} \mid \ d(z,p) \le d(z,q) \text{ for all } q \in \mathcal{P}\}$.

We then introduce the dual partition: the Delaunay cell of $z \in \mathfrak{h}$ is the convex hull of the points of $\mathcal{P}$ closest to z (for the Poincaré metric), hence it can figure the Voronoi domain $\mathcal{D}_z$. As one can imagine, the $\mathcal{F}$-perfect points are the vertices of the Dirichlet-Voronoi tiling, and the $\mathcal{F}$-eutactic points are those which lie in the interior of their Delaunay cell.

The 1-skeleton of the Dirichlet-Voronoi tiling is the graph of the neighbouring relation between perfect points. Bavard proved that it is connected, and finite modulo the convenient congruence subgroup. For a detailed description of the algorithm, we refer the reader to [Bav2], 1.5 and 1.6.

6.3 Some examples.

• For $M = \mathbb{D}_g$ $(g \geq 3)$, the algorithm only produces one perfect point $z = \frac{1+i}{2}$, corresponding to the so-called lattice $\mathbb{D}_{2g}^+$ (see next section).

• The choice $M = \mathbb{A}_g, g \geq 1$ is much less disappointing: it produces many symplectic-extreme lattices, among them $\mathbb{A}_2, \mathbb{D}_4, P_6, \mathbb{E}_8, K_{12}$.

• The densest lattices in the families attached to the Barnes lattices $M = P_g, g = 8, 12$ are the Barnes-Wall lattice BW_{16} and the Leech lattice.

• However, the union of the hyperbolic families of given dimension $2g$ has only dimension $g(g+1)/2 \ + 1$; hence it is no wonder that it misses some beautiful symplectic lattices, for instance the lattice $M(\mathbb{E}_6)$.

7. Other constructions

7.1 Hermitian lattices. Let K be a C.M. field or a totally definite quaternion algebra, and let $\mathfrak{M}$ be a maximal order of K. All the above-mentioned famous lattices (in even dimensions) can be constructed as $\mathfrak{M}$-modules of rank k equipped with the scalar product trace$(\alpha x.\overline{y})$, where trace is the reduced trace $K/\mathbb{Q}$, $x.\overline{y}$ the standard Hermitian inner product on $\mathbb{R} \otimes_{\mathbb{Q}} K^k$ and $\alpha \in K$ some convenient totally positive element (see for example [Bay]). We see in the following examples that such a construction often provides natural symplectic isodualities and automorphisms.

Lattices $\mathbb{D}_{2g}^+, g \geq 3$. Here $\mathfrak{M} = \mathbb{Z}[i] \subset \mathbb{C}$ is the ring of Gaussian integers. We consider in $\mathbb{C}^g$ equipped with the scalar product $\frac{1}{2}$ trace$(x.\overline{y})$ the lattice $\{x = (x_1, x_2, \cdots, x_g) \in \mathfrak{M}^g \mid x_1 + x_2 + \cdots + x_g \equiv 0 \mod (1+i)\}$ which is isometric to the root lattice $\mathbb{D}_{2g}$. Now we consider the conjugate elements $e = \frac{1}{1+i}(1, 1, \cdots, 1)$ and $\overline{e}$ $(= ie = (1, 1, \ldots, 1) - e)$ of $\mathbb{C}^g$; then the sets $\mathbb{D}_{2g}^+ = \mathbb{D}_{2g} \cup (e + \mathbb{D}_{2g})$ and $\mathbb{D}_{2g}^- = \mathbb{D}_{2g} \cup (\overline{e} + \mathbb{D}_{2g})$ turn out to be dual lattices, that coincide when g is even. In any case, the multiplication by i provides a symplectic isoduality. An obvious group of Hermitian automorphisms consist of permutations of the x_i's and even sign changes. Thus, comparing its order $2^g g!$ to the Hurwitz bound $(2)84(g-1)$, one sees that, except for $g = 3$, no lattice of the family is a Jacobian (in the opposite direction, all lattices, except for $g = 3$, are extreme in the Voronoi sense).

Barnes-Wall lattices $BW_{2^k}, k \geq 2$. Here, K is the quaternion field $\mathbb{Q}_{2,\infty}$ defined over $\mathbb{Q}$ by elements i, j such that $i^2 = j^2 = -1, ji = -ij$, $\mathfrak{M}$ is the Hurwitz order ($\mathbb{Z}$-module generated by $(1, i, j, \omega)$ where $\omega = 1/2(1 + i + j + ij)$), and we consider the two-sided ideal $\mathfrak{A} = (1+i)\mathfrak{M}$ of $\mathfrak{M}$. Starting from $M_0 = \mathfrak{A}$, we define inductively the right and left $\mathfrak{M}$-modules

$$M_{k+1} = \{(x, y) \in M_k \times M_k \mid x \equiv y \mod \mathfrak{A} M_k\} \quad \subset K^{2^{k+2}},$$

and for k odd (resp. even) we put $L_k = M_k$ (resp. $\mathfrak{A}^{-1} M_k$). For the scalar product $\frac{1}{2}$ trace$(x.\overline{y})$, we have $L_0 \sim \mathbb{D}_4$, $L_1 \sim \mathbb{E}_8$ and generally $L_k \sim BW_{2^{2k+2}}$. These lattices are alternatively 2-modular and unimodular: the right multiplication by i (resp. $j - i$) for k odd (resp. k even) provides a symplectic similarity σ from $L_k^* = L_k$ (resp. $\mathfrak{A}^{-1} L_k$) onto L_k. Using their logarithmic bound for the density of a period lattice, Buser and Sarnak proved that the Barnes-Wall lattices are certainly not Jacobians for $k \geq 5$. As usual, an argument of automorphisms extends this result for $1 \leq k \leq 4$: the group $\mathrm{Aut}_\sigma(L_k)$ embeds diagonally into $\mathrm{Aut}_\sigma(L_{k+2})$, and adding transpositions and sign changes, one obtains a subgroup of $\mathrm{Aut}_\sigma(L_{k+2})$ of order $2^7 \times \mathrm{Aut}_\sigma(L_k)$; starting from the subgroup of automorphisms of $\mathbb{D}_4$ given by left multiplication by the 24 units of $\mathfrak{M}$, or from the group $\mathrm{Aut}_\sigma(\mathbb{E}_8)$, one sees that $|\mathrm{Aut}_\sigma(L_k)|$ is largely over the Hurwitz bound.

Hermitian extensions of scalars. Let $\mathfrak{M}$ the ring of integers of a imaginary quadratic field. Following [G], Bachoc and Nebe show in [B-N] that by tensoring over $\mathfrak{M}$ a modular $\mathfrak{M}$-lattice, one can shift from one level to another; this construction preserves the symplectic nature of the isoduality, and hopefully, the minimum. For this last question, we refer to [Cou].

In [B-N], $\mathfrak{M}$ is the ring of integers of the quadratic field of discriminant -7. Let L_r be an $\mathfrak{M}$-lattice of rank r, unimodular with respect to its Hermitian structure, and consider the $\mathfrak{M}$-lattices $L_{2r} = (\mathbb{A}_2 \perp \mathbb{A}_2) \otimes_{\mathfrak{M}} L_r$ and $L_{4r} = \mathbb{E}_8 \otimes_{\mathfrak{M}} L_r$. By a determinant argument, one sees that (for the usual scalar product) the $\mathbb{Z}$-lattices L_r, L_{2r} and L_{4r} are symplectic modular lattices of respective levels $7, 3$ and 1. Starting from the Barnes lattice P_6, Gross obtained the Coxeter-Todd lattice K_{12} and the Leech lattice, of minimum 4. The same procedure was applied in [B-N] to a 20-dimensional lattice appearing in the ATLAS in connection with the Mathieu group M_{22}, and led to the first known extremal modular lattices of minimum 8, and respective dimensions 40 and 80. Note that while coding theory was involved in the original proof of the extremality of the unimodular lattices of dimension 80, an alternative "à la Kitaoka" proof is given in [Cou].

7.2 Exterior power. Let $1 \le k < n$ be two integers, and let E be a Euclidean space of dimension n; its exterior powers carry a natural scalar product which makes the canonical map $\sigma : \bigwedge^{n-k} E \to (\bigwedge^k E)^*$ an isometry $\bigwedge^{n-k} E \to (\bigwedge^k E)$ of square $(-1)^{k(n-k)}$. Let L be a lattice in E. It is shown in [Cou1] that σ maps the lattice $\bigwedge^{n-k} L$ onto $\sqrt{\det L}(\bigwedge^k L)^*$. In particular, when $n = 2k$, σ is a symplectic or orthogonal similarity of the $\binom{2k}{k}$-dimensional lattice $(\bigwedge^k L)$ onto its dual; if moreover L is integral, the lattice $(\bigwedge^k L)$ is modular of level $\det L$.

For instance, the lattice $\bigwedge^2 \mathbb{D}_4$ is isometric to $\mathbb{D}_6^+$, and the lattice $\bigwedge^3 \mathbb{E}_6$, in 20 dimensions, is 3-modular of symplectic type, with minimum 4, thus extremal.

Remark. Exterior even powers of unimodular lattices are unimodular lattices of special interest for the theory of group representations. Let us come back to the notation of this subsection. For even k, the canonical map $\mathrm{Aut}\, L \to \mathrm{Aut} \bigwedge^k L$ has kernel ± 1, and induces an embedding $\mathrm{Aut}\, L/(\pm 1) \hookrightarrow \mathrm{Aut} \bigwedge^k L$. Actually, the exterior squares of the lattice $\mathbb{E}_8$ and the Leech lattice provide faithful representations of minimal degrees of the group $O_8^+(2)$ and of the Conway group Co_1 respectively.

7.3 Group representations. Many important symplectic modular lattices were discovered by Nebe, Plesken ([N-P]) and Souvignier ([Sou]) while investigating finite rational matrix groups. In [S-T], Scharlau and Tiep, using symplectic groups over $\mathbb{F}_p$, construct large families of symplectic unimodular lattices, among them that of dimension 28 discovered by combinatorial devices by Bacher and Venkov in [B-V1].

I am indebted to C. Bavard and J. Martinet for helpful discussions when I was writing this survey. I am also grateful for the improvements they suggested after reading the first drafts of this paper.

References

[A] A. Ash, *On eutactic forms*, Can. J. Math. **29** (1977), 1040–1054.

[B-N] C. Bachoc, G. Nebe, *Extremal lattices of minimum 8 related to the Mathieu group M_{22}*, J. reine angew. Math. **494** (1998), 155–171.

[B-V1] R. Bacher, B. Venkov, *Lattices and association schemes: a unimodular example without roots in dimension 28*, Ann. Inst. Fourier, Grenoble **45** (1995), 1163–1176.

[B-V2] C. Bachoc, B. Venkov, *Modular forms, lattices and spherical designs*, preprint, Bordeaux, 1998.

[Bav1] C. Bavard, *Systole et invariant d'Hermite*, J. reine angew. Math. **482** (1997), 93–120.

[Bav2] C. Bavard, *Familles hyperboliques de réseaux symplectiques*, preprint, Bordeaux, 1997.

[Bay] E. Bayer-Fluckiger, *Definite unimodular lattices having an automorphism of given characteristic polynomial*, Comment. Math. Helvet. **59** (1984), 509–538.

[B-M] A.-M. Bergé, J. Martinet, *Densité dans des familles de réseaux. Application aux réseaux isoduaux*, L'Enseignement Mathématique **41** (1995), 335–365.

[B-M-S] A.-M. Bergé, J. Martinet, F. Sigrist, *Une généralisation de l'algorithme de Voronoï*, Astérisque **209** (1992), 137–158.

[Be-S] M. Bernstein and N.J.A. Sloane, *Some lattices obtained from Riemann surfaces*, Extremal Riemann Surfaces, J. R. Quine and P. Sarnak Eds. (Contemporary Math. Vol 201), Amer. Math. Soc., Providence, RI, 1997, pp. 29–32.

[B-S] P. Buser, P. Sarnak, *On the period matrix of a Riemann surface of large genus (with an Appendix by J.H. Conway and N.J.A. Sloane)*, Invent. math. **117** (1994), 27–56.

[C-S1] J.H. Conway, N.J.A. Sloane, *Sphere Packings, Lattices and Groups, Third edition*, Springer-Verlag , Grundlehren 290, Heidelberg, 1999.

[C-S2] J.H. Conway, N.J.A Sloane, *On Lattices Equivalent to Their Duals*, J. Number Theory **48** (1994), 373–382.

[Cou] R. Coulangeon, *Tensor products of Hermitian lattices*, Acta Arith. **92** (2000), 115–130.

[G] B. Gross, *Groups over $\mathbb{Z}$*, Invent. Math. **124** (1996), 263–279.

[M] J. Martinet, *Les réseaux parfaits des espaces euclidiens*, Masson, Paris, 1996.

[M-H] J. Milnor, D. Husemoller, *Symmetric Bilinear Forms*, Springer-Verlag, Ergebnisse 73, Heidelberg, 1973.

[N-P] G. Nebe, W. Plesken, *Finite rational matrix groups*, Memoirs A.M.S. **116** (1995), 1–144.

[N-V] G. Nebe, B. Venkov, *Nonexistence of Extremal Lattices in Certain Genera of Modular Lattices*, J. Number Theory **60** (1996), 310–317.

[Q1] H.-G. Quebbemann, *Modular Lattices in Euclidean Spaces*, J. Number Theory **54** (1995), 190–202.

[Q2] H.-G. Quebbemann, *Atkin-Lehner eigenforms and strongly modular lattices*, L'Enseignement Mathématique **43** (1997), 55–65.

[Qi] J.R. Quine, *Jacobian of the Picard Curve*, Contemp. Math. **201** (1997).

[Qi-Z] J.R. Quine, P. L. Zhang, *Extremal Symplectic Lattices*, Israel J. Math. **108** (1998), 237–251.

[S-T] R. Scharlau, P. H. Tiep, *Symplectic group lattices*, Trans. Amer. Math. Soc. **351** (1999), 2101–2131.

[S-SP] R. Scharlau, R. Schulze-Pillot, *Extremal lattices*, Algorithmic Algebra and Number Theory, B.H. Matzat, G.-M. Greul, G. Hiss ed., Springer-Verlag, Heidelberg, 1999, pp. 139–170.

[Sh] T. Shioda, *A Collection: Mordell-Weil Lattices*, Max-Planck Inst. Math., Bonn (1991).

[Sou] B. Souvignier, *Irreducible finite integral matrix groups of degree 8 and 10*, Math. Comp. **63** (1994), 335–350.

[V] G. Voronoï, *Nouvelles applications des paramètres continus à la théorie des formes quadratiques : 1 Sur quelques propriétés des formes quadratiques positives parfaites*, J. reine angew. Math **133** (1908), 97–178.

[Ve] B. Venkov, notes by J. Martinet, *Réseaux et "designs" sphériques*, preprint, Bordeaux, 1999.

Institut de Mathématiques
Université Bordeaux 1
351 cours de la Libération
33405 Talence Cedex, France

E-mail address: berge@math.u-bordeaux.fr

Contemporary Mathematics
Volume **272**, 2000

Universal Quadratic Forms and the Fifteen Theorem

J. H. Conway

ABSTRACT. This paper is an extended foreword to the paper of Manjul Bhargava [**1**] in these proceedings, which gives a short and elegant proof of the Conway-Schneeberger Fifteen Theorem on the representation of integers by quadratic forms.

The representation theory of quadratic forms has a long history, starting in the seventeenth century with Fermat's assertions of 1640 about the numbers represented by $x^2 + y^2$. In the next century, Euler gave proofs of these and some similar assertions about other simple binary quadratics, and although these proofs had some gaps, they contributed greatly to setting the theory on a firm foundation.

Lagrange started the theory of universal quadratic forms in 1770 by proving his celebrated Four Squares Theorem, which in current language is expressed by saying that the form $x^2+y^2+z^2+t^2$ is universal. The eighteenth century was closed by a considerably deeper statement – Legendre's Three Squares Theorem of 1798; this found exactly which numbers needed all four squares. In his *Theorie des Nombres* of 1830, Legendre also created a very general theory of binary quadratics.

The new century was opened by Gauss's *Disquisitiones Arithmeticae* of 1801, which brought that theory to essentially its modern state. Indeed, when Neil Sloane and I wanted to summarize the classification theory of binary forms for one of our books [**3**], we found that the only Number Theory textbook in the Cambridge Mathematical Library that handled every case was still the *Disquisitiones*! Gauss's initial exploration of ternary quadratics was continued by his great disciple Eisenstein, while Dirichlet started the analytic theory by his class number formula of 1839.

As the nineteenth century wore on, other investigators, notably H. J. S. Smith and Hermann Minkowski, explored the application of Gauss's concept of the genus to higher-dimensional forms, and introduced some invariants for the genus from which in this century Hasse was able to obtain a complete and very simple classification of rational quadratic forms based on Hensel's notion of "p-adic number", which has dominated the theory ever since.

In 1916, Ramanujan started the byway that concerns us here by asserting that

$$\begin{array}{lllllll}
[1,1,1,1], & [1,1,1,2], & [1,1,1,3], & [1,1,1,4], & [1,1,1,5], & [1,1,1,6], & [1,1,1,7], \\
[1,1,2,2], & [1,1,2,3], & [1,1,2,4], & [1,1,2,5], & [1,1,2,6], & [1,1,2,7], & [1,1,2,8], \\
[1,1,2,9], & [1,1,2,10], & [1,1,2,11], & [1,1,2,12], & [1,1,2,13], & [1,1,2,14], & [1,1,3,3], \\
[1,1,3,4], & [1,1,3,5], & [1,1,3,6], & [1,2,2,2], & [1,2,2,3], & [1,2,2,4], & [1,2,2,5], \\
[1,2,2,6], & [1,2,2,7], & [1,2,3,3], & [1,2,3,4], & [1,2,3,5], & [1,2,3,6], & [1,2,3,7], \\
[1,2,3,8], & [1,2,3,9], & [1,2,3,10], & [1,2,4,4], & [1,2,4,5], & [1,2,4,6], & [1,2,4,7], \\
[1,2,4,8], & [1,2,4,9], & [1,2,4,10], & [1,2,4,11], & [1,2,4,12], & [1,2,4,13], & [1,2,4,14], \\
[1,2,5,5], & [1,2,5,6], & [1,2,5,7], & [1,2,5,8], & [1,2,5,9], & [1,2,5,10] &
\end{array}$$

were all the diagonal quaternary forms that were universal in the sense appropriate to positive-definite forms, that is, represented every positive integer. In the rest of this paper, "form" will mean "positive-definite quadratic form", and "universal" will mean "universal in the above sense".

Although Ramanujan's assertion later had to be corrected slightly by the elision of the diagonal form $[1,2,5,5]$, it aroused great interest in the problem of enumerating all the universal quaternary forms, which was eagerly taken up, by Gordon Pall and his students in particular. In 1940, Pall also gave a complete system of invariants for the genus, while simultaneously Burton Jones found a system of canonical forms for it, so giving two equally definitive solutions for a problem raised by Smith in 1851.

There are actually two universal quadratic form problems, according to the definition of "integral" that one adopts. The easier one is that for Gauss's notion, according to which a form is integral only if not only are all its coefficients integers, but the off-diagonal ones are even. This is sometimes called "classically integral", but we prefer to use the more illuminating term "integer-matrix", since what is required is that the matrix of the form be comprised of integers. The difficult universality problem is that for the alternative notion introduced by Legendre, under which a form is integral merely if all its coefficients are. We describe such a form as "integer-valued", since the condition is precisely that all the values taken by the form are integers, and remark that this kind of integrality is the one most appropriate for the universality problem, since that is about the values of forms.

For nearly 50 years it has been supposed that the universality problem for quaternary integer-matrix forms had been solved by M. Willerding, who purported to list all such forms in 1948. However, the 15-theorem, which I proved with William Schneeberger in 1993, made it clear that Willerding's work had been unusually defective. In his paper in these proceedings, Manjul Bhargava [**1**] gives a very simple proof of the 15-theorem, and derives the complete list of universal quaternaries. As he remarks, of the 204 such forms, Willerding's purportedly complete list of 178 contains in fact only 168, because she missed 36 forms, listed 1 form twice, and listed 9 non-universal forms!

The 15-theorem closes the universality problem for integer-matrix forms by providing an extremely simple criterion. We no longer need a list of universal quaternaries, because a form is universal provided only that it represent the numbers up to 15. Moreover, this criterion works for larger numbers of variables, where the number of universal forms is no longer finite. (It is known that no form in three or fewer variables can be universal.)

I shall now briefly describe the history of the 15-theorem. In a 1993 Princeton graduate course on quadratic forms, I remarked that a reworking of Willerding's enumeration was very desirable, and could probably be achieved very easily in view of recent advances in the representation theory of quadratic forms, most particularly the work of Duke and Schultze-Pillot. Moreover, it was an easy consequence of this work that there must be a constant c with the property that if a matrix-integral form represented every positive integer up to c, then it was universal, and a similar but probably larger constant C for integer-valued forms. At that time, I feared that perhaps these constants would be very large indeed, but fortunately it appeared that they are quite small.

I started the next lecture by saying that we might try to find c, and wrote on the board a putative

THEOREM 0.1. *If an integer-matrix form represents every positive integer up to c (to be found!) then it is universal.*

We started to prove that theorem, and by the end of the lecture had found the 9 ternary "escalator" forms (see Bhargava's article [**1**] for their definition) and realised that we could almost as easily find the quaternary ones, and made it seem likely that c was much smaller than we had expected.

In the afternoon that followed, several class members, notably William Schneeberger and Christopher Simons, took the problem further by producing these forms and exploring their universality by machine. These calculations strongly suggested that c was in fact 15.

In subsequent lectures we proved that most of the 200+ quaternaries we had found were universal, so that when I had to leave for a meeting in Boston only nine particularly recalcitrant ones remained. In Boston I tackled seven of these, and when I returned to Princeton, Schneeberger and I managed to polish the remaining two off, and then complete this to a proof of the 15-theorem, modulo some computer calculations that were later done by Simons.

The arguments made heavy use of the notion of genus, which had enabled the nineteenth-century workers to extend Legendre's Three Squares theorem to other ternary forms. In fact the 15-theorem largely reduces to proving a number of such analogues of Legendre's theorem. Expressing the arguments was greatly simplified by my own symbol for the genus, which was originally derived by comparing Pall's invariants with Jones's canonical forms, although it has since been established more simply; see for instance my recent little book [**2**].

Our calculations also made it clear that the larger constant C for the integer-valued problem would almost certainly be 290, though obtaining a proof of the resulting "290-conjecture" would be very much harder indeed. Last year, in one of our semi-regular conversations I tempted Manjul Bhargava into trying his hand at the difficult job of proving the 290-conjecture.

Manjul started the task by reproving the 15-theorem, and now he has discovered the particularly simple proof he gives in the following paper, which has made it unnecessary for us to publish our rather more complicated proof. Manjul has also proved the "33-theorem" – much more difficult than the 15-theorem – which asserts that an integer-matrix form will represent all odd numbers provided only that it represents 1, 3, 5, 7, 11, 15, and 33. This result required the use of some very clever and subtle arithmetic arguments.

Finally, using these arithmetic arguments, as well as new analytic techniques, Manjul has made significant progress on the 290-conjecture, and I would not be surprised if the conjecture were to be finished off in the near future! He intends to publish these and other related results in a subsequent paper.

References

[1] M. Bhargava, *On the Conway-Schneeberger Fifteen Theorem*, these proceedings.

[2] J. H. Conway, *The sensual (quadratic) form*, Carus Mathematical Monographs **26**, MAA, 1997.

[3] J. H. Conway and N. A. Sloane, *Sphere packings, lattices and groups*, Grundlehren der Mathematischen Wissenschaften **290**, Springer-Verlag, New York, 1999.

Department of Mathematics
Princeton University
Princeton, NJ 08544
E-mail address: conway@math.princeton.edu

Contemporary Mathematics
Volume **272**, 2000

On the Conway-Schneeberger Fifteen Theorem

Manjul Bhargava

ABSTRACT. This paper gives a proof of the Conway-Schneeberger Fifteen Theorem on the representation of integers by quadratic forms, to which the paper of Conway [**1**] in these proceedings is an extended foreword.

1. Introduction. In 1993, Conway and Schneeberger announced the following remarkable result:

THEOREM 1 ("The Fifteen Theorem"). *If a positive-definite quadratic form having integer matrix represents every positive integer up to 15 then it represents every positive integer.*

The original proof of this theorem was never published, perhaps because several of the cases involved rather intricate arguments. A sketch of this original proof was given by Schneeberger in [**4**]; for further background and a brief history of the Fifteen Theorem, see Professor Conway's article [**1**] in these proceedings.

The purpose of this paper is to give a short and direct proof of the Fifteen Theorem. Our proof is in spirit much the same as that of the original unpublished arguments of Conway and Schneeberger; however, we are able to treat the various cases more uniformly, thereby obtaining a significantly simplified proof.

2. Preliminaries. The Fifteen Theorem deals with quadratic forms that are positive-definite and have integer matrix. As is well-known, there is a natural bijection between classes of such forms and lattices having integer inner products; precisely, a quadratic form f can be regarded as the inner product form for a corresponding lattice $L(f)$. Hence we shall oscillate freely between the language of forms and the language of lattices. For brevity, by a "form" we shall always mean a positive-definite quadratic form having integer matrix, and by a "lattice" we shall always mean a lattice having integer inner products.

A form (or its corresponding lattice) is said to be *universal* if it represents every positive integer. If a form f happens not to be universal, define the *truant* of f (or of its corresponding lattice $L(f)$) to be the smallest positive integer not represented by f.

Important in the proof of the Fifteen Theorem is the notion of "escalator lattice." An *escalation* of a nonuniversal lattice L is defined to be any lattice which is generated by L and a vector whose norm is equal to the truant of L. An *escalator lattice* is a lattice which can be obtained as the result of a sequence of successive escalations of the zero-dimensional lattice.

3. Small-dimensional Escalators. The unique escalation of the zero-dimensional lattice is the lattice generated by a single vector of norm 1. This lattice corresponds to the form x^2 (or, in matrix form, $[\,1\,]$) which fails to represent the number 2. Hence an escalation of $[\,1\,]$ has inner product matrix of the form

$$\begin{bmatrix} 1 & a \\ a & 2 \end{bmatrix}.$$

By the Cauchy-Schwartz inequality, $a^2 \leq 2$, so a equals either 0 or ± 1. The choices $a = \pm 1$ lead to isometric lattices, so we obtain only two nonisometric two-dimensional escalators, namely those lattices having Minkowski-reduced Gram matrices $\begin{bmatrix} 1 & 0 \\ 0 & 1 \end{bmatrix}$ and $\begin{bmatrix} 1 & 0 \\ 0 & 2 \end{bmatrix}$.

If we escalate each of these two-dimensional escalators in the same manner, we find that we obtain exactly 9 new nonisometric escalator lattices, namely those having Minkowski-reduced Gram matrices

$$\begin{bmatrix} 1 & 0 & 0 \\ 0 & 1 & 0 \\ 0 & 0 & 1 \end{bmatrix}, \begin{bmatrix} 1 & 0 & 0 \\ 0 & 1 & 0 \\ 0 & 0 & 2 \end{bmatrix}, \begin{bmatrix} 1 & 0 & 0 \\ 0 & 1 & 0 \\ 0 & 0 & 3 \end{bmatrix}, \begin{bmatrix} 1 & 0 & 0 \\ 0 & 2 & 0 \\ 0 & 0 & 2 \end{bmatrix}, \begin{bmatrix} 1 & 0 & 0 \\ 0 & 2 & 0 \\ 0 & 0 & 3 \end{bmatrix},$$
$$\begin{bmatrix} 1 & 0 & 0 \\ 0 & 2 & 1 \\ 0 & 1 & 4 \end{bmatrix}, \begin{bmatrix} 1 & 0 & 0 \\ 0 & 2 & 0 \\ 0 & 0 & 4 \end{bmatrix}, \begin{bmatrix} 1 & 0 & 0 \\ 0 & 2 & 1 \\ 0 & 1 & 5 \end{bmatrix}, \text{ and } \begin{bmatrix} 1 & 0 & 0 \\ 0 & 2 & 0 \\ 0 & 0 & 5 \end{bmatrix}.$$

Escalating now each of these nine three-dimensional escalators, we find exactly 207 nonisomorphic four-dimensional escalator lattices. All such lattices are of the form $[1] \oplus L$, and the 207 values of L are listed in Table 3.

When attempting to carry out the escalation process just once more, however, we find that many of the 207 four-dimensional lattices do not escalate (i.e., they are universal). For instance, one of the four-dimensional escalators turns out to be the lattice corresponding to the famous four squares form, $a^2 + b^2 + c^2 + d^2$, which is classically known to represent all integers. The question arises: how many of the four-dimensional escalators are universal?

4. Four-dimensional Escalators. In this section, we prove that in fact 201 of the 207 four-dimensional escalator lattices are universal; that is to say, only 6 of the four-dimensional escalators can be escalated once again.

The proof of universality of these 201 lattices proceeds as follows. In each such four-dimensional lattice L_4, we locate a 3-dimensional sublattice

L_3 which is known to represent some large set of integers. Typically, we simply choose L_3 to be unique in its genus; in that case, L_3 represents all integers that it represents locally (i.e., over each p-adic ring $\mathbb{Z}_p$). Armed with this knowledge of L_3, we then show that the direct sum of L_3 with its orthogonal complement in L_4 represents all sufficiently large integers $n \geq N$. A check of representability of n for all $n < N$ finally reveals that L_4 is indeed universal.

To see this argument in practice, we consider in detail the escalations L_4 of the escalator lattice $\begin{bmatrix} 1 & 0 & 0 \\ 0 & 2 & 0 \\ 0 & 0 & 2 \end{bmatrix}$ (labelled (4) in Table 1). The latter 3-dimensional lattice L_3 is unique in its genus, so a quick local calculation shows that it represents all positive integers not of the form $2^e(8k+7)$, where e is even. Let the orthogonal complement of L_3 in L_4 have Gram matrix $[m]$. We wish to show that $L_3 \oplus [m]$ represents all sufficiently large integers.

To this end, suppose L_4 is not universal, and let u be the first integer not represented by L_4. Then, in particular, u is not represented by L_3, so u must be of the form $2^e(8k+7)$. Moreover, u must be squarefree; for if $u = rt^2$ with $t > 1$, then $r = u/t^2$ is also not represented by L_4, contradicting the minimality of u. Therefore $e = 0$, and we have $u \equiv 7 \pmod 8$.

Now if $m \not\equiv 0$, 3 or 7 (mod 8), then clearly $u - m$ is not of the form $2^e(8k+7)$. Similarly, if $m \equiv 3$ or 7 (mod 8), then $u - 4m$ cannot be of the form $2^e(8k+7)$. Thus if $m \not\equiv 0 \pmod 8$, and given that $u \geq 4m$, then either $u - m$ or $u - 4m$ is represented by L_3; that is, u is represented by $L_3 \oplus [m]$ (a sublattice of L_4) for $u \geq 4m$. An explicit calculation shows that m never exceeds 28, and a computer check verifies that every escalation L_4 of L_3 represents all integers less than $4 \times 28 = 112$. It follows that any escalator L_4 arising from L_3, for which the value of m is not a multiple of 8, is universal.

Of course, the argument fails for those L_4 for which m is a multiple of 8. We call such an escalation "exceptional". Fortunately, such exceptional escalations are few and far between, and are easily handled. For instance, an explicit calculation shows that only two escalations of $L_3 = \begin{bmatrix} 1 & 0 & 0 \\ 0 & 2 & 0 \\ 0 & 0 & 2 \end{bmatrix}$ are exceptional (while the other 24 are not); these exceptional cases are listed in Table 2.1. As is also indicated in the table, although these lattices did escape our initial attempt at proof, the universality of these four-dimensional lattices L_4 is still not any more difficult to prove; we simply change the sublattice L_3 from the escalator lattice $\begin{bmatrix} 1 & 0 & 0 \\ 0 & 2 & 0 \\ 0 & 0 & 2 \end{bmatrix}$ to the ones listed in the table, and apply the same argument!

It turns out that all of the 3-dimensional escalator lattices listed in Table 1, except for the one labeled (6), are unique in their genus, so the universality of their escalations can be proved by essentially identical arguments, with just a few exceptions. As for escalator (6), although not unique in its genus, it does represent all numbers locally represented by it except possibly those which are 7 or 10 (mod 12). Indeed, this escalator contains the lattice $\begin{bmatrix} 1 & 0 & 0 \\ 0 & 4 & 2 \\ 0 & 2 & 8 \end{bmatrix}$, which is unique in its genus, and the lattices $\begin{bmatrix} 2 & -2 & 2 \\ -2 & 5 & 2 \\ 2 & 2 & 8 \end{bmatrix}$ and $\begin{bmatrix} 3 & 0 & 0 \\ 0 & 5 & 4 \\ 0 & 4 & 5 \end{bmatrix}$, which together form a genus; a local check shows that the first genus represents all numbers locally represented by escalator (6) which are not congruent to 2 or 3 (mod 4), while the second represents all such numbers not congruent to 1 (mod 3). The desired conclusion follows. (This fact has been independently proven by Kaplansky [**3**] using different methods.)

Knowing this, we may now proceed with essentially the same arguments on the escalations of $L_3 = \begin{bmatrix} 1 & 0 & 0 \\ 0 & 2 & 1 \\ 0 & 1 & 4 \end{bmatrix}$. The relevant portions of the proofs for all nonexceptional cases are summarized in Table 1.

"Exceptional" cases arise only for escalators (4) (as we have already seen), (6), and (7). Two arise for escalator (4). Although four arise for escalator (6), two of them turn out to be nonexceptional escalations of (1) and (8) respectively, and hence have already been handled. Similarly, two arise for escalator (7), but one is a nonexceptional escalation of (9). Thus only five truly exceptional four-dimensional escalators remain, and these are listed in Table 2. In these five exceptional cases, other three-dimensional sublattices unique in their genus are given for which essentially identical arguments work in proving universality. Again, all the relevant information is provided in Table 2.

5. Five-dimensional Escalators. As mentioned earlier, there are 6 four-dimensional escalators which escalate again; they have been italicized in Table 3 and are listed again in the first column of Table 4. A rather large calculation shows that these 6 four-dimensional lattices escalate to an additional 1630 five-dimensional escalators! With a bit of fear we may ask again whether any of these five-dimensional escalators escalate.

Fortunately, the answer is no; all five-dimensional escalators are universal. The proof is much the same as the proof of universality of the four-dimensional escalators, but easier. We simply observe that, for the 6 four-dimensional nonuniversal escalators, all parts of the proof of universality outlined in the second paragraph of Section 4 go through—except for the final check. The final check then reveals that each of these 6 lattices represent every positive integer *except for one single number* n. Hence once a single vector of norm n is inserted in such a lattice, the lattice must automatically become universal. Therefore all five-dimensional escalators are universal. A list of the 6 nonuniversal four-dimensional lattices, together with the single numbers they fail to represent, is given in Table 4.

Since no five-dimensional escalator can be escalated, it follows that there are only finitely many escalator lattices: 1 of dimension zero, 1 of dimension one, 2 of dimension two, 9 of dimension three, 207 of dimension four, and 1630 of dimension five, for a total of 1850.

6. Remarks on the Fifteen Theorem. It is now obvious that

(i) *Any universal lattice L contains a universal sublattice of dimension at most five.*

For we can construct an escalator sequence $0 = L_0 \subseteq L_1 \subseteq \ldots$ within L, and then from Sections 4 and 5, we see that either L_4 or (when defined) L_5 gives a universal escalator sublattice of L.

Our next remark includes the Fifteen Theorem.

(ii) *If a positive-definite quadratic form having integer matrix represents the nine critical numbers* 1, 2, 3, 5, 6, 7, 10, 14, *and* 15, *then it represents every positive integer.*
(Equivalently, the truant of any nonuniversal form must be one of these nine numbers.)

This is because examination of the proof shows that only these numbers arise as truants of escalator lattices.

We note that Remark (ii) is the best possible statement of the Fifteen Theorem, in the following sense.

(iii) *If t is any one of the above critical numbers, then there is a quaternary diagonal form that fails to represent t, but represents every other positive integer.*

Nine such forms of minimal determinant are $[2, 2, 3, 4]$ with truant 1, $[1, 3, 3, 5]$ with truant 2, $[1, 1, 4, 6]$ with truant 3, $[1, 2, 6, 6]$ with truant 5, $[1, 1, 3, 7]$ with truant 6, $[1, 1, 1, 9]$ with truant 7, $[1, 2, 3, 11]$ with truant 10, $[1, 1, 2, 15]$ with truant 14, and $[1, 2, 5, 5]$ with truant 15.

However, there is another slight strengthening of the Fifteen Theorem, which shows that the number 15 is rather special:

(iv) *If a positive-definite quadratic form having integer matrix represents every number below 15, then it represents every number above 15.*

This is because there are only four escalator lattices having truant 15, and as was shown in Section 5, each of these four escalators represents every number greater than 15.

Fifteen is the smallest number for which Remark (iv) holds. In fact:

(v) *There are forms which miss infinitely many integers starting from any of the eight critical numbers not equal to 15.*

Indeed, in each case one may simply take an appropriate escalator lattice of dimension one, two, or three.

(vi) *There are exactly 204 universal quaternary forms.*

An upper bound for the discriminant of such a form is easily determined; a systematic use of the Fifteen Theorem then yields the desired result. We note that the enumeration of universal quaternary forms was announced previously in the well-known work of Willerding [**5**], who found that there are exactly 178 universal quaternary forms; however, a comparison with our tables shows that she missed 36 universal forms, listed one universal form twice, and listed 9 non-universal forms. A list of all 204 universal quaternary forms is given in Table 5; the three entries not appearing among the list of escalators in Table 3 have been italicized.

Acknowledgments. The author wishes to thank Professors Conway and Kaplansky for many wonderful discussions, and for helpful comments on an earlier draft of this paper.

References

[1] J. H. Conway, *Universal quadratic forms and the Fifteen Theorem*, these proceedings.
[2] J. H. Conway and N. A. Sloane, *Sphere packings, lattices and groups*, Grundlehren der Mathematischen Wissenschaften **290**, Springer-Verlag, New York, 1999.
[3] I. Kaplansky, *The first non-trivial genus of positive definite ternary forms*, Math. Comp. **64** (1995), no. 209, 341–345.
[4] W. A. Schneeberger, *Arithmetic and Geometry of Integral Lattices*, Ph.D. Thesis, Princeton University, 1995.
[5] M. E. Willerding, *Determination of all classes of positive quaternary quadratic forms which represent all (positive) integers*, Ph.D. Thesis, St. Louis University, 1948.

	Three-dimensional escalator lattice	Truant	Represents nos. not of the form*	If m	Subtract	Check up to
(1)	$\begin{bmatrix} 1 & 0 & 0 \\ 0 & 1 & 0 \\ 0 & 0 & 1 \end{bmatrix}$	7	$2^e u_7$	$\not\equiv 0 \pmod 8$	m or $4m$	112
				$\equiv 0 \pmod 8$	does not arise	-
(2)	$\begin{bmatrix} 1 & 0 & 0 \\ 0 & 1 & 0 \\ 0 & 0 & 2 \end{bmatrix}$	14	$2^d u_7$	$\not\equiv 0 \pmod{16}$	m or $4m$	224
				$\equiv 0 \pmod{16}$	does not arise	-
(3)	$\begin{bmatrix} 1 & 0 & 0 \\ 0 & 1 & 0 \\ 0 & 0 & 3 \end{bmatrix}$	6	$3^d u_-$	$\not\equiv 0 \pmod 9$	m, $4m$, or $16m$	864
				$\equiv 0 \pmod 9$	does not arise	-
(4)	$\begin{bmatrix} 1 & 0 & 0 \\ 0 & 2 & 0 \\ 0 & 0 & 2 \end{bmatrix}$	7	$2^e u_7$	$\not\equiv 0 \pmod 8$	m or $4m$	112
				$\equiv 0 \pmod 8$	[See Table 2]	-
(5)	$\begin{bmatrix} 1 & 0 & 0 \\ 0 & 2 & 0 \\ 0 & 0 & 3 \end{bmatrix}$	10	$2^d u_5$	$\not\equiv 0 \pmod{16}$	m or $4m$	1440
				$\equiv 0 \pmod{16}$	does not arise	-
(6)	$\begin{bmatrix} 1 & 0 & 0 \\ 0 & 2 & 1 \\ 0 & 1 & 4 \end{bmatrix}$	7	$7^d u_-$ or $7, 10 \pmod{12}$	$\not\equiv 0, 3, 9 \pmod{12}$ $\& \not\equiv 0 \pmod{49}$	m, $4m$, or $9m$	3087
				$\equiv 0 \pmod{49}$	does not arise	-
				$\equiv 0, 3, 9 \pmod{12}$	[See Table 2]	-
(7)	$\begin{bmatrix} 1 & 0 & 0 \\ 0 & 2 & 0 \\ 0 & 0 & 4 \end{bmatrix}$	14	$2^d u_7$	$\not\equiv 0 \pmod{16}$	m or $4m$	224
				$\equiv 0 \pmod 8$	[See Table 2]	-
(8)	$\begin{bmatrix} 1 & 0 & 0 \\ 0 & 2 & 1 \\ 0 & 1 & 5 \end{bmatrix}$	7	$2^e u_7$	$\not\equiv 0 \pmod 8$	m or $4m$	252
				$\equiv 0 \pmod 8$	does not arise	-
(9)	$\begin{bmatrix} 1 & 0 & 0 \\ 0 & 2 & 0 \\ 0 & 0 & 5 \end{bmatrix}$	10	$5^d u_-$	$\not\equiv 0 \pmod{25}$	m or $4m$	4000
				$\equiv 0 \pmod{25}$	does not arise	-

Table 1. Proof of universality of four-dimensional escalators (nonexceptional cases)

*We follow the notation of Conway-Sloane [**2**]: p^d (resp. p^e) denotes an odd (resp. even) power of p; if $p = 2$, u_k denotes a number of the form $8n + k$ ($k = 1, 3, 5, 7$), and if p is odd, u_+ (resp. u_-) denotes a number which is a quadratic residue (resp. non-residue) modulo p.

"Exceptional" Lattice	New unique in genus sublattice	Unrepresented numbers	m	Subtract	Check up to
$\begin{bmatrix} 1 & 0 & 0 & 2 \\ 0 & 2 & 0 & 1 \\ 0 & 0 & 2 & 1 \\ 2 & 1 & 1 & 7 \end{bmatrix}$	$\begin{bmatrix} 1 & 0 & 0 \\ 0 & 2 & 1 \\ 0 & 1 & 3 \end{bmatrix}$	$5^d u_+$	40	m or $4m$	160
$\begin{bmatrix} 1 & 0 & 0 & 0 \\ 0 & 2 & 0 & 1 \\ 0 & 0 & 2 & 1 \\ 0 & 1 & 1 & 7 \end{bmatrix}$	$\begin{bmatrix} 2 & 0 & 1 \\ 0 & 2 & 1 \\ 1 & 1 & 7 \end{bmatrix}$	$2^e u_1, 2^e u_5,$ $2^d u_3, 2^d u_7, 3^d u_+$	1	m	14
$\begin{bmatrix} 1 & 0 & 0 & 1 \\ 0 & 2 & 1 & 0 \\ 0 & 1 & 4 & 3 \\ 1 & 0 & 3 & 7 \end{bmatrix}$	$\begin{bmatrix} 2 & 1 & 1 \\ 1 & 4 & 0 \\ 1 & 0 & 4 \end{bmatrix}$	$2^d u_7$	1	m, $4m$, or $9m$	9
$\begin{bmatrix} 1 & 0 & 0 & 0 \\ 0 & 2 & 1 & 1 \\ 0 & 1 & 4 & 0 \\ 0 & 1 & 0 & 7 \end{bmatrix}^{\dagger}$	$\begin{bmatrix} 2 & 0 & 0 \\ 0 & 4 & 2 \\ 0 & 2 & 10 \end{bmatrix}$	$2^d u_7$	90	m or $4m$	504
$\begin{bmatrix} 1 & 0 & 0 & 1 \\ 0 & 2 & 0 & 0 \\ 0 & 0 & 4 & 2 \\ 1 & 0 & 2 & 14 \end{bmatrix}$	$\begin{bmatrix} 1 & 0 & 0 \\ 0 & 4 & 2 \\ 0 & 2 & 13 \end{bmatrix}$	$2^d u_5, 2^e u_3$	2	m or $4m$	8

Table 2. Proof of universality of four-dimensional escalators (exceptional cases)

†In this exceptional case, the sublattice given here shows only that all even numbers are represented. However, the original argument of Table 1 (using escalator (6) as sublattice, with $m = 315$) shows that all odd numbers are represented, so the desired universality follows. [It turns out there is *no* sublattice unique in its genus that single-handedly proves universality in this case!]

1: 1 1 1 0 0 0
2: 1 1 2 0 0 0
3: 1 1 3 0 0 0
3: 1 2 2 2 0 0
4: 1 1 4 0 0 0
4: 1 2 2 0 0 0
4: 2 2 2 2 2 0
5: 1 1 5 0 0 0
5: 1 2 3 2 0 0
6: 1 1 6 0 0 0
6: 1 2 3 0 0 0
6: 2 2 2 2 0 0
7: 1 1 7 0 0 0
7: 1 2 4 2 0 0
7: 2 2 3 2 0 2
8: 1 2 4 0 0 0
8: 1 3 3 2 0 0
8: 2 2 2 0 0 0
8: 2 2 3 2 2 0
9: 1 2 5 2 0 0
9: 1 3 3 0 0 0
9: 2 2 3 0 0 2
10: 1 2 5 0 0 0
10: 2 2 3 2 0 0
10: 2 2 4 2 0 2
11: 1 2 6 2 0 0
11: 1 3 4 2 0 0
12: 1 2 6 0 0 0
12: 1 3 4 0 0 0
12: 2 2 3 0 0 0
12: 2 2 4 0 0 2
13: 2 2 5 2 0 2
13: 2 3 3 2 2 0
14: 1 2 7 0 0 0
14: 1 3 5 2 0 0
14: 2 2 4 2 0 0
15: 1 2 8 2 0 0
15: 1 3 5 0 0 0
15: 2 2 5 0 0 2
15: 2 3 3 0 2 0
16: 1 2 8 0 0 0
16: 2 2 4 0 0 0
16: 2 3 3 2 0 0
17: 1 2 9 2 0 0
17: 1 3 6 2 0 0
17: 2 3 4 0 2 2
18: 1 2 9 0 0 0
18: 1 3 6 0 0 0
18: 2 2 5 2 0 0
18: 2 3 3 0 0 0
18: 2 3 4 2 0 2
19: 1 2 10 2 0 0
19: 2 3 4 2 2 0
20: 1 2 10 0 0 0
20: 2 2 5 0 0 0
20: 2 2 6 2 2 0
20: 2 4 4 4 2 0
21: 2 3 4 0 2 0
22: 1 2 11 0 0 0
22: 2 2 6 2 0 0
22: 2 3 4 2 0 0
22: 2 3 5 0 2 2
23: 1 2 12 2 0 0
23: 2 3 5 2 0 2
24: 1 2 12 0 0 0
24: 2 2 6 0 0 0
24: 2 2 7 2 2 0
24: 2 3 4 0 0 0
24: 2 4 4 0 2 2
24: 2 4 4 4 0 0
25: 1 2 13 2 0 0
25: 2 3 5 2 2 0
26: 1 2 13 0 0 0
26: 2 2 7 2 0 0
26: 2 4 4 2 2 0
27: 1 2 14 2 0 0
27: 2 3 5 0 2 0
27: 2 4 5 4 0 2
28: 1 2 14 0 0 0
28: 2 2 7 0 0 0
28: 2 3 5 2 0 0
28: 2 4 4 0 2 0
28: 2 4 5 4 2 0
30: 2 3 5 0 0 0
30: 2 4 4 2 0 0
31: 2 3 6 2 2 0
31: 2 4 5 0 2 2
32: 2 4 4 0 0 0
32: 2 4 5 4 0 0
33: 2 3 6 0 2 0
33: 2 4 5 2 0 2
34: 2 3 6 2 0 0
34: 2 4 5 2 2 0
34: 2 4 6 4 0 2
35: 2 4 5 0 0 2
36: 2 3 6 0 0 0
36: 2 4 5 0 2 0
36: 2 4 6 4 2 0
36: 2 5 5 4 2 2
37: 2 5 5 4 2 0
38: 2 4 5 2 0 0
38: 2 4 6 0 2 2
39: 2 3 7 0 2 0
40: 2 3 7 2 0 0
40: 2 4 5 0 0 0
40: 2 4 6 2 0 2
40: 2 4 6 4 0 0
41: 2 4 7 4 0 2
42: 2 3 7 0 0 0
42: 2 4 6 0 0 2
42: 2 4 6 2 2 0
42: 2 5 5 4 0 0
43: 2 3 8 2 2 0
43: 2 5 5 2 0 2
44: 2 4 6 0 2 0
45: 2 4 7 0 2 2
45: 2 5 5 0 2 0
45: 2 5 6 4 2 2
46: 2 3 8 2 0 0
46: 2 4 6 2 0 0
46: 2 5 6 4 0 2
47: 2 4 7 2 0 2
47: 2 5 6 4 2 0
48: 2 3 8 0 0 0
48: 2 4 6 0 0 0
48: 2 5 5 2 0 0
49: 2 3 9 2 2 0
49: 2 4 7 0 0 2
49: 2 5 6 0 2 2
50: 2 4 7 2 2 0
50: 2 5 5 0 0 0
51: 2 3 9 0 2 0
52: 2 3 9 2 0 0
52: 2 5 6 2 0 2
52: 2 5 6 4 0 0
53: 2 5 6 2 2 0
54: 2 3 9 0 0 0
54: 2 4 7 2 0 0
54: 2 5 6 0 0 2
54: 2 5 7 4 2 2
55: 2 3 10 2 2 0
55: 2 5 6 0 2 0
55: 2 5 7 4 0 2
56: 2 4 7 0 0 0
56: 2 4 8 4 0 0
57: 2 3 10 0 2 0
58: 2 3 10 2 0 0
58: 2 4 8 2 2 0
58: 2 5 6 2 0 0
58: 2 5 7 0 2 2
60: 2 3 10 0 0 0
60: 2 4 9 4 2 0
60: 2 5 6 0 0 0
61: 2 5 7 2 0 2
62: 2 4 8 2 0 0
62: 2 5 7 4 0 0
63: 2 5 7 0 0 2
63: 2 5 7 2 2 0
64: 2 4 8 0 0 0
66: 2 4 9 2 2 0
67: 2 5 8 4 2 0
68: 2 4 9 0 2 0
68: 2 4 10 4 2 0
68: 2 5 7 2 0 0
70: 2 4 9 2 0 0
70: 2 5 7 0 0 0
72: 2 4 9 0 0 0
72: 2 4 10 4 0 0
72: 2 5 8 4 0 0
74: 2 4 10 2 2 0
76: 2 4 10 0 2 0
77: 2 5 9 4 2 0
78: 2 4 10 2 0 0
78: 2 5 8 2 0 0
80: 2 4 10 0 0 0
80: 2 4 11 4 0 0
80: 2 5 8 0 0 0
82: 2 4 11 2 2 0
82: 2 5 9 4 0 0
83: 2 5 9 2 2 0
85: 2 5 9 0 2 0
86: 2 4 11 2 0 0
87: 2 5 10 4 2 0
88: 2 4 11 0 0 0
88: 2 4 12 4 0 0
88: 2 5 9 2 0 0
90: 2 4 12 2 2 0
90: 2 5 9 0 0 0
92: 2 4 13 4 2 0
92: 2 5 10 4 0 0
93: 2 5 10 2 2 0
94: 2 4 12 2 0 0
95: 2 5 10 0 2 0
96: 2 4 12 0 0 0
96: 2 4 13 4 0 0
98: 2 4 13 2 2 0
98: 2 5 10 2 0 0
100: 2 4 13 0 2 0
100: 2 4 14 4 2 0
100: 2 5 10 0 0 0
102: 2 4 13 2 0 0
104: 2 4 13 0 0 0
104: 2 4 14 4 0 0
106: 2 4 14 2 2 0
108: 2 4 14 0 2 0
110: 2 4 14 2 0 0
112: 2 4 14 0 0 0

Table 3. Ternary forms[‡] L such that $[1] \oplus L$ is an escalator.
(The six entries not appearing in Table 5 have been italicized.)

[‡]We use the customary shorthand "D: a b c d e f" to represent the three-dimensional lattice $\begin{bmatrix} a & f/2 & e/2 \\ f/2 & b & d/2 \\ e/2 & d/2 & c \end{bmatrix}$ of determinant D.

Nonuniversal four-dimensional escalator	Unique number not represented
$\begin{bmatrix} 1 & 0 & 0 & 0 \\ 0 & 2 & 0 & 1 \\ 0 & 0 & 3 & 0 \\ 0 & 1 & 0 & 4 \end{bmatrix}$	10
$\begin{bmatrix} 1 & 0 & 0 & 0 \\ 0 & 2 & 1 & 0 \\ 0 & 1 & 4 & 1 \\ 0 & 0 & 1 & 5 \end{bmatrix}$	10
$\begin{bmatrix} 1 & 0 & 0 & 0 \\ 0 & 2 & 0 & 0 \\ 0 & 0 & 5 & 1 \\ 0 & 0 & 1 & 5 \end{bmatrix}$	15
$\begin{bmatrix} 1 & 0 & 0 & 0 \\ 0 & 2 & 0 & 0 \\ 0 & 0 & 5 & 0 \\ 0 & 0 & 0 & 5 \end{bmatrix}$	15
$\begin{bmatrix} 1 & 0 & 0 & 0 \\ 0 & 2 & 0 & 1 \\ 0 & 0 & 5 & 2 \\ 0 & 1 & 2 & 8 \end{bmatrix}$	15
$\begin{bmatrix} 1 & 0 & 0 & 0 \\ 0 & 2 & 0 & 1 \\ 0 & 0 & 5 & 1 \\ 0 & 1 & 1 & 9 \end{bmatrix}$	15

Table 4. Nonuniversal four-dimensional escalator lattices.
(The 1630 five-dimensional escalators are obtained from these.)

1: 1 1 1 0 0 0	16: 1 2 8 0 0 0	28: 2 3 5 2 0 0	48: 2 3 8 0 0 0	72: 2 4 9 0 0 0
2: 1 1 2 0 0 0	16: 2 2 4 0 0 0	28: 2 4 4 0 2 0	48: 2 4 6 0 0 0	72: 2 4 10 4 0 0
3: 1 1 3 0 0 0	16: 2 3 3 2 0 0	28: 2 4 5 4 2 0	48: 2 5 5 2 0 0	72: 2 5 8 4 0 0
3: 1 2 2 2 0 0	17: 1 2 9 2 0 0	30: 2 3 5 0 0 0	49: 2 3 9 2 2 0	74: 2 4 10 2 2 0
4: 1 1 4 0 0 0	17: 1 3 6 2 0 0	30: 2 4 4 2 0 0	49: 2 4 7 0 0 2	76: 2 4 10 0 2 0
4: 1 2 2 0 0 0	17: 2 3 4 0 2 2	31: 2 3 6 2 2 0	49: 2 5 6 0 2 2	77: 2 5 9 4 2 0
4: 2 2 2 2 2 0	18: 1 2 9 0 0 0	31: 2 4 5 0 2 2	50: 2 4 7 2 2 0	78: 2 4 10 2 0 0
5: 1 1 5 0 0 0	18: 1 3 6 0 0 0	32: 2 4 4 0 0 0	51: 2 3 9 0 2 0	78: 2 5 8 2 0 0
5: 1 2 3 2 0 0	18: 2 2 5 2 0 0	32: 2 4 5 4 0 0	52: 2 3 9 2 0 0	80: 2 4 10 0 0 0
6: 1 1 6 0 0 0	18: 2 3 3 0 0 0	33: 2 3 6 0 2 0	52: 2 5 6 2 0 2	80: 2 4 11 4 0 0
6: 1 2 3 0 0 0	18: 2 3 4 2 0 2	34: 2 3 6 2 0 0	52: 2 5 6 4 0 0	80: 2 5 8 0 0 0
6: 2 2 2 2 0 0	19: 1 2 10 2 0 0	34: 2 4 5 2 2 0	53: 2 5 6 2 2 0	82: 2 4 11 2 2 0
7: 1 1 7 0 0 0	19: 2 3 4 2 2 0	34: 2 4 6 4 0 2	54: 2 3 9 0 0 0	82: 2 5 9 4 0 0
7: 1 2 4 2 0 0	20: 1 2 10 0 0 0	35: 2 4 5 0 0 2	54: 2 4 7 2 0 0	85: 2 5 9 0 2 0
7: 2 2 3 2 0 2	20: 2 2 5 0 0 0	36: 2 3 6 0 0 0	54: 2 5 6 0 0 2	86: 2 4 11 2 0 0
8: 1 2 4 0 0 0	20: 2 2 6 2 2 0	36: 2 4 5 0 2 0	54: 2 5 7 4 2 2	87: 2 5 10 4 2 0
8: 1 3 3 2 0 0	*20: 2 3 4 0 0 2*	36: 2 4 6 4 2 0	55: 2 3 10 2 2 0	88: 2 4 11 0 0 0
8: 2 2 2 0 0 0	20: 2 4 4 4 2 0	36: 2 5 5 4 2 2	55: 2 5 6 0 2 0	88: 2 4 12 4 0 0
8: 2 2 3 2 2 0	22: 1 2 11 0 0 0	37: 2 5 5 4 2 0	55: 2 5 7 4 0 2	88: 2 5 9 2 0 0
9: 1 2 5 2 0 0	22: 2 2 6 2 0 0	38: 2 4 5 2 0 0	56: 2 4 7 0 0 0	90: 2 4 12 2 2 0
9: 1 3 3 0 0 0	22: 2 3 4 2 0 0	38: 2 4 6 0 2 2	56: 2 4 8 4 0 0	90: 2 5 9 0 0 0
9: 2 2 3 0 0 2	22: 2 3 5 0 2 2	39: 2 3 7 0 2 0	57: 2 3 10 0 2 0	92: 2 4 13 4 2 0
10: 1 2 5 0 0 0	23: 1 2 12 2 0 0	40: 2 3 7 2 0 0	58: 2 3 10 2 0 0	92: 2 5 10 4 0 0
10: 2 2 3 2 0 0	23: 2 3 5 2 0 2	40: 2 4 5 0 0 0	58: 2 4 8 2 2 0	93: 2 5 10 2 2 0
10: 2 2 4 2 0 2	24: 1 2 12 0 0 0	40: 2 4 6 2 0 2	58: 2 5 6 2 0 0	94: 2 4 12 2 0 0
11: 1 2 6 2 0 0	24: 2 2 6 0 0 0	40: 2 4 6 4 0 0	58: 2 5 7 0 2 2	95: 2 5 10 0 2 0
11: 1 3 4 2 0 0	24: 2 2 7 2 2 0	41: 2 4 7 4 0 2	60: 2 3 10 0 0 0	96: 2 4 12 0 0 0
12: 1 2 6 0 0 0	24: 2 3 4 0 0 0	42: 2 3 7 0 0 0	60: 2 4 9 4 2 0	96: 2 4 13 4 0 0
12: 1 3 4 0 0 0	24: 2 4 4 0 2 2	42: 2 4 6 0 0 2	60: 2 5 6 0 0 0	98: 2 4 13 2 2 0
12: 2 2 3 0 0 0	24: 2 4 4 4 0 0	42: 2 4 6 2 2 0	61: 2 5 7 2 0 2	98: 2 5 10 2 0 0
12: 2 2 4 0 0 2	25: 1 2 13 2 0 0	42: 2 5 5 4 0 0	62: 2 4 8 2 0 0	100: 2 4 13 0 2 0
12: 2 3 3 0 2 2	*25: 2 3 5 0 0 2*	43: 2 3 8 2 2 0	62: 2 5 7 4 0 0	100: 2 4 14 4 2 0
13: 2 2 5 2 0 2	25: 2 3 5 2 2 0	44: 2 4 6 0 2 0	63: 2 5 7 0 0 2	100: 2 5 10 0 0 0
13: 2 3 3 2 2 0	26: 1 2 13 0 0 0	45: 2 4 7 0 2 2	63: 2 5 7 2 2 0	102: 2 4 13 2 0 0
14: 1 2 7 0 0 0	26: 2 2 7 2 0 0	45: 2 5 5 0 2 0	64: 2 4 8 0 0 0	104: 2 4 13 0 0 0
14: 1 3 5 2 0 0	26: 2 4 4 2 2 0	45: 2 5 6 4 2 2	66: 2 4 9 2 2 0	104: 2 4 14 4 0 0
14: 2 2 4 2 0 0	27: 1 2 14 2 0 0	46: 2 3 8 2 0 0	68: 2 4 9 0 2 0	106: 2 4 14 2 2 0
15: 1 2 8 2 0 0	27: 2 3 5 0 2 0	46: 2 4 6 2 0 0	68: 2 4 10 4 2 0	108: 2 4 14 0 2 0
15: 1 3 5 0 0 0	27: 2 4 5 4 0 2	46: 2 5 6 4 0 2	68: 2 5 7 2 0 0	110: 2 4 14 2 0 0
15: 2 2 5 0 0 2	28: 1 2 14 0 0 0	47: 2 4 7 2 0 2	70: 2 4 9 2 0 0	112: 2 4 14 0 0 0
15: 2 3 3 0 2 0	28: 2 2 7 0 0 0	47: 2 5 6 4 2 0	70: 2 5 7 0 0 0	

Table 5. Ternary forms L such that $[1] \oplus L$ is universal.
(The three entries not appearing in Table 3 have been italicized.)

Department of Mathematics,
Princeton University,
Princeton, NJ 08544
E-mail address: bhargava@math.princeton.edu

Contemporary Mathematics
Volume **272**, 2000

On Trace Forms and the Burnside Ring

Martin Epkenhans

Abstract. We discuss the annihilating polynomials of trace forms arising in the literature. An improvement of these results is obtained by using a reduction to 2-groups. In several cases we determine the minimal monic annihilating polynomial of trace forms with given Galois action.

1. Introduction

Let L/K be a finite separable field extension, with $char(K) \neq 2$. Let N be a normal closure of L/K and let $G = G(N/K)$ be the Galois group of N/K. We are interested in a (monic) polynomial $p(X) \in \mathbb{Z}[X]$ of minimal degree, depending only on the Galois action of G on the set of left cosets of $G/G(N/L)$, such that $p(X)$ annihilates the trace form of L/K in the Witt ring $W(K)$ of K.

During the last decade several examples of annihilating polynomials have appeared in the literature. First D. Lewis [**Lew87**] defined a monic polynomial $L_n(X)$ which annihilates any quadratic form of dimension n. Hence there exists a monic annihilating polynomial of minimal degree in our situation. P.E. Conner [**Con87**] gave a polynomial of lower degree which annihilates trace forms. Beaulieu and Palfrey [**BP97**] and recently Lewis and McGarraghy [**LM00**] recover these results by giving annihilating polynomials which divide the corresponding Beaulieu-Palfrey polynomials. In [**Epk98**] we defined a polynomial $q(X)$, called the *signature polynomial,* which divides any annihilating polynomial. Further, there exists some integer $e \geq 0$ such that $2^e q(X)$ is an annihilating polynomial. Since the only torsion in the Witt ring is 2-torsion, we more like to get monic polynomials.

In this paper we discuss all these polynomials. We show that the Lewis-McGarraghy result improves the Beaulieu-Palfrey theorem. We improve the Lewis-McGarraghy theorem on trace forms of field extensions by using Springer's theorem on the lifting of quadratic forms to odd degree extensions.

In section 3 we give several examples which show that all the definitions of annihilating polynomials are essentially different.

In some cases we are able to give a further improvement by using some explicit calculations of the trace ideal $\mathcal{T}(G)$ of the Burnside ring $\mathcal{B}(G)$ of G.

Unfortunately we are not able to present a unified approach which yields a minimal monic annihilating polynomial. In section 5.2 we give an example of a

2000 *Mathematics Subject Classification.* 11E04.

group G acting transitively on a set S such that the signature polynomial is not an annihilating polynomial.

2. The Polynomials

We start by defining the different annihilating polynomials arising in the literature. The first polynomial is given by D. Lewis in [**Lew87**].

2.1. The Lewis polynomial $L_n(X)$.

DEFINITION 2.1. For $n \in \mathbb{N}$ the Lewis polynomial is defined as

$$\begin{array}{ll} L_n(X) = X(X^2-2^2)(X^2-4^2)\dots(X^2-n^2) & \text{if } n \text{ is even,} \\ L_n(X) = (X^2-1^2)(X^2-3^2)\dots(X^2-n^2) & \text{if } n \text{ is odd.} \end{array}$$

THEOREM 2.2. *Let ψ be a quadratic form of dimension n over the field K. Then*

$$L_n(\psi) = 0 \in W(K)$$

in the Witt ring $W(K)$ of K.

Observe, that the zeros of $L_n(X)$ are exactly those integers which arise as signatures of quadratic forms of dimension n.

We now restrict our attention to trace forms. Let A be an étale K-algebra of dimension n over the field K of characteristic $\neq 2$. Then

$$A \to K : x \mapsto trace_{A/K} x^2$$

defines a quadratic form of dimension n over K, called the *trace form of A/K*. We denote the trace form by $< A/K >$ or simply by $< A >$ if no confusion can arise. We know by Sylvester and Jacobi that trace forms have non-negative signatures.

2.2. The Conner polynomial $C_n(X)$.

DEFINITION 2.3. For $n \in \mathbb{N}$ the Conner polynomial $C_n(X)$ is defined to be

$$\begin{array}{ll} C_n(X) = X(X-2)(X-4)\dots(X-n) & \text{if } n \text{ is even,} \\ C_n(X) = (X-1)(X-3)\dots(X-n) & \text{if } n \text{ is odd.} \end{array}$$

Hence $C_n(X) = \prod_{k\geq 0, L_n(k)=0}(X-k)$ is the positive part of $L_n(X)$. From unpublished notes of P.E. Conner [**Con87**] we get

THEOREM 2.4. *Let L/K be a separable field extension of degree n. Then*

$$C_n(< L >) = 0 \in W(K).$$

The results of Lewis and Conner are optimal in the following sense. Let M be a class of quadratic forms. Then

$$I_M = \{f \in \mathbb{Z}[X] : f(\psi) = 0 \in W(K) \text{ for all } \psi \in M\}$$

is the *vanishing ideal of M*. Regarding the signatures of quadratic forms we observe the following result. $L_n(X)$ generates the principal ideal I_{Q_n}, where Q_n contains all quadratic forms of dimension n. Let T_n be the class of trace forms of dimension n of fields of characteristic $\neq 2$. Then $I_{T_n} = (C_n(X))$ (see [**Epk98**]). Hence the signatures are the only obstructions.

Let L/K be separable field extension of degree n. Choose a normal closure $N \supset L$ of L/K. Denote the Galois group of N/K by $G(N/K)$ and set $H = G(N/L)$. Let $f(X) \in K[X]$ be a polynomial with $L \simeq K[X]/(f)$. Then the transitive action of G on the roots of $f(X)$ is equivalent to the faithful action of G on the set of left cosets of H in G. Instead of fixing a field extension L/K we choose a finite group G and a subgroup $H < G$ such that the action

$$G \times G/H \to G/H$$

is faithful. We know that the action is transitive. Further G acts faithfully if and only if $\cap_{\sigma \in G} \sigma H \sigma^{-1} = 1$ if and only if H contains no subgroup $\neq 1$ which is normal in G.

DEFINITION 2.5. Let $H < G$ be finite groups with $\cap_{\sigma \in G} \sigma H \sigma^{-1} = 1$. Let $M(G, H)$ denotes the class of all quadratic forms ψ such that there is an irreducible and separable polynomial $f(X) \in K[K]$ with Galois group $Gal(f) \simeq G$ and such that

1. the action of $Gal(f)$ on the roots of $f(X)$ is equivalent to the action of G on G/H, and
2. ψ and the trace form of $K[X]/(f(X))$ over K are isometric.

We look for a polynomial $m_{G,H}(X) \in \mathbb{Z}[X]$ such that $m_{G,H}(X)$ annihilates any quadratic form in $M(G, H)$. Actually we are interested in a (monic) polynomial of minimal degree with the property cited above.

Beaulieu and Palfrey [**BP97**] gave an annihilating polynomial which divides Conner's polynomial. In general their polynomial has smaller degree.

2.3. The Beaulieu-Palfrey polynomial $B_{G,H}(X)$. Let G be a finite group acting on a finite set S. For $\sigma \in G$ let $\chi(\sigma) = \sharp S^\sigma$ be the number of fixed point of σ acting on S. Then the Galois number $t_{G,S}$ is defined to be the maximum value of $1 + \chi(\sigma)$ as σ runs over all elements of G which do not act as the identity on S. Beaulieu and Palfrey proved the following theorem.

THEOREM 2.6. *Let G be a finite group and let $H < G$ be a subgroup with $\cap_{\sigma \in G} \sigma H \sigma^{-1} = 1$. Let $t = t_{G,H}$ denote the Galois number of the action of G on G/H. Set*

$$B_{G,H}(X) = (X - n) \prod_{k=0, k \equiv 0 \bmod 2}^{t-1} (X - k),$$

where $n = [G : H]$. Then $B_{G,H}(X)$ annihilates any $\psi \in M(G, H)$.

In [**EG99**] we find a complete list of the Galois numbers of all doubly transitive permutation groups. Let us give some examples and consequences. We get $C_n(X) = B_n(X)$ in the case of the symmetric group $\mathfrak{S}_n$ acting on n letters.

A Frobenius group G of degree n has Galois number 2. Hence the Beaulieu-Palfrey polynomial is $X - n$, if n is odd and $X(X - n)$ if n is even.

Recently Lewis and McGarraghy [**LM00**] gave an annihilating polynomial for trace forms which divides the Beaulieu-Palfrey polynomial.

2.4. The Lewis-McGarraghy polynomial $p_{G,H}(X)$. Again let a finite group G act on a finite set S. For any subgroup $U < G$ let

$$S^U = \{s \in S : \ s^\sigma = s \text{ for all } \sigma \in U\}$$

be the set of fixed points of U. Set $\varphi_U(S) = \sharp S^U$.

DEFINITION 2.7. Let the finite group G act on set S. Let $n = \sharp S$ and set

$$\varphi(S) = \{\varphi_U(S) : \ U < G, \ \varphi_U(S) \equiv 0 \bmod 2\}.$$

Now define

$$p_{G,S} = \prod_{k \in \varphi(S)} (X - k).$$

Set $p_{G,H} = p_{G,G/H}$.

The result of Lewis and McGarraghy is as follows.

THEOREM 2.8. *Let $A = K[X]/(f(X))$ be an étale K-algebra. Consider the action of the Galois group G of $f(X)$ on the set S of roots of $f(X)$. Then $p_{G,S}(X)$ annihilates the trace form $< A >$.*

As we will see, this theorem improves the Beaulieu-Palfrey result in two directions. It also holds for étale algebras and, as we see later, there are examples where the degree of $p_{G,S}(X)$ is strictly smaller then the degree of $B_{G,H}(X)$. Observe the following. Let $\sigma \in G$ be a non-identity element. Then $\chi(\sigma) = \varphi_{<\sigma>}(S)$. Further $\varphi_U(S) \leq \chi(\sigma)$ for all $\sigma \in U$. Hence $p_{G,S}(X)$ divides $B_{G,H}(X)$ and the maximal root $\neq n$ of both polynomials coincide.

2.5. The dyadic annihilating polynomials $B^{(2)}_{G,H}(X)$and $p^{(2)}_{G,H}(X)$. The proofs of the trace form results cited above are in two steps. Consider the Burnside ring $\mathcal{B}(G)$ of G. Let χ_H be an element defined by the subgroup $H < G$. First we have to find a polynomial $g(X) \in \mathbb{Z}[X]$ such that $g(\chi_H) = 0$. Let N/K be a Galois extension with Galois group isomorphic to G. Then there is a homomorphism $h_{N/K} : \mathcal{B}(G) \to W(K)$ which maps χ_H onto $< N^H >$. Hence $g(X)$ annihilates the trace form of the fixed field N^H of H over K. By the fact that there are no non-trivial zero divisors of odd degree in $W(K)$, we can omit several roots of $g(X)$. We call it the even-odd trick. Hence the results of Conner, Beaulieu-Palfrey, and Lewis-McGarraghy follow from identities in the Burnside ring via a ring homomorphism together with the even-odd trick.

Next we show that we get better results by using Springer's theorem on the lifting of quadratic forms according to odd degree extensions.

DEFINITION 2.9. Let G be a finite group and let $H < G$ be a subgroup of index n such that $\cap_{\sigma \in G} \sigma H \sigma^{-1} = 1$. Let G_2 be a Sylow 2-group of G. Let $t = t_{G_2, G/H}$ be the Galois number of the action of G_2 on G/H. Set

$$B^{(2)}_{G,H}(X) = (X - n) \prod_{k=0, k \equiv n \bmod 2}^{t-1} (X - k)$$

and

$$p^{(2)}_{G,H} = p_{G_2, G/H}(X) = \prod_{k \in \varphi(G/H)} (X - k),$$

where $\varphi(G/H) = \{\varphi_U(G/H) : U < G_2\}$.

Observe, that $\varphi_U(G/H) \equiv n \bmod 2$ for any 2-group U. Hence the application of the even-odd trick in this situation does not yield polynomials of lower degree.

THEOREM 2.10. *Consider the situation of theorem 2.6 and set $H = G(N/L)$. We get*

$$B^{(2)}_{G,H}(< L >) = p^{(2)}_{G,H}(< L >) = 0 \in W(K).$$

PROOF. As above, we see that $p^{(2)}_{G,H}(X)$ divides $B^{(2)}_{G,H}(X)$. Let F be the fixed field of G_2 in N. Then $< L/K >$ lifts to the trace form of an étale algebra over F. By [**LM00**] $p^{(2)}_{G,H}(X)$ annihilates $< L/K > \otimes_K F$ in $W(F)$. Since F/K has odd degree, we are done by Springer's theorem [**Sch85**][I.5.5.9]. □

Next we define a polynomial which turns out to divide any annihilating polynomial.

2.6. The signature polynomial $q_{G,H}(X)$. We are interested in an optimal annihilating polynomial. In [**Epk98**] we defined the following polynomial. Let G act on S. For any $\sigma \in G$ with $\sigma^2 = 1$ let $\mathrm{sign}_\sigma(S) = \sharp S^\sigma$ be the signature of S defined by the involution σ. Set

$$q_{G,S}(X) = \prod_{k \in \{\mathrm{sign}_\sigma(S) : \sigma \in G, \sigma^2 = 1\}} (X - k).$$

For $S = G/H$ set $q_{G,H}(X) = q_{G,S}$. In [**Epk98**] we proved

THEOREM 2.11. *Let H, G be as above. Then*

$$I_{M(G,H)} \subset (q_{G,H}(X)).$$

There exists some integer $e \geq 0$, which only depends on G and H, such that

$$(2^e q_{G,H}(X)) \subset I_{M(G,H)}.$$

We call $\sharp S^\sigma$ a signature, since this value corresponds to signatures of trace forms via the homomorphism $h_{N/K}$.

3. The main result on annihilating polynomials

In this section we analyze the known vanishing results.

THEOREM 3.1. *Let G be a finite group with Sylow 2-group G_2 and let $H < G$ be a subgroup of index n with $\cap_{\sigma \in G} \sigma H \sigma^{-1} = 1$. Then*

$$q_{G,H}(X) \mid p_{G,H}(X) \mid B_{G,H}(X) \mid C_n(X) \mid L_n(X)$$

and

$$q_{G,H}(X) \mid p^{(2)}_{G,H}(X) \mid B^{(2)}_{G,H}(X).$$

All polynomials except the signature polynomial $q_{G,H}(X)$ are contained in $I_{M(G,H)}$. There are examples, where $q_{G,H}(X) \notin I_{M(G,H)}$. In general, these polynomials are different.

We already proved the results on the divisibility of the polynomials. In section 5.2 we give an example with $q_{G,H}(X) \notin I_{M(G,H)}$.

$C_n(X)$ and $L_n(X)$ are different by definition. Since the maximal root $\neq n$ of $C_n(X)$ is $n - 2$ we get

LEMMA 3.2. *$B_{G,H}(X) = C_n(X)$ if and only if G contains a transposition.*

In [**EG99**] we find a lot of example with $t_{G,H} \leq n - 3$. As already mentioned above, $B_{G,H}$ has degree ≤ 2 if G is a Frobenius group.

LEMMA 3.3. *Let $H < G$ be finite groups with $\cap_{\sigma\in G}\sigma H\sigma^{-1} = 1$.*

1. *Let $[G : H]$ be odd. If G acts doubly transitive, then*

$$X - 1 \mid p^{(2)}_{G,H}(X).$$

2. *Let $[G : H]$ be even. Then $X \mid p^{(2)}_{G,H}(X)$.*
   $X \mid q_{G,H}(X)$ if and only if G contains an involution which is not conjugate to any element in H.

Hence $p^{(2)}_{G,H}(X) \neq q_{G,H}(X)$ if $[G : H] \not\equiv \sharp H \equiv 1 \bmod 2$. And $X \not| q_{G,H}(X)$ if G contains only one class of involutions and H has even order. The Ree group $R(q)$ and any group with cyclic or generalized quaternion Sylow 2-group contain only one conjugacy class of involutions.

PROOF. 1. A two point stabilizer does not contain a Sylow 2-group. Hence $\varphi_{G_2}(G/H) = 1$.
2. A one point stabilizer contains a Sylow 2-group of G if and only if H has odd index in G. Hence $\varphi_{G_2}(G/H) = 0$. Now see corollary 11 in [**Epk99**]. □

We now prove that the Lewis-McGarraghy result improves the Beaulieu-Palfrey theorem.

PROPOSITION 3.4. *Let $R(q) = {}^2G_2(q), q = 3^{2n+1}$, where $n \geq 1$ be the Ree group. Consider $R(q)$ in its doubly transitive representation of degree $q^3 + 1$. Let H be a one-point stabilizer. Then*

$$B^{(2)}_{G,H}(X) = B_{G,H}(X) = (X - n) \cdot \prod_{k=0, k\equiv 0 \bmod 2}^{q+1} (X - k),$$
$$q_{G,H}(X) = (X - (q + 1)) \cdot (X - (q^3 + 1)),$$
$$p_{G,H}(X) = X(X - 2) \cdot q_{G,H}(X),$$
$$p^{(2)}_{G,H}(X) = X \cdot q_{G,H}(X).$$

Further

$$I_{M(G,H)} = (q_{G,H}).$$

PROOF. See [**HB82**][XI 13.2] for some basic facts on the Ree group. By [**EG99**][proposition 17] the Galois number is $q + 2$. Further $q + 1$ is the number of fixed points of a non-trivial involution in $R(q)$. This gives the result for both Beaulieu-Palfrey polynomials $B^{(2)}_{G,H}(X), B_{G,H}(X)$. Since all involutions are conjugate in $R(q)$, $q_{G,H}(X)$ has degree 2. By lemma 3.3 X divides $p^{(2)}_{G,H}$.

Now let $V < R(q)$ be a subgroup with $\varphi_V(S) \geq 3$, where $S = G/H$. Then V is contained in the stabilizer of three letters, which has order 2. Hence $\varphi_V(S) = n$ if $V = 1$ and $\varphi_V(S) = q + 1$ otherwise.

Finally, let V be the stabilizer of two letters. Then $\varphi_V(S) = 2$, since V has order $q - 1 \neq 2$. Since $0 \equiv q^3 + 1 \equiv 2 \equiv q + 1 \equiv n \bmod 2$, the even-odd trick does not reduce the number of roots. We later prove that $q_{G,H}(X)$ is already an annihilating polynomial (see proposition 4.3). □

Recently, McGarraghy found some other examples which show that the Lewis-McGarraghy-polynomial improves the Beaulieu-Palfrey result. Let us give another example.

PROPOSITION 3.5. *Let $G = AGL(n,q), n \geq 2$ be the affine linear group over $\mathbb{F}_q$. Consider the action of G on the vector space $V = \mathbb{F}_q^n$ by semilinear transformations and let $H = G_0$ be the stabilizer of the zero vector. Then $H = GL(n,q)$. We get*

$$B^{(2)}_{G,H}(X) = B_{G,H}(X) = (X - q^n) \cdot \prod_{k=0, k \equiv q \bmod 2}^{q^{n-1}} (X - k).$$

If q is odd, then

$$p^{(2)}_{G,H}(X) = p_{G,H}(X) = q_{G,H}(X) = \prod_{k=0}^{n} (X - q^k).$$

If q is even, we get

$$p^{(2)}_{G,H}(X) = p_{G,H}(X) = X \cdot \prod_{k=1}^{n} (X - q^k)$$

and

$$q_{G,H}(X) = X \cdot \prod_{k=1, 2k \geq n}^{n} (X - q^k).$$

PROOF. Let $U < G$ be a subgroup with $\varphi_U(V) \geq 1$. We can assume that U is a subgroup of $G_0 = GL(n,q)$. Hence the set of fixed points of U in V is the intersection of all eigenspaces of the eigenvalue 1 of all $\phi \in U$. Therefore $\varphi_U(V)$ is a power of q.

For any $k = 0, \ldots, n$ there is a linear map $\sigma \in GL(n,q)$ which has 1 as an eigenvalue of geometric multiplicity q^k. If q is odd, choose the diagonal matrix $E_k \oplus (-E_{n-k})$.

Let q be even. Then $x \mapsto x + {}^t(1, \ldots, 1)$ is a fixed point free involution. For $k = 2, \ldots, n$ set

$$J_k = \begin{pmatrix} 1 & 1 & & 0 \\ 0 & \ddots & \ddots & \\ \vdots & \ddots & \ddots & 1 \\ 0 & \ldots & 0 & 1 \end{pmatrix} \in M_k(\mathbb{F}_q)$$

and $B_k = J_k \oplus E_{n-k}$. Then B_k is an element of 2-power order having q^{n-k+1} fixed points. Any involution has a Jordan matrix of the form $r \cdot J_2 \oplus E_{n-2r}$. □

It remains to give an example with $B^{(2)}_{G,H}(X) \neq B_{G,H}(X)$. Note that $B^{(2)}_{G,H}(X) = B_{G,H}(X)$ if and only if the Galois number $t_{G,H}$ is given by an involution.

PROPOSITION 3.6. *Let G be a doubly transitive permutation group acting on $\mathbb{F}_3^6$ with a one point stabilizer $H = G_0 \simeq SL(2,13)$. Then*

$$B_{G,H}(X) = (X - 3^6) \prod_{k=0}^{4} (X - (2k+1)),$$
$$p_{G,H}(X) = (X - 3^6)(X - 9)(X - 1),$$
$$B^{(2)}_{G,H}(X) = (X - 3^6)(X - 1) = p^{(2)}_{G,H}(X) = q_{G,H}.$$

Hence

$$I_{M(G,H)} = (q_{G,H}).$$

PROOF. A two point stabilizer has order 3. Hence any non-trivial involution has exactly one fixed point. By proposition 9 in [**EG99**] the Galois number equals 10. □

The next example is due to Pierre Conner.

PROPOSITION 3.7. *Let G be a group of odd order and let H be a non trivial subgroup with $\cap_{\sigma\in G}\sigma H\sigma^{-1} = 1$. Then*

$$q_{G,H} = p^{(2)}_{G,H}(X) = B^{(2)}_{G,H}(X) = (X-n) \neq p_{G,H}(X), B_{G,H}(X).$$

PROOF. Observe, that $t_{G,H} \geq 2$, if $H \neq 1$. Now see [**Epk98**] proposition 4. □

4. The trace ideal

4.1. Definitions and basic properties. To make further progress on annihilating polynomials we introduce the Burnside ring $\mathcal{B}(G)$ and the trace ideal $\mathcal{T}(G)$ of a finite group G (see [**Epk98**], [**Epk99**], [**Hup98**][p. 159]). We briefly recall the definitions. For any subgroup H of G let $\chi_H = \chi_H^G$ denote the character induced by the representation of G on the left cosets of H.

DEFINITION 4.1. The *Burnside ring* $\mathcal{B}(G)$ is a free abelian group. The set

$$\{\chi_H : H \text{ runs over a set of representatives of conjugacy classes of subgroups of } G\}$$

is a free set of generators of $\mathcal{B}(G)$. The multiplication is induced by the tensor product of the underlying representations. Hence

$$\chi_H \cdot \chi_U = \bigoplus_{\sigma\in H\backslash G/U} \chi_{H\cap\sigma U\sigma^{-1}}.$$

Let $\sigma \in G$ be an element with $\sigma^2 = 1$. The number of fixed points $\text{sign}_\sigma \chi_H = \varphi_{<\sigma>}(G/H)$ of σ on G/H is called the signature of χ_H. This definition extends by linearity to a ring homomorphism $\text{sign}_\sigma : \mathcal{B}(G) \to \mathbb{Z}$. Let

$$L(G) = \cap_{\sigma\in G,\sigma^2=1} ker(\text{sign}_\sigma)$$

be the kernel of the total signature homomorphism.

Now let N/K be a Galois extension with Galois group $G(N/K) \simeq G$. Then there is a unique homomorphism

$$h_{N/K} : \mathcal{B}(G) \to W(K) : \; h_{N/K}(\chi_H) =< N^H >$$

(see [**Dre71**], [**BP97**], [**Epk98**], [**Epk99**]). Now the trace ideal $\mathcal{T}(G)$ in $\mathcal{B}(G)$ is defined to be the intersection of all kernels of homomorphisms $h_{N/K}$, where N/K runs over all Galois extensions with Galois group $\simeq G$ of fields of characteristic $\neq 2$. From [**Epk99**] theorem 6 we know that $L(G)/\mathcal{T}(G)$ is a finite 2-group. We calculated several examples in [**Epk98**], [**Epk99**]. Now we focus our attention on the polynomial $q_{G,H}(X)$.

PROPOSITION 4.2. *Let $H < G$ be finite groups with $\cap_{\sigma\in G}\sigma H\sigma^{-1} = 1$. Then*

$$q_{G,H}(\chi_H) \in L(G).$$

If G has even order, then $deg(q_{G,H}(X)) \geq 2$.

PROOF. The roots of $q_{G,H}(X)$ are by definition the possible signature values of $h_{N/K}(\chi_H) =< N^H >$. Any non-identity involution σ does not act as the identity on G/H. Hence $\text{sign}_\sigma \chi_H \neq [G:H] = \text{sign}_1 \chi_H$. □

We now illustrate how the concept of the trace ideal yields an optimal annihilating polynomial in some cases.

PROPOSITION 4.3. *Let G be a finite group with Sylow 2-group $G_2 \neq 1$. Let H be a subgroup with $\cap_{\sigma\in G}\sigma H\sigma^{-1} = 1$. Suppose, that any subgroup of G_2 is normal in G_2 and that $L(G_2)/\mathcal{T}(G_2)$ has exponent 2. Then*

$$q_{G,H}(\chi_H^G) \in \mathcal{T}(G)$$

Hence

$$I_{M(G,H)} = (q_{G,H}(X)).$$

PROOF. We prove that the image of $q_{G,H}(\chi_H^G)$ under the restriction homomorphism

$$res_{G_2}^G : \mathcal{B}(G) \to \mathcal{B}(G_2)$$

lies in the trace ideal $\mathcal{T}(G_2)$. By lemma 4 [**Epk98**]

$$res_{G_2}^G(q_{G,H}(\chi_H^G)) = q_{G,H}(res_{G_2}^G(\chi_H^G)) \in L(G_2).$$

Since $L(G_2)/\mathcal{T}(G_2)$ has exponent 2, the ideal $\mathcal{T}(G_2)$ is, regarded as a submodule of $L(G_2)$, given by a system of linear equations modulo 2. Hence it remains to consider the coefficients of $res_{G_2}^G(q_{G,H}(\chi_H^G))$ modulo 2. From $\text{sign}_\sigma\chi_H^G \equiv [G:H] \bmod 2$ we get

$$\begin{aligned} res_{G_2}^G(q_{G,H}(\chi_H^G)) &\equiv q_{G,H}(res_{G_2}^G(\chi_H^G)) \\ &\equiv (res_{G_2}^G(\chi_H^G) - [G:H]\chi_{G_2}^{G_2})^d \bmod 2\mathcal{B}(G_2), \end{aligned}$$

where $d = deg(q_{G,H})$. Let

$$res_{G_2}^G(\chi_H^G) = \sum_{\sigma\in G_2\backslash G/H} \chi_{G_2\cap\sigma H\sigma^{-1}}^{G_2} = \sum_{U<G_2} m_U\chi_U^{G_2}.$$

Then $m_{G_2} = \varphi_{G_2}(G/H) \equiv [G:H] \bmod 2$. By assumption any subgroup $U < G_2$ is normal in G_2. Therefore $(\chi_U^{G_2})^2 = [G_2:U]\chi_U^{G_2}$, which implies

$$\begin{aligned} (res_{G_2}^G(\chi_H^G) - [G:H]\chi_{G_2}^{G_2})^2 &\equiv \sum_{U\neq G_2} m_U^2(\chi_U^{G_2})^2 \equiv \sum_{U\neq G_2} m_U[G_2:U]\chi_U^{G_2} \\ &\equiv 0 \bmod 2\cdot\mathcal{B}(G_2). \end{aligned}$$

By proposition 4.2 $d \geq 2$. □

As an application we get $I_{M(G,H)} = (q_{G,H})$ in proposition 3.4, since the Sylow 2-group of $R(q)$ is an elementary abelian group of order 8. Now use proposition 6 [**Epk98**].

The assumption of the proposition above holds if G_2 is cyclic, or elementary abelian or a direct product of two cyclic groups. These examples yield the question.

Is $L(G_2)/\mathcal{T}(G_2)$ elementary abelian if G_2 is abelian?

4.2. A finiteness theorem. In this section we prove that $\mathcal{T}(G)$ is an intersection of finitely many kernels of homomorphisms $h_{N/K}$, where the field extensions can chosen to consist of Hilbertian fields.

LEMMA 4.4. *For any finite group G there exists a finite set $\mathfrak{M}$ of Galois extensions N/K with Galois group $G(N/K) \simeq G$ such that*

$$\mathcal{T}(G) = \cap_{N/K\in\mathfrak{M}} ker(h_{N/K}).$$

Proof. For any element $\sigma \in G$ of order ≤ 2 we can choose a Galois extension N_σ/K_σ of algebraic number fields such that $G(N_\sigma/K_\sigma) \simeq G$ and σ corresponds to the complex conjugation on N_σ (apply lemma 2 [**Epk99**]). We get

$$\mathcal{T}(G) = \cap_{N/K, G(N/K)\simeq G} \ker(h_{N/K}) \subset \cap_\sigma \ker(h_{N_\sigma/K_\sigma}) \subset L(G).$$

By theorem 2 [**Epk99**] the index of $\mathcal{T}(G)$ in $L(G)$ is finite. Hence we are done. □

Lemma 4.5. *Let N/K be a finite Galois extension and let L/K be any field extension such that $L \cap N = K$. Then the restriction homomorphism $G(NL/L) \to G(N/K)$ induces an isomorphism τ such that*

$$\begin{array}{ccc} \mathcal{B}(G(NL/L)) & \xrightarrow{h_{NL/L}} & W(L) \\ \uparrow \tau & & \uparrow s^\star \\ \mathcal{B}(G(N/K)) & \xrightarrow{h_{N/K}} & W(K) \end{array}$$

commutes. We get

$$ker(h_{N/K}) \subset ker(h_{NL/L}).$$

If $s^\star$ is injective, then equality holds.

Proof. The isomorphism τ is defined by

$$\tau(\chi_H^{G(N/K)}) = \chi_{G(N^H L/L)}^{G(NL/L)}$$

for $H \leq G(N/K)$. Note, that $L \cap N = K$ implies that $G(NL/L)$ and $G(N/K)$ are isomorphic. Hence τ is well defined. Therefore the results follows. □

Theorem 4.6. *For any finite group G there exists a finite set $\mathfrak{M}$ of Galois extensions N/K of Hilbertian fields with $G(N/K) \simeq G$ such that*

$$\mathcal{T}(G) = \cap_{N/K \in \mathfrak{M}} \ker(h_{N/K}).$$

Proof. Let N/K be a Galois extension with $G(N/K) \simeq G$. Choose

$$f(X) = X^m + \sum_{i=0}^{m-1} a_i X^i \in K[X]$$

with $N \simeq K[X]/(f(X))$. Set $K_1 = K_0(a_0, \dots, a_{m-1})$, where K_0 is the prime field of K. Let $N' \subset N$ be a splitting field of $f(X)$ over K_1 and set $K' = N' \cap K$. Then N'/K' is a Galois extension with Galois group isomorphic to G. Now lemma 4.5 implies $\ker(h_{N'/K'}) \subset \ker(h_{N/K})$. By theorem 2 p.155 [**Lan62**] the field K_1 is a finite field or a Hilbertian field.

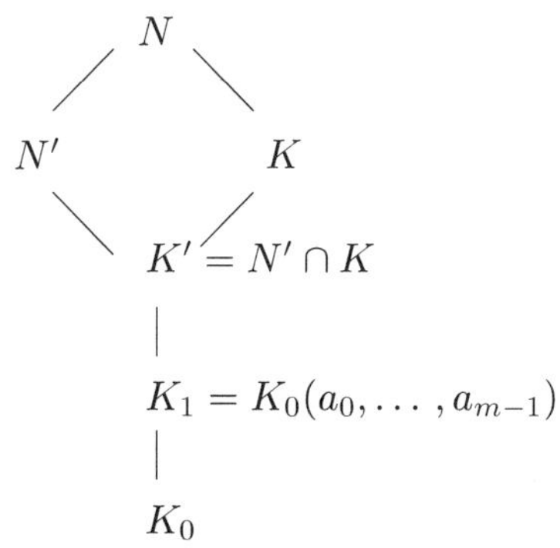

First suppose that K_1 is finite. Then G is cyclic. By proposition 5 in [**Epk98**] $\mathcal{T}(G)$ is the intersection of finitely many kernels of homomorphisms $h_{N/K}$ which are defined via algebraic number fields N/K.
If K_1 is a Hilbertian field then so is K' ([**FJ86**] Cor 11.7). By lemma 4.4 there exists a finite set $\mathfrak{M}$ of Galois extensions N/K with Galois group isomorphic to G and such that $\mathcal{T}(G) = \cap_{N/K \in \mathfrak{M}} \ker(h_{N/K})$. For any $N/K \in \mathfrak{M}$ choose some N'/K' as above. This defines a set $\mathfrak{M}'$. We conclude

$$\mathcal{T}(G) \subset \cap_{N'/K' \in \mathfrak{M}'} \ker(h_{N'/K'}) \subset \cap_{N/K \in \mathfrak{M}} \ker(h_{N/K}) = \mathcal{T}(G).$$

□

PROPOSITION 4.7. *Let G be a finite group with Sylow 2-group G_2. Suppose G_2 has a normal abelian complement A in G. Then $\chi \in \mathcal{T}(G)$ if and only if $res_G^{G_2}(\chi) \in \mathcal{T}(G_2)$.*

PROOF. Let N/K be a Galois extension of Hilbertian fields with $G(N/K) \simeq G_2$. By [**Mat87**] IV.3 Satz 2 the split embedding problem defined by N/K and the exact sequence $1 \longrightarrow A \longrightarrow G \longrightarrow G_2 \longrightarrow 1$ has a proper solution L/K. Lemma 4.5 together with Springer's theorem implies $\ker(h_{N/K}) = \ker(h_{L/L^{G_2}})$. Now the assertion follows from lemma 4.6 and [**Epk99**] proposition 21(2). □

Together with [**Epk99**] proposition 23 we get

COROLLARY 4.8. *Let G be an abelian group with Sylow 2-group G_2. Then $\chi \in \mathcal{T}(G)$ if and only if $res_G^{G_2}(\chi) \in \mathcal{T}(G_2)$. Further*

$$L(G)/\mathcal{T}(G) \simeq L(G_2)/\mathcal{T}(G_2).$$

4.3. The trace ideal of the dihedral group D_{2^n} of order 2^n. In [**Epk98**] we determined the trace ideal of an elementary abelian 2-group, of a cyclic 2-group and of the quaternion group of order 8. Now we consider the dihedral group D_{2^n} of order 2^n. We explicitly determine the trace ideal of D_8. In general, we only show that $L(D_{2^n})/\mathcal{T}(D_{2^n})$ has exponent 2. Let

$$D_{2^n} =< \sigma, \tau \mid \sigma^{2^{n-1}} = \tau^2 = 1, \tau^{-1}\sigma\tau = \sigma^{-1} >$$

be the dihedral group of order 2^n. Then

$$\mathcal{R}(D_8) = \{1, < \tau >, < \tau\sigma >, < \sigma^2 >, < \sigma >, V =< \tau, \sigma^2 >, W =< \tau\sigma, \sigma^2 >, D_8\}$$

is a complete set of representatives of the conjugacy classes of subgroups of D_8.

PROPOSITION 4.9. *We get*

$$\mathcal{T}(D_8) = \{\chi = \sum_{U\in\mathcal{R}(D_8)} m_U\chi_U : \chi \in L(G), m_{<\tau>} \equiv m_{<\tau\sigma>} \equiv 0 \bmod 2\}.$$

PROOF. Let

$$\begin{aligned}\chi &= m_G\chi_G + m_\tau\chi_{<\tau>} + m_{\sigma^2}\chi_{<\sigma^2>} + m_{\tau\sigma}\chi_{<\tau\sigma>} \\ &\quad + m_\sigma\chi_{<\sigma>} + m_V\chi_V + m_W\chi_W + m_1\chi_1 \in \mathcal{T}(D_8).\end{aligned}$$

The system of linear equations defining $L(D_8)$ is given by

$$\begin{array}{lllllllll} m_G & +2m_V & +2m_W & +2m_\sigma & +4m_\tau & +4m_{\sigma^2} & +4m_{\tau\sigma} & +8m_1 & =0 \\ m_G & +2m_V & & & +2m_\tau & & & & =0 \\ m_G & +2m_V & +2m_W & +2m_\sigma & & +4m_{\sigma^2} & & & =0 \\ m_G & & +2m_W & & & & +2m_{\tau\sigma} & & =0 \end{array}$$

Let N/K be a Galois extension with $G(N/K) \simeq D_8$. Set $L_\tau := N^{<\tau>}$. There is an irreducible polynomial $f = X^4 + aX^2 + b \in K[X]$ such that N is the splitting field of f over K. Assume $a \neq 0$. The relevant intermediate fields of N/K and their trace forms are as follows

H	N^H	$< N^H >$
G	K	$<1>$
$<\tau>$	$K(\alpha)$	$< 1, a^2 - 4b, -2a, -2ab(a^2 - 4b) >$
$<\sigma^2>$	$K(\sqrt{b}, \sqrt{a^2 - 4b})$	$2\times < 1, b(a^2 - 4b) >$
$<\tau\sigma>$	$K(\sqrt{b}, \sqrt{2\sqrt{b} - a})$	$< 1, b, -a, -ab(a^2 - 4b) >$
$<\sigma>$	$K(\sqrt{b(a^2 - 4b)})$	$< 2, 2b(a^2 - 4b) >$
V	$K(\sqrt{a^2 - 4b})$	$< 2, 2(a^2 - 4b) >$
W	$K(\sqrt{b})$	$< 2, 2b >$
$< id >$	N	$2\times < 1, b(a^2 - 4b), -a, -ab(a^2 - 4b) >$

Calculating the image of χ in $W(K)$ we get

$$h_{N/K}(\chi) = m_{\tau\sigma}\times < 1, -2 > \otimes < b > \otimes < 1, -ab, -a(a^2 - 4b), -b(a^2 - 4b) > .$$

If $m_{\tau\sigma}$ is even, then $m_{\tau\sigma}\times < 1, -2 >= 0$. Consider the generic situation: $K = \mathbb{Q}(X_1, X_2), f = X^4 + X_1X^2 + X_2$. Then $Gal(f) \simeq D_8$. Further

$$\psi =< 1, -X_1X_2, -X_1(X_1^2 - 4X_2), -X_2(X_1^2 - 4X_2) >$$

does not represents 2. Hence $< 1, -2 > \otimes\psi \neq 0$. The case $f = X^4 + b$ is left to the reader. □

PROPOSITION 4.10. *Let $n \geq 2$. Then $L(D_{2^n})/\mathcal{T}(D_{2^n})$ is an elementary abelian 2-group.*

PROOF. By proposition 6 [**Epk98**] and by proposition 4.9 we can assume $n \geq 4$. We only give a sketch of the proof. Set $G = D_{2^n}$, $V =< \tau, \sigma^2 >$, $\chi_\tau = \chi^G_{<\tau>}, \chi_{\tau\sigma} = \chi^G_{<\tau\sigma>}$, $\chi_\sigma = \chi^G_{<\sigma>}$ and $\chi_V = \chi^G_V$. We easily observe, that the following set B of elements in $\mathcal{B}(D_{2^n})$ defines a basis of $L(D_{2^n})$.

$$X_1 = \chi_1 - 2\chi_\tau + 2\chi_V - 2\chi_\sigma \in B.$$

$$X_{\tau\sigma} = \chi_{\tau\sigma} - \chi_\tau + 2\chi_V - \chi_\sigma - 2\chi_G \in B.$$

For any cyclic group $U \subset< \sigma^2 >, U \neq 1$ set

$$X_U = \chi_U - \frac{[G:U]}{2}\chi_\sigma \in B.$$

If $U \subset <\tau, \sigma^4>$ is a non-cyclic group, set

$$X_U = \chi_U - \chi_2 - \frac{[G:U]-2}{2}\chi_\sigma \in B.$$

If $U \neq G$ is a non-cyclic subgroup with $\tau\sigma \in U$, set

$$X_U = \chi_U - \frac{[G:U]}{2}\chi_\sigma + \chi_V - 2\chi_G \in B.$$

Now we have to prove $2\chi \in \mathcal{T}(D_{2^n})$ for any $\chi \in B$. This follows by induction using corollary 4 [**DEK97**] and a fact from the theory of embedding problems which we recall now (Theorem 6 [**Kim90**]).

Let $a, b \in K^\star$, such that a, b, ab are not squares in $K^\star$ and such the quadratic forms $<1, b>$ and $<a, ab>$ are isometric. Set $\Theta = a + \sqrt{a}$. Then $L = K(\sqrt{a}, \sqrt{b}, \sqrt{2q\Theta}), q \in K^\star$ parametrizes the D_8-extensions which contain $\sqrt{a}, \sqrt{b}$ and which are cyclic over $K(\sqrt{b})$. Further L/K is contained in a D_{16} extension, which is cyclic over $K(\sqrt{b})$ if and only if $<1,-2,-a,2a> \simeq <1,-q,b,-qb>$. We get

$$<K(\sqrt{2q\Theta})> \simeq <1, a, qa, qab> \simeq <1, a, q, qb>.$$

$2\times <1,-2,-a,2a> = 0 = 2\times <1,-q,b,-qb>$ implies $2\times <1,b> \simeq 2\times <q,qb>$. Hence $2\times <K(\sqrt{2q\theta})> \simeq 2\times <1,1,a,b>$, if $K(\sqrt{2q\theta})$ is contained in a D_{16}-extension. The case $b=-1$ in [**Kim90**] has to be treated separately. □

5. Some applications

5.1. Groups with dihedral Sylow 2-group. The complete classification of groups with dihedral Sylow 2-group can be found in [**Gor68**][16.3].

PROPOSITION 5.1. *Let G be a finite group with Sylow 2-group a dihedral group of order $2^n \geq 8$. Then for any subgroup $H < G$ with $\cap_{\sigma\in G}\sigma H\sigma^{-1} = 1$ we get*

$$I_{M(G,H)} = (q_{G,H}(X)).$$

PROOF. Let $n = 3$. Since $L(D_8)/\mathcal{T}(D_8)$ has exponent 2, we only have to consider the coefficients of $res^G_{G_2}(q_{G,H}(\chi^G_H))$ modulo 2. Let

$$res^G_{G_2}(\chi^G_H) = \sum_{U\in\mathcal{R}(D_8)} m_U\chi^{G_2}_U.$$

We get $q_{G,H}(X) \equiv (X - [G:H])^d \bmod \mathbb{F}_2[X]$, where $d = deg(q_{G,H}(X))$. We know that m_{G_2} is the number of fixed points of the action of G_2 on G/H. Hence $m_{G_2} \equiv [G:H] \bmod 2$. Therefore

$$\chi = (res^G_{G_2}(\chi^G_H) - [G:H]\chi^{G_2}_{G_2})^2 \equiv \sum_{U\in\mathcal{R}(D_8), U\neq G_2} m^2_U(\chi^{G_2}_U)^2 \bmod 2\mathcal{B}(G).$$

For any normal subgroup U we get $(\chi^{G_2}_U)^2 = [G:U]\chi^{G_2}_U$. Hence

$$\begin{aligned}\chi &\equiv m^2_\tau(\chi^{G_2}_\tau)^2 + m^2_{\tau\sigma}(\chi^{G_2}_{\tau\sigma})^2 \equiv m_\tau(2\chi^{G_2}_\tau + \chi^{G_2}_1) + m_{\tau\sigma}(2\chi^{G_2}_{\tau\sigma} + \chi^{G_2}_1)\\ &\equiv (m_\tau + m_{\tau\sigma})\chi^{G_2}_1 \equiv 0 \bmod 2\mathcal{B}(G),\end{aligned}$$

since $m_\tau \equiv m_{\tau\sigma} \bmod 2$ for any element in $L(D_8)$. Since $d = 2, 3, 4$, we are done.

Now consider $n \geq 4$. Then $\chi = \sum a_U\chi^{G_2}_U + 2\tilde{\chi}$, where U runs over all nontrivial cyclic subgroups of $<\sigma^2>$. Further $\tilde{\chi} \in \mathcal{B}(G_2)$. Let $U, V \in G_2$ be non-trivial subgroups of G_2 such that U is normal in G_2 and $V \neq G$. Then $\chi^{G_2}_U \cdot \chi^{G_2}_V \in 2\mathcal{B}(G_2)$.

We conclude $\chi \cdot (res^G_{G_2}(\chi^G_H) - [G : H]\chi^{G_2}_{G_2}) \in 2 \cdot \mathcal{B}(G_2)$. Hence we are done if $d = deg(q_{G,H}(X)) \geq 3$.
Let $d = 2$. Observe, that a_1 is even. By corollary 4 [**DEK97**] $\chi_U \equiv \frac{[G_2:U]}{2}\chi^{G_2}_\sigma \bmod \mathcal{T}(G_2)$ for any $U \subset <\sigma^2>, U \neq 1$. Hence $res^G_{G_2}(q_{G,H}(\chi^G_H)) = 2r\chi^{G_2}_\sigma + 2\tilde{\chi} + \tilde{\tilde{\chi}}$, where $r \in \mathbb{Z}, \tilde{\tilde{\chi}} \in \mathcal{T}(G_2) \subset L(G_2)$. Therefore $r\chi^{G_2}_\sigma + \tilde{\chi} \in L(G_2)$. □

5.2. Groups with quaternion Sylow 2-group. The aim of this section is to give an example where $q_{G,H}(X)$ does not annihilate $< L >$. Let us first recall the determination of the trace ideal of the quaternion group Q_8 of order 8. By [**Epk98**] proposition 7 we get

PROPOSITION 5.2.

$$\mathcal{T}(Q_8) = \{\chi = \sum_{H<Q_8} m_H\chi_H \ : \ \chi \in L(Q_8),\ m_{H_1} \equiv m_{H_2} \equiv m_{H_3} \bmod 4 \quad \textit{for all subgroups } H_1, H_2, H_3 \textit{ of order 4 in } Q_8\}.$$

PROPOSITION 5.3. *Let G be a finite group with Sylow 2-group G_2 a quaternion group of order 8. Let H be a subgroup of G such that G acts faithfully on G/H. Set $n := [G : H]$ and $s := \mathrm{sign}_\sigma\chi^G_H$. Then*

$$I_{M(G,H)} = (q_{G,H}(X)) = ((X - n)(X - s))$$

if one of the following cases occur

1. *H is of odd order.*
2. *$\sharp H \equiv 2 \bmod 4$.*
3. *$\sharp H \equiv 0 \bmod 4$ and the permutation representation defined by G on G/H contains only even permutations.*

PROOF. Since G_2 has a unique involution σ we get

$$q_{G,H}(X) = (X - n)(X - s).$$

We will prove $res^{G_2}_G(q_{G,H}(\chi^G_H)) \in \mathcal{T}(G_2)$ in the cases cited above. By proposition 5.2 we have to determine the coefficients of $\chi^{G_2}_{H_i}$ in $res^{G_2}_G(q_{G,H}(\chi^G_H))$. Let

$$res^{G_2}_G(\chi^G_H) = a\chi^{G_2}_{G_2} + c\chi^{G_2}_\sigma + d\chi^{G_2}_1 + \sum_{i=1}^3 b_i\chi^{G_2}_{H_i}.$$

Since all subgroups of G_2 are normal we get

$$\begin{array}{ll} \chi^{G_2}_U\chi^{G_2}_V = [G_2 : V]\chi^{G_2}_1 & \text{if } U \subset V \\ \chi^{G_2}_{H_i}\chi^{G_2}_{H_j} = \chi^{G_2}_\sigma & \text{if } i \neq j. \end{array}$$

Hence

$$res^{G_2}_G(q_{G,H}(\chi^G_H)) = a'\chi^{G_2}_{G_2} + c'\chi^{G_2}_\sigma + d'\chi^{G_2}_1 + \sum_{i=1}^3 b_i(2b_i + 2a - n - s)\chi^{G_2}_{H_i}.$$

Further $b_i = \sharp\{\tau \in G_2\backslash G/H : G_2 \cap \tau H\tau^{-1} = H_i\}$. Hence $b_1 = b_2 = b_3 = 0$ if $\sharp H \not\equiv 0 \bmod 4$.

Now suppose $\sharp H \equiv 0 \bmod 4$. We get $s = \text{sign}_\sigma \chi_H^G = a + 2b_1 + 2b_2 + 2b_3 + 4c \equiv n \bmod 8$. Hence $\frac{s+n}{2} \equiv n \bmod 4$. Let $i \neq j$. Then

$$\begin{aligned} & b_i(2b_i + 2a - n - s) - b_j(2b_j + 2a - n - s) \\ = \; & 2(b_i - b_j)(b_i + b_j + a + \frac{s+n}{2}) \\ \equiv \; & 0 \bmod 4 \end{aligned}$$

if and only if $(b_i - b_j)(b_i + b_j + a + n)$ is even. Since $a \equiv n \bmod 2$, we have to determine the parity of b_1, b_2, b_3. Let σ_i be a generator of H_i. The action of G_2 on G/H has the following types of orbits:

order of orbit	number of orbits	isotropy group
1	a	G_2
2	b_i	H_i
4	c	$<\sigma>$
8	d	1

Hence the sign of σ_i regarded as a permutation on G/H equals $(-1)^{b_j+b_k}$, where $\{i,j,k\} = \{1,2,3\}$. □

Observe, that G contains an odd permutation if and only if σ_i is odd for some $i = 1, 2, 3$. Now we would like to consider the following situation:
$\sharp H \equiv 0 \bmod 4$ and G contains an odd permutation.

LEMMA 5.4. *Under the condition above we get:*
G is a semidirect product of G_2 and a normal subgroup A of odd order. The conjugation of G_2 on A induces a monomorphism

$$\Phi : G_2 \longrightarrow \text{Aut}(A).$$

PROOF. Let U be the kernel of the homomorphism sign : $G \to \{1, -1\}$ defined by the action of G on G/H. Since U is a normal subgroup of index 2, its Sylow 2-group are cyclic of order 4. By [**Gor68**] 7.6.1 U contains a normal subgroup A of index 4 in U. By [**Hup67**] I.4.9 A is a normal subgroup of G. Hence G is a semidirect product of G_2 and A. Let

$$\Phi : G_2 \longrightarrow \text{Aut}(A) : \pi \mapsto (x \mapsto \pi x \pi^{-1})$$

Assume, that Φ is not injective. Then σ lies in the kernel of Φ, which implies $\sigma\pi = \pi\sigma$ for all $\pi \in A$. Since σ is commutes with every $\pi \in G_2$ it lies in the center of G. Hence σ is the unique involution in G. Since H has even order, G can not act faithfully on G/H, a contradiction. □

PROPOSITION 5.5. *Let G be a finite group with Sylow 2-group $G_2 \simeq Q_8$. Let $H < G$ be a subgroup with*

1. *$\sharp H \equiv 0 \bmod 4$*
2. *G acts faithfully on G/H.*
3. *the action of (2) contains odd permutations.*

Then

1. *G is a semidirect product of G_2 and a normal subgroup A of odd order.*
2. *$res_G^{G_2}(q_{G,H}(\chi_H^G)) \notin \mathcal{T}(G_2)$.*
3. *If A is abelian, then $q_{G,H}(\chi_H^G) \notin \mathcal{T}(G)$, but $2q_{G,H}(\chi_H^G) \in \mathcal{T}(G)$.*

PROOF. (2) follows from the remark following the proof of proposition 5.3. (3) is a consequence of (2) and proposition 4.7. □

Now choose a group A of odd order such that the automorphism group of A contains a quaternion group of order 8. Let

$$\Phi : Q_8 \to \mathrm{Aut}(A)$$

be a monomorphism and set $G = A \times_\Phi Q_8$. Let H be a subgroup of G with $\sharp H \equiv 4 \bmod 8$. We can assume that $H_1 < Q_8$ is a Sylow 2-group of H. We now determine the polynomial $p^{(2)}_{G,H}(X) = p(X)$. As above, let

$$res^{Q_8}_G(\chi^G_H) = a\chi^{Q_8}_{Q_8} + c\chi^{Q_8}_\sigma + d\chi^{Q_8}_1 + \sum_{i=1}^{3} b_i\chi^{Q_8}_{H_i} = \sum_{U<Q_8} m_U\chi^{Q_8}_U.$$

We know

$$m_U = \sharp\{\tau \in Q_8\backslash G/H : Q_8 \cap \tau H\tau^{-1} = U\}.$$

Hence $a = 0$.
Claim: $b_2 = b_3 = 0$. Assume $b_2 \neq 0$. Then $H_2 \subset \tau H\tau^{-1}$ for some $\tau \in G = AQ_8$. Since H_2 is normal in Q_8, we can choose $\tau \in A$. By Sylow's theorem we get $H_2 = \tau H_1\tau^{-1}$ for some $\tau \in A$, which is impossible.
Hence $res^{Q_8}_G(q_{G,H}(\chi^G_H)) = b_1\chi^{Q_8}_{H_1} + c\chi^{Q_8}_\sigma + d\chi^{Q_8}_1$. Therefore $\varphi_{Q_8}(S) = \varphi_{H_2}(S) = \varphi_{H_3}(S) = 0$, $\varphi_1(S) = n$, $\varphi_\sigma(S) = s = 2b_1 + 4c$ and $\varphi_{H_1}(S) = 2b_1$. We get

$$p^{(2)}_{G,H}(X) = \begin{cases} X(X-n)(X-s), & \text{if } c = 0 \\ X(X-n)(X-s)(X-2b_1), & \text{if } c \neq 0. \end{cases}$$

We conclude that b_1 is odd. Hence we are in the situation of proposition 5.5. Now we give a concrete example.

EXAMPLE 5.6. Let $A = \mathbb{Z}_3 \times \mathbb{Z}_3$ and set $G = A \times_\Phi Q_8$, where $\Phi : Q_8 \to \mathrm{Aut}(A) \simeq \widetilde{\mathfrak{S}_4}$ is a monomorphism. Let $H = H_1 < Q_8$ be a group of order 4. Then

$$\begin{aligned} I_{M(G,H)} &= (\ X(X-18)(X-2), 2(X-18)(X-2)\ ) \\ &= \{(X-18)(X-2)f(X) : f(X) \in \mathbb{Z}[X],\ f(0) \equiv 0 \bmod 2\}. \end{aligned}$$

PROOF. Observe, that the automorphism group of A is a double cover of the symmetric group $\mathfrak{S}_4$, which contains a quaternion group of order 8. We get $n = [G : A] = 18$.
Claim: $c = \sharp\{\tau \in H\backslash G/Q_8 : Q_8 \cap \tau H\tau^{-1} = <\sigma>\} = 0$.
Assume $\sigma \in \tau H\tau^{-1}$ for some non-trivial element $\tau \in A$. Then τ is an eigenvector according to the eigenvalue 1 of the linear map defined by $\Phi(\sigma)$. But $\Phi(\sigma) = -id$. Hence $\mathrm{sign}_\sigma\chi^G_H = s = 2b_1 = \phi_H(S)$. Since $18 = n = 2b_1 + 8d$, $d \neq 0$ we get $s = 2$, or $s = 10$. Lemma 11.5 in **[Hup98]** gives

$$s = \frac{18 \cdot \sharp(G\sigma \cap H)}{\sharp G\sigma} = \frac{18}{\sharp G\sigma} \neq 10.$$

Hence $s = 2$, which implies $q_{G,H}(X) = (X-18)(X-2)$ and $p_2(X) = X(X-18)(X-2)$. Now

$$(\ X(X-18)(X-2), 2(X-18)(X-2)) \subset I_{M(G,H)} \subset ((X-18)(X-2)).$$

The ideal of the left hand side has index 2 in $((X-18)(X-2))$. By proposition 5.5(3) we are done. □

6. Some examples

We close by giving some more examples.

PROPOSITION 6.1. *Let G be a Frobenius group of degree n.*

1. *If n is odd, then*

$$I_{M(G,H)} = (B_{G,H}) = (q_{G,H}) = (X-n).$$

2. *Let n be even. Then*

$$I_{M(G,H)} = (B_{G,H}) = (q_{G,H}) = X(X-n).$$

PROPOSITION 6.2. 1. *For $p = 7, 11$ consider the doubly transitive group $PSL(2,p)$ of degree p. Then $p^{(2)}_{G,H}(X) = (X-1)(X-3)(X-p)$ and*

$$I_{M(G,H)} = (q_{G,H}) = ((X-3)(X-p)).$$

2. *Consider $\mathfrak{A}_7$ in its doubly transitive representation of degree 15. Then $p^{(2)}_{G,H}(X) = (X-1)(X-3)(X-15)$ and*

$$I_{M(G,H)} = (q_{G,H}) = ((X-3)(X-15)).$$

PROOF. By lemma 3.3 $X-1$ divides $p^{(2)}_{G,H}(X)$. The Galois number is 4 in each case (see proposition 11,14,24 in [**EG99**]). We calculate $q_{G,H}(X)$ with the help of the character table in the ATLAS [**CCN+85**]. $q_{G,H}(X)$ is an annihilating polynomial since the Sylow 2-groups are elementary abelian or dihedral of order 8. □

PROPOSITION 6.3. *Let G be a Zassenhaus group of degree n.*

1. *If n is odd, then*

$$I_{M(G,H)} = (q_{G,H}) = (p^{(2)}_{G,H}) = ((X-1)(X-n)).$$

2. *Let n be even. Then $p^{(2)}_{G,H}(X) = X(X-2)(X-n)$.*
   We know $G = PSL(2,q)$, $PGL(2,q)$, $PML(2,q)$ with q odd.
   (a) *If $q \equiv 3 \bmod 4$ or $G = PGL(2,q)$, then*

$$I_{M(G,H)} = (q_{G,H}) = (p^{(2)}_{G,H}).$$

   (b) *If $q \equiv 1 \bmod 4$ and $G \neq PGL(2,q)$. Then $q_{G,H}(X) = (X-2)(X-n)$. If $G = PSL(2,q)$, then $I_{M(G,H)} = (q_{G,H})$.*

The result on $I_{M(PML(2,q),H)}$ is unknown at present.

References

[BP97] P. Beaulieu and T. Palfrey, *The Galois number*, Math. Ann. **309** (1997), 81–96.

[CCN+85] J.H. Conway, R.T. Curtis, S.P. Norton, R.A. Parker, and R.A. Wilson, *Atlas of finite Groups*, Clarendon Press London, 1985.

[Con87] P.E. Conner, *A proof of the conjecture concerning algebraic Witt classes*, unpublished, 1987.

[DEK97] Christof Drees, Martin Epkenhans, and Martin Krüskemper, *On the computation of the trace form of some Galois extensions*, J. Algebra **192** (1997), 209–234.

[Dre71] A.W.M. Dress, *The Grothendiek ring of monomial permutation representations*, J. Algebra **18** (1971), 137–157.

[EG99] Martin Epkenhans and Oliver Gerstengarbe, *On the Galois number and minimal degree of doubly transitive groups*, 1999, to appear.

[Epk98] Martin Epkenhans, *On Vanishing Theorems for Trace Forms*, Acta Mathematica et Informatica Universitatis Ostraviensis **6** (1998), 69–85.

[Epk99] Martin Epkenhans, *An analogue of Pfister's local-global principle in the Burnside ring*, J. Theorie des Nomb. Bordeaux **11** (1999), 31–44.

[FJ86] Michael D. Fried and Moshe Jarden, *Field arithmetic*, Ergebnisse der Mathematik und ihrer Grenzgebiete; 3.Folge, Bd. 11, Springer-Verlag, Berlin, Heidelberg, New York, London, Paris, Tokyo, 1986.

[Gor68] D. Gorenstein, *Finite Groups*, Harper and Row, New York, 1968.

[HB82] Bertram Huppert and Norman Blackburn, *Finite groups III*, Grundlehren der mathematischen Wissenschaften **243**, Springer-Verlag, Berlin, Heidelberg, New York, 1982.

[Hup67] Bertram Huppert, *Endliche Gruppen I*, Grundlehren der mathematischen Wissenschaften **134**, Springer-Verlag, Berlin, Heidelberg, New York, 1967.

[Hup98] Bertram Huppert, *Character theory of finite groups*, Expositions in Mathematics, Walter de Gruyter, Berlin, New York, 1998.

[Kim90] Ian Kiming, *Explicit classifications of some 2-extensions of a field of characteristic different from 2*, Can. J. Math. **XLII** (1990), 825–855.

[Lan62] Serge Lang, *Diophantine geometry*, Interscience, New York, London, 1962.

[Lew87] D. W. Lewis, *Witt rings as integral rings*, Invent. Math. **90** (1987), 631–633.

[LM00] D. W. Lewis and S. McGarraghy, *Annihilating polynomials, étale algebras, trace forms and the Galois number*, Arch. Mat. **75** (2000), 116–120.

[Mat87] B. Heinrich Matzat, *Konstruktive Galoistheorie*, Lecture Notes in Mathematics **1284**, Springer-Verlag, Berlin, Heidelberg, New York, London, Paris, Tokyo, 1987.

[Sch85] Winfried Scharlau, *Quadratic and hermitian forms*, Grundlehren der mathematischen Wissenschaften **270**, Springer-Verlag, Berlin, Heidelberg, New York, Tokyo, 1985.

FB MATHEMATIK
UNIVERSITY OF PADERBORN
D-33095 PADERBORN
GERMANY

E-mail address: martine@uni-paderborn.de

Contemporary Mathematics
Volume **272**, 2000

Equivariant Brauer groups

A. Fröhlich and C.T.C. Wall

Introduction

Suppose A an algebra over a field K (e.g. a group algebra $K[G]$), provided with an involutory anti-automorphism σ (e.g. induced by $g \mapsto g^{-1}$ for $g \in G$). There are natural definitions of σ-symmetric bilinear and quadratic forms (see e.g. [**6**]). In order to classify forms, we may seek to simplify A. Suppose A semi-simple: then we may decompose A as a sum of simple algebras, when the quadratic forms also decompose; and then use a Morita-type theory of algebras-with-involution to reduce to the case when A is a division ring. We are thus led to contemplate a Brauer group of central simple algebras with involution, and to seek to calculate such groups. This programme is performed in detail (for finite, local and global fields) in [**6**].

There are natural generalisations in several directions. We may replace K by an arbitrary commutative ring R (or even by an arbitrary scheme), and the single anti-automorphism of order 2 by a group Γ of automorphisms and anti-automorphisms. Also, along with the Brauer group, we may consider projective class groups and unit groups on one hand, or look at higher K-groups on the other.

A theory encompassing several of these generalisations was worked out by us about 30 years ago, and parts of it were published in [**1**], [**2**], [**3**] and [**7**]. However a further lengthy (1971) preprint 'Generalisations of the Brauer group I'[1] never attained final form, and this was the manuscript in which the extension to include anti-automorphisms was developed. It is our object here to outline the main features of this theory. We will omit the proofs, most of which involve verifications of identities; we also omit the discussion of automorphisms which was included in the earlier manuscript. Part of our motivation came from algebraic number theory: an example was developed in [**2**]: see also [**3**, §8].

The plan of the paper is as follows. We begin with general definitions and results about Γ-graded monoidal categories. The next section is devoted to the construction of categories of modules and of algebras which satisfy these conditions. We then obtain a number of exact sequences relating the groups and monoids we have defined. These form a particularly satisfying pattern when the grading group Γ is finite, and give explicit interpretations of several groups previously defined

[1] II - never written - was to include Brauer groups of graded - or, as they later became known, super-algebras.

abstractly. In §5 we return to the abstract theory to seek a justification for this pattern: this leads to some open questions. Up to this point the group Γ is fixed: in the final section we indicate the effect of varying it.

1. Graded monoidal categories

The concept of monoidal category is now classical (see e.g. [**4**]). A *monoidal category* consists of a category $\mathcal{C}$, a covariant functor $\nabla : \mathcal{C} \times \mathcal{C} \to \mathcal{C}$, an object E of $\mathcal{C}$, and natural equivalences $a : A\nabla(B\nabla C) \to (A\nabla B)\nabla C$, $c : A\nabla B \to B\nabla A$ and $e : E\nabla A \to A$ satisfying certain standard identities. In [**3**] we defined a category $\mathcal{MC}$ of monoidal categories. Here and below we will omit discussion of compatibility with the natural equivalences a, c and e.

If Γ is an abstract group, a Γ-*grading* on a category $\mathcal{C}$ is a functor $g : \mathcal{C} \to \Gamma$. For any morphism f, we refer to $g(f)$ as the *grade* of f. From now on we drop g from the notation, though we may make Γ explicit when needed. The grading is *stable* if for all $C \in \operatorname{ob}\mathcal{C}$, $\gamma \in \Gamma$ there is an equivalence $f \in \mathcal{C}$ with domain C and grade γ. A Γ-*graded monoidal category* consists of a stably Γ-graded category $(\mathcal{C}, g)$, a covariant Γ-functor $\nabla : \mathcal{C} \times_\Gamma \mathcal{C} \to \mathcal{C}$ (so only morphisms of the same grade can be 'added'), a covariant Γ-functor $E : \Gamma \to \mathcal{C}$ (whose image object is also denoted E), and natural equivalences (of grade 1) a, c, e satisfying the standard identities. We have a category $\Gamma - \mathcal{MC}$ of Γ-graded monoidal categories.

For $\mathcal{C}$ a Γ-graded monoidal category, we define $\mathcal{R}ep(\Gamma, \mathcal{C})$ to be the category of Γ-functors $F : \Gamma \to \mathcal{C}$ and natural transformations (of grade 1): we omit Γ from the notation if it is clear which group Γ is. An object of $\mathcal{R}ep(\mathcal{C})$ thus consists of an object C of $\mathcal{C}$ together with a representation of Γ by automorphisms of C. We have a functor $\mathcal{R}ep : \Gamma - \mathcal{MC} \to \mathcal{MC}$. More trivially, if Δ is a subgroup of Γ we have a functor $\Gamma - \mathcal{MC} \to \Delta - \mathcal{MC}$ defined by forgetting all morphisms other than those whose grade belongs to Δ. If Δ is trivial, we denote this by $\mathcal{K}er : \Gamma - \mathcal{MC} \to \mathcal{MC}$, and we have a (forgetful) natural transformation $T : \mathcal{R}ep \to \mathcal{K}er$.

For any monoidal category $\mathcal{C}$, we write $k(\mathcal{C})$ for the abelian monoid of isomorphism classes of objects of $\mathcal{C}$. An object A of $\mathcal{C}$ is said to be *invertible* if there exist an object B and an isomorphism $A\nabla B \to E$. Thus $k(\mathcal{C})$ is a group if and only if all objects of $\mathcal{C}$ are invertible. If $\mathcal{C}$ is stably graded, there is a natural action of Γ on $k(\mathcal{C})$ defined as follows. If $X \in \operatorname{ob}(\mathcal{C})$ and $\gamma \in \Gamma$, choose a morphism $f : X \to Y$ of grade γ and define $[X]^\gamma := [Y]$. The functor $T(\mathcal{C})$ induces a homomorphism $k\mathcal{R}ep(\mathcal{C}) \to k\mathcal{K}er(\mathcal{C})$, with image contained in the invariant part $H^0(\Gamma; k\mathcal{K}er(\mathcal{C}))$: we will denote the map $k\mathcal{R}ep(\mathcal{C}) \to H^0(\Gamma; k\mathcal{K}er(\mathcal{C}))$ by $T_\mathcal{C}$.

Write $U(\mathcal{C})$ for the abelian group (of 'units') of automorphisms (of grade 1) of the identity element E of $\mathcal{C}$. There is a natural action of Γ on $U(\mathcal{C})$ defined as follows. If $u \in U(\mathcal{C})$, choose a morphism $h : E \to E$ of grade γ and define $u^\gamma := h^{-1} \circ u \circ h$.

We have [**7**] algebraic K-theory groups $K_n(\mathcal{C}) := K_n(\mathcal{K}er(\mathcal{C}))$, and define the equivariant algebraic K-groups to be $K_n(\mathcal{C}, \Gamma) := K_n(\mathcal{R}ep(\Gamma, \mathcal{C}))$. We can identify $K_0(\mathcal{C})$ with the 'Grothendieck group' of the monoid $k(\mathcal{C})$, and $K_1(\mathcal{C})$ with the group $U(\mathcal{C})$. The equivariant Brauer group which is our main interest is an equivariant K_2 group (but of a 2-category: we return to this in §5). We aim to obtain exact sequences to calculate such groups.

The *projective category* (so-called because of the relation to projective representations) $P\mathcal{C}$ is defined as follows. For $u \in U(\mathcal{C})$ and $C \in \mathrm{ob}(\mathcal{C})$, the formula $\theta_C(u) := e_C \circ (u\nabla 1_C) \circ e_C^{-1}$ defines an automorphism (of grade 1) of C, and $\theta_C : U(\mathcal{C}) \to \mathrm{Aut}_{\mathcal{C}}(C)$ is a homomorphism, which is an isomorphism if C is invertible. Define two morphisms $s_1, s_2 : C \to D$ to be equivalent if, for some $u \in U(\mathcal{C})$, $s_2 = s_1 \circ \theta_C(u)$. Then $P\mathcal{C}$ has the same objects as $\mathcal{C}$ and its morphisms are the equivalence classes of those of $\mathcal{C}$ (equivalence respects the grade). Given a stable Γ-graded monoidal category $\mathcal{C}$, elementary arguments [**3**, 4.5] yield a sequence of abelian monoids

$$S(\mathcal{C}) : 1 \to H^1(\Gamma; U(\mathcal{C})) \xrightarrow{\phi_{\mathcal{C}}} k\mathcal{R}ep(\mathcal{C}) \xrightarrow{\mu_{\mathcal{C}}} k\mathcal{R}ep(P\mathcal{C}) \xrightarrow{\omega_{\mathcal{C}}} H^2(\Gamma; U(\mathcal{C}))$$

which is exact up to $k\mathcal{R}ep(\mathcal{C})$, and is exact if all objects are invertible. A functor $F : \mathcal{C} \to \mathcal{D}$ induces a morphism $S(F)$ of sequences.

A monoidal category is called *precise* if each of the natural equivalences a, c and e is equal to the corresponding identity. It is shown in [**3**, §5] that if, for all $C \in \mathrm{ob}(\mathcal{C})$, the map $c_{C,C} : C\nabla C \to C\nabla C$ is the identity — we then say that $\mathcal{C}$ is *strictly coherent* — then $\mathcal{C}$ is equivalent to a precise category. If $\mathcal{C}$ is a precise monoidal category we say that $\mathcal{C}$ is *group-like* if every object and every morphism is invertible. If the objects of $\mathcal{C}$ form a group under the composition law ∇, we call $\mathcal{C}$ a *group category*. Again, a strictly coherent group-like category is equivalent to a group category. These conditions on a monoidal category are very restrictive, but will be satisfied in important examples below.

If $\mathcal{C}$ is a Γ-graded group category, we can define cohomology groups $H^n(\Gamma; \mathcal{C})$ (zero in negative degrees), and establish an exact sequence [**3**, 7.2]

$$H(\mathcal{C}) : \dots H^n(\Gamma; U(\mathcal{C})) \xrightarrow{\alpha_{n,\mathcal{C}}} H^n(\Gamma; \mathcal{C}) \xrightarrow{\beta_{n,\mathcal{C}}} H^{n-1}(\Gamma; k(\mathcal{C})) \xrightarrow{\delta_{n,\mathcal{C}}} H^{n+1}(\Gamma; U(\mathcal{C})) \dots$$

which is functorial in an appropriate sense. Moreover we have an isomorphism $S(\mathcal{C}) \cong H^1(\mathcal{C})$, where $H^1(\mathcal{C})$ denotes the part of $H(\mathcal{C})$ between $H^1(\Gamma; U(\mathcal{C}))$ and $H^2(\Gamma; U(\mathcal{C}))$. In particular, we have isomorphisms

$$H^0(\Gamma; \mathcal{C}) \cong H^0(\Gamma; U(\mathcal{C})), \quad H^1(\Gamma; \mathcal{C}) \cong k\mathcal{R}ep(\mathcal{C}), \quad H^0(\Gamma; k(\mathcal{C})) \cong k\mathcal{R}ep(P\mathcal{C}).$$

See also [**5**] for a reinterpretation of the groups $H^n(\Gamma; \mathcal{C})$.

A more K-theoretic formulation of this was given in [**7**]. We will discuss it in more detail in §5.

In order to use all this, we need to construct some stable Γ-graded monoidal categories, and in particular some group categories. However, before proceeding to explicit examples, it is convenient here to recall the general construction of 'twisting' described in [**3**, §11]: we again omit details relating to compatibility with the equivalences a, c, e. Let $\mathcal{C}$ be a monoidal category, and $D : \mathcal{C} \to \mathcal{C}$ a morphism in $\mathcal{MC}$ such that there is a natural equivalence $j : I \to D \circ D$ with $Dj = jD$. We define a $\{\pm 1\}$-graded category $\mathcal{C}^*$ by setting $\mathcal{K}er\,\mathcal{C}^* := \mathcal{C}$ and letting the morphisms of grade -1 from P to P' correspond to $\mathcal{C}$-morphisms from DP to P'. Composition with a morphism $P' \to P''$ of grade $+1$ is defined in the natural way; if $g : DP' \to P''$ defines a morphism in $\mathcal{C}^*$ of grade -1, the composite with $f : P \to P'$ or with $f : DP \to P'$ is defined by $g \circ Df$ or by $g \circ Df \circ jP$ respectively. This composition is associative since $Dj = jD$.

If $\mathcal{C}$ is Γ-graded, the same construction works to give a $\Gamma \times \{\pm 1\}$-graded monoidal category $\mathcal{C}^*$. Now if $w : \Gamma \to \{\pm 1\}$ is a homomorphism, with graph

$\Gamma_w \subset \Gamma \times \{\pm 1\}$, we may restrict to the morphisms whose grades belong to Γ_w to obtain a Γ_w-graded monoidal category, which we say is obtained from $\mathcal{C}$ by twisting.

Similarly, given two triples $(\mathcal{C}, D, j)$ and $(\mathcal{C}', D', j')$ satisfying the conditions, we can enhance a functor $T : \mathcal{C} \to \mathcal{C}'$ to $T^* : \mathcal{C}^* \to \mathcal{C}'^*$ provided we are given a natural transformation $h : D' \circ T \to T \circ D$ such that, for each $P \in \text{ob}\mathcal{C}$, $TjA = hDA \circ D'hA \circ jTA$.

2. Construction of categories of modules and algebras

We use 'ring' to mean ring with identity element. If Γ is a group, a Γ-ring consists of a ring E together with an action of Γ by ring automorphisms. If E and F are Γ-rings, the category ${}_E\mathcal{M}od_F$ of E-F-bimodules is enhanced to a Γ-graded category by defining a morphism $M \to N$ of grade γ to be a pair (ϕ, γ) where $\phi : M \to N$ is a morphism of additive groups with $\phi(emf) = e^\gamma \phi(m) f^\gamma$ for all $e \in E$, $m \in M$, $f \in F$. Then $\mathcal{R}ep({}_E\mathcal{M}od_F)$ is equivalent to the category of bimodules with Γ-action. If G is a further Γ-ring there are functors

$$\otimes_F : ({}_E\mathcal{M}od_F) \times_\Gamma ({}_F\mathcal{M}od_G) \to {}_E\mathcal{M}od_G,$$
$$Hom_F : ({}_E\mathcal{M}od_F^{op}) \times_\Gamma ({}_G\mathcal{M}od_F) \to {}_G\mathcal{M}od_E$$

of Γ-graded monoidal categories.

Fix a group Γ and a homomorphism $w : \Gamma \to \{\pm 1\}$. A (Γ, w)-ring consists of a ring E and an action of Γ on the additive group such that $(e_1 e_2)^\gamma = e_1^\gamma e_2^\gamma$ if $w(\gamma) = +1$ and $e_2^\gamma e_1^\gamma$ if $w(\gamma) = -1$. If $w(\gamma) = -1$, a morphism of grade γ from an E-F-bimodule M to an F-E-bimodule M' is a group homomorphism ϕ such that $\phi(emf) = f^\gamma \phi(m) e^\gamma$ for all $e \in E, f \in F, m \in M$. The pair (Γ, w) will be fixed throughout §2-§5, and will usually be omitted from the notation. However, we add an affix + to the notation to signify if (Γ, w) is replaced by $(\Gamma, 1)$. A case of particular interest for the application to quadratic forms is when w is an isomorphism, so the action of Γ on E is given by an involution σ.

From now on we fix a commutative ring R, and consider only R-modules (and algebras) such that the left and right actions of R agree; write $\mathcal{M}od_R$ $(= {}_\mathbb{Z}\mathcal{M}od_R)$ for the category of R-modules. First suppose w trivial. Then $\mathcal{M}od_R$ is a Γ-graded monoidal category, where — as throughout §2-§4 — we take tensor product $\otimes$ as the operator ∇. Define $\mathcal{G}en_R$ to be the subcategory of $\mathcal{M}od_R$ whose objects are the R-progenerators, i.e. faithful, finitely generated projective R-modules, and whose morphisms are the invertible ones of all grades. This is a stable Γ-graded monoidal category.

To deal with the case when w is non-trivial, first observe that taking duals $M^* = Hom_R(M, R)$ of objects and *inverse* duals f^* for morphisms gives an equivalence $* : \mathcal{G}en_R \cong \mathcal{G}en_R$ of Γ-graded monoidal categories so that $** \cong 1_{\mathcal{G}en_R}$. We can now use the twisting construction defined earlier. We can interpret a morphism $M \to N$ of grade γ in $\mathcal{G}en_R$ with $w(\gamma) = -1$ is a morphism $M^* \to N$ of grade γ in $\mathcal{G}en_R^+$. There is a natural definition of composition in each case, leading to a well-defined Γ-graded monoidal category $\mathcal{G}en_R$.

The invertible objects of $\mathcal{G}en_R$ are the invertible R-modules, i.e. rank 1 projectives: these form a category $\mathcal{C}_R$. The category $\mathcal{C}_R$ is strictly coherent, hence equivalent to a group category. Write $i_R : \mathcal{C}_R \to \mathcal{G}en_R$ for the inclusion functor.

The unit group $U(\mathcal{C}_R) = U(\mathcal{G}en_R) = U(\mathcal{M}od_R)$ is identified with the group $U(R)$ of units of R, and $k(\mathcal{C}_R)$ is the Picard group or class group $C(R)$ of R. We inherit actions of Γ on $U(R)$ and $C(R)$. Note that the class $\mathrm{cl}\,(M) \in C(R)$ of $M \in \mathrm{ob}\,(\mathcal{C}_R)$ satisfies $\mathrm{cl}\,(M^*) = \mathrm{cl}\,(M)^{-1}$ and for $u \in U(\mathcal{G}en_R)$, $u^* = u^{-1}$. The induced action of Γ on both $C(R)$ and $U(R)$ is thus obtained from the corresponding action for $\mathcal{C}_R^+$ by tensoring by w: care is needed with signs in all cases when w is not trivial.

Since morphisms in $\mathcal{G}en_R$ of grade γ with $w(\gamma) = -1$ correspond to homomorphisms $M^* \to N$, or equivalently to pairings between M^* and N^*, we may reformulate the definitions as follows. First define the category ${}_E\mathcal{P}air_F$ (where E and F are rings) to have objects triples (M, M', μ) with M an E-F-bimodule, M' an F-E-bimodule and $\mu : M \times M' \to E$ a pairing that defines an E-E-bimodule map $M \otimes_F M' \to E$. A morphism $(M, M', \mu) \to (N, N', \nu)$ of grade γ is a triple (ϕ, ϕ', γ) where ϕ and ϕ' are morphisms of grade γ and

if $w(\gamma) = 1$, $\phi : M \to N$, $\phi' : M' \to N'$,

if $w(\gamma) = -1$, $\phi : M \to N'$, $\phi' : M' \to N$,

and in both cases $\nu(\phi(m), \phi'(m')) = \mu(m, m')^\gamma$. There is a functor

$$\otimes_F : ({}_E\mathcal{P}air_F) \times_\Gamma ({}_F\mathcal{P}air_G) \to {}_E\mathcal{P}air_G$$

of monoidal Γ-graded categories. Given $h : \Gamma \to \mathrm{Aut}_{{}_E\mathcal{P}air_F}(M, M', \mu)$, define $\mu_\sigma : M \times M \to E$ by $\mu_\sigma(m_1, m_2) = \mu(m_1, h_\sigma(m_2))$. Then μ_σ is sesquilinear and reflexive: $\mu_\sigma(m_1, m_2) = \mu_\sigma(h_{\sigma^2}(m_2), m_1)$; if $\sigma^2 = 1$, μ_σ is hermitian.

Now for R commutative define $\mathcal{P}air_R := {}_{\mathbb{Z}}\mathcal{P}air_R$, and let $\mathcal{G}en\mathcal{P}air_R$ denote the full subcategory of $\mathcal{P}air_R$ whose objects are triples (M, M', μ) with M and M' progenerators and μ non-singular. Then if $\langle , \rangle : M \times M^* \to R$ is the canonical pairing, the embedding functor $M \mapsto (M, M^*, \langle , \rangle)$ is an equivalence of Γ-graded monoidal categories from $\mathcal{G}en_R$ onto $\mathcal{G}en\mathcal{P}air_R$. We may thus work with the latter when this is more convenient.

We now define the category $\mathcal{A}lg_R$. The objects are associative R-algebras with identity and if $w = 1$, a morphism $(h, \gamma) : A \to B$ of grade γ is given by a ring homomorphism h which is a morphism of grade γ in $\mathcal{M}od_R$. Taking opposite algebras defines an equivalence $\mathrm{op} : \mathcal{A}lg_R \cong \mathcal{A}lg_R$ of Γ-graded monoidal categories, so that $\mathrm{op} \circ \mathrm{op} \cong 1$. We may thus again apply the twisting construction. Thus a morphism $(h, \gamma) : A \to B$ of grade γ with $w(\gamma) = -1$ is given by a ring homomorphism of A^{op} to B (equivalently, anti-homomorphism $h : A \to B$, i.e. satisfies $h(xy) = h(y)h(x)$), which is a morphism of grade γ in $\mathcal{M}od_R$.

We pick out the subcategory $\mathcal{A}z_R$ of $\mathcal{A}lg_R$ whose objects are the Azumaya (central separable) algebras and whose morphisms are the invertible ones only. Both $\mathcal{A}lg_R$ and $\mathcal{A}z_R$ are stable Γ-graded monoidal categories.

Write $\mathrm{Lin}_R : \mathcal{A}lg_R \to \mathcal{M}od_R$ for the forgetful functor from an algebra to the underlying module. If w is trivial, this induces functors $\mathrm{Lin}_R : \mathcal{A}lg_R \to \mathcal{M}od_R$ and $\mathrm{Lin}_R : \mathcal{A}z_R \to \mathcal{G}en_R$ of Γ-graded monoidal categories. To extend to the case when w is non-trivial we require a natural equivalence $h : (\mathrm{Lin}_R A)^* \to \mathrm{Lin}_R(A^{op})$. For this we use the reduced trace $\tau_A : A \to R$, which is defined for Azumaya algebras and has the properties:

(i) the pairing $T_A : A \times A \to R$ given by $T_A(a_1, a_2) = \tau_A(a_1 a_2)$ is non-singular,

(ii) we have $\tau_A(a_2 a_1) = \tau_A(a_1 a_2)$,

(iii) $\tau_{A^{op}}(a^{op}) = \tau_A(a)$,

(iv) if $f : A \to B$ is a Γ-ring isomorphism of grade γ, $\tau_B(f(a)) = \tau_A(a)^\gamma$,

(v) $\tau_{A \oplus B} = \tau_A \oplus \tau_B$.

Here (i) shows that the map $A \to A^*$ induced by T_A has an inverse, which we can take as h, (iv) yields the commutative diagram required for compatibility; (ii) and (iii) imply compatibility with the equivalences 'j', and (v) with sums.

More directly, we can define $\mathrm{Lin}_R : \mathcal{A}z_R \to \mathcal{G}en\mathcal{P}air_R$ by $\mathrm{Lin}_R(A) := (A, A, \tau_A)$, and check that this defines a functor of Γ-graded monoidal categories, and hence induces a morphism from $\mathcal{A}z_R$ to $\mathcal{G}en_R$.

Taking the endomorphism ring defines a morphism in $\Gamma - \mathcal{MC}$ in the opposite direction, $\mathrm{End}_R : \mathcal{G}en_R \to \mathcal{A}z_R$. This is clear if w is trivial, and there is an equivalence $\mathrm{End}_R \circ * \cong op \circ \mathrm{End}_R$ of functors of Γ-graded monoidal categories, which allows us to extend to the twisted case. There is now a $\Gamma - \mathcal{MC}$-equivalence $1_{\mathcal{A}z_R} \otimes_R op \cong \mathrm{End}_R \circ \mathrm{Lin}_R$ induced by the standard isomorphism $j_A : A \otimes_R A^{op} \cong \mathrm{End}_R(A)$ given by $j_A(a \otimes b^{op})(c) = acb$. Thus the functor End_R is cofinal.

If (X, h) is an object of $\mathcal{R}ep(\Gamma, \mathcal{G}en_R)$, so that $h : \Gamma \to \mathrm{Aut}_{\mathcal{G}en_R}(X)$, then $\mathcal{R}ep(\mathrm{End}_R)(X, h)$ is the object of $\mathcal{R}ep(\Gamma, \mathcal{A}z_R)$ given by the algebra $\mathrm{End}_R(X)$, where γ acts by conjugation with the semi-linear automorphism or anti-automorphism $h(\gamma)$ according as $w(\gamma) = 1$ or -1. The fibre of the functor End is described by

LEMMA 2.1. *(i) Let $M, N \in \mathrm{ob}(\mathcal{G}en_R)$. Then $\mathrm{End}_R(M) \cong \mathrm{End}_R(N)$ if and only if, for some $P \in \mathrm{ob}(\mathcal{C}_R)$, $M \cong N \otimes_R P$.*

(ii) Let (f, γ), (g, γ) be morphisms in $\mathcal{G}en_R$. Then $\mathrm{End}_R(f, \gamma) = \mathrm{End}_R(g, \gamma)$ if and only if, for some $u \in C(R)$, $(f, \gamma) = (u, 1) \circ (g, \gamma)$.

(iii) The functor $\mathrm{End}_R : \mathcal{G}en_R \to \mathcal{A}z_R$ factors as

$$\mathcal{G}en_R \xrightarrow{P_R} P\mathcal{G}en_R \xrightarrow{PEnd_R} \mathcal{A}z_R.$$

Taking inner automorphisms gives a homomorphism $A^\times \to \mathrm{Aut}(A)$ of the multiplicative group $A^\times$ of invertible elements of A, whose image $\mathrm{Inn}(A)$ consists of the inner automorphisms. We define new stable Γ-graded monoidal categories $Q\mathcal{A}lg_R$ and $Q\mathcal{A}z_R$ whose objects are those of $\mathcal{A}lg_R$ or $\mathcal{A}z_R$ respectively, but morphisms are equivalence classes of morphisms under the relation

$$f \sim g \; : \text{if } f = g \circ a \text{ for some } a \in A^\times$$

(the notation draws on the analogy with the construction of the projective category). There are quotient functors $QA_R : \mathcal{A}lg_R \to Q\mathcal{A}lg_R$, $QA_R : \mathcal{A}z_R \to Q\mathcal{A}z_R$, and since $QA_R \circ op \cong op \circ QA_R$, the construction passes to the twisted case. The category $Q\mathcal{A}z_R$ is strictly coherent.

An object of $\mathcal{R}ep(\Gamma, Q\mathcal{A}lg_R)$ is an R-algebra A together with a family $\{g_\gamma\}$ of ring automorphisms or antiautomorphisms according as $w(\gamma) = \pm 1$ such that g_γ restricts to γ on R, and for each $\gamma, \delta \in \Gamma$ there exists $a(\gamma, \delta) \in A^\times$ such that $g_\gamma \circ g_\delta = i(a(\gamma, \delta)) \circ g_{\gamma\delta}$.

We define the category $\mathcal{M}od - \mathcal{A}lg_R$, in the case when w is trivial, to have objects R-algebras, and a morphism $A \to B$ of grade γ is a $B - A$-bimodule M such that $mr = r^\gamma m$ for $r \in R, m \in M$. There are natural definitions of composition, identity and tensor product giving the structure of Γ-graded monoidal category; isomorphism in this category corresponds to Morita equivalence of algebras. An object A of the category is invertible if and only if there exists another object B with $A \otimes B$ equivalent to R. It follows that A must be an Azumaya algebra. Conversely, if A is Azumaya, the isomorphism $j_A : A \otimes_R A^{op} \cong \mathrm{End}_R(\mathrm{Lin}_R(A))$ shows that, since $\mathrm{End}_R(A)$ is equivalent in $\mathcal{M}od - \mathcal{A}lg_R$ to R, A^{op} provides an inverse object to A.

Now restrict to the subcategory of invertible morphisms. Then we have an endofunctor *op* which uses opposite algebras, viewing M as an $A^{op}-B^{op}$-bimodule $\overline{M}$, and defining M^{op} as the inverse to $\overline{M}$ so that e.g. $\overline{M_1} \otimes_R \overline{M_2} \cong \overline{M_1 \otimes_R M_2}$, $M_1^{op} \otimes_R M_2^{op} \cong (M_1 \otimes_R M_2)^{op}$, and if ${}_CN_B$, ${}_BM_A$ are invertible bimodules, then $\overline{N \otimes_B M} \cong \overline{M} \otimes_{B^{op}} \overline{N}$ and $N^{op} \otimes_{B^{op}} M^{op} \cong (N \otimes_B M)^{op}$. Then *op* is an equivalence of Γ-graded monoidal categories, and $op \circ op \cong 1_{\mathcal{M}od-\mathcal{A}lg_R}$. We now extend the definition of $\mathcal{M}od - \mathcal{A}lg_R$ by twisting to allow non-trivial w. Equivalently, a morphism $A \to B$ of grade γ with $w(\gamma) = -1$ is a $B^{op} - A$-bimodule M such that $mr = r^\gamma m$ for $r \in R, m \in M$.

Thus if w is trivial, an object of $\mathcal{R}ep(\Gamma, \mathcal{M}od - \mathcal{A}lg_R)$ is a pair (A, h) with A an R-algebra and, for each γ, $h(\gamma) = M_\gamma$ an $A - A$-bimodule such that $vr = r_\gamma v$ for all $r \in R$, $v \in V$ and there are isomorphisms $M_\gamma \otimes M_\delta \cong M_{\gamma\delta}$ of $A - A$-bimodules. In the case where w is an isomorphism, an object of $\mathcal{R}ep(\Gamma, \mathcal{M}od - \mathcal{A}lg_R)$ is a pair (A, M) with A an R-algebra and M an $A^{op} - A$-bimodule so that $vr = r_\gamma v$ as above, and $\overline{((M),\gamma)} \circ ((M),\gamma)$ is an identity morphism, i.e. $M^{op} \otimes_{A^{op}} M \cong A$ or equivalently, $M \cong \overline{M}$.

We now define a functor $W : \mathcal{A}lg_R \to \mathcal{M}od - \mathcal{A}lg_R$ of Γ-graded monoidal categories. If w is trivial, W is the identity on objects, and sends the morphism $(f, \gamma) : A \to B$ with $w(\gamma) = 1$ to the bimodule B_f whose left B-module structure is that of B and right A module structure given by $b_f.a = (bf(a))_f$. Then $W \circ op \cong op \circ W$, so the functor W extends by twisting to the case $w(\gamma) = -1$. Equivalently, if $w(\gamma) = -1$ we must replace B by B^{op}.

Finally we define the *Brauer category* $\mathcal{B}_R$ of R to be the subcategory of $\mathcal{M}od - \mathcal{A}lg_R$ consisting of Azumaya algebras and invertible morphisms. Then $\mathcal{B}_R$ is a stable Γ-graded monoidal category, and is strictly coherent. Thus $\mathcal{B}_R$ is equivalent to a group category. The functor W restricts to define a functor $W : \mathcal{A}z_R \to \mathcal{B}_R$, and this factors through $QA_R : \mathcal{A}z_R \to Q\mathcal{A}z_R$ to define $W' : Q\mathcal{A}z_R \to \mathcal{B}_R$.

3. Exact sequences of k groups and monoids

We now introduce notations for the monoids k of the categories $\mathcal{G}en, \mathcal{C}, \mathcal{A}z$ and $\mathcal{B}$, and their projective and equivariant versions (the categories $\mathcal{M}od, \mathcal{A}lg$ and $\mathcal{M}od - \mathcal{A}lg$ were merely used in the constructions). Set

$$\begin{array}{rclrcl}
Gen(R) & := & k(\mathcal{G}en_R) & Az(R) & := & k(\mathcal{A}z_R) \\
Gen(R,\Gamma) & := & k(\mathcal{R}ep(\Gamma, \mathcal{G}en_R)) & Az(R,\Gamma) & := & k(\mathcal{R}ep(\Gamma, \mathcal{A}z_R)) \\
PGen(R,\Gamma) & := & k(\mathcal{R}ep(\Gamma, P\mathcal{G}en_R)) & QAz(R,\Gamma) & := & k(\mathcal{R}ep(\Gamma, Q\mathcal{A}z_R)) \\
C(R) & := & k(\mathcal{C}_R) & B(R) & := & k(\mathcal{B}_R) \\
C(R,\Gamma) & := & k(\mathcal{R}ep(\Gamma, \mathcal{C}_R)) & RB(R,\Gamma) & := & k(\mathcal{R}ep(\Gamma, \mathcal{B}_R))
\end{array}$$

Then also $k(P\mathcal{G}en_R) = Gen(R)$ and $k(Q\mathcal{A}z_R) = Az(R)$; and the *equivariant class group* $C(R, \Gamma)$ is the maximal subgroup of the monoid $Gen(R, \Gamma)$. We have various maps induced by the functors

$$\begin{array}{ccccccc}
\mathcal{C}(R) & \xrightarrow{i_R} & \mathcal{G}en(R) & \xrightarrow{\mathrm{End}_R} & \mathcal{A}z(R) & \xrightarrow{W_R} & \mathcal{B}(R); \\
 & & \downarrow P_R & \nearrow PEnd_R & \downarrow QA_R & \nearrow W'_R & \\
 & & P\mathcal{G}en_R & & Q\mathcal{A}z_R & &
\end{array}$$

we denote the induced maps of k groups by the same letters; for the maps of equivariant k groups we add Γ to the subscript. However, the notations for maps to versions of $B(R)$ will be defined *ad hoc*.

The unit groups of the categories are trivial for $P\mathcal{G}en_R, \mathcal{A}z_R$ and $Q\mathcal{A}z_R$; we may identify

$$U(\mathcal{G}en_R) = U(\mathcal{C}_R) = U(R), \quad U(\mathcal{B}_R) \cong C(R).$$

We need more care in defining the equivariant versions of the Brauer group. First we define the *equivariant Brauer category* $\mathcal{B}(R,\Gamma)$. Its objects are those of $\mathcal{R}ep(\Gamma, \mathcal{A}z_R)$, which are $R-\Gamma$-algebras A with groups $f(\Gamma)$ of (anti)automorphisms. If w is trivial, a morphism $(A,f) \to (B,g)$ is an isomorphism class of pairs (M,h) where $M = {}_AM_B$ is an invertible bimodule and h a homomorphism of Γ to the automorphism group of the additive group M such that $h_\gamma(amb) = f_\gamma(a)h_\gamma(m)g_\gamma(b)$ for all $a \in A$, $b \in B$ and $m \in M$.

Here we cannot use the twisting construction since we are not defining a graded category. If w is non-trivial, then to define morphisms $A \to B$ we start with the category $\mathcal{R}ep(\Gamma, {}_A\mathcal{P}air_B)$ and consider isomorphism classes of quintuplets (M, M', ϕ, h, h') with bimodules ${}_AM_B$, ${}_BM'_A$, a non-singular pairing $\phi : M \times M' \to A$ over B with $\phi(h_\gamma m, h'_\gamma m') = f_\gamma(\phi(m,m'))$ and $h_\gamma(amb) = g_\gamma(b)h_\gamma(m)f_\gamma(a)$ for all a, b, m. After some (pages of) checking, we obtain

PROPOSITION 3.1. *The category $\mathcal{B}(R,\Gamma)$ is a strictly coherent group-like category. The functor W_R induces a functor $W_{R,\Gamma} : \mathcal{R}ep(\Gamma, \mathcal{A}z_R) \to \mathcal{B}(R,\Gamma)$ of monoidal categories such that the sequence*

$$Gen(R,\Gamma) \stackrel{End_{R,\Gamma}}{\longrightarrow} Az(R,\Gamma) \stackrel{W_{R,\Gamma}}{\longrightarrow} k(\mathcal{B}(R,\Gamma))$$

is exact. Moreover, $U(\mathcal{B}(R,\Gamma)) \cong C(R,\Gamma)$.

Define

$$B(R,\Gamma) := k(\mathcal{B}(R,\Gamma))) \cong \operatorname{Coker} End_{R,\Gamma} : Gen(R,\Gamma) \to Az(R,\Gamma).$$

Then $B(R,\Gamma)$ is a group, the *equivariant Brauer group*. Here the cokernel is the group of Brauer equivalence classes of (R,Γ)-Azumaya algebras, $k(\mathcal{B}(R,\Gamma))$ is the group of Morita equivalence classes, so the isomorphism expresses the fact that equivariant Brauer equivalence is the same as equivariant Morita equivalence. Recall that if w is an isomorphism, an equivariant Morita equivalence induces isomorphisms of categories of quadratic forms. Thus $B(R,\Gamma)$ is the group identified in the introduction as of particular importance.

If $M \in ob(\mathcal{G}en_R)$ represents a class in $H^0(\Gamma; Gen(R))$, there exists a section h to $\operatorname{Aut}_{Gen(R)}(M) \to \Gamma$. Define $\overline{h_\gamma} := End(h_\gamma)$. Since $h_\gamma h_\delta h_{\gamma\delta}^{-1}$ has grade 1, $\overline{h}_\gamma \overline{h}_\delta \overline{h}_{\gamma\delta}^{-1}$ is an inner automorphism of $End_R(M)$, so $(End_R(M), \overline{h})$ defines an object of $\mathcal{R}ep(\Gamma, Q\mathcal{A}z_R)$. This object depends only on M and is denoted $QEnd(M)$. This construction induces a homomorphism

$$QEnd_{R,\Gamma} : H^0(\Gamma; Gen(R)) \to QAz(R,\Gamma).$$

We denote the cokernel — which is a group — by $QB(R,\Gamma)$, and thus obtain a commutative diagram, whose maps we denote as follows

$$\begin{array}{ccccc}
Gen(R,\Gamma) & \stackrel{End_{R,\Gamma}}{\longrightarrow} & Az(R,\Gamma) & \stackrel{W_{R,\Gamma}}{\longrightarrow} & B(R,\Gamma) \\
\downarrow T_{Gen} & & \downarrow Q A_{R,\Gamma} & & \downarrow Q B_{R,\Gamma} \\
H^0(\Gamma; Gen(R)) & \stackrel{QEnd_{R,\Gamma}}{\longrightarrow} & QAz(R,\Gamma) & \stackrel{QW_{R,\Gamma}}{\longrightarrow} & QB(R,\Gamma)
\end{array}$$

Since $k\mathcal{R}ep\, W'_R : QAz(R,\Gamma) \to RB(R,\Gamma)$ vanishes on the image of $H^0(\Gamma;\, Gen(R))$, we have an induced homomorphism $\theta : QB(R,\Gamma) \to RB(R,\Gamma)$.

Forgetting the group Γ induces a homomorphism $QB(R,\Gamma) \to B(R)$ whose image is contained in the invariant subgroup $H^0(\Gamma; B(R))$. We write $Az_0(R,\Gamma)$, $QAz_0(R,\Gamma)$, $B_0(R,\Gamma)$ and $QB_0(R,\Gamma)$ for the kernels of the respective induced maps to $B(R)$. We also add a suffix 0 to denote the restrictions of various maps, e.g. $PEnd_{R,\Gamma,0} : PGen(R,\Gamma) \to Az_0(R,\Gamma)$.

THEOREM 3.2. *(i)* End_R *induces an exact sequence of* Γ*-modules*

$$1 \to C(R) \xrightarrow{i_R} Gen(R) \xrightarrow{\mathrm{End}_R} Az(R) \xrightarrow{W_R} B(R) \to 1.$$

(ii) The following diagram commutes and the top two rows are exact.

$$\begin{array}{ccccccccc}
1\to & C(R,\Gamma) & \xrightarrow{i_{R,\Gamma}} & Gen(R,\Gamma) & \xrightarrow{End_{R,\Gamma}} & Az(R,\Gamma) & \xrightarrow{W_{R,\Gamma}} & B(R,\Gamma) & \to 1 \\
 & \downarrow T_C & & \downarrow T_{Gen} & & \downarrow QA_{R,\Gamma} & & \downarrow QB_{R,\Gamma} & \\
1\to & H^0(\Gamma; C(R)) & \xrightarrow{H^0(i_R)} & H^0(\Gamma; Gen(R)) & \xrightarrow{QEnd_{R,\Gamma}} & QAz(R,\Gamma) & \xrightarrow{QW_{R,\Gamma}} & QB(R,\Gamma) & \to 1 \\
 & \| \downarrow & & \| \downarrow & & \downarrow T_{QAz} & & \downarrow T_{QB} & \\
 & H^0(\Gamma; C(R)) & \xrightarrow{H^0(i_R)} & H^0(\Gamma; Gen(R)) & \xrightarrow{H^0(End_R)} & H^0(\Gamma; Az(R)) & \xrightarrow{H^0(W_R)} & H^0(\Gamma; B(R)) &
\end{array}$$

Exactness for sequences of monoids is interpreted in accordance with the definitions in the appendix to [**3**]: here, in particular, the functor End_R is cofinal.

We also have exact sequences $S(\mathcal{C})$ with $\mathcal{C}$ any of $\mathcal{C}_R$, $\mathcal{G}en_R$ and $\mathcal{B}_R$ and, more importantly, $H(\mathcal{C})$ with $\mathcal{C}$ either of $\mathcal{C}_R$ and $\mathcal{B}_R$. Explicitly, these are

$$H(\mathcal{C}_R) \ldots H^n(\Gamma; U(R)) \xrightarrow{\alpha_{n,\mathcal{C}}} H^n(\Gamma; \mathcal{C}_R) \xrightarrow{\beta_{n,\mathcal{C}}} H^{n-1}(\Gamma; C(R)) \xrightarrow{\delta_{n,\mathcal{C}}} H^{n+1}(\Gamma; U(R)) \ldots$$
$$H(\mathcal{B}_R) \ldots H^n(\Gamma; C(R)) \xrightarrow{\alpha_{n,\mathcal{B}}} H^n(\Gamma; \mathcal{B}_R) \xrightarrow{\beta_{n,\mathcal{B}}} H^{n-1}(\Gamma; B(R)) \xrightarrow{\delta_{n,\mathcal{B}}} H^{n+1}(\Gamma; C(R)) \ldots$$

THEOREM 3.3. *There is a commutative diagram with exact rows and injective columns*

$$\begin{array}{ccccccc}
1 & \to & QB_0(R,\Gamma) & \longrightarrow & QB(R,\Gamma) & \longrightarrow & H^0(\Gamma; B(R)) \\
 & & \downarrow \xi & & \downarrow \theta & & \| \downarrow \\
1 & \to & H^1(\Gamma; C(R)) & \xrightarrow{\alpha_{1,\mathcal{B}}} & H^1(\Gamma; \mathcal{B}_R) & \xrightarrow{\beta_{1,\mathcal{B}}} & H^0(\Gamma; B(R))
\end{array}$$

To prove this, we first establish that an algebra representing an element of $\mathrm{Ker}\,\theta$ is in $\mathrm{Im}\,QEnd_{R,\Gamma}$, and then apply Theorem 3.2 to see that θ is injective. Now the upper row is trivially exact, and the lower is part of $H(\mathcal{B}_R)$. Since the right hand square clearly commutes, there is an induced map ξ, which is injective since θ is.

We now give two direct constructions of maps to $H^3(\Gamma; U(R))$. If G is a group, with centre $Z(G)$ and group $\mathrm{Out}\,(G) := \mathrm{Aut}\,(G)/\mathrm{Inn}\,(G)$ of outer automorphisms, then to any homomorphism $h : \Gamma \to \mathrm{Out}\,(G)$ is associated a cohomology class $c(h) \in H^3(\Gamma; Z(G))$, the obstruction to the existence of a group extension of G by Γ inducing h.

For A an R-algebra, write $\mathrm{Aut}^+(A)$ for its group of automorphisms and anti-automorphisms; recall that $A^\times$ denotes the multiplicative group of invertible elements. Define

$$\kappa_A : \mathrm{Aut}^+(A) \to \mathrm{Aut}\,(A^\times)$$

by $\kappa_A(t)(u) = t(u)$ if t is an automorphism, and $t(u)^{-1}$ if t is an anti-automorphism. This induces a homomorphism

$$\overline{\kappa}_A : \mathrm{Aut}^+(A)/\mathrm{Inn}\,(A) \to \mathrm{Out}\,(A^\times).$$

If (A, f) is an object of $\mathcal{R}ep(\Gamma, \mathcal{A}z_R)$, f is a homomorphism of Γ to the quotient $\mathrm{Aut}^+(A)/\mathrm{Inn}\,(A)$, hence we have a homomorphism

$$\overline{\kappa}_A \circ f : \Gamma \to \mathrm{Out}\,(A^\times)$$

and hence — since we can identify $Z(A^\times) = U(R)$ — a cohomology class
$$\rho(A,f) := c(\overline{\kappa}_A \circ f) \in H^3(\Gamma; U(R)).$$

Next let $(A,(M)) \in ob\mathcal{R}ep(\Gamma, \mathcal{B}_R)$. For each $\gamma \in \Gamma$ with $w(\gamma) = 1$, M_γ is an invertible $A-A$-bimodule of grade γ, and if $w(\gamma) = w(\delta) = 1$ we may choose isomorphisms $f_{\gamma,\delta} : M_\gamma \otimes_A M_\delta \cong M_{\gamma\delta}$. If however $w(\gamma) = 1$ but $w(\delta) = -1$, M_δ is an $A^{op} - A$-bimodule, and we get instead an isomorphism $f_{\gamma,\delta} : M_\gamma^{op} \otimes_{A^{op}} M_\delta \cong M_{\gamma\delta}$. Given 3 elements $\gamma_1, \gamma_2, \gamma_3 \in \Gamma$ we get two isomorphisms (of grade 1)
$$a(\gamma_1,\gamma_2,\gamma_3), b(\gamma_1,\gamma_2,\gamma_3) \ : \ M'_{\gamma_1} \otimes M'_{\gamma_2} \otimes M'_{\gamma_3} \longrightarrow M'_{\gamma_1\gamma_2\gamma_3},$$
where M'_γ denotes M_γ or M_γ^{op} and the tensor products are over A or A^{op} according to the values $w(\gamma_i)$.

Set $u(\gamma_1,\gamma_2,\gamma_3) := b(\gamma_1,\gamma_2,\gamma_3)\cdot a(\gamma_1,\gamma_2,\gamma_3)^{-1}$. This is a grade 1 automorphism of an invertible module, so can be identified with an element of $U(R)$. Further calculations show that u defines a 3-cocycle of Γ with values in $U(R)$, and that varying the choice of the isomorphisms $f(\gamma,\delta)$ by multiplying by a family $v(\gamma,\delta)$ of elements of $U(R)$ has the effect of multiplying u by the coboundary of v. We thus obtain a well defined cohomology class $\chi(A,(M)) \in H^3(\Gamma; U(R))$.

THEOREM 3.4. *The above constructions define*
(i) a map $\rho : QB(R,\Gamma) \to H^3(\Gamma; U(R))$ *which is zero on the image of* $B(R,\Gamma)$,
(ii) a homomorphism $\chi : RB(R,\Gamma) \to H^3(\Gamma; U(R))$, *such that*
(iii) we have $\rho = \chi \circ \theta : QB(R,\Gamma) \to RB(R,\Gamma) \to H^3(\Gamma; U(R))$,
(iv) we have $\delta_{\mathcal{C}_R} = \chi \circ \alpha_{\mathcal{B}_R} : H^1(\Gamma; C(R)) \to RB(R,\Gamma) \to H^3(\Gamma; U(R))$

up to eventual signs!

For A an Azumaya R-algebra, write ${}_A\mathcal{M}^*_R$ for the category of $A-R$-bimodules where R acts the same way on both sides; note that tensor product gives a functor $\otimes_R : {}_A\mathcal{M}^*_R \times \mathcal{C}_R \to {}_A\mathcal{M}^*_R$.

LEMMA 3.5. *(i)Let* $A_i \in \text{ob}(\mathcal{A}z_R)$, $M_i \in \text{ob}({}_{A_i}\mathcal{M}^*_R)$, $P_i \in \text{ob}(\mathcal{C}_R)$ *for* $i = 1,2$, *and let* $\alpha : A_1 \to A_2$ *be a morphism in* $\mathcal{A}z_R$ *and* $f : M_1 \to M_2$, $g : M_1 \otimes_R P_1 \to M_2 \otimes_R P_2$ *be isomorphisms of additive groups over the morphism* α. *Then there is a unique* R*-module isomorphism* $c : P_1 \to P_2$ *with* $g = f \otimes_R c$.

(ii) A corresponding conclusion holds if Γ *acts on* R *and* α *is a morphism of grade* γ, *also if* $w(\gamma) = -1$.

Let $(A,t) \in \text{ob}(\mathcal{R}ep(\Gamma, \mathcal{A}z_R))$, so $t : \Gamma \to \text{Aut}_{\mathcal{A}z_R}(A)$. Suppose we have an invertible $A-R$-bimodule M. Then if $w(\gamma) = 1$ we have $P(\gamma) \in \text{ob}(\mathcal{C}_R)$ and a group isomorphism $f_\gamma : M \to M \otimes_R P(\gamma)$ such that $f_\gamma(am) = t_\gamma(a)f_\gamma(m)$ for all $a \in A, m \in M$. If $w(\gamma) = -1$ we have $P(\gamma) \in \text{ob}(\mathcal{C}_R)$ and a group isomorphism $f_\gamma : M \to P(\gamma)^* \otimes_R M^*$ such that $f_\gamma(am) = f_\gamma(m)t_\gamma(a)$ for all $a \in A, m \in M$. From Lemma 3.5 we have unique additive isomorphisms
$$c(\gamma_1,\gamma_2) : p(\gamma_2) \to \begin{cases} P(\gamma_1)^* \otimes_R P(\gamma_1\gamma_2) & \text{if } w(\gamma_1) = 1 \\ (P(\gamma_1)^* \otimes_R P(\gamma_1\gamma_2))^* & \text{if } w(\gamma_1) = -1 \end{cases}$$
such that $f_{\gamma_1\gamma_2} \cdot f_{\gamma_2}^{-1} = [f_{\gamma_1}, c(\gamma_1,\gamma_2)]$. The elements $c(\gamma_1,\gamma_2)$ define a cocycle whose class in $H^2(\Gamma; \mathcal{C}_R)$ is independent of the choices of the P_γ and f_γ. The resulting map $\psi' : Az_0(R,\Gamma) \to H^2(\Gamma; \mathcal{C}_R)$ is a monoid homomorphism inducing an injective homomorphism $\psi : B_0(R,\Gamma) \to H^2(\Gamma; \mathcal{C}_R)$.

Denote by τ the composite map
$$PGen(R,\Gamma) \overset{PEnd_{R,\Gamma,0}}{\longrightarrow} Az_0(R,\Gamma) \overset{W_{R,\Gamma,0}}{\longrightarrow} B_0(R,\Gamma).$$

THEOREM 3.6. *The following diagram is commutative (up to signs) and the upper row is exact at* $B_0(R,\Gamma)$.

$$\begin{array}{ccccccc} PC(R,\Gamma) & \stackrel{Pi_{R,\Gamma}}{\longrightarrow} & PGen(R,\Gamma) & \stackrel{\tau}{\longrightarrow} & B_0(R,\Gamma) & \stackrel{QB_{R,\Gamma,0}}{\longrightarrow} & QB_0(R,\Gamma) \\ \downarrow\wr & & \downarrow \omega_{\mathcal{G}en} & & \downarrow \psi & & \downarrow \xi \\ H^0(\Gamma;C(R)) & \stackrel{\delta_{1,\mathcal{C}}}{\longrightarrow} & H^2(\Gamma;U(R) & \stackrel{\alpha_{2,\mathcal{C}}}{\longrightarrow} & H^2(\Gamma:\mathcal{C}_R) & \stackrel{\beta_{2,\mathcal{C}}}{\longrightarrow} & H^1(\Gamma;C(R)) \end{array}$$

4. Finite groups

We will now assume Γ to be finite. With this restriction we can strengthen the foregoing results.

THEOREM 4.1. *Let* Γ *be finite. Then*
(i) the map $\omega_{Gen} : PGen(R,\Gamma) \to H^2(\Gamma : U(R))$ *is surjective;*
(ii) the following maps are isomorphisms:

$$\theta : QB(R,\Gamma) \to k(\mathcal{R}ep(\Gamma,\mathcal{B}_R)),$$
$$\psi : B_0(R,\Gamma) \to H^2(\Gamma;\mathcal{C}_R),$$
$$\xi : QB_0(R,\Gamma) \to H^1(\Gamma;C(R)).$$

The proof of Theorem 4.1 uses the existence of a second operation besides $\otimes$: namely the direct sum $\oplus$, rather in the manner of Hilbert's 'Satz 90'. Indeed, (i) of Theorem 4.1 is a special case of [**3**, Prop 11.1]. We recall the construction. Given a cocycle $u : \Gamma \times \Gamma \to U(R)$, take M as the free R-module with basis $\{w_\gamma \,|\, \gamma \in \Gamma\}$, and define $f_\gamma(rw_\delta) := r^\gamma u(\gamma,\delta)w_{\gamma\delta}$. The cocycle property shows that this defines a projective action of Γ.

As to (ii), it suffices to establish the first assertion, as the rest will follow by diagram chasing using Theorem 3.3 and Theorem 3.6. We have already noted that the map is injective. Conversely, given an element of $k(\mathcal{R}ep(\Gamma,\mathcal{B}_R))$ represented by an Azumaya algebra A and invertible bimodules $M(\gamma)$ such that there exist isomorphisms $f_{\gamma,\delta} : M(\gamma)\otimes_A M(\delta) \cong M(\gamma\delta)$ satisfying the appropriate conditions, we set $X := \oplus_{\gamma\in\Gamma} M(\gamma)$ and $B := \mathrm{End}_A(X)^{op}$. We define isomorphisms $g_\sigma : M(\sigma)\otimes_A X \cong X$ by setting $g_\sigma(x(\sigma)\otimes x(\gamma)) := f_{\sigma,\gamma}(x(\sigma)\otimes x(\gamma))$. Then $M(\sigma)\otimes_A X$ is an invertible $A-B$-bimodule, so there is ring automorphism t_σ of B such that $g_\sigma(m(\sigma)\otimes xb) = g_\sigma(m(\sigma)\otimes x)t_\sigma(b)$. Now a few pages of checking show that t defines the desired representation (modulo inner automorphisms) of Γ on B.

Using these and the earlier results we obtain

THEOREM 4.2. *Let* Γ *be finite. Then we obtain a commutative diagram with exact rows*

$$\begin{array}{ccccccccccc} H^0(\Gamma;C(R)) & \stackrel{\delta_{1,\mathcal{C}}}{\longrightarrow} & H^2(\Gamma;U(R)) & \stackrel{\alpha_{2,\mathcal{C}}}{\longrightarrow} & H^2(\Gamma;\mathcal{C}_R) & \stackrel{\beta_{2,\mathcal{C}}}{\longrightarrow} & H^1(\Gamma;C(R)) & \stackrel{\delta_{2,\mathcal{C}}}{\longrightarrow} & H^3(\Gamma;U(R)) \\ \| & & \| & & \uparrow\wr\psi & & \uparrow\wr\xi & & \| \\ H^0(\Gamma;C(R)) & \longrightarrow & H^2(\Gamma;U(R)) & \longrightarrow & B_0(R;\Gamma) & \stackrel{QB_{R,\Gamma,0}}{\longrightarrow} & QB_0(R,\Gamma) & \stackrel{\rho_0}{\longrightarrow} & H^3(\Gamma;U(R)) \\ \downarrow & & \downarrow & & \downarrow & & \downarrow & & \downarrow \\ H^0(\Gamma;C(R)) & \longrightarrow & H^2(\Gamma;U(R)) & \longrightarrow & B(R;\Gamma) & \stackrel{QB_{R,\Gamma}}{\longrightarrow} & QB(R,\Gamma) & \stackrel{\rho}{\longrightarrow} & H^3(\Gamma;U(R)) \end{array}$$

Here the upper row is (part of) $H(\mathcal{C}_R)$. The second map in the second row is defined as $\psi^{-1}\circ\alpha_{2,\mathcal{C}}$: there is no direct construction.

For example, suppose $C(R)$ trivial — e.g. R a field. Then as θ and $\beta_{1,\mathcal{B}}$ are isomorphisms, the lower sequence of (4.2) reduces to

$$0 \to H^2(\Gamma;U(R)) \to B(R,\Gamma) \to H^0(\Gamma;B(R)) \to H^3(\Gamma;U(R)),$$

which we cited in [**6**, p.124] (with slight change in notation) for the case when w is an isomorphism.

The exact sequences $H(\mathcal{C}_R)$ and $H(\mathcal{B}_R)$ in low degrees fit into a commutative braid of exact sequences with the lower sequence of 4.2 and the sequence $0 \to B_0(R,\Gamma) \to B(R,\Gamma) \to H^0(\Gamma; B(R))$. These should form part of an infinite braid involving a new sequence of groups $H^n(\Gamma; E_R)$, as follows (we omit Γ from the notation to make the diagram more readable).

$$\begin{array}{ccccccccccccc}
 & \nearrow & \longrightarrow & \searrow & & \nearrow & \longrightarrow & \searrow & & \nearrow & \longrightarrow & \searrow & \\
H^{n-1}(C(R)) & & & & H^{n+1}(U(R)) & & & & H^{n+1}(E_R) & & & & H^{n-1}(B(R)) \\
 & \searrow & & \nearrow & & \searrow & & \nearrow & & \searrow & & \nearrow & \\
 & & H^{n-1}(\mathcal{B}_R) & & & & H^{n+1}(\mathcal{C}_R) & & & & H^n(\mathcal{B}_R) & & \\
 & \nearrow & & \searrow & & \nearrow & & \searrow & & \nearrow & & \searrow & \\
H^n(E_R) & & & & H^{n-2}(B(R)) & & & & H^n(C(R)) & & & & H^{n+2}(U(R)) \\
 & \searrow & & \nearrow & & \searrow & & \nearrow & & \searrow & & \nearrow & \\
 & & \longrightarrow & & & & \longrightarrow & & & & \longrightarrow & &
\end{array}$$

Here it follows from the diagram that we need to identify $H^0(E_R) \cong H^0(\mathcal{C}_R) \cong H^0(U(R))$ and $H^1(E_R) \cong H^1(\mathcal{C}_R)$, while $H^2(E_R) = B(R,\Gamma)$ is our central object of interest. We turn to the methods of [**7**] to seek a direct approach to this conjecture.

5. An abstract approach

In this section we recall and slightly amplify some results from [**7**]. First, suppose $\mathcal{C}$ is a Γ-graded group like category. Define a simplicial space $B\mathcal{C}$ whose 0-simplices are objects of $\mathcal{C}$, 1-simplices are morphisms (of all grades), 2-simplices are commutative triangles of morphisms, and higher simplices are filled in when possible. Then $B\mathcal{C}$ fibres over $B\Gamma$ and the fibre $B\mathcal{K}er\,\mathcal{C}$ is an infinite loop space with $K_0(\mathcal{C}) := \pi_0(B\mathcal{C}) = k(\mathcal{C})$; $K_1(\mathcal{C}) := \pi_1(B\mathcal{C}) = U(\mathcal{C})$, and vanishing higher groups. Further [**7**, Theorem] there is a spectral sequence with $E_2^{p,q} = H^p(\Gamma; K_{-q}(\mathcal{C}))$ and abutment $K_{-n}(\mathcal{C},\Gamma)$ (we have changed some signs to conform with standard notation for a cohomology spectral sequence). Since only two groups $K_q(\mathcal{C})$ are non-vanishing, the spectral sequence reduces to an exact sequence, which is naturally identified with $H(\mathcal{C})$.

For the present purpose we need a slightly more elaborate construction. Define a simplicial space E as follows. The 0-simplices are the Azumaya R-algebras A_i. The 1-simplices are the invertible bimodules ${}_iM_j = {}_{A_i}M_{A_j}$ (of all grades). The 2-simplices are the bimodule isomorphisms $f_{i,j,k} : {}_iM_j \otimes_{A_j} {}_jM_k \to {}_iM_k$. The 3-simplices are the commutative diagrams

$$\Delta_{ijkl} : \quad \begin{array}{ccc}
{}_iM_j \otimes {}_jM_k \otimes {}_kM_l & \xrightarrow{1\otimes f_{j,k,l}} & {}_iM_j \otimes {}_jM_l \\
\downarrow f_{i,j,k}\otimes 1 & & \downarrow f_{i,j,l} \\
{}_iM_k \otimes {}_kM_l & \xrightarrow{f_{i,k,l}} & {}_iM_l
\end{array}$$

and the n-simplices for $n > 3$ are the maps of the 3-skeleton of the standard n-simplex to the partial complex just constructed. There is a map to the classifying complex $B\Gamma$ which takes each 1-simplex to its grade.

We claim that E is a Kan complex, and hence that $E \to B\Gamma$ is a Kan fibration. We need to show, for each n, that any map into E of the complex Λ_{n-1} (obtained from the standard n-simplex Δ_n by deleting the interiors of the n-simplex and of one of its $(n-1)$-dimensional faces) extends to a map of Δ_n. Since Λ_{n-1} contains the $(n-2)$-skeleton of Δ_n, this is trivial for $n \geq 5$.

For $n = 1$ the extension is possible since the Γ-grading on $\mathcal{B}_R$ is stable - given A there exist invertible bimodules of all grades.

For $n = 2$ if the missing face is the third, we complete the map of Λ_1 defined by the modules ${}_iM_j$ and ${}_jM_k$ by the isomorphism ${}_iM_j \otimes_{A_j} {}_jM_k \to {}_iM_k$. If the missing face is different, the argument is essentially the same, since all the bimodules are invertible, and we may use their inverses.

For $n = 3$ a map of Λ_2 gives (perhaps after tensoring with some identity maps) the isomorphisms on 3 edges of the desired commutative diagram, and we may then complete the diagram uniquely.

Finally for $n = 4$ we are given 4 commutative diagrams and need to establish commutativity of a fifth. This follows from the following diagram, where we take the subscripts as $0, 1, 2, 3, 4$ and omit the symbol M for brevity:

$$\begin{array}{ccccccc}
(01)\otimes(12)\otimes(23)\otimes(34) & \longrightarrow & & \longrightarrow & & \longrightarrow & (01)\otimes(12)\otimes(24) \\
\downarrow & \searrow & & & & \swarrow & \downarrow \\
 & & (02)\otimes(23)\otimes(34) & \longrightarrow & (02)\otimes(24) & & \\
\downarrow & & \downarrow & & \downarrow & & \downarrow \\
 & & (03)\otimes(34) & \longrightarrow & (04) & & \\
\downarrow & \nearrow & & & & \nwarrow & \downarrow \\
(01)\otimes(13)\otimes(34) & \longrightarrow & & \longrightarrow & & \longrightarrow & (01)\otimes(14)
\end{array}$$

Here commutativity of the outer square follows from taking the tensor product of M_{01} with the simplex Δ_{1234}, of the left square follows from taking the tensor product of M_{34} with the simplex Δ_{0123}, of the centre square from Δ_{0234}, of the right square from Δ_{0124}, of the lower square from Δ_{0134}, and of the upper square from the identity $(f \otimes 1)(1 \otimes g) = (1 \otimes g)(f \otimes 1)$ where f and g are the morphisms corresponding to the 1-simplices 012 and 234. Since all the maps are isomorphisms, commutativity of any one square follows from that of the rest, as desired.

There is thus again a spectral sequence. If, as seems fairly certain, the fibre is an infinite loop space, the E^2 term is determined since we can identify $K_q(\mathcal{K}er\,\mathcal{C})$ with $U(R)$ if $q = 2$, $C(R)$ if $q = 1$, $B(R)$ if $q = 0$, and 0 otherwise. Unfortunately, the abutment is given only in terms of homotopy groups of the space of sections, and so is not easily interpreted.

We conjecture that, as in the case first described, there is an algebraic model for the spectral sequence, given in terms of some cochain model. We then expect that, up to chain homotopy equivalence, the spectral sequence is that of a filtered complex $0 \subset F_0 \subset F_1 \subset F_2$ of free Γ-modules, zero in negative degrees, such that F_0, F_1/F_0 and F_2/F_1 give free resolutions of $U(R), C(R)$ and $B(R)$ respectively; and the hypercohomology groups of F_1 and F_2/F_0 give the equivariant cohomology groups of the categories $\mathcal{C}_R$ and $\mathcal{B}_R$. Thus the two 'new' sequences at the end of §4 would be the exact cohomology sequences belonging to $F_0 \to F_2 \to F_2/F_0$ and $F_1 \to F_2 \to F_2/F_1$ respectively.

6. Change of groups

For the case of graded monoidal categories, a rather full discussion of constructions involving change of groups was given in [**3**, §9]. We content ourselves here with a brief summary.

The most general formulation is as follows. For Γ a group, and X a finite Γ-set, we form the Γ-graded category X_Γ whose object set is X and morphism set $\{(x, \gamma) : x \to x^\gamma \,|\, x \in X, \gamma \in \Gamma\}$. If $\mathcal{C}$ is a Γ-graded monoidal category, $\mathcal{H}(X) := \mathcal{H}om_\Gamma(X_\Gamma, \mathcal{C})$ can be considered a monoidal category, and $\mathcal{H}$ defines a Mackey functor from the category of finite Γ-sets to the homotopy category of $\mathcal{MC}$, and hence, composing e.g. with k, to the category of abelian monoids. If $\mathcal{C}$ is group-like we may compose with $H^i(\Gamma; -)$ to get a Mackey functor to abelian groups.

It follows, for example, that if Γ has finite order N, then [**3**, 7.7] N annihilates $H^n(\Gamma;\mathcal{C})$ for $n \geq 2$: this result is useful in some applications.

Let $\mathcal{C}$ be a Γ-graded group-like monoidal category, and $i : \Delta \to \Gamma$ the inclusion of a subgroup. Then there are restriction maps $i^* : H^i(\Gamma;\mathcal{C}) \to H^i(\Delta;\mathcal{C})$ and, if the subgroup Δ has finite index, corestriction maps i_* in the opposite direction: in particular, $i^* : Rep(\Gamma,\mathcal{C}) \to Rep(\Delta,\mathcal{C})$ and $i_* : Rep(\Delta,\mathcal{C}) \to Rep(\Gamma,\mathcal{C})$. Since $\mathcal{H}$ is a Mackey functor, these satisfy the standard relations: in fact [**1**] $K_0\mathcal{R}ep$ is a Frobenius functor.

If Δ is a normal subgroup of Γ, then $\mathcal{R}ep(\Delta,\mathcal{C})$ has the natural structure of Γ/Δ-graded monoidal category, and [**1**] $\mathcal{R}ep(\Gamma/\Delta,\mathcal{R}ep(\Delta,\mathcal{C})) \cong \mathcal{R}ep(\Gamma,\mathcal{C})$. Presumably if $\mathcal{C}$ is a group category there is a spectral sequence with E^2-term $H^p(\Gamma/\Delta; H^q(\Delta;\mathcal{C}))$ and abutment $H^n(\Gamma;\mathcal{C})$; at least we have [**3**, 9.7] the expected exact sequence of terms of low degree. This applies to the case $\mathcal{C} = \mathcal{C}_R$, and in view of the isomorphism ψ gives information about $B_0(R;\Gamma)$ if Γ is finite.

Finally we return to the twisting relation. Let $\mathcal{C}$ be a Γ-graded monoidal category, and $D : \mathcal{C} \to \mathcal{C}$ a morphism allowing us to extend the grading to $\Gamma \times \{\pm 1\}$. Suppose given two homomorphisms $w_i : \Gamma \to \{\pm 1\}$ $(i = 1, 2)$ with equaliser Δ: write $\Gamma_i \subset \Gamma \times \{\pm 1\}$ for the graph of w_i, and $\Delta := (\Delta, w_1|\Delta)$. Then [**3**, 10.2] there is a long exact sequence

$$\cdots H^i(\Gamma_1;\mathcal{C}) \longrightarrow H^i(\Delta;\mathcal{C}) \longrightarrow H^i(\Gamma_2;\mathcal{C}) \longrightarrow H^{i+1}(\Gamma_1;\mathcal{C}) \longrightarrow \cdots$$

where the first map is restriction and the second is corestriction. Specialising to $\mathcal{C}_R$, this gives

$$0 \to H^0(\Gamma_1;U(R)) \to H^0(\Delta;U(R)) \to H^0(\Gamma_2;U(R)) \to C(R,\Gamma_1) \to C(R,\Delta) \to$$
$$\to C(R,\Gamma_2) \to B_0(R,\Gamma_1) \to B_0(R,\Delta) \to B_0(R,\Gamma_2).$$

Perhaps the most interesting case is the simplest, when Γ has order 2, so w_1 is an isomorphism and w_2 is trivial (or vice-versa), and Δ is trivial. The exactness of these sequences may be verified directly.

We may also construct a braid with the exact sequences $H(\Gamma_1,\mathcal{C})$, $H(\Gamma_2,\mathcal{C})$, $H(\Delta,\mathcal{C})$, and exact cohomology sequences with coefficients $U(R)$ and $C(R)$.

References

[1] A. Fröhlich and C.T.C. Wall, Foundations of equivariant algebraic K-theory, Springer lecture notes in Math. **108** (1969) 12–27.

[2] A. Fröhlich and C.T.C. Wall, Equivariant Brauer groups in algebraic number theory, Bull. Math. Soc. France Mémoire **25** (1971) 91–96.

[3] A. Fröhlich and C.T.C. Wall, Graded monoidal categories, Compositio Math. **28** (1974) 229–286.

[4] S. MacLane, Natural associativity and commutativity, Rice Univ. Studies **49iv** (1963) 28-46.

[5] K.-H. Ulbrich, On cohomology of graded group categories, Compositio Math. **63** (1987) 409–417.

[6] C.T.C. Wall, On the classification of Hermitian forms II: semisimple rings, Invent. Math. **18** (1972) 119-141.

[7] C.T.C. Wall, Equivariant algebraic K-theory, pp 111–118 in *New developments in topology*, London Math. Soc. lecture notes **11** (1974).

Robinson College
Cambridge CB3 9AN
United Kingdom

Dept. of Mathematical Sciences
Liverpool University
Liverpool L69 3BX
United Kingdom
E-mail address: ctcw@liverpool.ac.uk

Contemporary Mathematics
Volume **272**, 2000

Isotropy of quadratic forms and field invariants

Detlev W. Hoffmann

Dedicated to the memory of Oleg Izhboldin

ABSTRACT. This paper is intended to give a survey for non-specialists in the algebraic theory of quadratic forms on the question of isotropy of quadratic forms and how certain answers to this question can be obtained by considering invariants of quadratic forms, such as the classical invariants dimension, signed discriminant, Clifford invariant, and invariants of fields pertaining to quadratic forms, such as the u-invariant, the Hasse number, the level, the Pythagoras number, and the l-invariant. We will interpret these field invariants as the supremum of the dimensions of certain types of anisotropic quadratic forms defined over the ground field. Particular emphasis is laid on the question of which values can be realized as invariants of a field, and on methods of how to construct such fields "generically."

1. Introduction

One of the central questions in the algebraic theory of quadratic forms is the following : When is a given quadratic form φ over a field F isotropic, i.e. when does the form φ represent zero nontrivially ?[1]

An answer to this question would automatically lead to an answer of another central problem, namely, when are two forms φ and ψ of the same dimension isometric (in which case we write $\varphi \cong \psi$), i.e. when does there exist a linear isomorphism $t : V \to W$ of the underlying vector spaces V of φ and W of ψ such that $\varphi(v) = \psi(tv)$ for all $v \in V$? This can be seen as follows. The diagonal form $\langle 1, -1 \rangle$ is called a hyperbolic plane, which we shall also denote by $\mathbb{H}$. A form is called hyperbolic if it has a diagonalization of the form $\langle 1, -1, 1, -1, \cdots, 1, -1 \rangle$, i.e. if it is isometric to an orthogonal sum of hyperbolic planes, $\mathbb{H} \perp \mathbb{H} \perp \cdots \perp \mathbb{H}$. Now φ being isotropic is equivalent to φ containing $\mathbb{H}$ as a subform.[2] If two forms φ and ψ are of the same dimension, then their isometry is equivalent to $\varphi \perp -\psi$ being hyperbolic. So first we have to check whether $\varphi \perp -\psi$ is isotropic, in which

2000 *Mathematics Subject Classification.* 11E04, 11E10, 11E25, 11E81, 12D15, 12F20.

[1]Throughout this paper, we only consider fields of characteristic different from 2, and quadratic forms are always assumed to be finite-dimensional and nondegenerate. We will often simply write "form" by which we shall refer to quadratic forms in the above sense.

[2]We say that η is a subform of μ if there exists a form τ such that $\mu \cong \eta \perp \tau$. In this situation, we write $\eta \subset \mu$ for short.

case we can write $\varphi \perp -\psi \cong \mathbb{H} \perp \tau$ for some form τ. By the Witt cancellation theorem, it then suffices to verify whether τ is hyperbolic in order to establish that $\varphi \perp -\psi$ is hyperbolic, and an induction argument on the dimension then yields that an answer to the isotropy question would provide an answer to the isometry question.

Rephrasing the initial isotropy question, suppose we have diagonalized the form φ, say, $\varphi \cong \langle a_1, \cdots, a_n \rangle$ with $a_i \in F^* = F \setminus \{0\}$. How can one tell, just by "looking" at the coefficients a_i, whether φ is isotropic, i.e. whether there exist $x_1, \cdots, x_n \in F$, not all equal to 0, such that $0 = \sum_{i=1}^n a_i x_i^2$? Now 1-dimensional forms are obviously anisotropic. A 2-dimensional form $\langle a, b \rangle$ is isotropic if and only if $\langle a, b \rangle \cong \langle 1, -1 \rangle$ if and only if the determinants ab of $\langle a, b \rangle$ and -1 of $\langle 1, -1 \rangle$ differ by a square if and only if $-ab \in F^2$. This is a criterion which is quite explicit and which in many cases can be readily verified. Are there criteria of that type for forms of dimension ≥ 3 ? To illustrate this question, let us look at some examples which will also serve us later.

EXAMPLE 1.1. If $\varphi \cong \langle a_1, \cdots, a_n \rangle$ is a form over $\mathbb{C}$, then since $\mathbb{C}$ is algebraically closed, φ will be isotropic if and only if $n \geq 2$. If the ground field is $\mathbb{R}$, then φ will be isotropic iff $n \geq 2$ and φ is indefinite, i.e. the a_i are not all of the same sign. Again, the isotropy can easily be checked simply by looking at the signs of the a_i.

EXAMPLE 1.2. Let $\mathbb{F}_q$ be a finite field with $q = p^n$ elements, p an odd prime. Then $\mathbb{F}_q$ has two nonzero square classes, say, 1 and a. An easy computation shows that any form of dimension 2 will represent both square classes, hence every form of dimension ≥ 3 will be isotropic over $\mathbb{F}_q$, and up to isometry the only anisotropic form over $\mathbb{F}_q$ will be $\langle 1, -a \rangle$.

EXAMPLE 1.3. The quadratic form theory over a p-adic field $\mathbb{Q}_p$, p a prime, is well understood, and again it is easy to tell from the coefficients whether a form over $\mathbb{Q}_p$ is isotropic or not (cf. [**S**, Ch. 5,§ 6]). In fact, suppose that p is odd. Now each square class of $\mathbb{Q}_p$ can be represented by some element in $\mathbb{Z}$, and if $a \in \mathbb{Z}$ is prime to p, then a is a square in $\mathbb{Q}_p$ if and only if a is a square modulo p by Hensel's Lemma. In particular, each form φ over $\mathbb{Q}_p$ has a diagonalization of the form $\langle a_1, \cdots, a_m \rangle \perp p\langle b_1, \cdots, b_n \rangle$, with a_i, b_j integers prime to p. Suppose that (after possibly scaling) $m \geq 3$. By the previous example, we know that there exist integers x, y such that $a_1 x^2 + a_2 y^2 \equiv -a_3 \bmod p$. By what was said before, there exists a $z \in \mathbb{Q}_p^*$ such that $-a_3^{-1}(a_1 x^2 + a_2 y^2) = z^2$, i.e. $a_1 x^2 + a_2 y^2 + a_3 z^2 = 0$, and hence φ is anisotropic. Thus, if $\dim \varphi \geq 5$, then φ is isotropic. It is also not too difficult to show (and we leave this to the reader) that up to isometry, there exists exactly one anisotropic 4-dimensional form, namely $\langle 1, -a, p, -ap \rangle$, where one can choose for a any integer prime to p which is not a quadratic residue modulo p.

The case $p = 2$ is a little more technical but can also be treated in a rather elementary way. Again, any form of dimension ≥ 5 over $\mathbb{Q}_2$ will be isotropic, and up to isometry, there exists exactly one anisotropic 4-dimensional form, $\langle 1, 1, 1, 1 \rangle$.

EXAMPLE 1.4. Over $\mathbb{Q}$, the situation is already more complicated. Clearly, φ would have to be indefinite in order to be isotropic. But this alone does not suffice as, for example, $\langle 1, 1, 1, -7 \rangle$ is anisotropic as 7 is not a sum of three squares in $\mathbb{Q}$. The Hasse-Minkowski theorem tells us that a form φ over $\mathbb{Q}$ is isotropic if and only if it is isotropic over $\mathbb{Q}_p$ for each prime p, and over $\mathbb{R}$ (see, e.g. [**L1**, Ch. VI, 3.1],

[**S**, Ch. 5, Theorem 7.2]). In particular, by the previous example, each indefinite form of dimension ≥ 5 over $\mathbb{Q}$ will be isotropic.

Using the previous example, we see that $\langle 1,1,1,-7\rangle$ is isotropic over each $\mathbb{Q}_p$, $p \neq 2$, and over $\mathbb{R}$. However, it is anisotropic over $\mathbb{Q}_2$ since $\langle 1,1,1,-7\rangle \cong \langle 1,1,1,1\rangle$ over $\mathbb{Q}_2$.

Before we consider rational function fields in one variable over the above fields, we shall mention a little lemma which we shall use quite often regarding forms over purely transcendental extensions.

LEMMA 1.5. (i) *Let φ and ψ be forms over F and let $K = F(T)$, the rational function field in one variable, or $K = F((T))$, the power series field in one variable over F. Then $\varphi \perp T\psi$ is anisotropic over K if and only if φ and ψ are anisotropic over F. In particular, anisotropic forms stay anisotropic over purely transcendental extensions.*

(ii) *If γ is a form over $K = F((T))$, then there exist forms φ and ψ over F such that $\gamma \cong \varphi \perp T\psi$.*

The first part can easily be proved by a degree argument. As for the second part, one may assume that γ is diagonalized and then use the fact that to each $x \in K^*$ there exists a $y \in F^*$ such that either $x \equiv y \bmod K^{*2}$ or $x \equiv yT \bmod K^{*2}$. We leave the details to the reader.

EXAMPLE 1.6. $\mathbb{C}$ is algebraically closed and therefore it is a so-called C_0 field.[3] Hence, $\mathbb{C}(T)$ is a C_1-field (cf. [**S**, Ch. 2, Th. 15.2]), which implies that every form of dimension ≥ 3 over $\mathbb{C}(T)$ will be isotropic.

EXAMPLE 1.7. As for $\mathbb{R}(T)$, the situation is more complicated. On $\mathbb{R}$, there exists exactly one ordering.[4] On $\mathbb{R}(T)$, there are many orderings which can be described explicitly (cf. [**KS**, Kap. II, § 9]). Let $X_{\mathbb{R}(T)}$ be the space of orderings. An obvious necessary condition for a form φ to be isotropic over $\mathbb{R}(T)$ is that φ be totally indefinite, i.e. if $\varphi \cong \langle a_1, \cdots, a_n\rangle$ then for each ordering $P \in X_{\mathbb{R}(T)}$ there exist a_i and a_j depending on P such that $a_i <_P 0 <_P a_j$. Using the explicit description of the orderings, the indefiniteness can in principal be checked using the coefficients of a diagonalization. It can also be shown that totally indefinite forms of dimension ≥ 3 over $\mathbb{R}(T)$ are always isotropic, a result essentially due to Witt [**Wi1**].

EXAMPLE 1.8. Now consider the rational function field $\mathbb{Q}(T)$ in one variable X over the rationals. Again, the field is formally real, i.e. there exist orderings on $\mathbb{Q}(T)$, and a necessary condition for a form φ to be isotropic is that φ be indefinite with respect to all orderings P in the space of orderings $X_{\mathbb{Q}(T)}$. However, this knowledge alone won't help us much as there exist anisotropic totally indefinite forms of any dimension ≥ 2 over $\mathbb{Q}(T)$. This can be seen as follows. Consider the anisotropic forms $\langle 1,-2\rangle$ and $\sigma_n = n \times \langle 1\rangle = \underbrace{\langle 1,\cdots,1\rangle}_{n}$ over $\mathbb{Q}$. Then 2 is a sum of squares and hence $2 \in P$ for all $P \in X_{\mathbb{Q}(T)}$. In particular, $\langle 1,-2\rangle \perp T\sigma_n$ is

[3] A field F is called a C_i-field if every homogeneous polynomial over F of degree d in at least $d^i + 1$ variables has a nontrivial zero.

[4] An ordering on a field F is a subset $P \subset F$ such that $P + P \subset P$, $P \cdot P \subset P$, $P \cup -P = F$, $P \cap -P = \{0\}$. It induces the order relation "$\geq_P$" defined by $x \geq_P y$ if and only if $x - y \in P$. Cf. [**S**, Ch. 3, § 1].

totally indefinite and anisotropic over $\mathbb{Q}(T)$ for all n (the anisotropy follows from Lemma 1.5).

One can justifiably say that we have no systematic method whatsoever to decide whether an arbitrarily given form over $\mathbb{Q}(T)$ is isotropic or not. One could say that with respect to many questions regarding quadratic forms, the field $\mathbb{Q}(T)$ is of a complexity which is beyond our current reach.

EXAMPLE 1.9. Although the quadratic form theory over $\mathbb{Q}_p$ is in a certain sense easier to handle than that over $\mathbb{Q}$, not much was known about the isotropy of forms over $\mathbb{Q}_p(T)$ until recently, when it was shown in [**HVG2**] that all φ of dimension > 22 over $\mathbb{Q}_p(T)$, $p \neq 2$, are isotropic. Later on, this was improved by Parimala and Suresh [**PS**], who proved that in fact all forms of dimension > 10 over $\mathbb{Q}_p(T)$, $p \neq 2$, are isotropic. The case $p = 2$ is still open.

As we have already remarked above, there does exist an anisotropic 4-dimensional form φ_p over $\mathbb{Q}_p$ for all primes p. This will yield the anisotropic 8-dimensional form $\varphi_p \perp T\varphi_p$ over $\mathbb{Q}_p(T)$. It is conjectured that all forms of dimension > 8 over $\mathbb{Q}_p(T)$ are isotropic.

All these examples give us already a glimpse of the difficulties which we encounter concerning the isotropy of forms. When dealing with forms over a particular field, one normally develops and uses a set of tools and methods to which quadratic forms over the field in question are amenable. These methods often come from number theory, algebraic and real algebraic geometry. For *general* fields, one approach to derive information on the isotropy of forms is to consider certain invariants of quadratic forms resp. of the underlying field and how these invariants influence the isotropy behaviour.

In section 2, we shall introduce quadratic form invariants and show how they can be used to classify forms resp. to tell us something about isotropy of forms. First, in section 2.1, we introduce the classical invariants dimension, determinant resp. signed discriminant, Clifford invariant and total signature. After saying a little about the Witt ring of a field and how these invariants can be interpreted as invariants of elements in this ring in section 2.2, we show in section 2.3 how the classical invariants lead us inevitably to higher cohomological invariants and the Milnor conjecture (see also Pfister's survey article [**P5**]). In section 2.4, we explain how these invariants lead to classification results on quadratic forms and mention how the invariants of a form φ (resp. the "place" where the form finds itself in the Witt ring) can provide information on isotropy.

In section 3, we introduce field invariants which can be defined as the suprema of the dimensions of certain types of anisotropic forms over a field F and which therefore, once these invariants have been determined for a certain field F, yield information on the (an)isotropy of forms over F. The invariants we shall introduce are the "old" and the "new" u-invariant (sections 3.1, 3.3), the Hasse number (section 3.2), the level (section 3.4), the Pythagoras number (section 5.2), and the length of a field, also called the l-invariant (section 3.6)

The main theme of this article will be to construct fields whose field invariants take prescribed values. The method we shall introduce may be called Merkurjev's method since it was Merkurjev who brought the method we shall describe to full fruition in his construction of fields with even u-invariant [**M2**], and the main idea behind this method as well as some of the necessary tools will be introduced in section 4. The basic method of construction will be sketched in section 4.1. The main tools will be function fields of quadrics and Pfister forms, and we shall collect some of their basic properties in sections 4.2, 4.3.

We then apply this technique in section 5 to prove Pfister's results on the level (section 5.1), the author's results on the Pythagoras number (section 5.2), and Merkurjev's results on the u-invariant (section 5.3).

Finally, in section 6, we add some remarks on the Hasse number and on how some of these field invariants relate to each other (section 6.1), and give some results on the l-invariant (section 6.2). We close by mentioning some further results concerning field invariants pertaining to quadratic forms and where Merkurjev's method had some impact in their proofs.

As a good general reference, in particular on field invariants, we recommend Pfister's beautiful book [**P4**], where the examples of fields given above (plus many more examples) and their invariants have been mentioned, and some of them have been treated in detail.

2. Invariants of quadratic forms

2.1. The classical invariants. The classical invariants of quadratic forms are the dimension, the determinant, and the Clifford invariant. The easiest to determine is obviously the dimension. A little more difficult is the determinant, as one has to know how to multiply in the field F. More precisely, let $\varphi = \langle a_1, \cdots, a_n\rangle$. Since our invariants should be invariants of the isometry class of a form, we define the determinant $\det(\varphi)$ to be the class of $\prod_{i=1}^{n} a_i$ in F^*/F^{*2}.

The Clifford invariant is defined via the Clifford algebra of a quadratic form. Let $T(V) = \bigoplus_{i=0}^{\infty} V^{\otimes n}$ be the tensor algebra of the underlying vector space of the form φ. Then the Clifford algebra $C(\varphi)$ is defined to be the quotient $T(V)/I$ where I is the 2-sided ideal generated by $\{x \otimes x - \varphi(x) \mid x \in V\}$. This can be shown to be a $\mathbb{Z}/2\mathbb{Z}$-graded 2^n-dimensional algebra over F ($n = \dim\varphi$), and the even part will be denoted by $C_0(\varphi)$. If $\dim\varphi$ is even (resp. odd) then $C(\varphi)$ (resp. $C_0(\varphi)$) is a central simple algebra over F whose class in the Brauer group $\operatorname{Br} F$ can be represented by a tensor product of quaternion algebras, its Brauer class is therefore an element in the exponent-2-part $\operatorname{Br}_2 F$ of the Brauer group. The Clifford invariant of φ, denoted by $c(\varphi)$, is defined to be the Brauer class of $C(\varphi)$ if $\dim\varphi$ is even (resp. $C_0(\varphi)$ if $\dim\varphi$ is odd).

We can already notice that these invariants (if we consider $\dim \bmod 2$ rather than the dimension itself) take values in certain Galois cohomology groups. Consider the Galois group G of a separable closure F_{sep} over F. Now let $H^nF = H^n(G, \mathbb{Z}/2\mathbb{Z})$ be the n-th Galois cohomology group with G acting trivially on $\mathbb{Z}/2\mathbb{Z}$. Identifying H^0F with $\mathbb{Z}/2\mathbb{Z}$, H^1F with F^*/F^{*2}, and H^2F with $\operatorname{Br}_2 F$, we see that the classical invariants take their values in the first three of these groups. A natural question is whether one can get invariants for quadratic forms also in the H^nF for $n \geq 3$. This will be made more precise below and it will lead to the famous Milnor conjecture, see also Pfister's article [**P5**].

An invariant of a somewhat different type exists if the field F is formally real, i.e. if there exist orderings on F. Let $\varphi = \langle a_1, \cdots, a_n\rangle$ and let P be an ordering on F. We define the signature of φ with respect to P, $\operatorname{sgn}_P \varphi$, to be $\#\{a_i \mid a_i >_P 0\} - \#\{a_i \mid a_i <_P 0\}$. Note that it is Sylvester's law of inertia which tells us that this is indeed an invariant of φ independent of the chosen diagonalization. If X

denotes the space of all orderings, then each form φ defines a map $\hat{\varphi} : X \to \mathbb{Z}$ by $\hat{\varphi}(P) = \operatorname{sgn}_P \varphi$. This map $\hat{\varphi}$ is called the total signature of φ.

2.2. The Witt ring. The Witt cancellation theorem states that if φ, ψ and η are forms over a field F and if $\varphi \perp \eta \cong \psi \perp \eta$, then $\varphi \cong \psi$. We call φ Witt equivalent to ψ, and we write $\varphi \sim \psi$, if $\varphi \perp -\psi$ is hyperbolic. The Witt cancellation theorem implies that this is indeed an equivalence relation. The class of a form φ with respect to this equivalence relation will be called Witt class of φ and be denoted for now by $[\varphi]$. These classes can be made into a ring as follows. The zero element is given by the class of hyperbolic forms, addition by $[\varphi]+[\psi] = [\varphi \perp \psi]$, and multiplication is induced by the tensor product of quadratic forms, $[\varphi]\cdot[\psi] = [\varphi\otimes\psi]$, where for diagonal forms $\varphi = \langle a_1, \cdots, a_n\rangle$ and $\psi = \langle b_1, \cdots, b_m\rangle$, this product is given by $\varphi \otimes \psi = \langle a_1b_1, \cdots, a_ib_j, \cdots, a_nb_m\rangle$. The ring thus obtained is called the Witt ring of F and denoted by WF. The Witt decomposition theorem states that each form φ decomposes into an orthogonal sum $\varphi_{\text{an}} \perp i\mathbb{H}$ where φ_{an} is anisotropic, and that φ_{an} is determined uniquely up to isometry. It is called the anisotropic part of φ. i is called the Witt index of φ denoted by $i_W(\varphi)$. We have that $[\varphi] = [\psi]$ iff $\varphi_{\text{an}} \cong \psi_{\text{an}}$, so as a set the Witt ring may be identified with isometry classes of anisotropic forms.

In the sequel, by abuse of notation, we shall simply write φ to denote the class $[\varphi]$ of a form φ over a field F. This does normally not cause any problems and makes the notations less clumsy. One should, however, carefully distinguish between isometry $\varphi \cong \psi$ and Witt equivalence $\varphi \sim \psi$.

Note that the dimension will not be an invariant for a Witt class $[\varphi]$, but we get an invariant if we replace it by the dimension index $\dim \bmod 2$ which takes values in $\mathbb{Z}/2\mathbb{Z}$. The same problem arises when we consider the determinant if -1 is not a square in F^* because $[\varphi] = [\varphi \perp \mathbb{H}]$, but $\det(\varphi) = -\det(\varphi \perp \mathbb{H})$. Hence, one introduces the signed discriminant $d_\pm\varphi = (-1)^{n(n-1)/2}\det(\varphi)$, where $n = \dim\varphi$. This signed discriminant is then an invariant of the Witt class of φ. The Clifford invariant is already an invariant of the Witt class of a form as a direct computation shows.

The classes of forms of even dimension form the so-called fundamental ideal IF of WF, which is maximal with $WF/IF \simeq \mathbb{Z}/2\mathbb{Z}$, the isomorphism being in fact induced by the dimension index map $\dim \bmod 2 : WF \to \mathbb{Z}/2\mathbb{Z}$. The higher powers I^nF of IF play a crucial role in the whole theory. I^nF is additively generated by the so-called n-fold Pfister forms. An n-fold Pfister form is a form of type $\langle 1, -a_1\rangle \otimes \cdots \otimes \langle 1, -a_n\rangle$, and we shall write $\langle\langle a_1, \cdots, a_n\rangle\rangle$ for short (note the sign convention which will become clearer later on). These Pfister forms and their many nice properties are of fundamental importance and we shall say more about them in section 4.3.

2.3. The Milnor conjecture. By a straightforward computation, the signed discriminant (resp. the Clifford invariant) induces a homomorphism of groups $d_\pm : IF \to F^*/F^{*2}$ (resp. $c : I^2F \to \operatorname{Br}_2 F$). It is also not too difficult to show that $d_\pm$ is surjective with kernel I^2F. The Clifford invariant map is rather more difficult to treat. That I^3F is in the kernel can still be readily shown. It is a deep theorem due to Merkurjev [**M1**] that the map c is surjective with kernel I^3F. Using the cohomology groups introduced above, we thus have isomorphisms $e_n : I^nF/I^{n+1}F \to H^nF$ for $n = 0, 1, 2$ induced by the classical invariants. These maps e_n send the class of an n-fold Pfister form $\langle\langle a_1, \cdots, a_n\rangle\rangle$ modulo $I^{n+1}F$ to the n-fold cup product $(a_1) \cup \cdots \cup (a_n)$ (this is one of the main reasons for the sign convention for Pfister forms). The Milnor conjecture states that for each n there exists a group isomorphism $e_n : I^nF/I^{n+1}F \to H^nF$ sending the class of an n-fold Pfister form to the corresponding n-fold cup product. This conjecture would then lead to a natural generalization of the classical invariants and in a certain sense to a complete set of invariants. Actually, in the introductory remarks, Question 4.3 and §6 of his article [**Mi**], Milnor stated his conjecture in terms of Milnor K-groups $K_nF/2K_nF$, and he asked whether the canonical homomorphisms $K_nF/2K_nF \to H^nF$ and $K_nF/2K_nF \to I^nF/I^{n+1}F$ are always isomorphisms.

For $n = 3$, the existence of such a homomorphism $e_n : I^nF/I^{n+1}F \to H^nF$ has been established by Arason [**A**], and for $n = 4$ by Jacob-Rost [**JR**] and, independently, by Szyjewski [**Sz**]. The fact that e_3 is an isomorphism was shown by Merkurjev-Suslin [**MSu**] and, independently, by Rost [**R1**]. A proof of the Milnor conjecture regarding the existence of isomorphisms $K_nF/2K_nF \to H^nF$ was announced by Voevodsky [**Vo**], and in collaboration with Orlov and Vishik [**OVV**] also the corresponding result on the maps e_n.

It should be noted that the classical invariants dim mod 2, $d_\pm$ and c are defined for all elements of the Witt ring, and they are functorial with respect to field extensions E/F, i.e. if $r^H_{E/F} : H^nF \to H^nE$ and $r^W_{E/F} : WF \to WE$ denote the restriction maps by passing from F to E, then for example $r^H_{E/F} \circ c = c \circ r^W_{E/F}$. Arason [**A**] showed that in general e_3 cannot be extended to the whole Witt ring and stay functorial with respect to field extensions. (His argument can be modified to yield explicit counterexamples for all e_n, $n \geq 3$ if one assumes the results of Voevodsky [**Vo**].)

2.4. Quadratic form invariants and the classification and isotropy problems. In a certain sense, one could say that the classification problem is solved once we have the above invariants e_n for all n. Indeed, let φ and ψ be forms over F. Classifying forms just means deciding whether φ is isometric to ψ, i.e. whether $\varphi \perp -\psi$ is hyperbolic. Now if this is the case, then clearly $\varphi \perp -\psi \in I^nF$ for all n and $e_n(\varphi \perp -\psi) = 0$. Conversely, suppose we can show that $\varphi \perp -\psi \in I^nF$ and $e_n(\varphi \perp -\psi) = 0$. Then $\varphi \perp -\psi \in I^{n+1}F$ and we can compute $e_{n+1}(\varphi \perp -\psi)$. If this equals 0, we can continue. If not, then $\varphi \not\cong \psi$. The Arason-Pfister Hauptsatz shows that it suffices to compute the e_n up to the smallest n such that $2^n > \dim(\varphi \perp -\psi)$ to decide whether $\varphi \cong \psi$.

THEOREM 2.1. (Arason-Pfister [**AP**, Hauptsatz and Kor. 3].) *Let φ be a form over F such that $\varphi \in I^nF$. If $\dim \varphi < 2^n$ then F is hyperbolic. If $\dim \varphi = 2^n$ then φ is similar to an n-fold Pfister form.*[5]

We will refer to this theorem simply by APH. Note that APH implies in particular that $\bigcap_{n=0}^{\infty} I^nF = 0$. It should be remarked that the only known proofs of APH always use function field techniques as will be introduced in section 4.2.

Returning to our classification problem, it remains to compute the e_n. Now e_0 and e_1 are easily computed, and it is equally easy to check whether their values in the target groups $H^0F = \mathbb{Z}/2\mathbb{Z}$ resp. $H^1F = F^*/F^{*2}$ are trivial or not. Suppose $\varphi \in I^2F$. Then $c(\varphi) = e_2(\varphi) \in H^2F = \operatorname{Br}_2 F$ can also explicitly be computed as the following lemma and its proof show.

---

[5]Two forms φ and ψ are similar if there exists an $a \in F^*$ such that $\varphi \cong a\psi$.

LEMMA 2.2. *Let φ be a form in I^2F of dimension $2n+2$, $n \geq 1$. Then there exist quaternion algebras $Q_1, \cdots, Q_n$ over F such that $c(\varphi) = [Q_1 \otimes \cdots \otimes Q_n] \in \operatorname{Br}_2 F$.*

PROOF. Suppose first that $n = 1$, so that φ is a 4-dimensional form in I^2F, i.e. with trivial signed discriminant. Then there are $a, b, x \in F^*$ such that $\varphi \cong x\langle 1, -a, -b, ab\rangle \cong x\langle\langle a, b\rangle\rangle$. Now $\langle\langle x, a, b\rangle\rangle \cong \langle\langle a, b\rangle\rangle \perp -x\langle\langle a, b\rangle\rangle \in I^3F$, hence $\langle\langle a, b\rangle\rangle \equiv x\langle\langle a, b\rangle\rangle \bmod I^3F$ and thus $c(\langle\langle a, b\rangle\rangle) = c(x\langle\langle a, b\rangle\rangle) \in \operatorname{Br}_2 F$. We have $c(\langle\langle a, b\rangle\rangle) = [(a, b)_F] \in \operatorname{Br}_2 F$, where $(a, b)_F$ denotes the quaternion algebra with F-basis 1, i, j, $ij = k$ and relations $i^2 = a$, $j^2 = b$, $ij = -ji$. (Under the identification $H^2F = \operatorname{Br}_2 F$, the cup product $(a) \cup (b)$ corresponds to the Brauer class of $(a, b)_F$.)

Now suppose that $n \geq 2$ and let $\varphi \in I^2F$, $\dim \varphi = 2n+2$. Then $\langle 1, -x\rangle\varphi \in I^3F$ and thus $\varphi \equiv x\varphi \bmod I^3F$, so that we may assume after scaling that $\varphi \cong \langle 1, -a, -b\rangle \perp \varphi'$. Then $\varphi \sim \langle\langle a, b\rangle\rangle \perp \psi$ in WF with $\psi \cong \langle -ab\rangle \perp \varphi'$. We have $\dim \psi = 2n$ and $\psi \in I^2F$. Thus, there exist by induction quaternion algebras $Q_1, \cdots, Q_{n-1}$ such that $c(\psi) = [Q_1 \otimes \cdots \otimes Q_{n-1}] \in \operatorname{Br}_2 F$. The homomorphism property of c on forms in I^2 implies $c(\langle\langle a, b\rangle\rangle \perp \psi) = c(\langle\langle a, b\rangle\rangle)c(\psi) = [Q_1 \otimes \cdots \otimes Q_n]$ with $Q_n = (a, b)_F$. □

Although we can explicitly write down an element in $\operatorname{Br}_2 F$ which represents the Clifford invariant of φ, it is still quite a different matter (if not impossible) to check whether this element is trivial in $\operatorname{Br}_2 F$. For the e_n, $n \geq 3$, the situation is even worse. Suppose we could show that $\varphi \in I^nF$ and suppose we have given φ in diagonal form $\langle a_1, \cdots, a_n\rangle$. There is no known formula for how to express $e_n(\varphi)$ as sum of n-fold cup products using the coefficients a_i in a way similar to what we did in the proof of the above lemma.

The following theorem of Elman-Lam tells us exactly for which fields the classical invariants (plus the signatures) are sufficient to classify quadratic forms.

THEOREM 2.3. (Elman-Lam [**EL3**, Classification Theorems 3′, 3].) *If F is not formally real, then quadratic forms over F are classified by dimension, signed discriminant, and Clifford invariant if and only if $I^3F = 0$.*

If F is formally real, then quadratic forms over F are classified by dimension, signed discriminant, Clifford invariant, and total signature if and only if I^3F is torsion free.

An element φ in WF is called torsion if for some $n \in \mathbb{N}$, $n \times \varphi = \underbrace{\varphi \perp \cdots \perp \varphi}_{n \text{ times}} \sim 0$ in WF. Pfister [**P3**] has shown that a torsion element in WF has always (additive) order a power of 2. If F is not formally real, i.e. if -1 is a sum of squares in F, then every element is torsion and its order will divide the level s of the field (see section 5.1). If F is formally real, i.e. -1 cannot be written as a sum of squares or, which is equivalent by the Artin-Schreier theorem, there exist orderings on F, then Pfister has also shown that the following sequence is exact :

$$0 \longrightarrow W_tF \longrightarrow WF \stackrel{(\operatorname{sgn}_P)}{\longrightarrow} \prod_{P \in X} \mathbb{Z} ,$$

where W_tF denotes the torsion part of the Witt ring and X the space of orderings of F. This is also referred to as Pfister's local-global principle and it essentially says that a form in WF is torsion if and only if its total signature is identically zero.

Fields to which the previous theorem by Elman-Lam can be applied include local and global fields, fields of transcendence degree ≤ 2 over finite fields or over the real numbers.

Let us conclude this section with a result which extends APH. Recall that APH tells us that anisotropic forms in I^nF must be of dimension $\geq 2^n$. Furthermore, up to similarity, the only anisotropic forms of dimension 2^n in I^nF will be anisotropic n-fold Pfister forms. But how about anisotropic forms of higher dimension in I^nF ? Pfister [**P3**] has shown that there are no anisotropic forms of dimension 10 in I^3F. It is not difficult to construct fields for which there exist anisotropic forms of dimension $2^n + 2^{n-1}$ in I^nF. So one might venture the following conjecture.

CONJECTURE 2.4. Let $n \in \mathbb{N}$, $n \geq 2$. Let φ be an anisotropic form in I^nF of dimension $> 2^n$. Then $\dim \varphi \geq 2^n + 2^{n-1}$.

For $n = 2$, this conjecture is trivially true. It is also true for $n = 3$ and 4.

THEOREM 2.5. (Pfister [**P3**, Satz 14, Zusatz] for $n = 3$, Hoffmann [**H4**, Main Theorem] for $n = 4$.) *Conjecture 2.4 is true for $n \leq 4$.*

B. Kahn informed me that A. Vishik [**Vi2**] has shown that Conjecture 2.4 is true for all n provided the field F is of characteristic 0. The proof is based on techniques developed by Voevodsky in his proof of the Milnor conjecture and which were further elaborated in Vishik's thesis [**Vi1**]. However, Vishik's proof does not rely on the Milnor conjecture.

APH and Theorem 2.5 (resp. Conjecture 2.4) can be interpreted as (an)isotropy results on forms whose invariants e_n are trivial up to a certain n (where we assume of course the Milnor conjecture): Suppose that Conjecture 2.4 is true and let φ be an anisotropic form over F such that $e_i(F) = 0$ for $0 \leq i \leq n$. Then $\dim \varphi = 2^{n+1}$ or $\dim \varphi \geq 2^{n+1} + 2^n$.

3. Field invariants

In the previous section, we have seen how invariants of quadratic forms yield information on the classification problem or on the isotropy problem. In this section, we want to exhibit certain field invariants which by their very definition tell us something about isotropy of quadratic forms (or certain types of quadratic forms).

Let F be a field and let $\mathcal{C}(F)$ be a set of quadratic forms over F, for example all forms which share a certain property. We define

$$\text{supdim}(\mathcal{C}(F)) = \sup\{\dim \varphi \mid \varphi \text{ anisotropic form over } F, \varphi \in \mathcal{C}(F)\} \ .$$

If $\mathcal{C}(F)$ is empty or if it only contains isotropic forms, we put $\text{supdim}(\mathcal{C}(F)) = 0$.

Let us consider some examples.

3.1. The old u-invariant. If we consider all quadratic forms over F,

$$\mathcal{C}_{all}(F) = \{\text{quadratic forms over } F\} \ ,$$

the field invariant $\text{supdim}(\mathcal{C}_{all}(F))$ coincides with the "old" u-invariant of F as originally defined by Kaplansky [**Ka**] (who calls it $C(F)$). It is just the supremum of the dimensions of anisotropic forms over F.

EXAMPLE 3.1. Consider $F = \mathbb{C}(t_1, \cdots, t_n)$, the function field in n variables over the complex numbers. Then it can be shown by an inductive argument that the form $\langle\langle t_1, \cdots, t_n\rangle\rangle$ is anisotropic (cf. Lemma 1.5). In particular, we have

$\text{supdim}(\mathcal{C}_{all}(F)) \geq \dim \langle\langle t_1, \cdots, t_n \rangle\rangle = 2^n$. On the other hand, F will be a C_n-field by Tsen-Lang theory. Hence, all forms of dimension $\geq 2^n + 1$ will be isotropic. This yields $\text{supdim}(\mathcal{C}_{all}(F)) = 2^n$.

3.2. The Hasse number. If F is formally real, then the form defined by a sum of n squares, $n \times \langle 1 \rangle$, will be anisotropic for all $n \in \mathbb{N}$. Hence, for formally real F the invariant $\text{supdim}(\mathcal{C}_{all}(F))$ contains no useful information. Therefore, it seems reasonable to replace $\mathcal{C}_{all}(F)$ by another class of forms which for nonformally real F coincides with $\mathcal{C}_{all}(F)$, but which for formally real F contains only forms which satisfy some necessary conditions for isotropy. So suppose that F is formally real and let P be an ordering on F. Let φ be a form over F. Let F_P be a real closure of F with respect to P and consider the form $\varphi_{F_P} = \varphi \otimes F_P$ obtained by passing from F to F_P via scalar extension. Then φ_{F_P} is isotropic if and only if φ_{F_P} is indefinite, i.e. $|\,\text{sgn}_P\, \varphi| < \dim \varphi$. Hence, for φ to be isotropic, a necessary condition is that φ be indefinite at each ordering P of F, in which case we say that φ is totally indefinite. If F is not formally real, there are no orderings and in this case we define each form to be totally indefinite to avoid case distinctions. We now put

$$\mathcal{C}_{ti}(F) = \{\text{totally indefinite quadratic forms over } F\}.$$

The invariant $\text{supdim}(\mathcal{C}_{ti}(F))$ we thus obtain is referred to as the Hasse number and denoted by $\tilde{u}(F)$. It coincides by definition with $\text{supdim}(\mathcal{C}_{all}(F))$ for nonformally real F. For formally real F, it yields useful further information.

EXAMPLE 3.2. For example, if $F = \mathbb{R}$ (or any real closed field), then $\tilde{u}(F) = 0$ (see also Example 1.1).

Consider the form $\varphi = \langle 1, -(1+T^2) \rangle$ over $\mathbb{R}(T)$. Then, since $1 + T^2$ is not a square in $\mathbb{R}(T)$, the form φ is anisotropic. It is also totally indefinite as 1 is totally positive (i.e. positive at each ordering P of $\mathbb{R}(T)$) and $-(1+T^2)$ is totally negative. In particular, $\tilde{u}(\mathbb{R}(T)) \geq 2$. By what we remarked in Example 1.7, we thus get $\tilde{u}(\mathbb{R}(T)) = 2$.

If we consider the function field $\mathbb{R}(T, X)$ in two variables, then $\langle 1, -(1+T^2) \rangle$ is still totally indefinite, hence the form $\langle 1, -(1+T^2) \rangle \perp X(n \times \langle 1 \rangle)$ will be totally indefinite, and it will be anisotropic for all $n \in \mathbb{N}$ as $\langle 1, -(1+T^2) \rangle$ and $\langle 1, \cdots, 1 \rangle$ are anisotropic over $\mathbb{R}(T)$, cf. Lemma 1.5. It follows that $\tilde{u}(\mathbb{R}(T, X)) = \infty$.

We have $\tilde{u}(\mathbb{Q}) = 4$, cf. Example 1.4.

3.3. The generalized u-invariant. As already remarked, if F is not formally real then all forms are torsion forms. If F is formally real, then torsion forms are exactly the forms with total signature zero, which then are necessarily totally indefinite forms. This leads to another modification of $\mathcal{C}_{all}(F)$ which yields meaningful information in the formally real case. We define

$$\mathcal{C}_{tor}(F) = \{\text{torsion quadratic forms over } F\}\ .$$

The invariant $u(F) = \text{supdim}(\mathcal{C}_{tor}(F))$ is called the (generalized) u-invariant as defined by Elman-Lam [**EL1**]. Since $\mathcal{C}_{tor}(F) \subset \mathcal{C}_{ti}(F)$, we obviously get $u(F) \leq \tilde{u}(F)$. Since $\dim \varphi - \text{sgn}_P\, \varphi \equiv 0 \bmod 2$ for all orderings P of a formally real F, we see that for formally real fields the u-invariant will always be even or infinite.

EXAMPLE 3.3. Clearly, $u(\mathbb{Q}) \leq \tilde{u}(\mathbb{Q}) = 4$. On the other hand, the form $\langle 1, 1, -7, -7 \rangle$, for example, has signature zero and is anisotropic. Hence $u(\mathbb{Q}) = 4$.

In fact, the Hasse-Minkowski theorem implies that for each global field K one has $\tilde{u}(K) = u(K) = 4$, cf. [**S**, Ch. 6, §6].

We have $u(\mathbb{R}) = \tilde{u}(\mathbb{R}) = 0$. The form $\langle 1, -(1+T^2)\rangle$ over $\mathbb{R}(T)$ is in fact a torsion form and thus $u(\mathbb{R}(T)) = \tilde{u}(\mathbb{R}(T)) = 2$. The form $\langle 1, -(1+T^2)\rangle \perp X\langle 1, -(1+T^2)\rangle$ is a torsion form and anisotropic over $\mathbb{R}(T,X)$, hence $u(\mathbb{R}(T,X)) \geq 4$. One can show that $u(\mathbb{R}(T,X)) \leq 6$, see, e.g., [**P4**, Ch. 8, Th. 2.12], so $u(\mathbb{R}(T,X)) = 4$ or 6. The precise value is not known at present.

3.4. The level. The level $s(F)$ (s for the German word "Stufe") of a field F is defined to be as follows. If -1 is not a sum of squares in F, i.e. if F is formally real, then we put $s(F) = \infty$. Otherwise, $s(F)$ is the smallest integer $n \geq 1$ such that -1 can be written as a sum of n squares in F. This definition can be reformulated as follows. Consider

$$\mathcal{C}_s(F) = \{\underbrace{\langle 1, \cdots, 1\rangle}_{n};\ n \in \mathbb{N}\}\ .$$

Then $s(F) = \operatorname{supdim}(\mathcal{C}_s(F))$, thus we have an interpretation of the level in terms of the supremum of the dimensions of certain anisotropic forms.

EXAMPLE 3.4. For finite fields $\mathbb{F}_p$, p an odd prime, we have $s(\mathbb{F}_p) = 1$ (resp. 2) if and only if -1 is a quadratic residue (resp. nonresidue) mod$p$ iff $p \equiv 1 \bmod 4$ (resp. $p \equiv 3 \bmod 4$).

For $\mathbb{Q}_p$ we have $s(\mathbb{Q}_p) = s(\mathbb{F}_p)$ for p odd, and $s(\mathbb{Q}_2) = 4$. Note that by passing to a rational function field, the level will not change, i.e. $s(F) = s(F(T))$.

3.5. The Pythagoras number. Another invariant which has been studied extensively is the so-called Pythagoras number $p(F)$ of a field. Define $\sum F^2$ to be the set of all elements in F^* which can be written as a sum of squares. Now let $p(F)$ be the smallest $n \in \mathbb{N}$ (if such an integer exists) such that each element in $\sum F^2$ can be written as a sum of $\leq n$ squares. If such an integer does not exist, i.e. if to each n there exists $x \in \sum F^2$ which cannot be written as sum of $\leq n$ squares, then we put $p(F) = \infty$. If we want to interpret $p(F)$ as the supremum of the dimensions of certain anisotropic forms, we have to consider

$$\mathcal{C}_p(F) = \{\underbrace{\langle 1, \cdots, 1\rangle}_{n} \perp \langle -x\rangle;\ n \in \mathbb{N} \cup \{0\},\ x \in \textstyle\sum F^2\}\ ,$$

and we get $p(F) = \operatorname{supdim}(\mathcal{C}_p(F))$.

If $s(F) = s < \infty$, i.e. if F is not formally real, then the form $(s+1) \times \langle 1\rangle$ will be isotropic and it will therefore contain $\mathbb{H} = \langle 1, -1\rangle$ as a subform. Now $\langle 1, -1\rangle$ is universal, i.e. it represents every element in F^*. This shows on the one hand that F is not formally real if and only if $F^* = \sum F^2$, on the other hand we get that $p(F) \leq s+1$. Since -1 is by definition of the level not a sum of $s-1$ squares, we have $p(F) \geq s$ and hence, for nonformally real F, $p(F) \in \{s(F), s(F)+1\}$.

For formally real F, $p(F)$ can be finite or infinite.

EXAMPLE 3.5. Fields with $p(F) = 1$ are called pythagorean fields. Each field F has a pythagorean closure F_{pyth} inside an algebraic closure F_{alg}, i.e. F_{pyth} is the smallest field inside F_{alg} with Pythagoras number 1. It can be obtained by taking the union of all fields K in F_{alg} such that there exists a tower $F = K_0 \subset K_1 \subset \cdots \subset K_n = K$, $n \in \mathbb{N}$, such that $K_{i+1} = K_i(\sqrt{a_i^2 + b_i^2})$, $a_i, b_i \in K_i$ (cf., for example, [**S**, p. 52]). $\mathbb{R}$ and $\mathbb{C}$ are pythagorean.

For odd primes p , one has $p(\mathbb{F}_p) = 2$, and $p(\mathbb{Q}_p) = 2$ (resp. 3) if $p \equiv 1 \bmod 4$ (resp. $p \equiv 3 \bmod 4$). Example 1.4 shows that $p(\mathbb{Q}) = 4$.

3.6. The l-invariant. A form $\varphi = \langle a_1, \cdots, a_n \rangle$ is called totally positive definite if all a_i are totally positive, i.e. all a_i are positive with respect to each ordering of F (if there are any), or, which is equivalent, all a_i are in $\sum F^2$. Note that by this definition, forms over nonformally real fields are always totally positive definite. The length $l(F)$ of a field F is defined to be the smallest $n \in \mathbb{N}$ (if such an integer exists) such that each totally positive definite form over F of dimension n represents all totally positive elements in F. This invariant is significant in the study of another invariant $g_F(n)$ which denotes the smallest $r \in \mathbb{N}$ such that every sum of squares of n-ary F-linear forms can be written as a sum of r squares of n-ary F-linear forms. This invariant $g_F(n)$ has been introduced in [**CDLR**], but it has implicitly already been studied by Mordell [**Mo**] for $F = \mathbb{Q}$, who showed that $g_{\mathbb{Q}}(n) = n + 3$. The invariant $l(F)$ has been introduced in [**BLOP**] where it is shown among many other things that $g_F(n) = n + l(F) - 1$ for all $n \geq l(F) - 1$, [**BLOP**, Th. 2.15].

Again, we want to interpret the invariant $l(F)$ as the supremum of the dimensions of certain anisotropic forms. We define

$$\mathcal{C}_l(F) = \{\langle a_1, \cdots, a_{n-1}, -a_n \rangle;\ n \in \mathbb{N},\ a_i \in \textstyle\sum F^2,\ 1 \leq i \leq n\}\ .$$

We thus get $l(F) = \operatorname{supdim}(\mathcal{C}_l(F))$. Note that $\mathcal{C}_p(F) \subset \mathcal{C}_l(F) \subset \mathcal{C}_{ti}(F)$, hence $p(F) \leq l(F) \leq \tilde{u}(F)$.

EXAMPLE 3.6. If F is not formally real, then $\sum F^2 = F^*$ and we have $\tilde{u}(F) = u(F) = l(F)$. We have $p(\mathbb{Q}) = 4 \leq l(\mathbb{Q}) \leq \tilde{u}(F) = 4$, hence $l(\mathbb{Q}) = 4$. Similarly, $l(\mathbb{R}(T)) = \tilde{u}(\mathbb{R}(T)) = 2$.

4. Construction of fields with prescribed invariants : Basic ideas and tools

Having introduced field invariants such as the level, the Pythagoras number, the l- and u-invariants and the Hasse number, it becomes a natural question to ask which values can appear for each of these invariants. Again, let us consider a certain class of forms $\mathcal{C}$ as in the examples above. For a given $n \in \mathbb{N}$, can there exist a field F such that $\operatorname{supdim}(\mathcal{C}(F)) = n$, and if so, how can one construct such a field ?

4.1. The basic idea of construction. We want to construct a field F such that $\operatorname{supdim}(\mathcal{C}(F)) = n$ for a certain $n \in \mathbb{N}$, i.e. we want to have an anisotropic form of dimension n in $\mathcal{C}(F)$, say φ, and we have to verify that all forms of dimension $> n$ in $\mathcal{C}(F)$ are anisotropic. The whole idea will be reminiscent of certain direct limit constructions which the reader might have encountered in other contexts, such as the construction of an algebraic closure of a field.

STEP 1. Choose a field F_0 such that there exists an anisotropic form φ of dimension n in $\mathcal{C}(F_0)$. If all forms in $\mathcal{C}(F_0)$ of dimension $> n$ are isotropic, F_0 is the desired field. In this case, we put $F_i = F_0$ for all $i \in \mathbb{N}$. If not, continue with step 2.

STEP 2. Construct a field extension F_1 of F_0 in such a way such that $\psi_{F_1} \in \mathcal{C}(F_1)$ for all $\psi \in \mathcal{C}(F_0)$, such that φ_{F_1} is anisotropic, and such that ψ_{F_1} is isotropic for all $\psi \in \mathcal{C}(F_0)$ of dimension $> n$. If all forms in $\mathcal{C}(F_1)$ of dimension $> n$ are isotropic, F_1 is the desired field and we put $F_i = F_1$ for all $i \in \mathbb{N}$, $i \geq 2$. If not, repeat this construction.

STEP 3. Repeating this construction, we get a tower of fields $F_0 \subset F_1 \subset \cdots \subset F_i \subset \cdots$ which may or may not become stationary. We let F be the direct limit of this tower of fields, i.e. $F = \bigcup_{i=0}^{\infty} F_i$. F is then again a field. We have to verify that for a form γ over F we have $\gamma \in \mathcal{C}(F)$ if and only if there exists some $i \in \mathbb{N}$ and some $\rho \in \mathcal{C}(F_i)$ such that $\gamma \cong \rho_F$. (For many types of $\mathcal{C}$, such as $\mathcal{C}_s$ = sums of squares, this is just a formality and often self-evident.)

STEP 4. We claim that F is the desired field. First, consider φ_F. Then $\varphi_F \in \mathcal{C}(F)$. Furthermore, φ_F is anisotropic. Indeed, if it were isotropic, there would be an $i \in \mathbb{N}$ such that already φ_{F_i} were isotropic, which is not possible because of the way the F_i were constructed. Hence, $\operatorname{supdim}(\mathcal{C}(F)) \geq n$.

Now let $\gamma \in \mathcal{C}(F)$ with $\dim \gamma > n$. Then, for some $i \in \mathbb{N}$ there exists $\rho \in \mathcal{C}(F_i)$ such that $\gamma \cong \rho_F$. By construction, $\rho_{F_{i+1}}$ is isotropic. Hence $\rho_F \cong \gamma$ is isotropic. This shows that $\operatorname{supdim}(\mathcal{C}(F)) \leq n$. Therefore, $\operatorname{supdim}(\mathcal{C}(F)) = n$.

Of course, it could a priori be impossible to have $\operatorname{supdim}(\mathcal{C}(F)) = n$ for certain $n \in \mathbb{N}$. For example, we have already seen that for formally real F, $u(F)$ will always be even. In section 5.1, we shall see that $s(F)$ will always be a 2-power or infinite. In many cases, we still don't know precisely which values can be realized.

Another obvious problem is how to control the (an)isotropy behaviour for many forms simultaneously by passing from F_i to F_{i+1}. If we consider a field extension L/K such that an anisotropic form μ over K becomes isotropic over L, then in general many more anisotropic forms over K will also become isotropic over L. So the aim is to choose L in such a way that as few anisotropic forms over K become isotropic over L as possible, but that μ should be one of the forms which do become isotropic. This can be achieved "generically" by passing to function fields as we shall explain in the next section. We will refer to the above step by step procedure using function field extensions as Merkurjev's method, as it was Merkurjev who demonstrated the power of this method most spectacularly by constructing fields with prescribed even u-invariant. This result will be the theme of section 5.3.

4.2. Function fields of quadratic forms. Let φ be a form over F of dimension $n \geq 3$. Then the function field of φ, denoted by $F(\varphi)$, is the function field of the projective quadric defined by the equation $\varphi = 0$.

More explicitly, consider $\varphi = \langle a_1, \cdots, a_n \rangle$ as a homogeneous polynomial in the polynomial ring $F[x_1, \cdots, x_n]$, $\varphi(x_1, \cdots, x_n) = \sum_{i=1}^{n} a_i x_i^2$. Then it can readily be shown that the polynomial $\varphi(x_1, \cdots, x_{n-1}, 1)$ is irreducible, and if we denote by I the ideal generated by $\varphi(x_1, \cdots, x_{n-1}, 1)$ in $F[x_1, \cdots, x_{n-1}]$, then $F(\varphi)$ is the quotient field of the integral domain $F[x_1, \cdots, x_{n-1}]/I$.

If we put $\varphi' = \langle a_2, \cdots, a_n \rangle$, then we have

$$F(\varphi) = F(x_2, \cdots, x_{n-1})(\sqrt{-a_1^{-1}\varphi'(x_2, \cdots, x_{n-1}, 1)})\ .$$

If $\varphi = \langle a, b \rangle$ is a binary form, we put $F(\varphi) = F(\sqrt{-ab})$. This is consistent with the previous expression and it gives a quadratic extension if $\langle a, b \rangle$ is anisotropic (i.e. $-ab \notin F^{*2}$), and it is F itself if $\langle a, b \rangle$ is isotropic. To avoid case distinctions, we put $F(\varphi) = F$ for $\dim \varphi \leq 1$ (sometimes it is useful to consider 0-dimensional quadratic forms). If $\dim\varphi \geq 2$, then φ will be isotropic over $F(\varphi)$ by definition of the function field.

One of the first systematic studies of function fields of quadratic forms and the behaviour of quadratic forms over such function fields has been undertaken by Knebusch [**K1**], [**K2**]. But these function fields have already appeared earlier, for example in [**P1**] and [**AP**]. We collect some useful facts which we will use later. We refer to [**S**, Ch. 4, §5] for details.

Let φ be a form over F with $\dim\varphi = n \geq 2$. Then $F(\varphi)$ is of transcendence degree $n-2$ over F, and $F(\varphi)/F$ is purely transcendental if and only if φ is isotropic (here, we consider F to be purely transcendental of transcendence degree 0 over itself in order to include the case $\varphi \cong \mathbb{H}$). This shows in particular that if K/F is an extension such that φ_K is isotropic, then $K(\varphi)/K$ is purely transcendental. It follows that if ψ is another form over F, then if ψ is anisotropic over K it will be anisotropic over $F(\varphi)$ (go up to $K(\varphi)$ and then down to $F(\varphi)$).

$F(\varphi)$ is also called a generic zero field for the form φ as it has the property that if L is any field over F such that φ_L is isotropic, then there exists an F-place $\lambda : F(\varphi) \to L \cup \infty$.[6]

Among the most important problems in the theory of function fields of quadrics are the following questions :

QUESTION 4.1. (i) Let φ be an anisotropic form over F. Which anisotropic forms ψ become isotropic over $F(\varphi)$?

(ii) Let φ be an anisotropic form over F. For which anisotropic forms ψ over F is $\varphi_{F(\psi)}$ isotropic ?

A complete answer to the first question is known for $\dim\varphi = 2$. In fact, the following is easy to show (cf. [**S**, Ch. 2, Lemma 5.1], [**L1**, Ch. VII, Lemma 3.1]) :

LEMMA 4.2. *Let ψ be a form over F such that $\dim\psi \geq 3$ or $\dim\psi = 2$ and anisotropic. Let $d \in F^* \setminus F^{*2}$. Then $\psi_{F(\sqrt{d})}$ is isotropic if and only if there exists $a \in F^*$ such that $a\langle 1, -d\rangle \subset \psi$.*

The case of $\dim\varphi = 3$ has been studied by various authors, among others by Rost [**R2**], Hoffmann, Lewis and Van Geel [**LVG**], [**HLVG**], [**HVG1**]. In [**HLVG**], a reasonably complete answer is given in terms of so-called splitting sequences and minimal forms. The case of $\dim\varphi \geq 4$ is largely open and only some partial results are known.

The second question has attracted a lot of attention over the last few years. A complete answer is known if $\dim\varphi \leq 5$. An almost complete answer is also known for $\dim\varphi = 6$. Partial results have been obtained in dimensions 7 and 8 (e.g., a rather complete description is known if φ is an 8-dimensional form in I^2F). Among the authors who contributed to these results on forms of small dimension are Wadsworth [**W**], Leep [**Le3**], Hoffmann [**H1**], [**H2**], Laghribi [**Lag1**], [**Lag2**], [**Lag3**], Izhboldin and Karpenko [**IK1**], [**IK2**], [**IK3**].

If in Question 4.1 we replace the word "isotropic" by "hyperbolic", the problem becomes somewhat easier but by no means trivial. Part (i) will then become the problem of determining the kernel of the ring homomorphism $WF \to WF(\varphi)$. This

[6] A place $\lambda : K \to L\cup\infty$ is a pseudo-homomorphism with the properties $\lambda(a+b) = \lambda(a)+\lambda(b)$, $\lambda(ab) = \lambda(a)\lambda(b)$ whenever the right hand sides are defined, where one applies the obvious rules $a + \infty = \infty$ for $a \in L$, $a\infty = \infty$ for $a \in L^* \cup \infty$, $1/\infty = 0$, $1/0 = \infty$, and where the expressions $\infty + \infty$ and 0∞ are not defined.

is known for Pfister forms as we shall see in the next section, but also for other types of forms, so for example in the case $\dim \varphi \leq 5$ (cf. [**F**]).

One of the most general and most useful results on the "hyperbolic" question is the following subform theorem due to Wadsworth [**W**, Th. 2] and, independently, Knebusch [**K1**, Lemma 4.5]. See also [**S**, Ch. 4, Th. 5.4].

THEOREM 4.3. *Let φ and ψ be forms over F such that* $\dim \varphi \geq 2$*, $\varphi \not\cong \mathbb{H}$, and ψ nonhyperbolic. Suppose that $\psi_{F(\varphi)}$ is hyperbolic. Then for each $a \in F^*$ represented by φ and each $b \in F^*$ represented by ψ, we have that $ab\varphi \subset \psi$. In particular,* $\dim \psi \geq \dim \varphi$.

This theorem is often referred to (as we shall do) as the Cassels-Pfister subform theorem, CPST for short, although the original theorem which goes by this name (and which is the main ingredient in the proof of the theorem above) states the following : Let φ be a form of dimension n over F and let ψ be another form over F. Then ψ represents $\varphi(x_1, \cdots, x_n)$ over the rational function field in n variables $F(x_1, \cdots, x_n)$ if and only if $\varphi \subset \psi$. Cf. [**P2**], see also [**L1**, Ch. IX, Th. 2.8], [**S**, Ch. 4, Th. 3.7].

General results on the isotropy problem are even harder to obtain or so it seems, although great progress has been made recently in the work of Vishik, Karpenko and Izhboldin who invoke highly sophisticated machinery such as Chow groups and Chow motives of quadrics, unramified cohomology and graded Grothendieck groups of quadrics to tackle this and related problems. Further progress in the algebraic theory of quadratic forms in general and on the isotropy problem in particular will most likely necessitate more and more the use of such highly developed techniques originating in algebraic geometry, K-theory and cohomology theory.

An important general result which nevertheless can be proved in a fairly elementary fashion using properties of Pfister forms (see section below) is the following :

THEOREM 4.4. (Hoffmann [**H3**, Th. 1].) *Let φ and ψ be anisotropic forms over F such that* $\dim \varphi \leq 2^n < \dim \psi$. *Then φ stays anisotropic over $F(\psi)$.*

A somewhat different proof from the one in [**H3**] was given by Hurrelbrink-Rehmann [**HuR**].

4.3. Pfister forms. An n-fold Pfister form over F is a form of type $\langle 1, -a\rangle \otimes \cdots \otimes \langle 1, -a_n\rangle$, $a_i \in F^*$, and we write $\langle\langle a_1, \cdots, a_n\rangle\rangle$ for short. We have already recognized them in section 2.2 as being the generators of I^nF. Pfister forms share many nice and useful properties which make them into a powerful tool and place them at the core of the whole algebraic theory of quadratic forms. They have first been studied systematically by Pfister [**P2**] who proved their basic properties. Many of these proofs have been simplified later by Witt [**Wi2**] (see also [**S**, Ch. 2, §10]). We will denote the set of forms isometric (resp. similar) to n-fold Pfister forms by P_nF (resp. GP_nF).

We collect some of these properties which will be useful later on. First of all, if $\varphi \in P_nF$, then $a \in F^*$ is represented by φ if and only if $\varphi \cong a\varphi$. If $D_F(\varphi)$ denotes the elements in F^* represented by φ, and if $G_F(\varphi)$ represents the similarity factors of φ over F, then the above just means $G_F(\varphi) = D_F(\varphi)$ for $\varphi \in P_nF$. In particular, $D_F(\varphi)$ is a group. A form ψ with the property $G_F(\psi) = D_F(\psi)$ is called round (this definition is due to Witt [**Wi2**] who also provided an elegant proof of the roundness of Pfister forms). An n-dimensional form ψ over F is called multiplicative if for the n-tuples of variables $X = (x_1, \cdots, x_n)$, $Y = (y_1, \cdots, y_n)$ there exist an n-tuple of rational functions $Z = (z_1, \cdots, z_n)$, $z_k \in F(x_i, y_j; 1 \le i, j \le n)$ such that $\psi(X)\psi(Y) = \psi(Z)$.[7] Since isotropic forms are universal, they are always multiplicative. Pfister showed that an anisotropic form φ over F is a Pfister form if and only if φ is multiplicative if and only if $D_L(\varphi)$ is a group for all field extensions L/F iff φ_L is round for all field extensions L/F (cf. Pfister [**P2**, Satz 5, Th. 2], see also Lam's book [**L1**, Ch. X § 2] and Scharlau's book [**S**, Ch. 2, § 10; Ch. 4, Th. 4.4]).

If F is formally real and P is an ordering of F, then for $\varphi = \langle\langle a_1, \cdots, a_n\rangle\rangle$ we have $\operatorname{sgn}_P \varphi = 2^n = \dim \varphi$ if all $a_i <_P 0$, otherwise $\operatorname{sgn}_P \varphi = 0$.

Pfister forms are either anisotropic or hyperbolic. This shows that if $\varphi \in GP_nF$ is anisotropic, $n \ge 1$, then $\varphi_{F(\varphi)}$ is isotropic and therefore hyperbolic. The converse of this statement is also true : Let φ be an anisotropic form over F of dimension ≥ 2 such that $\varphi_{F(\varphi)}$ is hyperbolic, then $\varphi \in GP_nF$ for some $n \ge 1$. This has been shown by Wadsworth [**W**, Th. 5] and, independently, by Knebusch [**K1**, Th. 5.8] (see also [**S**, Ch. 4, Th. 5.4]).

We have that if $\varphi \in GP_nF$ and ψ are anisotropic forms over F, and if $\psi_{F(\varphi)}$ is hyperbolic, then there exists a form τ over F such that $\psi \cong \varphi \otimes \tau$. This can easily be shown using the fact that Pfister forms become hyperbolic over their own function field, invoking CPST and doing an induction on the dimension of ψ. We leave the details as an exercise to the reader. This result also leads to a proof of APH.

PROOF OF APH. Let φ be an anisotropic form in I^nF and write $\varphi \sim \sum_{i=1}^r \pi_i$ with anisotropic $\pi_i \in GP_nF$. If $r = 0$ then $\varphi \sim 0$, i.e. $\dim \varphi = 0$. If $r = 1$ then $\varphi \cong \pi_1$ (as φ and π_1 are anisotropic), and we have $\dim \varphi = 2^n$. If $r \ge 2$, consider $K = F(\pi_r)$. If φ_K is hyperbolic then $\varphi \cong \tau \otimes \pi_r$ for a certain form τ over F by the preceding paragraph. In particular $\dim \varphi = 2^n \dim \tau \ge 2^n$. If φ_K is not hyperbolic, then $(\varphi_K)_{\text{an}} = \sum_{i=1}^s (\pi_i)_K$, where the $(\pi_i)_K$ are anisotropic, $1 \le s < r$ (after deleting all hyperbolic $(\pi_i)_K$ and rearranging the remaining ones : $s \ge 1$ because φ_K is not hyperbolic, $s < r$ because $(\pi_r)_K$ is hyperbolic). Induction on r shows that $\dim \varphi \ge \dim(\varphi_K)_{\text{an}} \ge 2^n$.

If $\dim \varphi = 2^n$, then $\dim(\varphi_{F(\varphi)})_{\text{an}} < 2^n$ and $(\varphi_{F(\varphi)})_{\text{an}} \in I^nF(\varphi)$, which by the above shows that $\varphi_{F(\varphi)}$ is hyperbolic, and thus, by what we mentioned above, $\varphi \in GP_nF$. □

An important notion is that of a Pfister neighbor. φ is called a Pfister neighbor if there exists a form $\pi \in P_nF$ for some $n \ge 0$ and an $a \in F^*$ such that $a\varphi \subset \pi$ and $\dim \varphi > \frac{1}{2} \dim \pi = 2^{n-1}$. In this case, we say that φ is a Pfister neighbor of the Pfister form π. In this situation, if φ is isotropic then π is isotropic and hence hyperbolic. Conversely, if π is hyperbolic than φ is isotropic as follows readily from the fact that $\dim \varphi > \frac{1}{2} \dim \pi$ and the following rather obvious but useful observation (whose proof is left as an easy exercise to the reader) :

LEMMA 4.5. *Let ψ and τ be forms over F such that $\tau \subset \psi$. If $\dim \tau > \dim \psi - i_W(\psi)$, then τ is isotropic.*

---

[7] Some authors call a form multiplicative if it is round. In Scharlau's book [**S**, Ch. 2, Def. 10.1], a form is called multiplicative if it is either hyperbolic, or anisotropic and round. With this definition, Pfister forms are multiplicative.

If φ is a Pfister neighbor of the Pfister forms π_1 and π_2, then $\pi_1 \cong \pi_2$. Indeed, for dimension reasons, there exists $n \in \mathbb{N}$ such that $\pi_1, \pi_2 \in P_nF$. Let $a_i \in F^*$ and τ_i such that $a_i\varphi \perp \tau_i \cong \pi_i$, $i = 1, 2$. Note that $\dim \tau_i < \frac{1}{2} \dim \pi_i = 2^{n-1}$. Then, in WF, $a_2\pi_1 - a_1\pi_2 \sim a_2\tau_1 - a_1\tau_2$. The left hand side is in I^nF, the right hand side is of dimension $< 2^n$, hence, by APH, $a_2\pi_1 - a_1\pi_2 \sim 0$, i.e. $a_2\pi_1 \cong a_1\pi_2$. Mutliplicativity of Pfister forms then readily implies that $\pi_1 \cong \pi_2$.

5. Construction of fields with prescribed invariants : Examples

5.1. The level. Recall that by definition the level $s(F)$ is the supremum of the dimensions of anisotropic forms of type $\langle 1, \cdots, 1\rangle$. This is one of the first field invariants pertaining to quadratic forms which has been studied, in the beginning in the context of number fields, see the historical remarks in [**P4**, p. 42]. Let us just mention that H. Kneser [**Kn**] proved in 1934 that the level of a field F, if finite (i.e. if F is not formally real) is always of the form 1, 2, 4, 8 or a multiple of 16. Obviously unaware of Kneser's result, Kaplansky [**Ka**] proved in 1953 the weaker result that the level for nonformally real F is 1, 2, 4 or a multiple of 8. (Kaplansky does not use the word level and calls this invariant $B(F)$.) Back then, there were no examples known of nonformally real F with $s(F) > 4$.

The aim of this section is to prove Pfister's famous results on the level, published in [**P1**].

THEOREM 5.1. (Pfister)

(i) *Let F be a nonformally real field. Then there exists $n \in \mathbb{N} \cup \{0\}$ such that $s(F) = 2^n$.*

(ii) *For each $n \in \mathbb{N} \cup \{0\}$ there exists a field F with $s(F) = 2^n$.*

PROOF. (i) Define $\sigma_n = 2^n \times \langle 1\rangle = \langle\langle -1, \cdots, -1\rangle\rangle \in P_nF$. Let $m \in \mathbb{N}$ and suppose that $s(F) \geq m$. Let $\varphi = m \times \langle 1\rangle$. Then by definition of the level as $\operatorname{supdim} \mathcal{C}_s(F)$, where $\mathcal{C}_s(F) = \{n \times \langle 1\rangle; n \in \mathbb{N}\}$, we have that φ is anisotropic. Let n be such that $2^{n-1} < m \leq 2^n$. Then φ is a Pfister neighbor of σ_n, and the anisotropy of φ implies that of σ_n. Hence $s(F) \geq 2^n$, which immediately yields the desired result.

(ii) Let K be any formally real field and consider the n-fold Pfister form σ_n defined as above, and $\psi = (2^n+1)\times\langle 1\rangle$. Let $F = K(\psi)$. Then ψ_K is isotropic, hence $m \times \langle 1\rangle$ is isotropic for all $m > 2^n$. On the other hand, σ_n will stay anisotropic over $F = K(\psi)$ as follows from Theorem 4.4. Of course, one can also invoke CPST : If $(\sigma_n)_F$ were isotropic, it would be hyperbolic as it is a Pfister form. Hence ψ would be similar to a subform of σ_n, which is impossible for dimension reasons. □

In the notations of section 4.1, K corresponds to the field F_0, $K(\psi)$ to F_1, σ_n to φ, and our construction stops after the second step.

5.2. The Pythagoras number. We have already seen that if F is not formally real, then $p(F) \in \{s(F), s(F) + 1\}$, and thus $p(F) \in \{2^n, 2^n + 1\}$ for some integer $n \geq 0$. If F is any field with $s(F) = 2^n$, then consider the rational function field in one variable $K = F(T)$. Now for $a \in F^*$, $a + T^2$ is a sum of m squares in K if and only if -1 or a is a sum of $m - 1$ squares in F. This results has been shown by Cassels [**Ca**] and has been generalized by Pfister [**P2**, Satz 2] (see also the "Second Representation Theorem" [**L1**, Ch. IX, Th. 2.1]). This readily implies that $-1 + T^2$ cannot be written as a sum of 2^n squares and we have $p(K) = 2^n + 1$.

To get a field K' with $p(K') = 2^n$, let K_{alg} be an algebraic closure of K and consider all intermediate fields $K \subset L \subset K_{alg}$ with the property that $2^n \times \langle 1 \rangle$ is anisotropic over L. These fields are inductively ordered by inclusion, so there exists a intermediate field K' which is maximal with respect to this property. Then it is not to difficult to show that $2^n \times \langle 1 \rangle$ will be universal over K'. In particular, each element of K'^* will be a sum of $\leq 2^n$ squares, but by construction -1 will not be a sum of $2^n - 1$ squares. Hence, $p(K') = 2^n$. For more details, cf. [**P4**, Ch. 7, Prop. 1.5].

We can also construct such fields using our method. Indeed, let F_0 be any field of level 2^n. We construct a tower of fields $F_0 \subset F_1 \subset \cdots$ as follows. Suppose we have constructed F_i, $i \geq 0$, such that $2^n \times \langle 1 \rangle$ is isotropic over F_i (this is the case for F_0 as $s(F_0) = 2^n$). Then let F_{i+1} be the free compositum over F_i of all function fields $F_i(\varphi)$, where φ runs over all forms of dimension $2^n + 1$ over F_i. Let $F = \bigcup_{i=0}^{\infty}$. Then, by construction, all forms over F of dimension 2^n+1 are isotropic. In particular, $u(F) \leq 2^n$. On the other hand, since anisotropic forms of dimension $\leq 2^n$ stay anisotropic over function fields of forms of dimension $\geq 2^n+1$, we see that $2^n \times \langle 1 \rangle$ stays anisotropic over F. We immediately get $p(F) = s(F) = u(F) = 2^n$.

Let us summarize the above.

THEOREM 5.2. *Let F be a nonformally real field. Then there exists an integer $n \geq 0$ such that $s(F) = 2^n$ and $p(F) \in \{2^n, 2^n + 1\}$. Conversely, to each integer $n \geq 0$ there exist nonformally real fields F, F' such that $s(F) = p(F) = s(F') = 2^n$ and $p(F') = 2^n + 1$.*

For formally real F, the situation is more complicated. Prestel [**Pr2**] showed that to each integer $n \geq 0$, there exist formally real fields F, F' such that $p(F) = 2^n$, $p(F') = 2^n + 1$, and fields F'' with $p(F'') = \infty$. In fact, he could construct such fields inside $\mathbb{R}$ using a technique called "intersection of henselian fields". In [**H6**], it was finally shown that in fact all $n \in \mathbb{N}$ can appear as Pythagoras number of formally real fields. Let us give the proof which is essentially the one to be found in [**H6**], except that there some of the auxiliary results are proved in greater generality. First, some notations.

Let $n \in \mathbb{N}$ and let $m \in \mathbb{N} \cup \{0\}$ such that $2^{m-1} < n \leq 2^m$. Let us define $n(0) = n$ and $n(1) = 2^m - n$, and inductively, if $n(i) \geq 1$ then $n(i+1) = n(i)(1)$. Note that if $n(i) > 1$, then there exists $k \in \mathbb{N} \cup \{0\}$ such that $n(i) > 2^k > n(i+1)$. We put $\sigma_n = n \times \langle 1 \rangle$ and $\pi_n = \sigma_n \perp \sigma_{n(1)}$. In other words, with n and m as above, $\pi_n = \langle\langle -1, \cdots, -1 \rangle\rangle \in P_m F$, and σ_n is a Pfister neighbor of π_n.

We begin with a lemma which is essentially due to Izhboldin [**I1**].

LEMMA 5.3. *Let $m, n, r \in \mathbb{N}$ such that $n = 2^m + r$ and $1 \leq r \leq 2^m - 1$, and let $x \in F^*$. Put $\varphi = \sigma_n \perp \langle x \rangle$ and $\psi = \sigma_{n(1)} \perp \langle -x \rangle$. Then the following holds :*

(i) *Let L/F be any field extension. Then φ_L is a Pfister neighbor if and only if $\varphi_L \subset (\pi_n)_L$ iff ψ_L is isotropic.*

(ii) *If φ is anisotropic, then $\varphi_{F(\psi)}$ is an anisotropic Pfister neighbor. In particular, $\varphi_{F(\psi)} \subset (\pi_n)_{F(\psi)}$ and $(\pi_n)_{F(\psi)}$ is anisotropic.*

PROOF. (i) We have $\pi_n = \sigma_n \perp \sigma_{n(1)} = 2^{m+1} \times \langle 1 \rangle$, hence φ contains the Pfister neighbor $(2^m + 1) \times \langle 1 \rangle$ of π_n. If φ_L is a Pfister neighbor, then it must be a neighbor of the same Pfister form as $(2^m + 1) \times \langle 1 \rangle$, hence of $(\pi_n)_L$. Since both φ_L and $(\pi_n)_L$ represent 1, multiplicativity of Pfister forms implies $\varphi_L \subset (\pi_n)_L$. The converse is obvious.

By Witt cancellation, we have $(\sigma_n \perp \langle x\rangle)_L \subset (\pi_n)_L = (\sigma_n \perp \sigma_{n(1)})_L$ iff $(\sigma_{n(1)})_L$ represents x if and only if $(\sigma_{n(1)} \perp \langle -x\rangle)_L = \psi_L$ is isotropic.

(ii) $\psi_{F(\psi)}$ is isotropic and thus, by part (i), $\varphi_{F(\psi)}$ is a Pfister neighbor of $(\pi_n)_{F(\psi)}$. If φ is anisotropic, then σ_n is anisotropic, and since σ_n is a Pfister neighbor of π_n, we have that π_n is anisotropic. Suppose that $\varphi_{F(\psi)}$ is isotropic. Then $(\pi_n)_{F(\psi)}$ is hyperbolic, and by CPST and because π_n and ψ represent 1, we have $\psi \subset \pi_n$. By a similar argument as before, $\psi = \sigma_{n(1)} \perp \langle -x\rangle \subset \pi_n$ implies that σ_n represents $-x$ and hence that φ is isotropic, a contradiction. □

PROPOSITION 5.4. *Let $1 \leq n' < n$ be integers, $x, y \in F^*$, and let $\varphi = \sigma_{n'} \perp \langle x\rangle$, $\psi = \sigma_n \perp \langle y\rangle$. Let $m \geq 0$ be an integer such that $2^m + 1 \leq \dim\psi \leq 2^{m+1}$. Suppose that φ is anisotropic and that one of the following conditions holds :*

(i) $\dim\varphi \leq 2^m$, *or*

(ii) $2^m + 1 \leq \dim\varphi < \dim\psi \leq 2^{m+1}$ *and ψ is not a Pfister neighbor.*

Then $\varphi_{F(\psi)}$ is anisotropic.

PROOF. The proof is by induction on m. In the situation of (i), in particular if $m = 0$, the statement follows from Theorem 4.4.

So suppose that (ii) holds. If ψ is isotropic, then $F(\psi)/F$ is purely transcendental and φ stays therefore anisotropic over $F(\psi)$. Hence we may assume that ψ is anisotropic.

Suppose first that φ is a Pfister neighbor, say, of $\gamma \in P_{m+1}F$. Then γ is anisotropic as φ is supposed to be anisotropic. Suppose $\varphi_{F(\psi)}$ is isotropic. Then γ becomes hyperbolic over $F(\psi)$, which, by CPST implies that ψ is similar to a subform of γ and hence, for dimension reasons, that ψ is a Pfister neighbor, contrary to our assumption. Thus, if φ is a Pfister neighbor, $\varphi_{F(\psi)}$ is anisotropic.

Note that if $\dim\varphi = 2^m+1$, then φ is a Pfister neighbor of $\langle\langle -1, \cdots, -1, -y\rangle\rangle \in P_{m+1}F$. Hence, what remains to check is the case $2^m + 2 \leq \dim\varphi < \dim\psi \leq 2^{m+1}$ with φ and ψ not being Pfister neighbors. In particular, this implies $m \geq 2$.

In this situation, we have $1 \leq \dim\sigma_{n(1)} < \dim\sigma_{n'(1)} \leq 2^m - 1$. Let $\rho = \sigma_{n'(1)} \perp \langle -x\rangle$, $\tau = \sigma_{n(1)} \perp \langle -y\rangle$. Recall that in this situation, $\pi_n = \pi_{n'} = \langle\langle -1, \cdots, -1\rangle\rangle \in P_{m+1}F$.

Since φ and ψ are not Pfister neighbors, it follows from Lemma 5.3 that ρ and τ are anisotropic and that $\varphi_{F(\rho)} \subset (\pi_n)_{F(\rho)}$, and that $(\pi_n)_{F(\rho)}$ is anisotropic.

Suppose that $\varphi_{F(\psi)}$ is isotropic. Then $\varphi_{F(\psi)(\rho)}$ is isotropic. Now $F(\psi)(\rho)$ and $F(\rho)(\psi)$ are F-isomorphic. Hence, $\varphi_{F(\rho)(\psi)}$ is isotropic and thus $(\pi_n)_{F(\rho)(\psi)}$ is hyperbolic. By CPST and since both ψ and π_n represent 1, we have $\psi_{F(\rho)} \subset (\pi_n)_{F(\rho)}$ and hence, $\psi_{F(\rho)}$ is a Pfister neighbor which, by Lemma 5.3, implies that $\tau_{F(\rho)}$ is isotropic.

Note that $2 \leq \dim\tau < \dim\rho \leq 2^m$ with $m \geq 2$. Let $k \geq 1$ such that $2^k + 1 \leq \dim\rho \leq 2^{k+1} \leq 2^m$. Recall that $\tau = \sigma_{n(1)} \perp \langle -y\rangle$ and $\rho = \sigma_{n'(1)} \perp \langle -x\rangle$ are anisotropic with $\dim\tau < \dim\rho$. We will show that $\tau_{F(\rho)}$ is anisotropic, in contradiction to what we have shown above under the assumption that $\varphi_{F(\psi)}$ is isotropic.

If $\dim\tau \leq 2^k$ then we are in case (i) which by the above implies that $\tau_{F(\rho)}$ is anisotropic. So suppose that $\dim\tau \geq 2^k + 1$. Then $2^k + 2 \leq \dim\rho \leq 2^{k+1}$, and if we can show that ρ is not a Pfister neighbor then we are in case (ii) and the anisotropy of $\tau_{F(\rho)}$ follows by induction because $k < m$.

So suppose $\rho = \sigma_{n'(1)} \perp \langle -x \rangle$ is a Pfister neighbor. By Lemma 5.3, this implies that $\sigma_{n'(2)} \perp \langle x \rangle$ is isotropic. But $\sigma_{n'(2)} \perp \langle x \rangle \subset \sigma_{n'} \perp \langle x \rangle = \varphi$, contradicting the anisotropy of φ. Hence ρ is not a Pfister neighbor and the proof is complete. $\square$

THEOREM 5.5. *Let $n \in \mathbb{N} \cup \{\infty\}$. Let E be a formally real field. Then there exists a formally real field extension F of E such that $p(F) = n$.*

PROOF. Let $F_0 = E(x_1, x_2, \cdots)$ be the rational function field in an infinite number of variables x_i over F_0. Let $a_n = 1 + x_1^2 + \cdots + x_n^2$. Then a_n is a sum of $n+1$ squares, but it cannot be written as a sum of n squares in F_0 (in $E(x_1, \cdots, x_n)$, this follows by induction from the "Second Representation Theorem" [**L1**, Ch. IX, Th. 2.1] which we already mentioned above, and for F_0 this follows from the fact that $F_0/E(x_1, \cdots, x_n)$ is purely transcendental). In particular, this shows that $p(F_0) = \infty$.

Now let $n \in \mathbb{N}$ and let $\varphi = (n-1) \times \langle 1 \rangle \perp \langle -a_{n-1} \rangle$, which is anisotropic by the above. For a field K, let us define

$$\mathcal{P}(K) = \{n \times \langle 1 \rangle \perp \langle -b \rangle;\ b \in \sum K^2\} .$$

We have $\mathcal{P}(K) \subset \mathcal{C}_p(K)$. We construct a tower of fields $F_0 \subset F_1 \subset F_2 \subset \cdots$ by putting F_{i+1} to be the free compositum of all function fields $F_i(\psi)$ over F_i with $\psi \in \mathcal{P}(F_i)$. Then we put $F = \bigcup_{i=0}^{\infty} F_i$. We claim that F is formally real and $p(F) = n$.

Suppose we have shown that F_i is formally real. Then all forms in $\mathcal{P}(F_i)$ are by construction totally indefinite. By [**ELW**, Th. 3.5], this implies that all orderings on F_i will extend to orderings on F_{i+1}, hence F_{i+1} will be formally real. This shows that all orderings on F_0 extend to orderings on F and F is therefore formally real.

By construction, all forms of type $n \times \langle 1 \rangle \perp \langle -b \rangle$, $b \in \sum F^2$, are isotropic. Hence, $p(F) \leq n$. To show equality, it suffices to verify that φ_F is anisotropic. This follows if we can show that if K is a formally real extension of F with φ_K anisotropic, and if $\psi \in \mathcal{P}(K)$, then $\varphi_{K(\psi)}$ is anisotropic. But this follows from the previous proposition if we can show that if $m \in \mathbb{N}$ is such that $2^m+1 < n+1 = \dim \psi \leq 2^{m+1}$, then ψ is not a Pfister neighbor (note that this case can only occur if $n+1 \geq 4$ and thus $m \geq 1$). With the notations as above and by Lemma 5.3, $\psi = \sigma_n \perp \langle -b \rangle$, $b \in \sum K^2$, $2^m + 1 \leq n \leq 2^{m+1} - 1$, is a Pfister neighbor iff $\sigma_{n(1)} \perp \langle b \rangle$ is isotropic. But $\sigma_{n(1)} \perp \langle b \rangle$ is positive definite at each ordering of K as $b \in \sum K^2$, hence $\sigma_{n(1)} \perp \langle b \rangle$ is anisotropic and ψ is not a Pfister neighbor. $\square$

5.3. The u-invariant. In this section, we want to describe the main ideas behind Merkurjev's construction of fields whose u-invariants have as value any given even $n \in \mathbb{N}$, cf. [**M2**]. This result was a major breakthrough in the algebraic theory of quadratic forms and it triggered a renewed interest in the isotropy problem for quadratic forms over function fields of quadrics. Up to then, the only known values for the u-invariant were powers of 2, and Kaplansky [**Ka**] conjectured that this would always be the case. However, up to Merkurjev's results, the only values which could be ruled out were $3, 5, 7$ (note that if F is formally real, then u will necessarily be even) :

PROPOSITION 5.6. *Let F be nonformally real and $I^3 F = 0$. If $\infty > u(F) > 1$ then $u(F)$ is even. In particular, $u(F) \notin \{3, 5, 7\}$.*

PROOF. If $u(F) < 8$, then , by APH, $I^3 F = 0$, so the second part follows from the first.

Now suppose that $I^3F = 0$. Let $n \in \mathbb{N}$ and let φ be form over F of dimension $2n+1$. Let $d \in F^*$ such that $d_\pm(\varphi \perp \langle d\rangle) = 1$, i.e. $\varphi \perp \langle d\rangle \in I^2F$. If $\varphi \perp \langle d\rangle$ is anisotropic, then $u(F) \geq 2n+2$. If $\varphi \perp \langle d\rangle$ is isotropic, then $\varphi \cong \psi \perp \langle -d\rangle$, and by comparing signed discriminants, we have $\psi \in I^2F$. After scaling, we may assume that ψ represents 1. Then $\varphi \cong \psi \perp \langle -d\rangle \subset \langle 1,-d\rangle \otimes \psi \in I^3F = 0$, hence $\langle 1,-d\rangle \otimes \psi$ is hyperbolic and φ is therefore isotropic as $\varphi \subset \langle 1,-d\rangle \otimes \psi$ and $\dim\varphi > \frac{1}{2}\dim(\langle 1,-d\rangle \otimes \psi)$ (see Lemma 4.5). All this shows that either φ is isotropic or $u(F) \geq 2n+2$. Hence, $u(F)$ is even. □

We have already encountered examples of fields with $u(F) = 2^n$ in Example 3.1. Another possibility to construct such fields is as follows. Let F be a field with $u(F) = m$. Then $u(F((T))) = 2m$ by Lemma 1.5. Hence, starting with any field with $u = 1$, we get all 2-powers by taking iterated power series extensions.

Let us now turn to Merkurjev's construction. At the core lie the index reduction formulas which tell us when a central division algebra D over F will have zero divisors over $F(\psi)$, the function field of a quadric ψ over F. Merkurjev's original proofs use Swan's computation of the K-theory of quadrics. A more elementary proof has been given by Tignol [**T2**]. In Tignol's formulation, the result on index reduction reads as follows.

THEOREM 5.7. *Let D be a central division algebra over F and let ψ be a form over F of dimension ≥ 2. Then $D_{F(\psi)} = D \otimes_F F(\psi)$ is not a division algebra if and only if D contains a homomorphic image of $C_0(\psi)$, the even part of the Clifford algebra of ψ.*

COROLLARY 5.8. *Let $n \geq 1$ and let $D = Q_1 \otimes \cdots \otimes Q_n$ be a central division algebra over F, where the Q_i's are quaternion algebras over F.*

(i) *Let ψ be a form over F of dimension $2n+3$. Then $D_{F(\psi)}$ is a division algebra.*
(ii) *Let ψ be a form over F in I^3F. Then $D_{F(\psi)}$ is a division algebra.*

PROOF. (i) We have $\dim_F D = 4^n$ and $\dim C_0(\psi) = \frac{1}{2}\dim C(\psi) = 2^{2n+2} = 4^{n+1}$. Since $\dim\psi$ is odd, $C_0(\psi)$ is a central simple F-algebra, and any homomorphic image of $C_0(\psi)$ is therefore isomorphic to $C_0(\psi)$. Thus, for dimension reasons, D cannot contain a homomorphic image of $C_0(\psi)$. By the above theorem, $D_{F(\psi)}$ is a division algebra.

(ii) If ψ is isotropic, then $F(\psi)/F$ is purely transcendental, and since division algebras stay division over purely transcendental extensions, $D_{F(\psi)}$ is a division algebra.

So we may assume that ψ is anisotropic and thus, by APH, $\dim\psi \geq 8$. Now $\psi \in I^3F$ implies that $[C(\psi)] = 0 \in \operatorname{Br}_2 F$. It is known that since $\psi \in I^2F$, there exists a central simple F-algebra A such that $C_0(\psi) \simeq A \times A$ and $C(\psi) \simeq A \otimes_F M_2(F)$ (see, e.g., [**L1**, Ch. V, Th. 2,5]). Since $\dim_F C(\psi) \geq 2^8$, and since $C(\psi)$ is isomorphic to a matrix algebra over F, it follows that A is a matrix algebra over F of rank ≥ 8. In particular, A has zero divisors. On the other hand, in our situation, D contains a homomorphic image of $C_0(\psi)$ iff D contains a subalgebra isomorphic to A. But this is impossible if D is division as A has zero divisors. □

For the construction of the fields themselves, we shall need the following lemma which can be considered as some sort of converse of Lemma 2.2.

LEMMA 5.9. *Let $n \in \mathbb{N}$. Let $Q_i = (a_i, b_i)_F$, $1 \le i \le n$, be quaternion algebras over F, and let $A = \bigotimes_{i=1}^n Q_i$. Then there exist $r_i \in F^*$, $1 \le i \le n$, and a form φ over F such that $\dim \varphi = 2n+2$ and $\varphi \sim \sum_{i=1}^n r_i \langle\langle a_i, b_i \rangle\rangle$ in WF. In particular, $c(\varphi) = [A] \in \mathrm{Br}_2 F$. Furthermore, if A is a division algebra, then φ is anisotropic.*

PROOF. The proof is by induction on n. If $n = 1$, we put $\varphi = \langle\langle a_i, b_i \rangle\rangle$. Suppose $n \ge 2$ and we have constructed $r_i \in F^*$, $1 \le i \le n-1$, and $\psi \in I^2F$, $\dim \psi = 2n$, such that $\psi \sim \sum_{i=1}^{n-1} r_i \langle\langle a_i, b_i \rangle\rangle$ in WF. Write $\psi \cong \langle r \rangle \perp \psi'$ and consider $\varphi = \psi' \perp -r\langle -a_n, -b_n, a_n b_n \rangle$. Then, in WF, $\varphi \sim \psi - r\langle\langle a_n, b_n \rangle\rangle$, in particular, $\varphi \in I^2F$. Also, $\dim \varphi = 2n+2$, and with $r_n = -r$ we have the desired form.

$c(\varphi) = [A] \in \mathrm{Br}_2 F$ follows from the fact that c yields an isomorphism from I^2F/I^3F to $\mathrm{Br}_2 F$ mapping $r\langle\langle a, b \rangle\rangle \bmod I^3F$ to $[(a,b)_F] \in \mathrm{Br}_2 F$.

Suppose that φ is isotropic, say, $\varphi \cong \varphi_0 \perp \mathbb{H}$. Then $\varphi_0 \in I^2F$, $c(\varphi) = c(\varphi_0)$, and $\dim \varphi_0 = 2n$. By Lemma 2.2, there exist quaternion algebras $B_1, \cdots, B_{n-1}$ over F such that for $B = B_1 \otimes \cdots \otimes B_{n-1}$ we have $[B] = c(\varphi_0) = c(\varphi) = [A]$. Now $\dim_F B = 4^{n-1} < \dim_F A = 4^n$, hence $A \cong M_2(B)$ is not division. □

In the situation of the above lemma, we call φ an Albert form associated with A.

THEOREM 5.10. (Merkurjev) *Let $m \in \mathbb{N}$ be even. Let E be any field. Then there exists a (nonformally real) field F over E with $u(F) = m$ and $I^3F = 0$.*

PROOF. Let E_{alg} be an algebraic closure of E. Then we have $u(E_{alg}) = 1$ and $u(E_{alg}(T)) = 2$ (cf. section 3.1).

Now let $m = 2n+2$ with $n \ge 1$, and let $F_0 = E(x_1, y_1, \cdots, x_n, y_n)$ be the rational function field in $2n$ variables over E. Let $Q_i = (x_i, y_i)_{F_0}$ and $A = \bigoplus_{i=1}^n Q_i$. Then A is a division algebra (see, e.g., [**T1**, 2.12]). Let φ be an Albert form associated with A as in Lemma 5.9, which is anisotropic as A is division.

We now construct a tower of fields $F_0 \subset F_1 \subset \cdots$ as follows. Let K be a field extension of F_0 such that A_K is division, and define

$$\begin{aligned} \mathcal{U}_1(K) &= \{\psi \text{ form over } K,\ \dim \psi = 2n+3\}, \\ \mathcal{U}_2(K) &= \{\psi \text{ form over } K,\ \psi \in I^3K\}. \end{aligned}$$

Then, by Corollary 5.8, $A_{K(\psi)}$ will be division if $\psi \in \mathcal{U}_1(K) \cup \mathcal{U}_2(K)$. Having constructed F_i, let F_{i+1} be the free compositum over F_i of all function fields $F_i(\psi)$, $\psi \in \mathcal{U}_1(F_1) \cup \mathcal{U}_2(F_2)$. Put $F = \bigcup_{i=0}^\infty F_i$.

By construction, all forms in I^3F are isotropic, hence $I^3F = 0$. Also, all forms of dimension $2n+3$ are isotropic over F, hence $u(F) \le 2n+2$.

By Corollary 5.8, A_F is a division algebra. Hence, φ_F is anisotropic by Lemma 5.9 and we have $u(F) \ge 2n+2$. Thus, $u(F) = 2n+2$. □

In view of Proposition 5.6, we see that for $m \in \mathbb{N}$, a necessary and sufficient condition for a field F with $I^3F = 0$ and $u(F) = m$ to exist is that m be even.

6. Some final examples and remarks

6.1. Values of $\tilde{u}$ and u and relations between invariants. If F is nonformally real, then $\tilde{u}(F) = u(F)$. This equality no longer holds in general when F in formally real. We have the following result which gives the analogue of Proposition 5.6 and Theorem 5.10 for formally real fields. Recall that for formally real F, $u(F)$ is either even or infinite, and that $u(F) \le \tilde{u}(F)$.

THEOREM 6.1. *Let F be a formally real field.*

(i) *If I^3F is torsion free and $\tilde{u}(F) < \infty$, then $\tilde{u}(F)$ is even. In particular, $\tilde{u}(F) \notin \{1, 3, 5, 7\}$.*
(ii) *Suppose that I^3F is torsion free and $u(F) < \infty$. If $u(F) \leq 2$ then $\tilde{u}(F) \in \{u(F), \infty\}$. If $u(F) = 2n$ for an integer $n \geq 2$, then $\tilde{u}(F) \in \{u(F), u(F) + 2, \infty\}$.*
(iii) *There exist formally real fields F_{2n}, F'_{2n}, $n \in \mathbb{N} \cup \{0\}$, and F''_{2n} for $n \geq 2$, such that for all these fields I^3 is torsion free and such that for $n \in \mathbb{N} \cup \{0\}$ we have $u(F_{2n}) = u(F'_{2n}) = \tilde{u}(F_{2n}) = 2n$ and $\tilde{u}(F'_{2n}) = \infty$, and for $n \geq 2$ we have $u(F''_{2n}) = 2n$ and $\tilde{u}(F''_{2n}) = 2n + 2$.*

We will not prove this theorem but only give some relevant references. Part (i) has been proved in [**ELP**, Th. H]. The fact that $u(F) \leq 2$ implies $\tilde{u}(F) \leq 2$ (provided $\tilde{u}(F) < \infty$) was shown in [**ELP**, Ths. E,F]. Since binary totally indefinite forms are torsion, this implies that if $u(F) \in \{0, 2\}$ and if $\tilde{u}(F) < \infty$, then $u(F) = \tilde{u}(F)$. If $u(F) = 2n$ for some integer $n \geq 2$, and if $\tilde{u}(F) < \infty$, then it will be shown in a forthcoming paper [**H7**] that I^3F being torsion free implies $\tilde{u}(F) \in \{2n, 2n+2\}$.

As for the existence, examples of fields F_0, F'_0 with the desired properties are given by $F_0 = \mathbb{R}$ and $F'_0 = \mathbb{R}((X))((Y))$. By applying Lemma 1.5 twice, we see that $u(F'_0) = 4u(\mathbb{R}) = 0$. Consider the form $\varphi_m = m \times \langle -1, X, Y, XY \rangle$. Since $m \times \langle 1 \rangle$ is anisotropic over $\mathbb{R}$ for all $m \in \mathbb{N}$, we have again by Lemma 1.5 that φ_m is anisotropic. Also $\det \langle -1, X, Y, XY \rangle = -1$, which shows that the 4-dimensional form $\langle -1, X, Y, XY \rangle$ is totally indefinite with respect to each ordering on $\mathbb{R}((X))((Y))$. Hence, φ_m is totally indefinite. All this together yields that $\tilde{u}(\mathbb{R}((X))((Y))) = \infty$.

Examples of fields F_{2n} and F''_{2n} for $n \geq 2$ have been constructed by Hornix [**Hor5**] and by Lam [**L3**]. The methods of construction are a variation of Merkurjev's method of constructing fields with even u-invariant. (Lam actually doesn't construct the above fields having the additional property of I^3 being torsion free, but this can readily be achieved by a slight generalization of his construction. He shows for his F''_{2n} only that $u(F''_{2n}) \leq 2n$, but it can be shown that also for his example one gets $u(F''_{2n}) = 2n$.) Finally, examples of the type F'_{2n} can be found in [**H7**].

Using this result, we can conclude that if $n \in \mathbb{N} \cup \{0\}$, then there exists a formally real field F with $u(F) = n$ if and only if n is even. The question remains which values can occur for u in the nonformally real case (resp. for $\tilde{u}$ in the formally real case). Now 1 and all even $n \in \mathbb{N}$ can be realized as $u(F)$ for a suitable nonformally real F. We also know that $3, 5, 7$ are not possible. The first open case was 9, until recently when Izhboldin announced the construction of a (necessarily nonformally real) field F with $u(F) = 9$, [**I2**]. The method of construction is again of Merkurjev type, however, the techniques and auxiliary results needed to show that a certain 9-dimensional form stays anisotropic after taking successively function fields of 10-dimensional forms are highly sophisticated and go far beyond the scope of this article (see also the remarks preceding Theorem 4.4). So one might conjecture that there exists a nonformally real F with $u(F) = n$ if and only if $n \notin \{3, 5, 7\}$. But it should be noted that it seems that Izhboldin's methods cannot easily be generalized to yield odd values ≥ 11.

It should also be noted that Izhboldin's results can readily be used to construct formally real F with $\tilde{u}(F) = 9$. We get a conjecture analogous to the u-invariant case. Let us summarize these conjectures.

CONJECTURE 6.2. (i) Let $n \in \mathbb{N}$. Then there exists a nonformally real field F with $u(F) = n$ if and only if $n \notin \{3, 5, 7\}$.
(ii) Let $n \in \mathbb{N} \cup \{0\}$. Then there exists a formally real field F with $\tilde{u}(F) = n$ if and only if $n \notin \{1, 3, 5, 7\}$.

A natural question to ask is how much $\tilde{u}(F)$ (if finite) can actually differ from $u(F)$ for a formally real F. The above examples show that $\tilde{u}(F) - u(F) = 2$ is possible. Using a result announced by Izhboldin [**I3**], one can construct a formally real field F with $u(F) = 8$ and $\tilde{u}(F) = 12$.

THEOREM 6.3. (Izhboldin, announced.) *Let φ and ψ be forms over F such that φ is anisotropic, $\dim \varphi = 12$, $\varphi \in I^3F$ and $\dim \psi \geq 9$. Then $\varphi_{F(\psi)}$ is isotropic if and only if ψ is similar to a subform of φ.*

COROLLARY 6.4. *There exists a formally real field F with $u(F) = 8$ and $\tilde{u}(F) = 12$.*

PROOF. Let $F_0 = \mathbb{Q}(T)$. Then $\langle 1, 1\rangle \otimes \langle 1, 1, 1, 7\rangle$ and $\langle 1, 1, -7, -7\rangle$ are anisotropic over $\mathbb{Q}$, hence, if we put $\alpha = \langle 1, 1, 1, 7\rangle \perp T\langle 1, -7\rangle$, then $\varphi = \langle 1, 1\rangle \otimes \alpha$ is anisotropic over F_0 (cf. Lemma 1.5) and in I^3F_0 as $\alpha \in I^2F_0$ and $\langle 1, 1\rangle \in IF_0$. Now $\langle 1, 1\rangle \otimes \langle 1, 1, 1, 7\rangle$ is positive definite and $\langle 1, 1, -7, -7\rangle$ is torsion. Hence, $\mathrm{sgn}_P\, \varphi = 8$ for all orderings P on F_0.

Suppose that K is a field extension of F_0 such that all orderings on F_0 extend to orderings on K, and such that φ_K is anisotropic. Define

$$\begin{aligned} \mathcal{U}_1(K) &= \{\psi \text{ torsion form over } K, \dim \psi \in \{10, 12\}\}, \\ \mathcal{U}_2(K) &= \{\psi \text{ totally indefinite form over } K, \dim \psi \geq 13\}. \end{aligned}$$

Let $\psi \in \mathcal{U}_1(K) \cup \mathcal{U}_2(K)$. Since ψ is indefinite at each ordering on K, it follows that each ordering on K extends to an ordering on $K(\psi)$. Also, ψ is not similar to a subform of φ. This is obvious for dimension reasons if $\psi \in \mathcal{U}_2(K)$, and if $\psi \in \mathcal{U}_1(K)$, then by a simple signature argument it is clear that a 12-dimensional form with signature 8 at each ordering cannot contain a 10- or 12-dimensional subform with total signature 0. Hence, by Izhboldin's result, $\varphi_{K(\psi)}$ will be anisotropic.

Now we construct a tower of fields $F_0 \subset F_1 \subset \cdots$ with F_{i+1} being the free compositum over F_i of all function fields $F_i(\psi)$ with $\psi \in \mathcal{U}_1(F_i) \cup \mathcal{U}_2(F_i)$, and we put $F = \bigcup_{i=0}^{\infty} F_i$.

By construction, all orderings on F_i, $i \geq 0$, extend to orderings on F. All totally indefinite forms over F of dimension ≥ 13 are isotropic, hence $\tilde{u}(F) \leq 12$. φ_F will be anisotropic. Note that $\mathrm{sgn}_P\, \varphi_K = 8 < \dim \varphi = 12$ for all orderings on F, hence φ_F is totally indefinite and thus $\tilde{u}(F) = 12$. Clearly, $u(F) \leq \tilde{u}(F) = 12$. But all torsion forms of dimension 10 or 12 are isotropic, hence $u(F) \leq 8$. Now any 8-dimensional anisotropic form over F_0 will stay anisotropic over F by Theorem 4.4, in particular the anisotropic torsion form $\langle 1, 1, -7, -7\rangle \perp T\langle 1, 1, -7, -7\rangle$. Hence, $u(F) = 8$ □

An obvious question is when $u(F) < \infty$ implies $\tilde{u}(F) < \infty$. This can be answered in terms of certain properties of the space of orderings of F (equipped with a suitable topology), namely the properties SAP (for strong approximation property) and S_1. There are quadratic form characterizations of these properties. A field is SAP if and only if for all $x, y \in F^*$ there exists $n \in \mathbb{N}$ such that $n \times \langle -1, x, y, xy \rangle$ is isotropic (we then say that $\langle -1, x, y, xy \rangle$ is weakly isotropic), cf. [**Pr1**, Satz 3.1], [**ELP**, Th. C]. A field has property S_1 if and only if for each binary torsion form β over F one has $D_F(\beta) \cap \sum F^2 \neq \emptyset$, cf. [**EP**]. The properties SAP and S_1 together are equivalent to the property ED (effective diagonalization), which states that each form φ over F has a diagonalization $\langle a_1, \cdots, a_n \rangle$ such that for each ordering P on F, $a_i >_P 0$ implies $a_{i+1} >_P 0$ for $1 \leq i < n$. The equivalence has been shown in [**PrW**, Th. 2].

The following was proved by Elman-Prestel [**EP**, Th. 2.5].

THEOREM 6.5. *Let F be formally real. Then $\tilde{u}(F) < \infty$ if and only if F satisfies $u(F) < \infty$, SAP and S_1.*

In this situation, they also derive bounds on $\tilde{u}$ in terms of u and the Pythagoras number. These bounds are improved in [**H7**], where the following is shown.

THEOREM 6.6. *Let F be formally real and ED. Then*

$$\tilde{u}(F) \leq \frac{p(F)}{2}(u(F) + 2)\ .$$

It should be noted that $p(F) \leq u(F)$. In fact, suppose that $p(F) \geq 2^m + 1$. Then there exists $x \in \sum F^2$ such that $2^m \times \langle 1 \rangle \perp \langle -x \rangle$ is anisotropic. But this is a Pfister neighbor of $\langle\langle -1, \cdots, -1, x \rangle\rangle \in P_{m+1}F$, which is therefore also anisotropic, and which is torsion as it is totally indefinite. Hence, $u(F) \geq 2^{m+1}$ and it follows immediately that $p(F) \leq u(F)$.

For some types of fields, equality of u and $\tilde{u}$ could be established. One class of such fields is the class of so-called linked fields. A field is called linked if the classes of quaternion algebras form a subgroup of the Brauer group (by Merkurjev's theorem, this just means that the classes of quaternion algebras constitute the whole $\mathrm{Br}_2 F$). Finite, local and global fields belong into this class, as well as fields of transcendence degree ≤ 1 (resp. ≤ 2) over $\mathbb{R}$ (resp. $\mathbb{C}$), or the field $\mathbb{C}((X))((Y))((Z))$, the latter being a field of u-invariant 8. It is not difficult to show that fields with $\tilde{u}(F) \leq 4$ are always linked. The following has been shown by Elman-Lam [**EL2**] and Elman [**E**].

THEOREM 6.7. *Let F be a linked field. Then $u(F) = \tilde{u}(F) \in \{0, 1, 2, 4, 8\}$. Each of these values can be realized.*

The examples of linked fields we mentioned above show that all these values can indeed be realized. Note that the value 0 (resp. 1) necessarily means that F is formally real (resp. not formally real).

6.2. Some remarks on the l-invariant. Recall that for nonformally real F, we have $l(F) = u(F)$, so the interesting case is the formally real one. The following is known.

THEOREM 6.8. *Let F be formally real.*

(i) *Suppose that I^3F is torsion free and $1 < l(F) < \infty$. Then $l(F) \not\equiv 1 \bmod 4$.*
(ii) *To each integer $n \geq 1$ with $n \not\equiv 1 \bmod 4$ if $n \geq 2$, there exists a formally real F such that $l(F) = n$ and I^3F is torsion free.*

The first part has been proved in [**BLOP**, Lemma 4.11]. The second part was shown by Hornix for $n \geq 4$, cf. [**Hor5**, Remark 3.8]. Again, the construction is a variation of Merkurjev's method. Now $l(\mathbb{R}) = 1$ and $l(\mathbb{R}(T)) = 2$ as is rather obvious, and in both cases I^3 is clearly torsion free. Let us show how to get the case $l(F) = 3$.

PROPOSITION 6.9. *There exists a formally real F with $l(F) = 3$ and I^3F torsion free.*

PROOF. Let $F_0 = \mathbb{Q}$ and consider $\varphi = \langle 1, 1, -7, -7 \rangle$ which is torsion and anisotropic. If K is any extension of F_0 such that the unique ordering of F_0 extends to K and such that φ_K is anisotropic, let $\psi \cong \langle a_1, a_2, a_3, -a_4 \rangle$ with $a_i \in \sum K^2$. Note that ψ has signature 2 at each ordering and that therefore ψ is not similar to a subform of (and hence similar to) φ_K. Since $\varphi_{K(\psi)}$ is a Pfister form and hence anisotropic or hyperbolic, it follows from CPST that $\varphi_{K(\psi)}$ is anisotropic. This is still the case if ψ is a torsion form in I^3K. For if ψ is isotropic then $K(\psi)/K$ is purely transcendental, and if ψ is anisotropic then $\dim \psi \geq 8$ by APH. Note that in any case, ψ will be totally indefinite, hence all orderings on K will extend to $K(\psi)$.

Having constructed F_i such that the ordering on F_0 extends to F_i and such that φ_{F_i} is anisotropic, let F_{i+1} be the free compositum over F_i of all function fields $F_i(\psi)$ where $\psi \cong \langle a_1, a_2, a_3, -a_4 \rangle$ with $a_i \in \sum F_i^2$, or ψ is a torsion form in I^3F_i. We put $F = \bigcup_{i=0}^{\infty} F_i$. By construction, the ordering on F_0 extends to F. Furthermore, forms $\langle a_1, a_2, a_3, -a_4 \rangle$ with $a_i \in \sum F^2$ are isotropic. By the definition of l this yields $l(F) \leq 3$. On the other hand, φ_F is anisotropic, hence $\langle 1, 1, -7 \rangle$ is anisotropic over F, which shows that $l(F) = 3$. Also, there are no anisotropic torsion forms in I^3F, hence I^3F is torsion free. □

6.3. Related results. Merkurjev's method has been used and modified in many ways to yield examples of fields with prescribed invariants. For instance, in [**EL1**], a certain filtration $u^{(0)} \leq u^{(1)} \leq \cdots \leq u$ of the u-invariant has been defined and the question was asked whether this sequence is actually always constant (cf. also [**P4**, Ch. 8, §2]). A variation of Merkurjev's method was used in [**H5**] to construct various types of nonconstant such sequences.

There are many other field invariants pertaining to dimensions of anisotropic quadratic forms with additional properties. We have only mentioned a few of them. Several others have been defined and studied, most notably in the work of Hornix [**Hor1**]–[**Hor5**], but some of these invariants are of a rather technical nature, so we will not give the definitions here. Merkurjev's method has been put to good use in [**Hor5**] in the construction of fields where these invariants attain prescribed values.

Another important problem is the behaviour of field invariants under algebraic extensions of the ground field, in particular under quadratic extensions. In other words, if L/K is a field extension, then how do $u(L)$ and $u(K)$ relate to each other ? For example, if K is not formally real, then Leep [**Le1**] proves that $u(L) \leq \frac{1}{2}[L:K]u(K)$. Questions of this type have been studied for instance in [**E**], [**EL1**], [**EL4**], [**EP**], [**Hor3**], [**Le1**], [**Le2**]. Also in this context, Merkurjev's method could be modified suitably to show that in Leep's estimate above, equality can occur in the cases $[L:K] \in \{2, 3\}$. This was done by Leep-Merkurjev [**LeM**], with further generalizations by Mináč [**Min**], Mináč-Wadsworth [**MinW**].

References

[A] ARASON, J.KR.: *Cohomologische Invarianten quadratischer Formen.* J. Algebra **36** (1975) 448–491.

[AP] ARASON, J.KR.; PFISTER, A.: *Beweis des Krullschen Durschnittsatzes für den Wittring.* Invent. Math. **12** (1971) 173–176.

[BLOP] BAEZA, R.; LEEP, D.; O'RYAN, M.; PRIETO, J.P.: *Sums of squares of linear forms.* Math. Z. **193** (1986) 297–306.

[Ca] CASSELS, J.W.S.: *On the representations of rational functions as sums of squares.* Acta Arith. **9** (1964) 79–82.

[CDLR] CHOI, M.D.; DAI, Z.D.; LAM, T.Y.; REZNICK, B.: *The Pythagoras number of some affine algebras and local algebras.* J. Reine Angew. Math. **336** (1982) 45–82.

[E] ELMAN, R.: *Quadratic forms and the u-invariant*, III. Proc. of Quadratic Forms Conference (ed. G. Orzech). Queen's Papers in Pure and Applied Mathematics No. **46** (1977) 422–444.

[EL1] ELMAN, R.; LAM, T.Y.: *Quadratic forms and the u-invariant* I. Math. Z. **131** (1973) 283–304.

[EL2] ELMAN, R.; LAM, T.Y.: *Quadratic forms and the u-invariant* II. Invent. Math. **21** (1973) 125–137.

[EL3] ELMAN, R.; LAM, T.Y.: *Classification theorems for quadratic forms over fields.* Comm. Math. Helv. **49** (1974) 373–381.

[EL4] ELMAN, R.; LAM, T.Y.: *Quadratic forms under algebraic extensions.* Math. Ann. **219** (1976) 21–42.

[ELP] ELMAN, R.; LAM, T.Y.; PRESTEL, A.: *On some Hasse principles over formally real fields.* Math. Z. **134** (1973) 291–301.

[ELW] ELMAN, R.; LAM, T.Y.; WADSWORTH, A.R.: *Orderings under field extensions.* J. Reine Angew. Math. **306** (1979) 7–27.

[EP] ELMAN, R.; PRESTEL, A.: *Reduced stability of the Witt ring of a field and its Pythagorean closure.* Amer. J. Math. **106** (1983), 1237–1260.

[F] FITZGERALD, R.W.: *Witt kernels of function field extensions.* Pacific J. Math. **109** (1983) 89–106.

[H1] HOFFMANN, D.W.: *Isotropy of* 5*-dimensional quadratic forms over the function field of a quadric.* K-Theory and Algebraic Geometry: Connections with Quadratic Forms and Division Algebras, Proceedings of the 1992 Santa Barbara Summer Research Institute (eds. B. Jacob, A. Rosenberg). Proc. Symp. Pure Math. **58.2** (1995) 217–225.

[H2] HOFFMANN, D.W.: *On 6-dimensional quadratic forms isotropic over the function field of a quadric.* Comm. Alg. **22** (1994) 1999–2014.

[H3] HOFFMANN, D.W.: *Isotropy of quadratic forms over the function field of a quadric.* Math. Z. **220** (1995) 461–476.

[H4] HOFFMANN, D.W.: *On the dimensions of anisotropic quadratic forms in* I^4.

[H5] HOFFMANN, D.W.: *On Elman and Lam's filtration of the u-invariant.* J. Reine Angew. Math. **495** (1998) 175–186.

[H6] HOFFMANN, D.W.: *Pythagoras numbers of fields.* J. Amer. Math. Soc. **12** (1999), 839–848.

[H7] HOFFMANN, D.W.: *Quadratic forms and field invariants.* Preprint (2000).

[HLVG] HOFFMANN, D.W.; LEWIS, D.W.; VAN GEEL, J.: *Minimal forms for function fields of conics.* K-Theory and Algebraic Geometry: Connections with Quadratic Forms and Division Algebras, Proceedings of the 1992 Santa Barbara Summer Research Institute (eds. B. Jacob, A. Rosenberg). Proc. Symp. Pure Math. **58.2** (1995) 227–237.

[HVG1] HOFFMANN, D.W.; VAN GEEL, J.: *Minimal forms with respect to function fields of conics.* Manuscripta Math. **86** (1995) 23–48.

[HVG2] HOFFMANN, D.W.; VAN GEEL, J.: *Zeros and norm groups of quadratic forms over function fields in one variable over a local non-dyadic field.* J. Ramanujan Math. Soc. **13** (1998) 85–110.

[Hor1] HORNIX, E.A.M.: *Totally indefinite quadratic forms over formally real fields.* Proc. Kon. Akad. van Wetensch., Series A, **88** (1985) 305–312.

[Hor2] HORNIX, E.A.M.: *Invariants of formally real fields and quadratic forms.* Proc. Kon. Akad. van Wetensch., Series A, **89** (1986) 47–60.

[Hor3] HORNIX, E.A.M.: *Field invariants under totally positive quadratic extensions.* Proc. Kon. Akad. van Wetensch., Series A, **91** (1988) 277-291.

[Hor4] HORNIX, E.A.M.: *Fed-linked fields and formally real fields satisfying B_3.* Indag. Math. **1** (1990) 303–321.

[Hor5] HORNIX, E.A.M.: *Formally real fields with prescribed invariants in the theory of quadratic forms.* Indag. Math. **2** (1991) 65–78.

[HuR] HURRELBRINK, J.; REHMANN, U.: *Splitting patterns of quadratic forms.* Math. Nachr. **176** (1995) 111–127.

[I1] IZHBOLDIN, O.T.: *On the isotropy of quadratic forms over the function field of a quadric*, Algebra i Analiz. **10** (1998) 32–57. (English transl.: St. Petersburg Math. J. **10** (1999) 25–43.)

[I2] IZHBOLDIN, O.T.: *Field with u-invariant* 9. Preprint (1999).

[I3] IZHBOLDIN, O.T.: *Some new results concerning isotropy of low-dimensional forms.* Preprint (1999).

[IK1] IZHBOLDIN, O.T.; KARPENKO, N.A.: *Isotropy of six-dimensional quadratic forms over function fields of quadrics.* J. Algebra **209** (1998) 65–93.

[IK2] IZHBOLDIN, O.T.; KARPENKO, N.A.: *Isotropy of* 8*-dimensional quadratic forms over function fields of quadrics.* Comm. Alg. **27** (1999) 1823–1841.

[IK3] IZHBOLDIN, O.T.; KARPENKO, N.A.: *Isotropy of virtual Albert forms over function fields of quadrics.* Math. Nachr. **206** (1999) 111–122.

[JR] JACOB, B.; ROST, M.: *Degree four cohomological invariants for quadratic forms.* Invent. Math. **96** (1989) 551–570.

[Ka] KAPLANSKY, I.: *Quadratic forms.* J. Math. Soc. Japan **6** (1953) 200–207.

[K1] KNEBUSCH, M.: *Generic splitting of quadratic forms* I. Proc. London Math. Soc. **33** (1976) 65–93.

[K2] KNEBUSCH, M.: *Generic splitting of quadratic forms* II. Proc. London Math. Soc. **34** (1977) 1–31.

[KS] KNEBUSCH, M.; SCHEIDERER, C.: Einführung in die reelle Algebra. Braunschweig, Wiesbaden: Vieweg 1989.

[Kn] KNESER, H.: *Verschwindende Quadratsummen in Körpern.* Jber. Dt. Math. Ver. **44** (1934) 143–146.

[Lag1] LAGHRIBI, A.: *Isotropie de certaines formes quadratiques de dimension* 7 *et* 8 *sur le corps des fonctions d'une quadrique.* Duke Math. J. **85** (1996) 397–410.

[Lag2] LAGHRIBI, A.: *Formes quadratiques en 8 variables dont l'algèbre de Clifford est d'indice 8.* K-theory **12** (1997) 371–383.

[Lag3] LAGHRIBI, A.: *Formes quadratiques de dimension 6.* Math. Nachr. **204** (1999) 125–135.

[L1] LAM, T.Y.: The Algebraic Theory of Quadratic Forms. Reading, Massachusetts: Benjamin 1973 (revised printing 1980).

[L2] LAM, T.Y.: *Fields of u-invariant* 6 *after A. Merkurjev.* Israel Math. Conf. Proc. Vol. 1: Ring Theory 1989 (in honor of S.A. Amitsur) (ed. L. Rowen). pp.12–31. Jerusalem: Weizmann Science Press 1989.

[L3] LAM, T.Y.: *Some consequences of Merkurjev's work on function fields.* Preprint 1989.

[Le1] LEEP, D.: *Systems of quadratic forms.* J. Reine Angew. Math. **350** (1984) 109–116.

[Le2] LEEP, D.: *Invariants of quadratic forms under algebraic extensions.* Preprint (1988).

[Le3] LEEP, D.: *Function field results.* Handwritten notes taken by T.Y. Lam (1989).

[LeM] LEEP, D.B., MERKURJEV, A.S.: *Growth of the u-invariant under algebraic extensions.* Contemp. Math. **155** (1994) 327–332.

[LVG] LEWIS, D.W.; VAN GEEL, J.: *Quadratic forms isotropic over the function field of a conic.* Indag. Math. **5** (1994) 325–339.

[M1] MERKURJEV, A.S.: *On the norm residue symbol of degree 2.* Dokladi Akad. Nauk. SSSR **261** (1981) 542–547. (English translation: Soviet Math. Doklady **24** (1981) 546–551.)

[M2] MERKURJEV, A.S.: *Simple algebras and quadratic forms.* Izv. Akad. Nauk. SSSR **55** (1991) 218–224. (English translation: Math. USSR Izvestiya **38** (1992) 215–221.)

[MSu] MERKURJEV, A.S.; SUSLIN, A.A.: *On the norm residue homomorphism of degree three.* Izv. Akad. Nauk. SSSR **54** (1990) 339–356. (English translation: Math. USSR Izvestiya **36** (1991) 349–368.)

[Mi] MILNOR, J.: *Algebraic K-theory and quadratic forms.* Invent. Math. **9** (1970) 318–344.

[Min] MINÁČ, J.: *Remarks on Merkurjev's investigations on the u-invariant.* Contemp. Math. **155** (1994) 333–338.

[MinW] MINÁČ, J.; WADSWORTH, A.R.: *The u-invariant for algebraic extensions.* K-Theory and Algebraic Geometry: Connections with Quadratic Forms and Division Algebras, Proceedings of the 1992 Santa Barbara Summer Research Institute (eds. B. Jacob, A. Rosenberg). Proc. Symp. Pure Math. **58.2** (1995) 333–358.

[Mo] MORDELL, J.L.: *On the representation of a binary quadratic form as a sum of squares of linear forms.* Math. Z. **35** (1932) 1–15.

[OVV] ORLOV, D.; VISHIK, A.; VOEVODSKY, V.: *Motivic cohomology of Pfister quadrics and Milnor's conjecture one quadratic forms.* Preprint.

[PS] PARIMALA, R.; SURESH, V.: *Isotropy of quadratic forms over function fields of curves over p-adic fields.* Publ. Math., Inst. Hautes Etud. Sci. **88** (1998) 129–150.

[P1] PFISTER, A.: *Darstellung von -1 als Summe von Quadraten in einem Körper.* J. London Math. Soc. **40** (1965) 159–165.

[P2] PFISTER, A.: *Multiplikative quadratische Formen.* Arch. Math. **16** (1965) 363–370.

[P3] PFISTER, A.: *Quadratische Formen in beliebigen Körpern.* Invent. Math. **1** (1966) 116–132.

[P4] PFISTER, A.: Quadratic forms with applications to algebraic geometry and topology. London Math. Soc. Lect. Notes **217**, Cambridge University Press 1995.

[P5] PFISTER, A.: *On the Milnor Conjectures: History, Influence, Applications.* Jber. d. Dt. Math.-Verein. **102** (2000) 15–41.

[Pr1] PRESTEL, A: *Quadratische Semi-Ordnungen und quadratische Formen.* Math. Z. **133** (1973), 319–342.

[Pr2] PRESTEL, A.: *Remarks on the Pythagoras and Hasse number of real fields*, J. Reine Angew. Math. **303/304** (1978) 284–294.

[PrW] PRESTEL, A; WARE, R.: *Almost isotropic quadratic forms.* J. London Math. Soc. **19** (1979), 241–244.

[R1] ROST, M.: *Hilbert 90 for K_3 for degree-two extensions.* Preprint (1986).

[R2] ROST, M.: *Quadratic forms isotropic over the function field of a conic.* Math. Ann. **288** (1990) 511–513.

[S] SCHARLAU, W.: Quadratic and Hermitian Forms. Grundlehren **270**, Berlin, Heidelberg, New York, Tokyo : Springer 1985.

[Sz] SZYJEWSKI, M.: *The fifth invariant of quadratic forms.* Algebra Anal. **2** (1990) 213–234. (English translation: Leningrad Math. J. **2** (1991) 179–198.)

[T1] TIGNOL, J.-P.: *Algèbres indécomposables d'exposant premier.* Adv. Math. **65** (1987) 205–228.

[T2] TIGNOL, J.-P.: *Réduction de l'indice d'une algèbre simple centrale sur le corps des fonctions d'une quadrique.* Bull. Soc. Math. Belgique Sér. A **42** (1990) 735–745.

[Vi1] VISHIK, A.: *Integral motives of quadrics.* Max-Planck-Institut für Mathematik, preprint **13** (1998).

[Vi2] VISHIK, A.: *On the dimensions of anisotropic forms in I^n.* Preprint (1999).

[Vo] VOEVODSKY, V.: *The Milnor conjecture.* Preprint (1996).

[W] WADSWORTH, A.R.: *Similarity of quadratic forms and isomorphisms of their function fields.* Trans. Amer. Math. Soc. **208** (1975) 352–358.

[Wi1] WITT, E.: *Theorie der quadratischen Formen in beliebigen Körpern.* J. Reine Angew. Math. **176** (1937) 31–44 (= Collected Papers, pp. 2–15).

[Wi2] WITT, E.: *Über quadratische Formen in Körpern.* Unpublished manuscript (1967) (= Collected Papers pp. 35–39).

[Wi3] WITT, E.: Collected Papers/Gesammelte Abhandlungen (ed. I. Kersten), Berlin, Heidelberg, New York : Springer (1998)

Département de Mathématiques
UMR 6623 du CNRS
Université de Franche-Comté
16, Route de Gray
F-25030 Besançon Cedex
France

E-mail address: detlev@math.univ-fcomte.fr

Contemporary Mathematics
Volume **272**, 2000

Quadratic forms with absolutely maximal splitting

Oleg Izhboldin and Alexander Vishik

ABSTRACT. Let F be a field and ϕ be a quadratic form over F. The higher Witt indices of ϕ are defined recursively by the rule $i_{k+1}(\phi) = i_k((\phi_{an})_{F(\phi_{an})})$, where $i_0(\phi) = i_W(\phi)$ is the usual Witt index of the form ϕ. We say that anisotropic form ϕ has *absolutely maximal splitting* if $i_1(\phi) > i_k(\phi)$ for all $k > 1$.

One of the main results of this paper claims that for all anisotropic forms ϕ satisfying the condition $2^{n-1} + 2^{n-3} < \dim \phi \leq 2^n$, the following three conditions are equivalent: (i) the kernel of the natural homomorphism $H^n(F, \mathbb{Z}/2\mathbb{Z}) \to H^n(F(\phi), \mathbb{Z}/2\mathbb{Z})$ is nontrivial, (ii) ϕ has absolutely maximal splitting, (iii) ϕ has maximal splitting (i.e., $i_1(\phi) = \dim \phi - 2^{n-1}$). Moreover, we show that if we assume additionally that $\dim \phi \geq 2^n - 7$, then these three conditions hold if and only if ϕ is an anisotropic n-fold Pfister neighbor. In our proof we use the technique developed by V. Voevodsky in his proof of Milnor's conjecture.

1. Introduction

Let F be a field of characteristic $\neq 2$ and let $H^n(F)$ be the Galois cohomology group of F with $\mathbf{Z}/2\mathbf{Z}$-coefficients. For a given extension L/F, we denote by $H^n(L/F)$ the kernel of the natural homomorphism $H^n(F) \to H^n(L)$. Now, let ϕ be a quadratic form over F. An important part of the algebraic theory of quadratic forms deals with the behavior of the groups $H^n(F)$ under the field extension $F(\phi)/F$. Of particular interest is the group

$$H^n(F(\phi)/F) = \ker(H^n(F) \to H^n(F(\phi))).$$

The computation of this group is connected to Milnor's conjecture and plays an important role in K-theory and in the theory of quadratic forms.

The first nontrivial result in this direction is due to J. K. Arason. In [**1**], he computed the group $H^n(F(\phi)/F)$ for the case $n \leq 3$. The case $n = 4$ was completely studied by Kahn, Rost and Sujatha ([**13**]). In the cases where $n \geq 5$, there are only partial results depending on Milnor's conjecture: the group $H^n(F(\phi)/F)$ was computed for all Pfister neighbors ([**26**]) and for all 4-dimensional forms ([**33**]). All known results make natural the following conjecture.

The first-named author was supported by Alexander von Humboldt Stiftung.
The first-named author died on 17th April, 2000.
The second-named author was supported by Max-Planck Institut für Mathematik.

CONJECTURE 1.1. *Let F be a field, n be a positive integer, and let ϕ be an F-form of dimension $> 2^{n-1}$. Then the following conditions are equivalent:*

(1) *the group $H^n(F(\phi)/F) = \ker(H^n(F) \to H^n(F(\phi)))$ is nonzero,*
(2) *the form ϕ is an anisotropic n-fold Pfister neighbor.*

Moreover, if these conditions hold, then the group $H^n(F(\phi)/F)$ is isomorphic to $\mathbf{Z}/2\mathbf{Z}$ and is generated by $e^n(\pi)$, where π is the n-fold Pfister form associated with ϕ.

The proof of the implication (2)⇒(1) follows from the fact that the Pfister quadric is isotropic if and only if the corresponding *pure symbol* $\alpha \in \mathrm{K}_n^M(k)/2$ is zero, and the fact that the *norm-residue* homomorphism is injective on α (see [**36**]). The implication (1)⇒(2) seems much more difficult. In this paper, we give only a partial answer to the conjecture.

1.1. Forms with maximal splitting. It turns out that Conjecture 1.1 is closely related to a conjecture concerning so-called forms with maximal splitting. Let us recall some basic definitions and results. By $i_W(\phi)$ we denote the Witt index of ϕ. For an anisotropic quadratic form ϕ, the *first higher Witt index of ϕ* is defined as follows: $i_1(\phi) = i_W(\phi_{F(\phi)})$. Since $\phi_{F(\phi)}$ is isotropic, we obviously have $i_1(\phi) \geq 1$. In [**4**] Hoffmann proved the following

THEOREM 1.2. *Let ϕ be an anisotropic quadratic form. Let n be such that $2^{n-1} < \dim\phi \leq 2^n$ and m be such that $\dim\phi = 2^{n-1} + m$. Then*

- *$i_1(\phi) \leq m$,*
- *if ϕ is a Pfister neighbor, then $i_1(\phi) = m$.*

This theorem gives rise to the following

DEFINITION 1.3 (see [**4**]). Let ϕ be an anisotropic quadratic form. Let us write $\dim\phi$ in the form $\dim\phi = 2^{n-1} + m$, where $0 < m \leq 2^{n-1}$. We say that ϕ has maximal splitting if $i_1(\phi) = m$.

Our interest in forms with maximal splitting is motivated (in particular) by the following observation (which depends on the Milnor conjecture, see Proposition 7.5, and requires $char(F) = 0$): *Let ϕ and n be as in Conjecture 1.1. If $H^n(F(\phi)/F) \neq 0$, then ϕ has maximal splitting and $H^n(F(\phi)/F) \simeq \mathbf{Z}/2\mathbf{Z}$.* Therefore, the problem of classification of forms with maximal splitting is closely related to Conjecture 1.1. On the other hand, there are many other problems depending on the classification of forms with maximal splitting.

Let us explain some known results concerning this classification. By Theorem 1.2, all Pfister neighbors and all forms of dimension $2^n + 1$ have maximal splitting. By [**5**], these examples present an exhaustive list of forms with maximal splitting of dimension ≤ 9. The case $\dim\phi = 10$ is much more complicated. In [**9**], it was proved that a 10-dimensional form ϕ has maximal splitting only in the following cases:

- ϕ is a Pfister neighbor,
- ϕ can be written in the form $\phi = \langle\langle a\rangle\rangle\, q$, where q is a 5-dimensional form.

The structure of quadratic forms with maximal splitting of dimensions 11, 12, 13, 14, 15, and 16 is very simple: they are Pfister neighbors (see [**5**] or [**7**]). Since $17 = 2^4 + 1$, it follows that any 17-dimensional form has maximal splitting. The previous discussion shows that we have a complete classification of forms with maximal splitting of dimensions ≤ 17.

Conjecture 1.1 together with our previous discussion make the following problem natural:

PROBLEM 1.4. *Find the condition on the positive integer d such that each d-dimensional form ϕ with maximal splitting is necessarily a Pfister neighbor.*

The following example is due to Hoffmann. Let F be the field of rational functions $k(x_1, \dots, x_{n-3}, y_1, \dots, y_5)$ and let

$$q = \langle\langle x_1, \dots, x_{n-3}\rangle\rangle \otimes \langle y_1, y_2, y_3, y_4, y_5\rangle .$$

Then q has maximal splitting and is not a Pfister neighbor. We obviously have $\dim q = 2^{n-1} + 2^{n-3}$. This example gives rise to the following

PROPOSITION 1.5. *Let d be an integer satisfying $2^{n-1} \leq d \leq 2^{n-1} + 2^{n-3}$ for some $n \geq 4$. Then there exists a field F and a d-dimensional F-from ϕ with maximal splitting which is not a Pfister neighbor.*

For the proof, we can define ϕ as an arbitrary d-dimensional subform of the form q constructed above. We remind that if $\phi \subset q$ is a subform, where $2^{n-1} \leq \dim \phi \leq \dim q \leq 2^n$, and q has maximal splitting , then ϕ also has maximal splitting (see [**4**], Prop.4).

Let us return to Problem 1.4. By Proposition 1.5, it suffices to study Problem 1.4 only in the case where $2^{n-1} + 2^{n-3} < d \leq 2^n$. Here, we state the following

CONJECTURE 1.6. *Let $n \geq 3$ and F be an arbitrary field. For any anisotropic quadratic F-form with maximal splitting the condition $2^{n-1} + 2^{n-3} < \dim \phi \leq 2^n$ implies that ϕ is a Pfister neighbor.*

This conjecture is true in the cases $n = 3$ and $n = 4$ (see [**5**], [**7**]). At the time we cannot prove Conjecture 1.6 in the case $n \geq 5$. However, we prove the following partial case of the conjecture.

THEOREM 1.7. *Let $n \geq 5$ and q be an anisotropic form such that $2^n - 7 \leq \dim q \leq 2^n$. Then the following conditions are equivalent:*

(i) *q has maximal splitting,*
(ii) *q is a Pfister neighbor.*

Moreover, we show that for any form ϕ satisfying the condition $2^{n-1} + 2^{n-3} < \dim \phi \leq 2^n$, Conjectures 1.1 and 1.6 are equivalent for all fields of characteristic zero (here we use the Milnor conjecture). The equivalence of the conjectures follows readily from the following theorem.

THEOREM 1.8. ((*M*), see the end of Section 1) *Let n be an integer ≥ 4 and F be a field of characteristic 0. Let ϕ be an anisotropic form such that $2^{n-1} + 2^{n-3} < \dim \phi \leq 2^n$. Then the following conditions are equivalent:*

(1) *ϕ has maximal splitting,*
(2) *$H^n(F(\phi)/F) \neq 0$.*

On the other hand, Theorems 1.7 and 1.8 give rise to the proof of the following partial case of Conjecture 1.1.

COROLLARY 1.9. ((*M*), see the end of Section 1) *Let F be a field of characteristic zero and let $n \geq 5$. Then for any F-form ϕ satisfying the condition $\dim \phi \geq 2^n - 7$, the following conditions are equivalent:*

(1) *the group* $H^n(F(\phi)/F) = \ker(H^n(F) \to H^n(F(\phi)))$ *is nonzero,*
(2) *the form* ϕ *is an anisotropic* n*-fold Pfister neighbor.*

1.2. Plan of works. In section 3, we prove Theorem 1.7. Our proof is based on the following ideas of Bruno Kahn ([**11**]): First of all, we recall some result of M. Knebusch: *let* q *be an anisotropic* F*-form. If the* $F(q)$*-form* $(q_{F(q)})_{an}$ *is defined over* F*, then* q *is a Pfister neighbor.* Now, let q be an F-form satisfying the hypotheses of Theorem 1.7 (in particular, dim $q > 16$). Let us consider the $F(q)$-form $\phi = (q_{F(q)}))_{an}$. By the definition of forms with maximal splitting, we obviously have $\dim\phi \leq 7$. By the construction, the form ϕ belongs to the image of the homomorphism $W(F) \to W(F(q))$. This implies that ϕ belongs to the unramified part $W_{nr}(F(q))$ of the Witt group $W(F(q))$. Using some deep results concerning the group $W_{nr}(F(q))$, we prove that all forms of dimension ≤ 7 belonging to $W_{nr}(F(q))$ are necessarily defined over F (provided that $\dim q > 16$). In particular, this implies that $\phi = (q_{F(q)})_{an}$ is defined over F. Then Knebusch's theorem says that q is a Pfister neighbor. This completes the proof of Theorem 1.7.

To prove Theorem 1.8, we need the Milnor conjecture. The implication (1)⇒(2) is the most difficult part of the theorem. To explain the plan, we introduce the notion of "forms with absolutely maximal splitting". First, we recall that for any F-form ϕ, we can define the higher Witt indices by the following recursive rule: $i_{s+1}(\phi) = i_s((\phi_{an})_{F(\phi_{an})})$.

DEFINITION 1.10. Let ϕ be an anisotropic quadratic form. We say that ϕ has absolutely maximal splitting, or that ϕ is an AMS-form, if $i_1(\phi) > i_r(\phi)$ for all $r > 1$.

Such terminology is justified by the fact that at least in the case $char(F) = 0$, AMS implies maximal splitting (see Theorem 7.1). It is not difficult to show, that if the form ϕ has maximal splitting and satisfies the condition $2^{n-1} + 2^{n-3} < \dim\phi \leq 2^n$, then ϕ is an AMS-form (see Lemma 4.1). This shows, that the form ϕ satisfying the condition (1) of Theorem 1.8, is necessarily an AMS-form. Therefore, it suffices to prove the following theorem.

THEOREM 1.11. ((*M*), see the end of Section 1) *Let* F *be a field of characteristic zero. Let* ϕ *be an AMS-form satisfying the condition* $2^{n-1} < \dim\phi \leq 2^n$*. Then* $H^n(F(\phi)/F) \neq 0$.

To prove this theorem, we study the motive of the projective quadric Q corresponding to a subform q of ϕ of codimension $i_1(\phi) - 1$. It is well known that the function fields of the forms ϕ and q are stably equivalent. Hence, it suffices to prove that $H^n(F(q)/F) \neq 0$. In §5, we show that the motive $M(Q)$ of the quadric Q has some specific endomorphism $\omega : M(Q) \to M(Q)$ which we call the *Rost projector.* Let us give the definition of the latter. First, we recall that the set of endomorphisms $M(Q) \to M(Q)$ is defined as $\mathrm{CH}^d(Q \times Q)$, where $d = \dim Q$. We say that $\omega \in \mathrm{End}(M(Q))$ is a Rost projector, if ω is an idempotent ($\omega \circ \omega = \omega$), and the identity $\omega_{\bar{F}} = pt \times Q_{\bar{F}} + Q_{\bar{F}} \times pt$ holds over the algebraic closure $\bar{F}$ of F. The existence of the *Rost projector* means that $M(Q)$ contains a direct summand N such that $N_{\bar{k}}$ is isomorphic to the direct sum of two so-called *Tate-motives* $\mathbb{Z} \oplus \mathbb{Z}(d)[2d]$ (since the mutually orthogonal projectors $Q_{\bar{F}} \times pt$ and $pt \times Q_{\bar{F}}$ give direct summands isomorphic to $\mathbb{Z}$ and $\mathbb{Z}(d)[2d]$, respectively). The final step in the proof of theorem 1.8 is based on the following theorem.

THEOREM 1.12. ((*M*), see the end of Section 1) *Let F be a field of characteristic zero. Let Q be the projective quadric corresponding to an anisotropic F-form q. Assume that Q admits a Rost projector. Then* $\dim q = 2^{m-1} + 1$ *for suitable m. Moreover,* $H^m(F(q)/F) \neq 0$.

Now, it is very easy to complete the proof of the implication (1)⇒(2) of Theorem 1.8. Since $2^{m-1} < \dim q \leq 2^m$, $2^{n-1} < \dim \phi \leq 2^n$, and the extensions $F(\phi)/F$ and $F(q)/F$ are stably equivalent, it follows that $n = m$ (this follows easily from Hoffmann's theorem [**4**]). Therefore, $H^n(F(\phi)/F) = H^m(F(q)/F) \neq 0$. This completes the proof of the implication (1)⇒(2).

The proof of Theorem 1.12 is given in section 6. It is based on the technique developed by V.Voevodsky for the proof of Milnor's conjecture (see [**36**]). All needed results of Voevodsky's preprints are collected in Appendix A. Aside from the Appendix we also use the main results of [**26**] (in Theorem 7.3). Here we should point out that these results can be obtained from those of the Appendix in a rather simple way (the recipe is given, for example, in [**14**, Remark 3.3.]). All the major statements of the current paper which are using the abovementioned unpublished results are marked with (*M*) with the reference to this page.

Acknowledgements. An essential part of this work was done while the first author was visiting the Bielefeld University and the second author was visiting the Max-Plank Institute für Mathematik. We would like to express our gratitude to these institutes for their support and hospitality. The support of the Alexander von Humboldt Foundation for the first author is gratefully acknowledged. Also, we would like to thank Prof. Ulf Rehmann who made possible the visit of the second author to Bielefeld. Finally, we are grateful to the referee for the numerous suggestions and remarks which improved the text substantially.

2. Notation and background

In this article we use the standard quadratic form terminology from [**19**],[**30**]. We use the notation $\langle\langle a_1, \ldots, a_n\rangle\rangle$ for the *Pfister form* $\langle 1, -a_1\rangle \otimes \cdots \otimes \langle 1, -a_n\rangle$. Under $GP_n(F)$ we mean the set of forms over F which are similar to n-fold Pfister forms. The n-fold Pfister forms provide a system of generators for the abelian group $I^n(F)$. We recall that the Arason–Pfister Hauptsatz (APH in what follows) states that: *every quadratic form over F of dimension $< 2^n$ which lies in $I^n(F)$ is necessarily hyperbolic; if $\phi \in I^n(F)$ and $\dim \phi = 2^n$, then the form ϕ is necessarily similar to a Pfister form.* We use the notation e^n for the *generalized Arason invariant* [1]

$$I^n(F)/I^{n+1}(F) \to H^n(F), \quad \text{where} \quad \langle\langle a_1, \ldots, a_n\rangle\rangle \mapsto (a_1, \ldots, a_n).$$

The following statements describe the relationship between the Witt ring $W(F)$ and the cohomology $H^n(F)$. They will be used extensively in the next two sections.

THEOREM 2.1. ([**24**],[**1**],[**10**],[**32**],[**21**],[**23**],[**27**],Rost-unpublished)
For $n \leq 4$, we have canonical isomorphisms $e^n : I^n(F)/I^{n+1}(F) \to H^n(F)$

THEOREM 2.2. (Arason[**1**], Kahn-Rost-Sujatha[**13**], Merkurjev[**22**])
Let $0 \leq m \leq n \leq 4$ and π be an m-fold Pfister form over F. Then $H^n(F(\pi)/F) = e^m(\pi)H^{n-m}(F)$.

[1]The existence of e^n was proven by Arason for $n \leq 3$, and by Jacob-Rost/Szyjewski for $n = 4$.

THEOREM 2.3. (Arason[**1**], Kahn-Rost-Sujatha[**13**])
Let $n \leq 4$ and ρ be a form over F of dimension $> 2^n$. Then $H^n(F(\rho)/F) = 0$.

The following statement is an evident corollary of the theorems above.

COROLLARY 2.4. *Let ρ be a form over F and n be a positive integer ≤ 5. Let ξ be a form over F such that $\xi_{F(\rho)} \in I^n(F(\rho))$. Then*

- *if ρ is a Pfister neighbour of a Pfister form π, then $\xi \in \pi W(F) + I^n(F)$,*
- *if $\dim \rho > 2^{n-1}$, then $\xi \in I^n(F)$.*

REMARK 2.5. Actually, the restriction on the integer n here is unnecessary (at least in characteristic 0) - see Theorem 7.3.

In section 5 we use the notation $\mathbb{Z}$ for the trivial *Tate-motive* (which is just the motive of a point $M(\mathrm{Spec}(k))$), and $\mathbb{Z}(m)[2m]$ for the tensor power $\mathbb{Z}(1)[2]^{\otimes m}$ of the *Tate-motive* $\mathbb{Z}(1)[2]$, where the latter is defined as a complementary direct summand to $\mathbb{Z}$ in $M(\mathbb{P}^1)$ ($M(\mathbb{P}^1) = \mathbb{Z} \oplus \mathbb{Z}(1)[2]$). For this reason, we use the notation **Z** for all groups and rings $\mathbb{Z}$ throughout the text. In section 5 we work in the classical Chow-motivic category of Grothendieck (see [**3**],[**20**],[**31**],[**29**]). We remind that in this category the group $\mathrm{Hom}(M(P), M(Q))$ is naturally identified with $\mathrm{CH}^{\dim(Q)}(P \times Q)$, for any smooth connected projective varieties P and Q over k.

In this connection we should mention that the motive of a completely split quadric P (of dimension d) is a direct sum of *Tate-motives*:

$M(P) = \oplus_{0 \leq i \leq d} \mathbb{Z}(i)[2i]$, if d is odd;
$M(P) = (\oplus_{0 \leq i \leq d} \mathbb{Z}(i)[2i]) \oplus \mathbb{Z}(d/2)[d]$, if d is even.

The corresponding mutually orthogonal projectors in $\mathrm{End}(M(P))$ are given by $h^i \times l_i$, and $l_i \times h^i$, where $0 \leq i < d/2$, and $h^i \subset P$ is a plane section of codimension i, and $l_i \subset P$ is a projective subspace of dimension i (in the case d even, we also have $l^1_{d/2} \times l^2_{d/2}$ and $l^2_{d/2} \times l^1_{d/2}$, where $l^1_{d/2}, l^2_{d/2} \subset P$ are the projective subspaces of half the dimension from the two different families).

In section 6 we work in the bigger triangulated category of motives $DM^{eff}_{-}(k)$ constructed by V.Voevodsky (see [**34**]). This category contains the category of Chow-motives as a full additive subcategory closed with respect to direct summands. All the necessary facts and references are given in the Appendix.

3. Descent problem and forms with maximal splitting

The main goal of this section is to prove Theorem 1.7. It should be noticed that in all cases except for $\dim q = 2^n - 7$ this theorem was proved earlier:

- if $\dim q = 2^n$ or $2^n - 1$, the theorem was proved by M. Knebusch and A. Wadsworth (independently);
- if $\dim q = 2^n - 2$ or $2^n - 3$, the theorem was proved by D. Hoffmann [**4**];
- if $\dim q = 2^n - 4$ or $2^n - 5$, the theorem was proved by B. Kahn [**11**, remark after Th.4] (see also more elementary proofs in [**5**] or [**7**]);
- In the case $\dim q = 2^n - 6$, the theorem follow easily from a result of A. Laghribi [**18**].

To prove the theorem in the case $\dim q = 2^n - 7$ we use the same method as in the paper of Bruno Kahn [**11**]. Namely, we reduce Theorem 1.7 to the study of *a descent problem for quadratic forms* (see Proposition 3.8 and Theorem 3.9). As in the paper of B. Kahn, we work modulo a suitable power $I^n(F)$ of the fundamental ideal $I(F)$.

We start with the following notation.

DEFINITION 3.1. Let ψ be a form over F and $n \geq 0$ be an integer. We define $\dim_n \psi$ as follows:

$$\dim_n \psi = \min\{\dim \phi \mid \phi \equiv \psi \pmod{I^n(F)}\}$$

LEMMA 3.2. *Let ψ be a form over F and L/F be some field extension. Then* $\dim_n \psi_L \leq \dim_n \psi$. *If L/F is unirational, then* $\dim_n \psi_L = \dim_n \psi$.

PROOF. The inequality $\dim_n \psi_L \leq \dim_n \psi$ is obvious. If L/F is unirational, the identity $\dim_n \psi_L = \dim_n \psi$ follows easily from standard specialization arguments. □

COROLLARY 3.3. *Let ψ and $q_0 \subset q$ be forms over F. Then* $\dim_n \psi_{F(q_0)} \leq \dim_n \psi_{F(q)}$.

PROOF. Since $q_0 \subset q$, it follows that $q_{F(q_0)}$ is isotropic and hence the extension $F(q, q_0)/F(q_0)$ is purely transcendental. By Lemma 3.2, we have $\dim_n \psi_{F(q_0)} = \dim_n \psi_{F(q,q_0)} \leq \dim_n \psi_{F(q)}$. □

Now, we recall an evident consequence of Merkurjev's index reduction formula: *if A is a central simple algebra of index 2^n and q is a form of dimension $> 2n+2$, then* $\operatorname{ind} A_{F(q)} = \operatorname{ind} A$. The following lemma is an obvious generalization of this statement.

LEMMA 3.4. *Let A be a central simple F-algebra of index 2^n and q be a quadratic form over F. Let $F_0 = F, F_1, \ldots, F_h$ be the generic splitting tower of q. Let $i \geq 1$ be an integer such that* $\dim((q_{F_{i-1}})_{an}) > 2n+2$. *Then* $\operatorname{ind} A_{F_i} = \operatorname{ind} A$.

LEMMA 3.5. *Let A be a central simple F-algebra of index 2^n and q be a quadratic form of dimension $> 2n+4$. Then there exists a unirational extension E/F and a 3-dimensional form $q_0 \subset q_E$ such that* $\operatorname{ind}(A_E \otimes C_0(q_0)) = 2^{n+1}$

PROOF. Let $\tilde{F} = F(X,Y,Z)$, $\tilde{q} = q_{\tilde{F}} \perp -X\langle\langle Y, Z\rangle\rangle$, and $\tilde{A} = A_{\tilde{F}} \otimes (Y,Z)$. Clearly, $\operatorname{ind} \tilde{A} = 2 \operatorname{ind} A = 2^{n+1}$. Let $\tilde{F}_0 = \tilde{F}, \tilde{F}_1, \ldots, \tilde{F}_h$ be the generic splitting tower for $\tilde{q}$. Let $\tilde{q}_i = (\tilde{q}_{\tilde{F}_i})_{an}$ for $i = 0, \ldots, h$. Let s be the minimal integer such that $\dim \tilde{q}_s \leq \dim q - 2$. We have $\dim \tilde{q}_{s-1} \geq \dim q > 2(n+1)+2$. By Lemma 3.4, we have $\operatorname{ind} A_{\tilde{F}_s} = \operatorname{ind} \tilde{A} = 2^{n+1}$.

We set $E = \tilde{F}_s$. Since $\tilde{q}_E = q_E \perp -X\langle\langle Y, Z\rangle\rangle$, the forms q_E and $X\langle\langle Y, Z\rangle\rangle$ contain a common subform of dimension

$$\frac{1}{2}(\dim q + \dim(X\langle\langle Y,Z\rangle\rangle) - \dim(\tilde{q}_E)_{an}) = \frac{1}{2}(\dim q + 4 - \dim \tilde{q}_s)$$
$$\geq \frac{1}{2}(\dim q + 4 - (\dim q - 2)) = 3.$$

Hence, there exists a 3-dimensional E-form q_0 such that $q_0 \subset q_E$ and $q_0 \subset X\langle\langle Y, Z\rangle\rangle_E$. Clearly, $C_0(q_0) = (Y,Z)$. Hence, $\operatorname{ind}(A_E \otimes_E C_0(q_0)) = \operatorname{ind} \tilde{A}_E = 2^{n+1}$.

To complete the proof, it suffices to show that E/F is unirational. To prove this, let us write q in the form $q = x\langle 1, -y, -z\rangle \perp q_0$ with $x, y, z \in F^*$. Let us consider the field

$$K = \tilde{F}(\sqrt{X/x}, \sqrt{Y/y}, \sqrt{Z/z}) = F(X,Y,Z)(\sqrt{X/x}, \sqrt{Y/y}, \sqrt{Z/z}).$$

Clearly, K/F is purely transcendental. In the Witt ring $W(K)$, we have $\tilde{q}_K = q_K - X\langle\langle Y,Z\rangle\rangle_K = x\langle 1,-y,-z\rangle_K + q_0 - X\langle 1,-Y,-Z,YZ\rangle = q_0 - \langle XYZ\rangle$. Hence $\dim(\tilde{q}_K)_{an} \leq \dim q_0 + 1 = \dim q - 3 + 1 = \dim q - 2$. Since s is the minimal integer such that $\dim \tilde{q}_s \leq \dim q - 2$, it follows that the extension $(K \cdot \tilde{F}_s)/K$ is purely transcendental (see, e.g., [**16**, Cor. 3.9 and Prop. 5.13]), where $K \cdot \tilde{F}_s$ is the free composite of K and $\tilde{F}_s$ over $\tilde{F}$. Since K/F is purely transcendental, it follows that $(K\cdot\tilde{F}_s)/F$ is also purely transcendental. Hence $\tilde{F}_s/F$ is unirational. Since $E = \tilde{F}_s$, we are done. □

LEMMA 3.6. *Let ρ be a Pfister neighbor of $\langle\langle a,b\rangle\rangle$ and n be a positive integer. Let ψ be a form such that $\dim_n \psi_{F(\rho)} < 2^{n-1}$. Then there exists an F-form μ such that $\dim\mu = \dim_n \psi_{F(\rho)}$ and $\psi_{F(\rho)} \equiv \mu_{F(\rho)} \pmod{I^n(F)}$.*

PROOF. Let ξ be an $F(\rho)$-form such that $\dim\xi = \dim_n \psi_{F(\rho)}$ and $\psi_{F(\rho)} \equiv \xi \pmod{I^n(F(\rho))}$. By [**18**, Lemme 3.1], we have $\xi \in W_{nr}(F(\rho)/F)$. By the excellence property of $F(\rho)/F$ (see [**2**, Lemma 3.1]), there exists an F-form μ such that $\xi = \mu_{F(\rho)}$. □

COROLLARY 3.7. *Let ρ be a Pfister neighbor of $\langle\langle a,b\rangle\rangle$ and n be a positive integer such that $n \leq 5$. Let ψ be a form such that $\dim_n \psi_{F(\rho)} < 2^{n-1}$. Then there exist F-forms μ and γ such that $\dim\mu = \dim_n \psi_{F(\rho)}$ and $\psi \equiv \mu + \langle\langle a,b\rangle\rangle\gamma \pmod{I^n(F)}$.*

PROOF. Let μ be a form as in Lemma 3.6. We have $(\psi-\mu)_{F(\rho)} \in I^n(F(\rho))$. By Corollary 2.4, we have $\psi - \mu \in \langle\langle a,b\rangle\rangle W(F) + I^n(F)$. Hence, there exists γ such that $\psi - \mu \in \langle\langle a,b\rangle\rangle\gamma + I^n(F)$. □

PROPOSITION 3.8. *Let q be an F-form of dimension > 16 and ψ be a form over F such that $\dim_5 \psi_{F(q)} \leq 7$. Then $\dim_5\psi = \dim_5\psi_{F(q)}$. In particular, $\dim_5\psi \leq 7$.*

PROOF. By Lemma 3.2, we have $\dim_5\psi \geq \dim_5\psi_{F(q)}$. Hence, it suffices to verify that $\dim_5\psi \leq \dim_5\psi_{F(q)}$. As usually, we denote as $C_0(\psi)$ the even part of the Clifford algebra and as $c(\psi)$ the Clifford invariant of ψ. We start with the following case:

Case 1. either $\dim_5\psi_{F(q)} \leq 6$ or $\dim_5\psi_{F(q)} = 7$ and $\operatorname{ind} C_0(\psi) \neq 8$.

By the definition of $\dim_5\psi_{F(q)}$, there exists an $F(q)$-form ϕ such that $\dim\phi = \dim_5\psi_{F(q)} \leq 7$ and $\phi \equiv \psi_{F(q)} \pmod{I^5(F(q))}$. In particular, we have $c(\phi) = c(\psi_{F(q)})$. Since $\dim\phi \leq 7$ and $\dim q > 16$, the index reduction formula shows that $\operatorname{ind} C_0(\phi) = \operatorname{ind} C_0(\psi)$. By the assumption of Case 1, we see that

- either $\dim\phi \leq 6$,
- or $\dim\phi = 7$ and $\operatorname{ind} C_0(\phi) \neq 8$.

Since $\phi \equiv \psi_{F(q)} \pmod{I^5(F(q))}$, it follows that $\phi \in im(W(F) \to W(F(q))) + I^5(F(q))$. The principal theorem of [**18**] shows that ϕ is defined over F. In other words, there exists an F-form μ such that $\phi = \mu_{F(q)}$. Therefore, $\psi_{F(q)} \equiv \phi \equiv \mu_{F(q)} \pmod{I^5(F(q))}$. By Corollary 2.4, we see that $\psi \equiv \mu \pmod{I^5(F)}$. Hence, $\dim_5(\psi) \leq \dim\mu = \dim\phi = \dim_5(\psi_{F(q)})$. This completes the proof in Case 1.

Case 2. $\dim_5\psi_{F(q)} = 7$ and $\operatorname{ind} C_0(\psi) = 8$.

Lemma 3.2 shows that we can change the ground field by an arbitrary unirational extension. After this, Lemma 3.5 (applied to $A = C_0(\psi)$, $n = 3$ and q) shows, that we can assume that *there exists a 3-dimensional subform $q_0 \subset q$ such that* $\operatorname{ind}(C_0(\psi) \otimes C_0(q_0)) = 16$.

Let $a, b \in F^*$ be such that q_0 is a Pfister neighbor of $\langle\langle a,b\rangle\rangle$.

By Corollary 3.3, we have $\dim_5 \psi_{F(q_0)} \leq 7$. By Corollary 3.7, there exists a form μ of dimension ≤ 7 and a form λ such that $\psi \equiv \mu + \langle\langle a,b\rangle\rangle\,\lambda \pmod{I^5(F)}$.

First, consider the case where $\dim\lambda$ is odd. Then $c(\psi) = c(\mu)+(a,b)$. Therefore $\operatorname{ind} C_0(\mu) = \operatorname{ind}(C_0(\psi)\otimes(a,b)) = \operatorname{ind}(C_0(\psi)\otimes C_0(q_0)) = 16$. On the other hand, $\dim\mu \leq 7$ and hence $\operatorname{ind} C_0(\mu) \leq 8$. We get a contradiction.

Now, we can assume that $\dim\lambda$ is even. Then $\lambda \equiv \langle\langle c\rangle\rangle \pmod{I^2(F)}$, where $c = d_{\pm}\lambda$. Hence, $\langle\langle a,b\rangle\rangle\,\lambda \equiv \langle\langle a,b,c\rangle\rangle \pmod{I^4(F)}$. Hence, $\psi - \mu \equiv \langle\langle a,b\rangle\rangle\,\lambda \equiv \langle\langle a,b,c\rangle\rangle \pmod{I^4(F)}$. Let $\pi = \langle\langle a,b,c\rangle\rangle$. We have $\psi \equiv \mu + \pi \pmod{I^4(F)}$.

Since $\pi_{F(\pi)}$ is hyperbolic, it follows that $\psi_{F(q,\pi)} \equiv \mu_{F(q,\pi)} \pmod{I^4(F(q,\pi))}$. Since $\dim_5 \psi_{F(q)} = 7$, there exists a 7-dimensional $F(q)$-form ξ such that $\psi_{F(q)} \equiv \xi \pmod{I^5(F(q))}$. This implies that $\mu_{F(q,\pi)} \equiv \psi_{F(q,\pi)} \equiv \xi_{F(q,\pi)} \pmod{I^4(F(q,\pi))}$. Since $\dim\mu + \dim\xi \leq 7+7 = 14 < 2^4$, APH shows that $\mu_{F(q,\pi)} = \xi_{F(q,\pi)}$. Hence, $\psi_{F(q,\pi)} \equiv \xi_{F(q,\pi)} \equiv \mu_{F(q,\pi)} \pmod{I^5(F(q,\pi))}$. Since $\dim q > 16$, Corollary 2.4 shows that $\psi_{F(\pi)} \equiv \mu_{F(\pi)} \pmod{I^5(F(\pi))}$. Hence $(\psi-\mu)_{F(\pi)} \in I^5(F(\pi))$.

By Corollary 2.4, there exists an F-form γ such that $\psi - \mu \equiv \pi\gamma \pmod{I^5(F)}$. Since $\psi - \mu \equiv \pi \pmod{I^4(F)}$, it follows that either π is hyperbolic or $\dim\gamma$ is odd. In any case, we can assume that $\dim\gamma$ is odd. Then $\gamma \equiv \langle k\rangle \pmod{I^2(F)}$, where $k = d_{\pm}\gamma$. Hence, $\psi - \mu \equiv \pi\gamma \equiv k\pi \pmod{I^5(F)}$. Therefore, $\xi \equiv \psi_{F(q)} \equiv (\mu + k\pi)_{F(q)} \pmod{I^5(F(q))}$. Since $\dim\xi + \dim\mu + \dim\pi = 7+7+8 < 2^5$, APH shows that $\xi = (\zeta_{F(q)})_{an}$, where $\zeta = (\mu \perp k\pi)_{an}$. Since $\dim\zeta \leq \dim(\mu\perp k\pi) \leq 7+8 < 2^4 < \dim q$, Hoffmann's theorem shows that $\zeta_{F(q)}$ is anisotropic. Hence, $\xi = \zeta_{F(q)}$. In particular, $\dim\zeta = 7$. We have $\psi_{F(q)} \equiv \xi \equiv \zeta_{F(q)} \pmod{I^5(F(q))}$. Since $\dim q > 16$, Corollary 2.4 shows that $\psi \equiv \zeta \pmod{I^5(F)}$. Hence, $\dim_5\psi \leq \dim\zeta = 7$. On the other hand, $\dim_5\psi \geq \dim_5\psi_{F(q)} = 7$. The proof is complete. □

The essential part of the following theorem was proved by Ahmed Laghribi in [**18**].

THEOREM 3.9. (cf. [**18**, Théorème principal]). *Let q be a form of dimension > 16. Let ϕ be a form of dimension ≤ 7 over the field $F(q)$. Then the following conditions are equivalent.*

(1) *ϕ is defined over F,*
(2) *$\phi \in im(W(F) \to W(F(q)))$,*
(3) *$\phi \in im(W(F) \to W(F(q))) + I^5(F(q))$,*
(4) *$\phi \in W_{nr}(F(q)/F)$.*

PROOF. This theorem is proved in [**18**] except for the case where $\dim\phi = 7$ and $\operatorname{ind} C_0(\phi) = 8$. Implications (1)⇒(2)⇒(3) ⟺ (4) are also proved in [**18**]. It suffices to prove implication (3)⇒(1).

Condition (3) shows that there exists a form ψ over F such that $\psi_{F(q)} \equiv \phi \pmod{I^5(F(q))}$. Therefore $\dim_5\psi_{F(q)} \leq \dim\phi \leq 7$. By Proposition 3.8, we have $\dim_5\psi \leq 7$. Hence there exists an anisotropic F-form μ of dimension ≤ 7 such that $\psi \equiv \mu \pmod{I^5(F)}$. Thus $\phi \equiv \psi_{F(q)} \equiv \mu_{F(q)} \pmod{I^5(F(q))}$. Since $\dim\phi + \dim\mu = 7+7 < 2^5$, APH shows that $\phi_{an} = (\mu_{F(q)})_{an}$. Since $\dim\mu < 8 < \dim q$, Hoffmann's theorem shows that $\mu_{F(q)}$ is anisotropic. Hence $\phi_{an} = \mu_{F(q)}$. Therefore ϕ_{an} is defined over F. Hence, ϕ is defined over F. □

PROOF OF THEOREM 1.7. (i)⇒(ii). Let $\phi = (q_{F(q)})_{an}$. By [**17**, Th. 7.13], it suffices to prove that ϕ is defined over F. Since $n \geq 5$, we have $\dim q \geq 2^n - 7 > 16$.

Clearly, $\phi \in im(W(F) \to W(F(q)))$. Since q has maximal splitting, it follows that $\dim \phi = 2^n - \dim q \le 7$. By Theorem 3.9, we see that ϕ is defined over F.

(ii)⇒(i). Obvious. □

4. Elementary properties of AMS-forms

In this section we start studying forms with absolutely maximal splitting (AMS-forms) defined in the introduction (Definition 1.10).

LEMMA 4.1. *Let ϕ be an anisotropic form and n be an integer such that $2^{n-1} + 2^{n-3} < \dim \phi \le 2^n$. Suppose that ϕ has maximal splitting. Then ϕ has absolutely maximal splitting.*

PROOF. Let $m = \dim \phi - 2^{n-1}$. Clearly, $\dim \phi = 2^{n-1} + m$ and $2^{n-3} < m \le 2^{n-1}$. Since ϕ has maximal splitting, we have $i_1(\phi) = m$. Let $F = F_0, F_1, \dots, F_h$ be the generic splitting tower of ϕ. Let $\phi_i = (\phi_{F_i})_{an}$ for $i = 0, \dots, h$. Let us fix $r > 1$. To prove that ϕ has absolutely maximal splitting, we need to verify that $i_r(\phi) < m$. Clearly, $i_r(\phi) = i_1(\phi_r)$. Thus, we need to verify that $i_1(\phi_r) < m$. In the case where $\dim \phi_r \le 2^{n-2}$, we have $i_1(\phi_r) \le \frac{1}{2} \dim \phi_r \le \frac{1}{2} 2^{n-2} = 2^{n-3} < m$.

Thus, we can suppose that $\dim \phi_r > 2^{n-2}$. Since $r \ge 1$, we have $\dim \phi_r \le \dim \phi_1 = \dim \phi - 2i_1(\phi) = 2^{n-1} + m - 2m = 2^{n-1} - m$. Hence, $2^{n-2} < \dim \phi_r \le 2^{n-2} + (2^{n-2} - m)$. By Theorem 1.2, we have $i_1(\phi_r) \le 2^{n-2} - m$. Since $m > 2^{n-3}$, we have $2^{n-2} - m < m$. Hence $i_1(\phi_r) \le 2^{n-2} - m < m$. □

From the results proven in the next sections (see Theorem 7.1) it follows that in the dimension range we are interested in ($2^{n-1} + 2^{n-3} < \dim \phi \le 2^n$), the form has maximal splitting if and only if it has absolutely maximal splitting.

REMARK 4.2. We cannot change the strict inequality $2^{n-1} + 2^{n-3} < \dim \phi$ by $2^{n-1} + 2^{n-3} \le \dim \phi$ in the formulation of the lemma. Indeed, for any $n \ge 3$ there exists an example of $(2^{n-1}+2^{n-3})$-dimensional form ϕ with maximal splitting which is not an AMS-form. The simplest example is the following:

$$\phi = \langle\langle x_1, x_2, \dots, x_{n-3} \rangle\rangle \otimes \langle 1, 1, 1, 1, 1 \rangle \quad \text{over the field} \quad \mathbb{R}(x_1, \dots, x_{n-3}).$$

In this case $i_1(\phi) = i_2(\phi) = 2^{n-3}$.

5. Motivic decomposition of AMS-Quadrics

In this section we will produce some "binary" motive related to an AMS-quadric.

Let X, Y and Z be smooth projective varieties over k of dimensions l, m and n, respectively. Then we have a natural (associative) pairing:

$$\circ : \mathrm{CH}^{n+b}(Y \times Z) \otimes \mathrm{CH}^{m+a}(X \times Y) \to \mathrm{CH}^{n+a+b}(X \times Z),$$

where $v \circ u := (\pi_{X,Z})_*(\pi^*_{X,Y}(u) \cap \pi^*_{Y,Z}(v))$, and $\pi_{X,Y} : X \times Y \times Z \to X \times Y$, $\pi_{Y,Z} : X \times Y \times Z \to Y \times Z$, $\pi_{X,Z} : X \times Y \times Z \to X \times Z$ are the natural projections. In particular, taking $X = \mathrm{Spec}(k)$, we get a pairing:

$$\mathrm{CH}^{n+b}(Y \times Z) \otimes \mathrm{CH}_r(Y) \to \mathrm{CH}_{r-b}(Z).$$

In this case, we will denote $v \circ u$ as $v(u)$.

THEOREM 5.1. (cf. [**33**, Proof of Statement 6.1]) *Let Q be an AMS-quadric. Let $P \subset Q$ be any subquadric of codimension $= i_1(q) - 1$. Then P possesses a Rost projector (in other words, $M(P)$ contains a direct summand N such that $N|_{\overline{k}} \simeq \mathbb{Z} \oplus \mathbb{Z}(\dim(P))[2\dim(P)]$).*

PROOF. We say that "we are in the situation $(*)$", if we have the following data:

Q - some quadric; $P \subset Q$ - some subquadric of codimension d;
$\Phi \in \mathrm{CH}^m(Q \times Q)$, where $m := \dim(P)$.

In this case, let $\Psi \in \mathrm{CH}^{m+d}(P \times Q)$ denote the class of the graph of the natural embedding $P \subset Q$, and let $\Psi^\vee \in \mathrm{CH}^{m+d}(Q \times P)$ denote the dual cycle. We define $\varepsilon := \Psi^\vee \circ \Phi \circ \Psi \in \mathrm{CH}^m(P \times P)$.

The action on $\mathrm{CH}_*(P_{\overline{k}})$ identifies: $\mathrm{CH}^m(P_{\overline{k}} \times P_{\overline{k}}) = \prod_r \mathrm{End}(\mathrm{CH}_r(P_{\overline{k}}))$ (see [**29**, Lemma 7]), and we will denote as $\varepsilon_{(r)} \in \mathrm{End}(\mathrm{CH}_r(P_{\overline{k}}))$ the corresponding coordinate of $\varepsilon_{\overline{k}}$.

- If $0 \le s < m/2$, then $\mathrm{CH}_s(P_{\overline{k}}) = \mathbf{Z}$ with the generator l_s - the class of projective subspace of dimension s on $P_{\overline{k}}$;
- if $m/2 < s \le m$, then $\mathrm{CH}_s(P_{\overline{k}}) = \mathbf{Z}$ with the generator h^{m-s} - the class of plane section of codimension $m - s$ on $P_{\overline{k}}$;
- if $s = m/2$, then $\mathrm{CH}_s(P_{\overline{k}}) = \mathbf{Z} \oplus \mathbf{Z}$ with the generators $l^1_{m/2}$ and $l^2_{m/2}$ - the classes of $m/2$-dimensional projective subspaces from two different families.

This permits to identify $\mathrm{End}(\mathrm{CH}_s(P_{\overline{k}}))$ with $\mathbf{Z}$ if $0 \le s \le m$, $s \ne m/2$, and with $\mathrm{Mat}_{2\times 2}(\mathbf{Z})$, if $s = m/2$. We should mention, that since for an arbitrary field extension E/k, the natural map $\mathrm{CH}_s(P|_{\overline{k}}) \to \mathrm{CH}_s(P|_{\overline{E}})$ is an isomorphism (preserving the generators above), we have an equality: $(\varepsilon_E)_{(s)} = \varepsilon_{(s)}$ (in $\mathbf{Z}$, resp. $\mathrm{Mat}_{2\times 2}(\mathbf{Z})$).

We will need the following easy corollary of Springer's theorem. Under the degree of the cycle $A \in \mathrm{CH}_s(Q)$ we will understand the degree of the 0-cycle $A \cap h^s$.

LEMMA 5.2. *Let $0 \le s \le \dim(Q)/2$. Then the following conditions are equivalent:*

(1) *$q = (s+1) \cdot \mathbb{H} \perp q'$, for some form q';*
(2) *Q contains (projective subspace) $\mathbb{P}^s$ as a subvariety;*
(3) *$\mathrm{CH}_s(Q)$ contains cycle of odd degree.*

PROOF. (1) $\Rightarrow$ (2) and (2) $\Rightarrow$ (3) are evident. (3) $\Rightarrow$ (1) Use induction on s. For $s = 0$ the statement is equivalent to the Theorem of Springer. Now if $s > 0$, then $\mathrm{CH}_0(Q)$ also contains a cycle of odd degree (obtained via intersection with h^s). So, $q = \mathbb{H} \perp q''$. And we have the natural degree preserving isomorphism: $\mathrm{CH}_s(Q) = \mathrm{CH}_{s-1}(Q'')$. By induction, $q'' = s \cdot \mathbb{H} \perp q'$. □

LEMMA 5.3. *In the situation of $(*)$, suppose, for some $0 \le s < m/2$, that $\varepsilon_{(s)} \in \mathbf{Z}$ is odd. Then for an arbitrary field extension E/k, if Q_E contains a projective space of dimension s, then it contains a projective space of dimension $s + d$.*

PROOF. Let E/k be such an extension that $l_s \in image(\mathrm{CH}_s(Q_E) \to \mathrm{CH}_s(Q_{\overline{E}}))$, and suppose that $\varepsilon_{(s)}$ is odd. We have: $\Psi \circ \Psi^\vee \circ \Phi(l_s) = \lambda \cdot l_s \subset \mathrm{CH}_s(Q_{\overline{E}})$, where $\lambda \in \mathbf{Z}$ is odd (since $\Psi : \mathrm{CH}_s(P_{\overline{E}}) \to \mathrm{CH}_s(Q_{\overline{E}})$ is an isomorphism). On the other hand, the composition $\Psi \circ \Psi^\vee : \mathrm{CH}_{s+d}(Q_{\overline{E}}) \to \mathrm{CH}_s(Q_{\overline{E}})$ is given by the intersection with the plane section of codimension d, so it preserves the *degree* of the cycle. This implies that $\Phi(l_s) \in image(\mathrm{CH}_{s+d}(Q_E) \to \mathrm{CH}_{s+d}(Q_{\overline{E}}))$ has odd degree. By Lemma 5.2, Q_E contains a projective space of dimension $s+d$. □

LEMMA 5.4. *In the situation of* $(*)$, *suppose, for some* $m/2 < s \leq m$, *that* $\varepsilon_{(s)} \in \mathbf{Z}$ *is odd. Then for an arbitrary field extension* E/k, *if* Q_E *contains a projective space of dimension* $(m-s)$, *then it contains a projective space of dimension* $(m-s+d)$.

PROOF. Consider the cycle $\varepsilon^\vee \in \mathrm{CH}^m(P \times P)$ dual to ε. Since $(A \circ B)^\vee = B^\vee \circ A^\vee$, we have: $\varepsilon^\vee = \Psi^\vee \circ \Phi^\vee \circ \Psi$. On the other hand, $(\varepsilon^\vee)_{(s)} = \varepsilon_{(m-s)}$. Now, the statement follows from Lemma 5.3. □

LEMMA 5.5. *In the situation of* $(*)$, *if* $d > 0$, *then* $\varepsilon_{\overline{k}}(l^1_{m/2}) = \varepsilon_{\overline{k}}(l^2_{m/2}) = c \cdot h^{m/2}$, *where* $c \in \mathbf{Z}$.

PROOF. Clearly, $\Psi(l^i_{m/2}) = l_{m/2} \in \mathrm{CH}_{m/2}(Q_{\overline{k}})$. On the other hand, $\Phi \circ \Psi(l^i_{m/2}) \in \mathrm{CH}_{m/2+d}(Q_{\overline{k}})$, the later group is generated by $h^{m/2}$ (since $(m/2)+d> (m+d)/2$), and $\Psi^\vee(h^{m/2}) = h^{m/2}$. □

Let now Q be an AMS-quadric, and $P \subset Q$ be a subquadric of codimension $i_1(q)-1$. By the definition of AMS-quadrics, either $\dim(Q) = 0$, or $i_1(q) > 1$ and P is a proper subform of Q. Clearly, it is enough to consider the second possibility.

By the definition of $i_1(q)$, we have: $q_{k(Q)} = i_1(q) \cdot \mathbb{H} \perp q_1$. So, the quadric $Q_{k(Q)}$ contains an $(i_1(q)-1)$-dimensional projective subspace $l_{(i_1(q)-1)}$. Denote: $d := i_1(q)-1$, and $m := \dim(P)$. Let $\Phi \in \mathrm{CH}^m(Q \times Q)$ be the class of the closure of $l_d \subset \mathrm{Spec}(k(Q)) \times Q \subset Q \times Q$. Let us denote this particular case of $(*)$ as $(**)$.

LEMMA 5.6. *In the situation of* $(**)$, $\varepsilon_{\overline{k}} = P_{\overline{k}} \times l_0 + \sum_{0<i<m} b_i \cdot (h^{m-i} \times h^i) + a \cdot l_0 \times P_{\overline{k}}$, *where* $b_1, \ldots, b_{m-1}, a \in \mathbf{Z}$.

PROOF. If for some $0 \leq i < m/2$, the coordinate $\varepsilon_{(i)}$ is odd, then by Lemma 5.3, in the *generalized splitting tower* $k = F_0 \subset F_1 \subset \cdots \subset F_h$ for the quadric Q (see [16]), there exists $0 \leq t < h$ such that $i_W(q_{F_t}) \leq i < i + i_1(q) - 1 < i_W(q_{F_{t+1}})$. Since q is an AMS-form, this can happen only if $i = 0$. In the same way, using Lemma 5.4, we get that for all $m/2 < i < m$, the coordinates $\varepsilon_{(i)}$ are even.

This implies that on the group $\mathrm{CH}_i(P_{\overline{k}})$, where $0 < i < m$, $i \neq m/2$, the map $\varepsilon_{\overline{k}}$ acts as some (integral) multiple of $h^{m-i} \times h^i$ (notice also that $h^{m-i} \times h^i$ acts trivially on all $\mathrm{CH}_j(P_{\overline{k}})$, $j \neq i$). The same holds for $i = m/2$ by Lemma 5.5.

Clearly, $P_{\overline{k}} \times l_0$ (resp. $l_0 \times P_{\overline{k}}$) acts on $\mathrm{CH}_0(P_{\overline{k}})$ (resp. $\mathrm{CH}_m(P_{\overline{k}})$) as a generator of $\mathrm{End}(\mathrm{CH}_0(P_{\overline{k}})) = \mathbf{Z}$ (resp. $\mathrm{End}(\mathrm{CH}_m(P_{\overline{k}})) = \mathbf{Z}$), and acts trivially on $\mathrm{CH}_j(P_{\overline{k}})$, $j \neq 0$ (resp. $j \neq m$). So, we need only to observe that $\varepsilon_{(0)} = 1$ (since $\Psi^\vee \circ \Phi \circ \Psi(l_0) = \Psi^\vee \circ \Phi(l_0) = \Psi^\vee(l_d) = l_0$)(this is the only place where we use the specifics of Φ). □

Now we can use ε to construct the desired projector in $\mathrm{End}(M(P))$, where $M(P)$ is a *motive* of the quadric P, considered as an object of the classical Chow-motivic category of Grothendieck $Chow^{eff}(k)$ (see [3],[20],[31],[29]). We remind that $\mathrm{End}(M(P))$ is naturally identified with $\mathrm{CH}^m(P \times P)$ with the composition given by the pairing $\circ$.

Take $\omega := \varepsilon - \sum_{0<i<m} b_i \cdot (h^i \times h^{m-i}) - [a/2] \cdot (h^m \times P) \in \mathrm{End}(M(P))$. Then $\omega_{\overline{k}}$ is a projector equal to either $(P_{\overline{k}} \times l_0 + l_0 \times P_{\overline{k}})$, or to $P_{\overline{k}} \times l_0$ (depending on the parity of a).

We have the following easy consequence of the Rost Nilpotence Theorem ([**29**, Corollary 10]):

LEMMA 5.7. ([**33**, Lemma 3.12]) *If for some $\omega \in \mathrm{End}(M(P))$, $\omega_{\overline{k}}$ is an idempotent, then for some r, ω^{2^r} is an idempotent.*

The mutually orthogonal idempotents $P_{\overline{k}} \times l_0$ and $l_0 \times P_{\overline{k}}$ give the direct summands $\mathbb{Z}$ and $\mathbb{Z}(m)[2m]$ in $M(P_{\overline{k}})$. By lemma 5.7, we get a direct summand L in $M(P)$ such that either $L_{\overline{k}} = \mathbb{Z} \oplus \mathbb{Z}(m)[2m]$, or $L_{\overline{k}} = \mathbb{Z}$. The latter possibility is excluded by the following lemma.

LEMMA 5.8. *Let L be a direct summand of $M(P)$ such that $L_{\overline{k}} \simeq \mathbb{Z}$. Then P is isotropic.*

PROOF. Let $w \in \mathrm{CH}^{\dim(P)}(P \times P)$ be the projector, corresponding to L. We have: $\mathrm{End}(M(P_{\overline{k}})) = \prod_r \mathrm{End}(\mathrm{CH}_r(P_{\overline{k}}))$. So, if $\dim(P) > 0$, then the restriction $w_{\overline{k}}$ of our projector to $\overline{k}$ has no choice but to be $P_{\overline{k}} \times l_0 \in \mathrm{CH}^{\dim(P)}(P_{\overline{k}} \times P_{\overline{k}})$. Then, evidently, $degree(w \cap \Delta_P) = 1$, and on $P \times P$, and therefore also on P, we get a point of odd degree. By Springer's Theorem, P is isotropic. If $\dim(P) = 0$, then $\mathrm{End}(M(P))$ has a nontrivial projector if and only if $\det_{\pm}(p) = 1$ ($\Leftrightarrow$ p is isotropic). □

From Lemma 5.7 it follows that ω^{2^r} is an idempotent, and by Lemma 5.8, $\omega^{2^r}|_{\overline{k}} = P_{\overline{k}} \times l_0 + l_0 \times P_{\overline{k}}$. Theorem 5.1 is proven. □

6. Binary direct summands in the motives of quadrics

The following result was proven (but not formulated) by the second author in his thesis (see the proof of Statement 6.1 in [**33**]). We will reproduce its proof here for the reader's convenience.

THEOREM 6.1. ([**33**]) ((*M*), see the end of Section 1) *Let k be a field of characteristic 0, and P be smooth anisotropic projective quadric of dimension n whose Chow-motive $M(P)$ contains a direct summand N such that $N|_{\overline{k}} = \mathbb{Z} \oplus \mathbb{Z}(n)[2n]$. Then $n = 2^s - 1$ for some s.*

PROOF OF THEOREM 6.1. The construction we use here is very close to that used by V.Voevodsky in [**36**].

The category of *Chow-motives* $Chow^{eff}(k)$ which we used in the previous section is a full additive subcategory (closed under taking direct summands) in the triangulated category $DM_{-}^{eff}(k)$ - see [**34**]. The category $DM_{-}^{eff}(k)$ contains the "motives" of all smooth simplicial schemes over k. If P is a smooth projective variety over k. we denote as $\check{C}(P)^{\bullet}$ the *standard simplicial scheme* corresponding to the pair $P \to \mathrm{Spec}(k)$ (see Definition A.8). We will denote its motive by $\mathcal{X}_P$.

From the natural projection: $\check{C}(P)^{\bullet} \xrightarrow{pr} \mathrm{Spec}(k)$, we get a map: $\mathcal{X}_P \xrightarrow{M(pr)} \mathbb{Z}$. By Theorem A.9, $M(pr)_{\overline{k}}$ is an isomorphism. From this point, we will denote $M(pr)$ simply as pr (since we will not use simplicial schemes themselves anymore).

By Theorem A.11, we get that in $DM_{-}^{eff}(k)$,

$$N := Cone[-1](\mathcal{X}_P \xrightarrow{\mu'} \mathcal{X}_P(n)[2n+1]),$$

where μ' is some (actually, the only) nontrivial [2] element from

$$\mathrm{Hom}(\mathcal{X}_P, \mathcal{X}_P(n)[2n+1]).$$

By Theorem A.15, $pr : \mathcal{X}_P \to \mathbb{Z}$ induces the natural isomorphism for all a, b:

$$pr_* : \mathrm{Hom}(\mathcal{X}_P, \mathcal{X}_P(a)[b]) \to \mathrm{Hom}(\mathcal{X}_P, \mathbb{Z}(a)[b]).$$

Denote: $\mu := pr_*(\mu') \in \mathrm{Hom}(\mathcal{X}_P, \mathbb{Z}(n)[2n+1])$.

SUBLEMMA 6.2. *The map*

$$(\mu')^* : \mathrm{Hom}(\mathcal{X}_P, \mathbb{Z}(c)[d]) \to \mathrm{Hom}(\mathcal{X}_P, \mathbb{Z}(c+n)[d+2n+1])$$

coincides with the multiplication by $\mu \in \mathrm{Hom}(\mathcal{X}_P, \mathbb{Z}(n)[2n+1])$.

PROOF. The maps $\Delta_{\mathcal{X}_P} : \mathcal{X}_P \to \mathcal{X}_P \otimes \mathcal{X}_P$, and $\pi_i : \mathcal{X}_P \otimes \mathcal{X}_P \to \mathcal{X}_P$ are mutually inverse isomorphisms (by Theorem A.13). Clearly, $\mu \cdot u = \Delta_{\mathcal{X}_P}(\mu \otimes u)$.

The map $\mu \otimes u : \mathcal{X}_P \otimes \mathcal{X}_P \to \mathbb{Z}(n)[2n+1] \otimes \mathbb{Z}(c)[d]$ coincides with the composition:

$$\begin{array}{ccccc} \mathcal{X}_P & \xrightarrow{\mu'} & \mathcal{X}_P(n)[2n+1] & \xrightarrow{pr(n)[2n+1]} & \mathbb{Z}(n)[2n+1] \\ \otimes & & \otimes & & \otimes \\ \mathcal{X}_P & \xrightarrow{id} & \mathcal{X}_P & \xrightarrow{u} & \mathbb{Z}(c)[d] \end{array}$$

which can be identified with the composition:

$$\mathcal{X}_P \xrightarrow{\mu'} \mathcal{X}_P(n)[2n+1] \xrightarrow{u} \mathbb{Z}(n+c)[2n+1+d],$$

which is equal to $(\mu')^*(u)$. □

SUBLEMMA 6.3. *Multiplication by* μ *induces a homomorphism*

$$\mathrm{Hom}(\mathcal{X}_P, \mathbb{Z}(c)[d]) \to \mathrm{Hom}(\mathcal{X}_P, \mathbb{Z}(c+n)[d+2n+1])$$

which is an isomorphism if $d - c > 0$, *and which is surjective if* $d = c$. *The same holds for cohomology with* $\mathbb{Z}/2$*-coefficients.*

PROOF. Since N is a direct summand in $M(P)$, $\mathrm{Hom}(N, \mathbb{Z}(a)[b]) = 0$, for $b - a > n = \dim(P)$, by Theorem A.2(1).

Consider Hom's from the exact triangle $N \to \mathcal{X}_P \xrightarrow{\mu'} \mathcal{X}_P(n)[2n+1] \to N[1]$ to $\mathbb{Z}(n+c)[2n+d+1]$. We have: $\mathrm{Hom}(N, \mathbb{Z}(n+c)[2n+d+1]) = 0$, if $d - c \geq 0$, and $\mathrm{Hom}(N, \mathbb{Z}(n+c)[2n+d]) = 0$, if $d - c > 0$. This, combined with Sublemma 6.2, implies the statement for $\mathbb{Z}$-coefficients. The case of $\mathbb{Z}/2$-coefficients follows from the five-lemma. □

We can also consider $\tilde{\mathcal{X}}_P := \mathrm{Cone}[-1](\mathcal{X}_P \xrightarrow{pr} \mathbb{Z})$.

SUBLEMMA 6.4. *Let* a *and* b *be integers such that* $b > a$. *Then*

- $\mathrm{Hom}(\mathcal{X}_P, \mathbb{Z}(a)[b])$ *is a* 2*-torsion group,*
- $\mathrm{Hom}(\mathcal{X}_P, \mathbb{Z}(a)[b])$ *embeds into* $\mathrm{Hom}(\mathcal{X}_P, \mathbb{Z}/2(a)[b])$,
- *the natural map* $\tilde{\mathcal{X}}_P \xrightarrow{\delta} \mathcal{X}_P$ *induces an isomorphism*

$$\mathrm{Hom}(\mathcal{X}_P, \mathbb{Z}/2(a)[b]) \xrightarrow{=} \mathrm{Hom}(\tilde{\mathcal{X}}_P, \mathbb{Z}/2(a)[b]).$$

[2]Since, otherwise, $\mathcal{X}_P$ would be a direct summand of N and hence also of $M(P)$, and by Lemma 5.8, P would be isotropic

PROOF. For a finite field extension E/k we have the action of *transfers* on motivic cohomology:

$$\mathrm{Tr} : \mathrm{Hom}(X|_E, \mathbb{Z}(a)[b]) \to \mathrm{Hom}(X, \mathbb{Z}(a)[b]),$$

which is induced by the natural map $\mathbb{Z} \to M(\mathrm{Spec}(E))$ (given by the generic cycle on $\mathrm{Spec}(k) \times \mathrm{Spec}(E) = \mathrm{Spec}(E)$). The main property of the transfer is that $\mathrm{Tr} \circ j$ acts as multiplication by the degree $[E : k]$, where

$$j : \mathrm{Hom}(X, \mathbb{Z}(a)[b]) \to \mathrm{Hom}(X|_E, \mathbb{Z}(a)[b])$$

is the natural restriction.

A quadric P has a point E of degree 2, and over E, $\mathcal{X}_P$ becomes $\mathbb{Z}$ (by Theorem A.9), so we have that $\mathrm{Hom}(\mathcal{X}_P|_E, \mathbb{Z}(a)[b]) = 0$ for $b - a > 0$ (by Theorem A.2(1)).

Considering the composition $\mathrm{Tr} \circ j = \cdot[E : k]$ we get that $\mathrm{Hom}(\mathcal{X}_P, \mathbb{Z}(a)[b])$ is a 2-torsion group for $b > a$. In particular, the natural map $\mathrm{Hom}(\mathcal{X}_P, \mathbb{Z}(a)[b]) \to \mathrm{Hom}(\mathcal{X}_P, \mathbb{Z}/2(a)[b])$ is injective for $b > a$.

Since $\mathrm{Hom}(\mathbb{Z}, \mathbb{Z}/2(a)[b]) = 0$ for any $b > a$ (see Theorem A.2(1)), we also have that for $b > a$, $\delta^* : \mathrm{Hom}(\mathcal{X}_P, \mathbb{Z}/2(a)[b]) \to \mathrm{Hom}(\tilde{\mathcal{X}}_P, \mathbb{Z}/2(a)[b])$ is an isomorphism. □

We have the action of motivic cohomological operations Q_i on $\mathrm{Hom}(\mathcal{X}_P, \mathbb{Z}(*)[*'])$ and $\mathrm{Hom}(\tilde{\mathcal{X}}_P, \mathbb{Z}(*)[*'])$ (see Theorems A.5 and A.6). The differential Q_i acts without cohomology on $\mathrm{Hom}(\tilde{\mathcal{X}}_P, \mathbb{Z}/2(*)[*'])$ for any $i \leq [\log_2(n+1)]$ (see Theorem A.16).

Denote $\eta := \mu(mod2)$, i.e. the image of μ in the cohomology with $\mathbb{Z}/2$ coefficients. From Sublemma 6.4 it follows that $\eta \neq 0$.

Denote $r = [\log_2(n)]$.

SUBLEMMA 6.5. *$Q_i(\eta) = 0$, for all $i \leq r$.*

PROOF. In fact, $Q_i(\eta) \in \mathrm{Hom}(\mathcal{X}_P, \mathbb{Z}/2(n + 2^i - 1)[2n + 2^{i+1}])$, and the latter group is an extension of 2-cotorsion in $\mathrm{Hom}(\mathcal{X}_P, \mathbb{Z}(n + 2^i - 1)[2n + 2^{i+1}])$, and 2-torsion in $\mathrm{Hom}(\mathcal{X}_P, \mathbb{Z}(n + 2^i - 1)[2n + 2^{i+1} + 1])$.

But, by Sublemma 6.3, the multiplication by μ induces surjections $\mathrm{Hom}(\mathcal{X}_P, \mathbb{Z}(2^i - 1)[2^{i+1} - 1]) \to \mathrm{Hom}(\mathcal{X}_P, \mathbb{Z}(n + 2^i - 1)[2n + 2^{i+1}])$ and $\mathrm{Hom}(\mathcal{X}_P, \mathbb{Z}(2^i - 1)[2^{i+1}]) \to \mathrm{Hom}(\mathcal{X}_P, \mathbb{Z}(n + 2^i - 1)[2n + 2^{i+1} + 1])$. Furthermore, the groups: $\mathrm{Hom}(\mathcal{X}_P, \mathbb{Z}(2^i - 1)[2^{i+1} - 1])$, $\mathrm{Hom}(\mathcal{X}_P, \mathbb{Z}(2^i - 1)[2^{i+1}])$ are zero.

In fact, from the exact triangle $N \to \mathcal{X}_P \to \mathcal{X}_P(n)[2n + 1] \to N[1]$, we get an exact sequence: $\mathrm{Hom}(N, \mathbb{Z}(2^i - 1)[2^{i+1} - 1]) \leftarrow \mathrm{Hom}(\mathcal{X}_P, \mathbb{Z}(2^i - 1)[2^{i+1} - 1]) \leftarrow \mathrm{Hom}(\mathcal{X}_P(n)[2n + 1], \mathbb{Z}(2^i - 1)[2^{i+1} - 1])$. The first group is zero since N is a direct summand in the motive of a smooth projective variety, and (consequently) $\mathrm{Hom}(N, \mathbb{Z}(a)[b]) = 0$ for $b > 2a$ (see Theorem A.2(2)). The third group is zero, since $n > 2^i - 1$ (see Theorem A.1). Hence the second is zero as well. The case of $\mathrm{Hom}(\mathcal{X}_P, \mathbb{Z}(2^i - 1)[2^{i+1}])$ follows in an analogous manner.

Thus, $Q_i(\eta) = 0$. □

SUBLEMMA 6.6. *Let $0 \leq j \leq r$. Then Q_j is injective on $\mathrm{Hom}(\tilde{\mathcal{X}}_P, \mathbb{Z}/2(c)[d])$, if $d - c = n + 1 + 2^j$.*

PROOF. Let $\tilde{v} \in \mathrm{Hom}(\tilde{\mathcal{X}}_P, \mathbb{Z}/2(c)[d])$, where $d - c = n + 1 + 2^j$. If $Q_j(\tilde{v}) = 0$, then $\tilde{v} = Q_j(\tilde{w})$, for some $\tilde{w} \in \mathrm{Hom}(\tilde{\mathcal{X}}_P, \mathbb{Z}/2(c - 2^j + 1)[d - 2^{j+1} + 1])$ (since Q_j acts without cohomology on $\mathrm{Hom}(\tilde{\mathcal{X}}_P, \mathbb{Z}/2(*)[*'])$, by Theorem A.16). Since

$(d-2^{j+1}+1)-(c-2^j+1)=n+1>0$, we have that $\delta^*:\mathrm{Hom}(\mathcal{X}_P,\mathbb{Z}/2(c-2^j+1)[d-2^{j+1}+1])\to\mathrm{Hom}(\tilde{\mathcal{X}}_P,\mathbb{Z}/2(c-2^j+1)[d-2^{j+1}+1])$ is an isomorphism, and there exists $w\in\mathrm{Hom}(\mathcal{X}_P,\mathbb{Z}/2(c-2^j+1)[d-2^{j+1}+1])$ such that $\tilde{w}=\delta^*(w)$.

By Sublemma 6.3, $w=\eta\cdot u$, for some $u\in\mathrm{Hom}(\mathcal{X}_P,\mathbb{Z}/2(c-2^j+1-n)[d-2^{j+1}-2n])$. By Theorem A.6(2), $Q_j(\eta\cdot u)=Q_j(\eta)\cdot u+\eta\cdot Q_j(u)+\sum\{-1\}^{x_i}\phi_i(\eta)\cdot\psi_i(u)$, where $x_i>0$, and ϕ_i,ψ_i are cohomological operations of some bidegree $(*)[*']$, where $*'>2*\geq 0$.

Note that $c-2^j+1-n=d-2^{j+1}-2n=:s$. But by Theorem A.18, we have that $\mathrm{Hom}(\mathbb{Z},\mathbb{Z}/2(a)[b])\xrightarrow{pr^*}\mathrm{Hom}(\mathcal{X}_P,\mathbb{Z}/2(a)[b])$ is an isomorphism for $a\geq b$. Hence, $u=pr^*(u_0)$, where $u_0\in\mathrm{Hom}(\mathbb{Z},\mathbb{Z}/2(s)[s])=\mathrm{K}^M_s(k)/2$. We have: $Q_j(u_0)=0$ and $\psi_i(u_0)=0$ (since $\mathrm{Hom}(\mathbb{Z},\mathbb{Z}/2(a)[b])=0$ for $b>a$).

But pr^* commutes with Q_j and ψ_i. So, $Q_j(u)=0$ and $\psi_i(u)=0$. That means: $Q_j(w)=Q_j(\eta\cdot u)=Q_j(\eta)\cdot u=0$, by Sublemma 6.5.

We get: $\tilde{v}=Q_j(\tilde{w})=Q_j\circ\delta^*(w)=\delta^*\circ Q_j(w)=0$. I.e., Q_j is injective on $\mathrm{Hom}(\tilde{\mathcal{X}}_P,\mathbb{Z}/2(c)[d])$. □

Denote $\tilde{\eta}:=\delta^*(\eta)\in\mathrm{Hom}(\tilde{\mathcal{X}}_P,\mathbb{Z}/2(n)[2n+1])$. Since $\eta\neq 0$, we have $\tilde{\eta}\neq 0$ (by Sublemma 6.4).

SUBLEMMA 6.7. *Let $0\leq m<r$, and $\tilde{\eta}=Q_m\circ\cdots\circ Q_1\circ Q_0(\tilde{\eta}_m)$ for some $\tilde{\eta}_m\in\mathrm{Hom}(\tilde{\mathcal{X}}_P,\mathbb{Z}/2(n-2^{m+1}+m+2)[2n-2^{m+2}+m+4])$. Then there exists $\tilde{\eta}_{m+1}$ such that $\tilde{\eta}_m=Q_{m+1}(\tilde{\eta}_{m+1})$.*

PROOF. Since Q_{m+1} acts without cohomology on $\mathrm{Hom}(\tilde{\mathcal{X}}_P,\mathbb{Z}/2(*)[*'])$, it is enough to show that $Q_{m+1}(\tilde{\eta}_m)=0$.

Denote $\tilde{v}:=Q_{m+1}(\tilde{\eta}_m)$. We have $\tilde{v}\in\mathrm{Hom}(\tilde{\mathcal{X}}_P,\mathbb{Z}/2(n+m+1)[2n+m+3])$. Since Q_i commutes with Q_j (by Theorem A.6(1)), we have: $Q_m\circ Q_{m-1}\circ\cdots\circ Q_0(\tilde{v})=Q_m\circ Q_{m-1}\circ\cdots\circ Q_0\circ Q_{m+1}(\tilde{\eta}_m)=Q_{m+1}\circ Q_m\circ Q_{m-1}\circ\cdots\circ Q_0(\tilde{\eta}_m)=Q_{m+1}(\tilde{\eta})=0$, by Sublemma 6.5.

But, for any $0\leq t\leq m$, $Q_{t-1}\circ\cdots\circ Q_0(\tilde{v})\in\mathrm{Hom}(\tilde{\mathcal{X}}_P,\mathbb{Z}(c)[d])$, where $d-c=n+1+2^t$, and Q_t is injective on $\mathrm{Hom}(\tilde{\mathcal{X}}_P,\mathbb{Z}(c)[d])$, by Sublemma 6.6. So, from the equality $Q_m\circ Q_{m-1}\circ\cdots\circ Q_0(\tilde{v})=0$, we get $\tilde{v}=0$. □

From Sublemma 6.7 it follows that $\tilde{\eta}=Q_r\circ\cdots\circ Q_1\circ Q_0(\tilde{\eta}_r)$. Denote $\tilde{\gamma}:=\tilde{\eta}_r$. We have $\tilde{\gamma}\in\mathrm{Hom}(\tilde{\mathcal{X}}_P,\mathbb{Z}/2(n-2^{r+1}+2+r)[2n-2^{r+2}+4+r])$.

But $(2n-2^{r+2}+4+r)-(n-2^{r+1}+2+r)=n-2^{r+1}+2$, and $r=[\log_2(n)]$, hence $2^r\leq n<2^{r+1}$. Since we know that $\mathrm{Hom}(\tilde{\mathcal{X}}_P,\mathbb{Z}/2(a)[b])=0$ for $a\geq b$ (by Theorem A.18), and $\tilde{\eta}\neq 0$, the only possible choice for n is $n=2^{r+1}-1$.

Theorem 6.1 is proven. □

LEMMA 6.8. *Let $0\leq j\leq r$. Then Q_j is injective on $\mathrm{Hom}(\tilde{\mathcal{X}}_P,\mathbb{Z}/2(c)[d])$ provided $d-c=2^j$.*

PROOF. Let $\tilde{v}\in\mathrm{Hom}(\tilde{\mathcal{X}}_P,\mathbb{Z}/2(c)[d])$, where $d-c=2^j$. If $Q_j(\tilde{v})=0$, then $\tilde{v}=Q_j(\tilde{w})$ for some $\tilde{w}\in\mathrm{Hom}(\tilde{\mathcal{X}}_P,\mathbb{Z}/2(c-2^j+1)[d-2^{j+1}+1])$ (since Q_j acts without cohomology on $\mathrm{Hom}(\tilde{\mathcal{X}}_P,\mathbb{Z}/2(*)[*'])$, by Theorem A.16). But $(c-2^j+1)=(d-2^{j+1}+1)$, and $\mathrm{Hom}(\tilde{\mathcal{X}}_P,\mathbb{Z}/2(c-2^j+1)[d-2^{j+1}+1])=0$, by Theorem A.18. □

THEOREM 6.9. (compare with [**8**, Theorem 3.1]) ((*M*), see the end of Section 1) *Let k be a field of characteristic 0, P be a smooth n-dimensional anisotropic projective quadric over k, and N be a direct summand in $M(P)$ such that $N|_{\overline{k}}=$*

$\mathbb{Z} \oplus \mathbb{Z}(n)[2n]$. *Then* $n = 2^{s-1} - 1$, *and there exists* $\alpha \in \mathrm{K}_s^M(k)/2$ *such that for any field extension* E/k, *the following conditions are equivalent:*
1) $\alpha|_E = 0$; *2)* $P|_E$ *is isotropic.*

In particular, $\alpha \in \mathrm{Ker}(\mathrm{K}_s^M(k)/2 \to \mathrm{K}_s^M(k(P))/2) \neq 0$.

PROOF. It follows from Theorem 6.1 that $n = 2^{r+1} - 1$ for some r, and there exists $\tilde{\gamma} \in \mathrm{Hom}(\tilde{\mathcal{X}}_P, \mathbb{Z}/2(r+1)[r+2])$ such that $\tilde{\eta} = Q_r \circ \cdots \circ Q_1 \circ Q_0(\tilde{\gamma})$. By Sublemma 6.4, the map $\delta^* : \mathrm{Hom}(\mathcal{X}_P, \mathbb{Z}/2(r+1)[r+2]) \to \mathrm{Hom}(\tilde{\mathcal{X}}_P, \mathbb{Z}/2(r+1)[r+2])$ is an isomorphism, and $\tilde{\gamma} = \delta^*(\gamma)$ for some $\gamma \in \mathrm{Hom}(\mathcal{X}_P, \mathbb{Z}/2(r+1)[r+2])$. Let τ be the only nontrivial element of $\mathrm{Hom}(\mathbb{Z}/2, \mathbb{Z}/2(1)) = \mathbf{Z}/2$. Denote as α the element corresponding to $\tau \circ \gamma$ via identification (by Theorem A.18) $\mathrm{Hom}(\mathcal{X}_P, \mathbb{Z}/2(r+2)[r+2]) = \mathrm{Hom}(\mathbb{Z}, \mathbb{Z}/2(r+2)[r+2]) = \mathrm{K}_{r+2}^M(k)/2$. Then, by Theorem A.20, for any field extension E/k, $\alpha|_E = 0$ if and only if $\gamma|_E = 0$. But $\gamma|_E = 0 \Leftrightarrow \tilde{\gamma}|_E = 0$. By Lemma 6.8, $\tilde{\gamma}|_E = 0 \Leftrightarrow \tilde{\eta}|_E = 0$. By Sublemma 6.4, $\tilde{\eta}|_E = 0 \Leftrightarrow \eta|_E = 0 \Leftrightarrow \mu|_E = 0$. Finally, $\mu|_E = 0$ if and only if $\mathcal{X}_P|_E$ is a direct summand in N and, consequently, in $M(P)$, which by Lemma 5.8, is equivalent to $P|_E$ being isotropic. □

REMARK 6.10. 1) Theorem 6.9 basically says that under the mentioned conditions, the quadric P is a *norm-variety* for $\alpha \in \mathrm{K}_s^M(k)/2$.
2) Taking into account the Milnor conjecture ([**36**]) and the definition of the *Rost projector*, we see that Theorem 6.9 implies Theorem 1.12.
3) It should be mentioned that in small-dimensional cases it is possible to prove the result (in arbitrary characteristic $\neq 2$) without the use of Voevodsky's technique. For example, the case $n = 7$ was considered in [**15**].

7. Properties of forms with absolutely maximal splitting

In this section we work with fields satisfying the condition $\mathrm{char}\, F = 0$. We begin with the following modification of Theorem 1.11.

THEOREM 7.1. ((*M*), see the end of Section 1) *Let* ϕ *be an anisotropic quadratic form over a field* F *of characteristic* 0. *Suppose that* ϕ *is an AMS-form. Then*

(1) ϕ *has maximal splitting,*
(2) *the group* $H^s(F(\phi)/F)$ *is nontrivial, where* s *is the integer such that* $2^{s-1} < \dim\phi \leq 2^s$.

PROOF. (1) Let ψ be subform of ϕ of codimension $i_1(q) - 1$. Let X be the projective quadric corresponding to ψ. By Theorem 5.1, X possesses a Rost projector. Theorem 6.1 shows that $\dim(X) = 2^{s-1} - 1$ for suitable s. Hence $\dim\psi = 2^{s-1} + 1$. By the definition of ψ, we have $\dim\phi - i_1(\phi) = \dim\psi - 1 = 2^{s-1}$. Therefore $\dim\phi = 2^{s-1} + m$, where $m = i_1(\phi)$. To prove that ϕ has maximal splitting, it suffices to verify that $m \leq 2^{s-1}$. This is obvious because $2^{s-1} + m = \dim\phi \geq 2i_1(\phi) = 2m$.
(2) Obvious in view of Theorem 6.9 and the isomorphism $k_s(F) \simeq H^s(F)$. □

Theorem 7.1 and Conjecture 1.1 make natural the following

CONJECTURE 7.2. *If an anisotropic quadratic form has absolutely maximal splitting, then it is a Pfister neighbor.*

In the proof of Theorem 1.8 we will need some deep results related to the Milnor conjecture.

THEOREM 7.3. (see [**36**],[**26**]). ((*M*), see the end of Section 1) *Let F be a field of characteristic 0. Then for any $n \geq 0$*

(1) *there exists an isomorphism $e^n : I^n(F)/I^{n+1}(F) \stackrel{\simeq}{\to} H^n(F)$ such that*

$$e^n(\langle\langle a_1, \ldots, a_n\rangle\rangle) = (a_1, \ldots, a_n).$$

(2) *If ϕ is a Pfister neighbor of $\pi \in GP_n(F)$. Then $H^n(F(\phi)/F)$ is generated by $e^n(\pi)$.*
(3) *If $\dim \tau > 2^n$, then $H^n(F(\tau)/F) = 0$.*
(4) *The ideal $I^n(F)$ coincides with Knebusch's ideal $J_n(F)$. In other words, for any $\tau \in I^n(F)\backslash I^{n+1}(F)$, we have $\deg \tau = n$.*

We need also the following easy consequence of a result by Hoffmann.

LEMMA 7.4. *Let ϕ be an anisotropic form such that $\dim \phi \leq 2^n$. Let τ be an anisotropic quadratic form and $F_0 = F, F_1, \ldots, F_h$ be the generic splitting tower of τ. Let j be such that $\dim(\tau_{F_{j-1}})_{an} > 2^n$. Suppose that ϕ_{F_j} has maximal splitting. Then ϕ has maximal splitting.*

PROOF. Obvious in view of [**4**, Lemma 5]. □

PROPOSITION 7.5. ((*M*), see the end of Section 1) *Let F be a field of characteristic 0. Let ϕ be a quadratic form over F and n be such that $2^{n-1} < \dim \phi \leq 2^n$. Suppose that $H^n(F(\phi)/F) \neq 0$. Then $H^n(F(\phi)/F) \simeq \mathbf{Z}/2\mathbf{Z}$ and ϕ has maximal splitting.*

PROOF. Let u be an arbitrary nonzero element of the group $H^n(F(\phi)/F)$. Since the homomorphism $e^n : I^n(F)/I^{n+1}(F) \to H^n(F)$ is an isomorphism, there exists an anisotropic $\tau \in I^n(F)$ such that $\tau \notin I^{n+1}(F)$ and $e^n(\tau) = u \in H^n(F(\phi)/F)$. Let $F_0 = F, F_1, \ldots, F_h$ be the generic splitting tower of τ. Let $\tau_i = (\tau_{F_i})_{an}$. Since $\tau \in I^n(F)\backslash I^{n+1}(F)$, Item (4) of Theorem 7.3 shows that $\deg \tau = n$. Therefore, τ_{h-1} is a nonhyperbolic form in $GP_n(F_{h-1})$. Since $e^n(\tau) \in H^n(F(\phi)/F)$, we have $e^n((\tau_{h-1})_{F_{h-1}(\phi)}) = 0$. Hence, τ_{h-1} is hyperbolic over the function field of $\phi_{F_{h-1}}$. Since τ_{h-1} is an anisotropic form in $GP_n(F_{h-1})$, the Cassels–Pfister subform theorem shows that $\phi_{F_{h-1}}$ is a Pfister neighbor of $\tau_{F_{h-1}}$. Hence $\phi_{F_{h-1}}$ has maximal splitting. Lemma 7.4 shows that ϕ has maximal splitting.

Since $\phi_{F_{h-1}}$ is a Pfister neighbor of $\tau_{F_{h-1}}$, Item (2) of Theorem 7.3 shows that $|H^n(F_{h-1}(\phi)/F_{h-1})| \leq 2$. By Item (3) of Theorem 7.3, we have $H^n(F_{h-1}/F) = 0$. Hence $|H^n(F_{h-1}(\phi)/F)| \leq 2$. Since $H^n(F(\phi)/F) \subset H^n(F_{h-1}(\phi)/F)$, we get $|H^n(F(\phi)/F)| \leq 2$. Now, since $H^n(F(\phi)/F) \neq 0$, we have $H^n(F(\phi)/F) \simeq \mathbf{Z}/2\mathbf{Z}$. □

COROLLARY 7.6. ((*M*), see the end of Section 1) *Let $n \geq 5$ and ϕ be an anisotropic form such that $2^n - 7 \leq \dim \phi \leq 2^n$ Then the following conditions are equivalent:*

(a) *ϕ has maximal splitting,*
(b) *ϕ is a Pfister neighbor,*
(c) *$H^n(F(\phi)/F) \simeq \mathbf{Z}/2\mathbf{Z}$.*
(d) *$H^n(F(\phi)/F) \neq 0$.*

PROOF. (a)⇒(b) follows from Theorem 1.7. (b)⇒(c) follows from Theorem 7.3; (c)⇒(d) is obvious; (d)⇒(a) is proved in Proposition 7.5. □

PROOF OF THEOREM 1.8. Let ϕ and n be as in Theorem 1.8. If ϕ has maximal splitting, then Lemma 4.1 shows that ϕ has absolutely maximal splitting. Then Theorem 7.1 shows that $H^n(F(\phi)/F) \neq 0$. Conversely, if we suppose that $H^n(F(\phi)/F) \neq 0$, then Proposition 7.5 shows that ϕ has maximal splitting. □

Appendix A.

In this section we will list some results of V.Voevodsky, which we use in the proof of Theorems 6.1 and 6.9.

We will assume everywhere that $char(k) = 0$.

First of all, we need some facts about triviality of motivic cohomology of smooth simplicial schemes. If not specified otherwise, under $\mathrm{Hom}(-,-)$ we will mean $\mathrm{Hom}_{DM_-^{eff}(k)}(-,-)$. We remind that $\mathrm{Hom}_{DM_-^{eff}(k)}(M(X), \mathbb{Z}(a)[b])$ is naturally identified with $H_B^{b,a}(X, \mathbb{Z})$ (see [**36**]).

THEOREM A.1. ([**36**, Corollary 2.2(1)]) *Let $\mathcal{X}$ be smooth simplicial scheme over k. Then* $\mathrm{Hom}(M(\mathcal{X})(a)[b], \mathbb{Z}(c)[d]) = 0$, *for any $a > c$.*

In the case of a smooth variety we have further restrictions on motivic cohomology:

THEOREM A.2. ([**36**, Corollary 2.3]) *Let N be a direct summand in $M(X)$, where X is a smooth scheme over k. Then* $\mathrm{Hom}(N, \mathbb{Z}(a)[b]) = 0$ *in the following cases:*

1 *If* $b - a > \dim(X)$;
2 *If* $b > 2a$.

The same is true about cohomology with $\mathbb{Z}/2$-coefficients.

In [**36**] the *Stable homotopy category of schemes over* $\mathrm{Spec}(k)$, $\mathcal{SH}(k)$ was defined (see also [**25**]). $\mathcal{SH}(k)$ is a *triangulated* category, and there is a functor $S : SmSimpl/k \to \mathcal{SH}(k)$, and a triangulated functor $G : \mathcal{SH}(k) \to DM_-^{eff}(k)$ such that the composition $G \circ S : SmSimpl/k \to DM_-^{eff}(k)$ coincides with the usual *motivic functor*: $X \mapsto M(X)$ (here $SmSimpl/k$ is the category of smooth simplicial schemes over $\mathrm{Spec}(k)$). In [**36**], Section 3.3, the *Eilenberg-MacLane spectrum* $\mathbf{H}_{\mathbb{Z}/2}$ (as an object of $\mathcal{SH}(k)$) is defined, together with its shifts $\mathbf{H}_{\mathbb{Z}/2}(a)[b]$, for $a, b \in \mathbf{Z}$.

THEOREM A.3. ([**36**, Theorem 3.12]) *If X is a smooth simplicial scheme, then there exist canonical isomorphisms*

$$\mathrm{Hom}_{\mathcal{SH}(k)}(S(X), \mathbf{H}_{\mathbb{Z}/2}(a)[b]) = \mathrm{Hom}_{DM_-^{eff}(k)}(M(X), \mathbb{Z}/2(a)[b]).$$

DEFINITION A.4. ([**36**, p.31]) The *motivic Steenrod algebra* is the algebra of endomorphisms of $\mathbf{H}_{\mathbb{Z}/2}$ in $\mathcal{SH}(k)$, i.e.:

$$\mathcal{A}^{b,a}(k, \mathbb{Z}/2) = \mathrm{Hom}_{\mathcal{SH}(k)}(\mathbf{H}_{\mathbb{Z}/2}, \mathbf{H}_{\mathbb{Z}/2}(a)[b]).$$

The composition gives a pairing:

$$\mathrm{Hom}_{\mathcal{SH}(k)}(U, \mathbf{H}_{\mathbb{Z}/2}(c)[d]) \otimes \mathcal{A}^{b,a}(k, \mathbb{Z}/2) \to \mathrm{Hom}_{\mathcal{SH}(k)}(U, \mathbf{H}_{\mathbb{Z}/2}(c+a)[d+b]),$$

which is natural on U.

Let now $f : X \to Y$ be a morphism in $SmSimpl/k$. In $\mathcal{SH}(k)$ we have an *exact triangle*: $\mathrm{cone}(S(f))[-1] \xrightarrow{\delta'} S(X) \xrightarrow{S(f)} S(Y) \to \mathrm{cone}(f)$.

Theorem A.3 implies:

THEOREM A.5. *We have an action of the motivic Steenrod algebra* $\mathcal{A}^{*,*}(k,\mathbb{Z}/2)$ *on* $\oplus_{a,b}\operatorname{Hom}_{DM_-^{eff}(k)}(M(X),\mathbb{Z}/2(a)[b])$, $\oplus_{a,b}\operatorname{Hom}_{DM_-^{eff}(k)}(M(Y),\mathbb{Z}/2(a)[b])$, *and* $\oplus_{a,b}\operatorname{Hom}(\operatorname{cone}(M(f))[-1],\mathbb{Z}/2(a)[b])$, *which is compatible with* $M(f)^*$ *and* δ^*.

We have some special elements $Q_i\in\mathcal{A}^{2^{i+1}-1,2^i-1}(k,\mathbb{Z}/2)$ (see [**36**, p.32]).

THEOREM A.6. ([**36**, Theorems 3.17 and 3.14])

1) $Q_i^2=0$, *and* $Q_iQ_j+Q_jQ_i=0$.
2) *Let* $u,v\in\operatorname{Hom}_{DM_-^{eff}(k)}(M(X),\mathbb{Z}(*)[*'])$, *for a smooth simplicial scheme* X. *Then* $Q_i(u\cdot v)=Q_i(u)\cdot v+u\cdot Q_i(v)+\sum\{-1\}^{n_j}\phi_j(u)\cdot\psi_j(v)$, *where* $n_j>0$, *and* $\phi_j,\psi_j\in\mathcal{A}(k,\mathbb{Z}/2)$ *are some (homogeneous) elements of bidegree* (b,a), *where* $b>2a\geq 0$.
3) $Q_i=[\beta,q_i]$, *where* β *is Bockstein, and* $q_i\in\mathcal{A}(k,\mathbb{Z}/2)$.

Following [**36**], we define:

DEFINITION A.7. ([**36**, p.32]) Margolis motivic cohomology $\tilde{H}M_i^{b,a}(U)$ are cohomology groups of the complex: $\operatorname{Hom}_{\mathcal{SH}(k)}(U,\mathbf{H}_{\mathbb{Z}/2}(a-2^i+1)[b-2^{i+1}+1])\stackrel{Q_i}{\to}\operatorname{Hom}_{\mathcal{SH}(k)}(U,\mathbf{H}_{\mathbb{Z}/2}(a)[b])\stackrel{Q_i}{\to}\operatorname{Hom}_{\mathcal{SH}(k)}(U,\mathbf{H}_{\mathbb{Z}/2}(a+2^i-1)[b+2^{i+1}-1])$, for any $U\in Ob(\mathcal{SH}(k))$.

If U is $\operatorname{Cone}[-1](S(f))$, for some morphism $f:X\to Y$ of simplicial schemes, then by Theorems A.3 and A.5, $\tilde{H}M_i^{b,a}(U)$ coincides with the cohomology of the complex

$$\mathrm{H}_{\mathcal{M}}^{b-2^{i+1}+1,a-2^i+1}(M(U),\mathbb{Z}/2)\stackrel{Q_i}{\to}\mathrm{H}_{\mathcal{M}}^{b,a}(M(U),\mathbb{Z}/2)\stackrel{Q_i}{\to}\mathrm{H}_{\mathcal{M}}^{b+2^{i+1}-1,a+2^i-1}(M(U),\mathbb{Z}/2),$$

where $\mathrm{H}_{\mathcal{M}}^{d,c}(*,\mathbb{Z}/2):=\operatorname{Hom}_{DM_-^{eff}(k)}(*,\mathbb{Z}/2(c)[d])$, and $M(U)=\operatorname{Cone}[-1](M(f))$. Since $\tilde{H}M_i^{b,a}(\operatorname{Cone}[-1](S(f)))$ depends only on $M(f)$, we can denote it simply as $\tilde{H}M_i^{b,a}(\operatorname{Cone}[-1](M(f)))$.

Let P be some smooth projective variety over $\operatorname{Spec}(k)$.

DEFINITION A.8. The *standard simplicial scheme* $\check{C}(P)^\bullet$, corresponding to the pair $P\to\operatorname{Spec}(k)$ is the simplicial scheme such that $\check{C}(P)^n=P\times\cdots\times P$ ($n+1$-times), with faces and degeneration maps given by partial projections and diagonals.

In $SmSimpl/k$ we have a natural projection: $pr:\check{C}(P)^\bullet\to\operatorname{Spec}(k)$. Let us denote $\mathcal{X}_P:=M(\check{C}(P)^\bullet)$. We get the natural map $M(pr):\mathcal{X}_P\to\mathbb{Z}$.

THEOREM A.9. ([**36**, Lemma 3.8]) *If* P *has a* k*-rational point, then* $M(pr):\mathcal{X}_P\to\mathbb{Z}$ *is an isomorphism.*

REMARK A.10. 1) In the notations of [**36**, Lemma 3.8], one should take $X=P$, $Y=\operatorname{Spec}(k)$, and observe that the simplicial weak equivalence gives an isomorphism on the level of *motives*. 2) Actually, $M(pr)$ is an isomorphism if and only if P has a 0-cycle of degree 1 (see [**33**, Theorem 2.3.4]).

Theorem A.9 shows that $\mathcal{X}_P|_{\overline{k}}=\mathbb{Z}$, which means that $\mathcal{X}_P$ is a *form* of the Tate-motive.

THEOREM A.11. ([**36**, Theorem 4.4]) *Let* P *be an anisotropic projective quadric of dimension* n. *Let* N *be a direct summand in* $M(P)$ *such that* $N|_{\overline{k}}=\mathbb{Z}\oplus\mathbb{Z}(n)[2n]$. *Then in* $DM_-^{eff}(k)$ *there exists a distinguished triangle of the form:*

$$\mathcal{X}_P(n)[2n]\to N\to\mathcal{X}_P\stackrel{\mu'}{\to}\mathcal{X}_P(n)[2n+1].$$

REMARK A.12. Actually, Theorem 4.4 of [**36**] is formulated only for the case of the *Rost motive* (as a direct summand in the motive of the *minimal Pfister neighbour*). But the proof does not use any specifics of the Pfister form case, and works with any "binary" direct summand of dimension $= dim(P)$. At the same time, Theorem A.11 is a very particular case of [**33**, Lemma 3.23].

THEOREM A.13. ([**36**, Lemma 3.8]) *The natural diagonal map* $\Delta_{\mathcal{X}_P} : \mathcal{X}_P \to \mathcal{X}_P \otimes \mathcal{X}_P$ *is an isomorphism.*

REMARK A.14. One should observe that the same proof as in [**36**, Lemma 3.8] gives the *simplicial weak equivalence* $\check{C}(P)^\bullet \times \check{C}(P)^\bullet \stackrel{pr_1}{\to} \check{C}(P)^\bullet$ with the diagonal map as inverse.

THEOREM A.15. ([**36**, Lemma 4.7])
$M(pr)_* : \mathrm{Hom}(\mathcal{X}_P, \mathcal{X}_P(a)[b]) \to \mathrm{Hom}(\mathcal{X}_P, \mathbb{Z}(a)[b])$ *is an isomorphism, for any* a, b.

Let us denote: $\tilde{\mathcal{X}}_P := \mathrm{Cone}[-1](M(pr))$. Since $\tilde{\mathcal{X}}_P$ comes from $\mathcal{SH}(k)$, it makes sense to speak of the *Margolis cohomology* $\tilde{H}M_i^{b,a}(\tilde{\mathcal{X}}_P)$ of $\tilde{\mathcal{X}}_P$.

Suppose now that P be a smooth projective quadric of dimension $\geq 2^i - 1$.

The following result of V.Voevodsky is the main tool in studying motivic cohomology of quadrics:

THEOREM A.16. ([**36**, Theorem 3.25 and Lemma 4.11]) *Let* P *be a smooth projective quadric of dimension* $\geq 2^i - 1$*, then* $\tilde{H}M_i^{b,a}(\tilde{\mathcal{X}}_P) = 0$*, for any* a, b.

REMARK A.17. In [**36**, Lemma 4.11], the result is formulated only for the case of a $(2^i - 1)$-dimensional *Pfister quadric* (corresponding to the form $\langle\langle a_1, \ldots, a_i\rangle\rangle \perp -\langle a_{i+1}\rangle)$. But the proof does not use any specifics of the *Pfister case* (the only thing which is used is: for any $j \leq i$, P has a plane section of dimension $2^j - 1$, which is again a quadric). Thus, for any quadric P of dimension $\geq 2^i - 1$, the ideal I_P contains a $(v_i, 2)$-element (notations from [**36**, Lemma 4.11]).

The following statement is a consequence of the Beilinson-Lichtenbaum Conjecture for $\mathbb{Z}/2$-coefficients.

THEOREM A.18. ([**36**, Proposition 2.7, Corollary 2.13(2) and Theorem 4.1]) *The map* $M(pr)^* : \mathrm{Hom}(\mathbb{Z}, \mathbb{Z}/2(a)[b]) \to \mathrm{Hom}(\mathcal{X}_P, \mathbb{Z}/2(a)[b])$ *is an isomorphism for any* $b \leq a$.

REMARK A.19. We should add that in [**36**, Theorem 4.1] it is proven that the condition $H90(n, 2)$ is satisfied for all n and all fields of characteristic 0 - see p.11 of [**36**]. Also, $\mathrm{Hom}(M(X), \mathbb{Z}/2(a)[b])$ can be identified with $H_B^{b,a}(X, \mathbb{Z}/2)$.

Motivic cohomology of $\mathcal{X}_P$ can be used to compute the kernel on Milnor's K-theory ($mod\,2$):

THEOREM A.20. ([**35**, Lemma 6.4], [**36**, Theorem 4.1] ; or [**12**, Theorem A.1]) *Let* $\tau \in \mathrm{Hom}_{DM_-^{eff}(k)}(\mathbb{Z}/2, \mathbb{Z}/2(1)) = \mathbf{Z}/2$ *be the only nontrivial element. Let* P *be a smooth projective quadric over* k*. Then the composition*

$$\mathrm{Hom}(\mathcal{X}_P, \mathbb{Z}/2(m-1)[m]) \stackrel{\tau\circ}{\to} \mathrm{Hom}(\mathcal{X}_P, \mathbb{Z}/2(m)[m]) = \mathrm{Hom}(\mathbb{Z}, \mathbb{Z}/2(m)[m]) = \mathrm{K}_m^M(k)/2$$

identifies the first group with the $\ker(\mathrm{K}_m^M(k)/2 \to \mathrm{K}_m^M(k(P))/2)$.

References

[1] Arason, J. K. *Cohomologische invarianten quadratischer Formen.* J. Algebra **36** (1975), no. 3, 448–491.

[2] Colliot-Thélène, J.-L.; Sujatha, R. *Unramified Witt groups of real anisotropic quadrics.* K-theory and algebraic geometry: connections with quadratic forms and division algebras (Santa Barbara, CA, 1992), 127–147, Proc. Sympos. Pure Math., **58**, Part 2, Amer. Math. Soc., Providence, RI, 1995.

[3] Fulton, W. *Intersection Theory.* Springer-Verlag, 1984.

[4] Hoffmann, D. W. *Isotropy of quadratic forms over the function field of a quadric.* Math. Z. **220** (1995), 461–476.

[5] Hoffmann, D. W. *Splitting patterns and invariants of quadratic forms.* Math. Nachr. **190** (1998), 149-168.

[6] Hurrelbrink, J., Rehmann, U. *Splitting patterns of quadratic forms.* Math. Nachr. **176** (1995), 111–127.

[7] Izhboldin, O. T. *Quadratic forms with maximal splitting.* Algebra i Analiz, vol. **9**, no. 2, 51–57. (Russian). English transl.: in St. Petersburg Math J. Vol. **9** (1998), No. 2.

[8] Izhboldin, O. T. *Quadratic Forms with Maximal Splitting, II.*, Preprint, February 1999.

[9] Izhboldin, O. T. *Fields of u-invariant 9.* Preprint, Bielefeld University, 1999, 1-50.

[10] Jacob, W.; Rost, M. *Degree four cohomological invariants for quadratic forms*, Invent. Math. **96** (1989), 551-570.

[11] Kahn, B. *A descent problem for quadratic forms.* Duke Math. J. **80** (1995), no. 1, 139–155.

[12] Kahn, B. *Motivic cohomology of smooth geometrically cellular varieties.*, Algebraic K-theory (Seattle, WA, 1997), 149–174, Amer. Math. Soc., Providence, RI., 1999.

[13] Kahn, B.; Rost, M.; Sujatha, R. *Unramified cohomology of quadrics I.* Amer. J. Math. **120** (1998), no. 4, 841–891.

[14] Kahn, B. and Sujatha, R. *Motivic cohomology and unramified cohomology of quadrics*, J. of the Eur. Math. Soc. **2** (2000), 145–177.

[15] Karpenko, N. A. *Characterization of minimal Pfister neighbors via Rost projectors.* Universität Münster, Preprintreihe SFB 478 - Geometrische Strukturen in der Mathematik, 1999 Heft 65

[16] Knebusch, M. *Generic splitting of quadratic forms, I.* Proc. London Math. Soc **33** (1976), 65–93.

[17] Knebusch, M. *Generic splitting of quadratic forms, II.* Proc. London Math. Soc **34** (1977), 1–31.

[18] Laghribi, A. *Sur le problème de descente des formes quadratiques.* Arch. Math. **73** (1999), no. 1, 18–24.

[19] Lam, T. Y. *The Algebraic Theory of Quadratic Forms.* Massachusetts: Benjamin 1973 (revised printing: 1980).

[20] Manin, Yu. I. *Correspondences, motifs and monoidal transformations.*, Mat. Sb. **77** (1968), no. 4, 475-507; English transl. in Math. USSR-Sb. **6** (1968), no. 4, 439-470.

[21] Merkurjev, A. S. *The norm residue homomorphism of degree* 2 (in Russian), Dokl. Akad. Nauk SSSR **261** (1981), 542-547. English translation: Soviet Mathematics Doklady **24** (1981), 546-551.

[22] Merkurjev, A. S. *On the norm residue homomorphism for fields*, Mathematics in St. Petersburg, 49-71, A.M.S. Transl. Ser. 2, **174**, A.M.S., Providence, RI, 1996.

[23] Merkurjev, A. S.; Suslin, A. A. *The norm residue homomorphism of degree 3* (in Russian), Izv. Akad. Nauk SSSR **54** (1990), 339-356. English translation: Math. USSR Izv. **36** (1991), 349-368.

[24] Milnor, J. W. *Algebraic K-theory and quadratic forms*, Invent. Math. **9** (1969/70), 318-344.

[25] Morel, F.; Voevodsky, V. *A^1-homotopy theory of schemes.*, Preprint, Oct. 1998 (www.math.uiuc.edu/K-theory/0305/)

[26] Orlov, D.; Vishik, A.; Voevodsky, V. *Motivic cohomology of Pfister quadrics and Milnor's conjecture on quadratic forms.* Preprint in preparation.

[27] Rost, M. *Hilbert theorem 90 for K_3^M for degree-two extensions*, Preprint, Regensburg, 1986.

[28] Rost, M. *Some new results on the Chow-groups of quadrics.*, Preprint, Regensburg, 1990.

[29] Rost, M. *The motive of a Pfister form.*, Preprint, 1998 (see www.physik.uniregensburg.de/ rom03516/papers.html).

[30] Scharlau, W. *Quadratic and Hermitian Forms* Springer, Berlin, Heidelberg, New York, Tokyo (1985).

[31] Scholl, A. J. *Classical motives.* Proc. Symp. Pure Math. **55.1** (1994) 163–187.

[32] Szyjewski, M. *The fifth invariant of quadratic forms* (in Russian), Algebra i Analiz **2** (1990), 213-224. English translation: Leningrad Math. J. **2** (1991), 179-198.

[33] Vishik, A. *Integral motives of quadrics.* Max Planck Institut Für Mathematik, Bonn, preprint MPI-1998-13, 1–82.

[34] Voevodsky, V. *Triangulated category of motives over a field*, K-Theory Preprint Archives, Preprint 74, 1995 (see www.math.uiuc.edu/K-theory/0074/).

[35] Voevodsky, V. *Bloch-Kato conjecture for $\mathbb{Z}/2$-coefficients and algebraic Morava K-theories.*, Preprint, June 1995. (www.math.uiuc.edu/K-theory/0076/)

[36] Voevodsky, V. *Milnor conjecture.*, preprint, December 1996. (www.math.uiuc.edu/K-theory/0170/)

Fakultät für Mathematik
Universität Bielefeld
Postfach 100131
D-33501 Bielefeld, GERMANY

School of Mathematics
Institute for Advanced Study
Olden Lane
Princeton, New Jersey 08540, USA
E-mail address: vishik@mccme.ru

Contemporary Mathematics
Volume **272**, 2000

2-Regularity and Reversibility of Quadratic Mappings

Alexey F. Izmailov

ABSTRACT. Some particular answers to the following question are presented: does a surjective quadratic mapping between Banach spaces have a bounded right inverse or not?

These notes are concerned with some particular answers to one question connected with quadratic mappings. Let X and Y be linear spaces. We refer to a mapping $Q : X \to Y$ as *quadratic* if there exists a bilinear mapping $B : X \times X \to Y$ such that

$$Q(x) = B(x, x) \quad \forall x \in X.$$

To say it in other words, a quadratic mapping is a homogeneous of degree 2 polynomial mapping. In the case of a finite-dimensional Y, a quadratic mapping is a mapping whose components are quadratic forms. Recall that for any quadratic mapping Q, there exists a unique symmetric bilinear mapping B related to Q in the sense mentioned above. Hence in the sequel, we shall denote quadratic mapping and associated symmetric bilinear mapping by the same symbol. Several mathematicians, e.g. R. Aron, B. Cole, S. Dineen, T. Gamelin, R. Gonzalo, J. Jaramillo, R. Ryan, have studied polynomials in Banach spaces. See the book of Dineen [**D**], especially chapters 1 and 2, for more information including a comprehensive list of references.

When studying singular points of smooth nonlinear mappings, quadratic mappings analysis is of great importance. Let us mention two different and rather productive approaches to the study of irregular problems which have been actively developed for the last two decades. The first approach is based on the so-called 2-normality concept [**Ar1**], the other one is based on the 2-regularity construction (see [**IT**] and the bibliography there). But for any approach of such a kind it is typical that second derivatives are taken into account. When the first derivative is onto (that is the regular case), the first derivative is a good local approximation to the mapping under consideration. Applying the highly developed theory of linear operators to linear approximation one can obtain the most important facts of nonlinear analysis, such as the implicit function theorem and its numerous corollaries.

2000 *Mathematics Subject Classification.* Primary 47H60; Secondary 58C15.

The author was supported by the Russian Foundation for Basic Research Grant #97-01-00188. The author also thanks IMPA, where he was a visiting professor during the completion of this paper.

Naturally, when the first derivative is not onto (that is the singular case), linear approximation is not enough for a description of the nonlinear mapping local structure, and one has to take into account the quadratic term of the Taylor formula.

On the other hand, quadratic mappings are of great interest by themselves because a quadratic mapping is the most simple model of a substantially nonlinear mapping. When we are to construct an example for some theorem on singular points, we consider quadratic mappings first of all.

But in contrast with the theory of linear operators, quadratic mappings theory is not really developed so far. There are no answers to some basic questions, and here we would like to discuss one such question connected with the existence of a bounded right inverse to a quadratic mapping. For linear operators, a complete answer to this question is given by the classical Banach open mapping theorem. Let us recall this result.

Suppose now that X and Y are Banach spaces, and $A: X \to Y$ is a continuous linear operator. The *right inverse* to A is the mapping

$$A^{-1}: Y \to 2^X, \quad A^{-1}(y) = \{x \in X \,|\, Ax = y\}.$$

Hence for any $y \in Y$, $A^{-1}y$ is the complete pre-image of y with respect to A. Let us define the "norm"

$$\|A^{-1}\| = \sup_{\substack{y \in Y,\\ \|y\|=1}} \inf_{x \in A^{-1}y} \|x\|$$

(if A is a one-to-one operator, then this "norm" is a classical operator norm of its classical inverse). Let us refer to the right inverse as *bounded* if $\|A^{-1}\|$ is finite. By the open mapping theorem, A^{-1} is bounded if and only if A is onto, i.e.

$$\operatorname{Im} A = Y.$$

Now let us replace here the continuous linear operator A by a continuous quadratic mapping Q and consider the same question (all the notions and notations remain without any modification). In topological terms the question is: determine the conditions under which the image of a neighborhood of zero element in X is a neighborhood of zero element in Y?

First, let us mention that in contrast with theory of linear operators, in the general case the right inverse to Q can be unbounded when Q is onto, as illustrated by the following example.

EXAMPLE 1. Consider the mapping

$$Q: l_2 \to l_2, \quad Q(x) = \left(\sum_{i=2}^{\infty} \frac{x_i^2}{i} - x_1^2,\, x_1x_2,\, x_1x_3,\, \ldots \right).$$

One can show (by means of some fairly straightforward computations) that this mapping is onto, but its right inverse is not bounded.

We next consider some sufficient conditions for boundedness of a right inverse, but for that purpose we need some more terminology. For any element $h \in X$ let us define the linear operator

$$Q_h: X \to Y, \quad Q_h x = Q(h, x).$$

Recall that Q here is the symmetric bilinear mapping associated with the quadratic mapping under consideration. Note that $2Q_h$ is exactly the Fréchet derivative of Q

at the point h:

$$Q'(h) = 2Q_h, \quad h \in X.$$

By $N(Q)$ let us denote the null-set of Q:

$$N(Q) = \{x \in X | \, Q(x) = 0\}.$$

Note that $N(Q)$ is always a closed cone but normally this cone is not convex.

A quadratic mapping Q is referred to as:

- *2-regular with respect to an element* $h \in X$ if $\operatorname{Im} Q_h = Y$ (to say it in other words, if Q is regular, or normal, at the point h);
- *2-regular* if it is 2-regular with respect to every element $h \in N(Q) \setminus \{0\}$;
- *strongly 2-regular* if there exists a number $\alpha > 0$ such that

$$\sup_{\substack{h \in N_\alpha(Q), \\ \|h\|=1}} \|Q_h^{-1}\| < \infty,$$

where

$$N_\alpha(Q) = \{x \in X | \, \|Q(x)\| \leq \alpha\}$$

is the α-extension of the null-set of Q.

It is easy to see that for the finite-dimensional case, strong 2-regularity is equivalent to 2-regularity, but for the general case strong 2-regularity is somewhat stronger (for instance, it is possible that $N(Q) = \{0\}$, and the 2-regularity condition is trivially satisfied, but at the same time $N_\alpha(Q) \neq \{0\} \; \forall \, \alpha > 0$). Note also that all this terminology can be applied to a general nonlinear mapping at a fixed point, and this results in the basic notions of 2-regularity theory (for that purpose one has to consider the invariant second differential of the mapping at the point under consideration as Q). Some necessary and sufficient conditions for strong 2-regularity of quadratic mappings were proposed in [**Ar2**].

THEOREM 1. *Let X and Y be Banach spaces. Assume that $Q : X \to Y$ is a continuous quadratic mapping, and one of the following conditions is satisfied:*

1. *$N(Q) = \{0\}$, Q is strongly 2-regular, and $Q(X) = Y$;*
2. *Q is 2-regular with respect to some $h \in N(Q)$.*

Then Q^{-1} is bounded.

Certainly, conditions 1) and 2) cannot be satisfied simultaneously. If the condition 2) holds, the assertion of Theorem 1 is fairly standard as it follows immediately from the standard implicit function theorem (note that in this case Q is surjective automatically, so one does not have to assume that Q is surjective). The assertion corresponding to condition 1) is not so standard as it follows from a special theorem on distance estimates for strongly 2-regular mappings. This theorem can be considered as the generalization of the classical Lyusternik's theorem on a tangent subspace. This generalization was proposed in the concurrent papers [**BT**], [**Av2**] (see also [**T**], [**Av1**], [**IT**], [**Ar1**]).

Now let us consider the finite-dimensional case. Clearly, in this case the strong 2-regularity condition in Theorem 1 can be omitted. Note that Example 1 has a strong infinite-dimensional specificity. For the mapping Q from Example 1 the conditions $N(Q) = \{0\}$ and $Q(X) = Y$ hold, and the assertion of Theorem 1 is not valid only for one reason: Q is not strongly 2-regular. We do not know any finite-dimensional example of a surjective quadratic mapping with unbounded right inverse, and we have to admit that we do not know so far if such an example is possible or not. However, we would like to discuss some particular answers to this question. We are in an algebraic situation now, and it seems reasonable that in order to find to find such answers, one has to take into account both analytical and algebraic arguments.

To begin with, note that for the finite-dimensional case Theorem 1 (without the strong 2-regularity assumption) provides a complete description for a typical quadratic mapping.

PROPOSITION 1. *For any positive integers n and m, the set of 2-regular quadratic mappings is open and dense in the set of all quadratic mappings from $\mathbf{R}^n$ to $\mathbf{R}^m$.*

The set of all quadratic mappings from $\mathbf{R}^n$ to $\mathbf{R}^m$ is considered here with the standard norm topology. It is well known that this set has a natural linear structure and can be normed in a natural way.

Proposition 1 follows from Thom's transversality theorem. It was stated in [**Ag**] in a different form.

We see now that for almost any finite-dimensional quadratic mapping, the following alternative is true: the null-set is trivial, or the mapping is 2-regular with respect to any nonzero element from the null-set. Hence by Theorem 1, for almost any finite-dimensional quadratic mapping the following implication holds: if it is surjective then its right inverse is bounded. The question is: can one replace here the words "for almost any" by "for any" or not? It is clear that the answer is positive for $m = 1$ (i.e. for the case of one quadratic form, and this case is fairly trivial). The answer turns out to be positive for two forms as well, but this case is already not trivial at all.

PROPOSITION 2. *For any positive integer n and any quadratic mapping $Q : \mathbf{R}^n \to \mathbf{R}^2$ condition $Q(X) = \mathbf{R}^2$ is equivalent to the boundedness of Q^{-1}.*

PROOF. If the conclusion fails to hold, then there exists a sequence $\{y^k\} \subset \mathbf{R}^2$ such that

$$\inf_{x \in Q^{-1}(y^k)} \|x\| \to \infty \text{ as } k \to \infty, \tag{1}$$

$$\|y^k\| = 1 \quad \forall\, k = 1,\, 2,\, \ldots$$

It is possible to assume that $\{y^k\} \to \eta$ as $(k \to \infty)$, $\|\eta\| = 1$. Then taking into account that Q is surjective, we can claim the existence of an element $x(\eta) \in \mathbf{R}^n$ such that $Q(x(\eta)) = \eta$. Obviously, the case $Q_{x(\eta)} = 0$ can be omitted as $\eta \in \operatorname{Im} Q_{x(\eta)}$. Hence, we have to consider two cases:

- $\operatorname{Im} Q_{x(\eta)} = \mathbf{R}^2$, then a contradiction can be obtained immediately by taking advantage of the implicit function theorem;
- $\operatorname{Im} Q_{x(\eta)} = \operatorname{span}\{\eta\}$ is a straight line.

Obviously, one of the open half-planes defined in $\mathbf{R}^2$ by this straight line contains infinitely many elements of the sequence $\{y^k\}$. Hence, we can assume that

$$\langle \zeta,\, y^k \rangle > 0 \quad \forall\, k = 1,\, 2,\, \ldots$$

where $\zeta \in \mathbf{R}^2$ is a fixed element such that

$$\langle \eta,\, \zeta \rangle = 0, \quad \|\zeta\| = 1.$$

Again we take into account that Q is surjective; this implies that there exists $x(\zeta) \in \mathbf{R}^n$ such that $Q(x(\zeta)) = \zeta$.

Without loss of generality, we can assume that

$$\langle \eta, Q(x(\eta), x(\zeta)) \rangle \geq 0$$

(if this inequality does not hold, one should replace $x(\eta)$ by $-x(\eta)$).

Any element $y \in \mathbf{R}^2$ can be uniquely represented as $y = y_1\eta + y_2\zeta$, with $y_1, y_2 \in \mathbf{R}$. Assume that

$$\langle \zeta, y \rangle > 0, \quad \|y\| = 1$$

(the inequality means that $y_2 > 0$) and consider the equation

$$Q(x(\eta) + tx(\zeta)) = \tau y$$

with respect to $(t, \tau) \in \mathbf{R} \times \mathbf{R}$. Under our assumptions, this equation can be reformulated as the system of equations

$$1 + 2t\|Q(x(\eta), x(\zeta))\| = \tau y_1,$$

$$t^2 = \tau y_2.$$

It is easy to verify that if $y_1 > 0$ then this system has two solutions, and both tend to $(0, 1)$ as $y_1 \to 1$, $y_2 \to 0$. Hence for any sufficiently large k, one can find $t_k, \tau_k \in \mathbf{R}$ such that

$$y^k = Q\left(\frac{x(\eta) + t_k x(\zeta)}{\sqrt{\tau_k}}\right),$$

and

$$\left\{\frac{x(\eta) + t_k x(\zeta)}{\sqrt{\tau_k}}\right\} \to x(\eta) \text{ as } k \to \infty.$$

But this is in contradiction with (1). □

We are not aware so far whether or not a similar result holds for $m > 2$. Note that two cases are of most importance in applications: $n \gg m$ and $n = m$. The first case is important in view of constrained optimization problems. Some special results for this case were obtained by Agrachyov in [**Ag**]. Let us discuss here the second case.

PROPOSITION 3. *For $n = 2$ or 3 and for any quadratic mapping $Q : \mathbf{R}^n \to \mathbf{R}^n$ condition $Q(X) = \mathbf{R}^n$ is sufficient for the equality $N(Q) = \{0\}$.*

PROOF. Assume that $N(Q) \neq \{0\}$. Let us take advantage of the following procedure which is sometimes useful for a reduction to lower dimension.

Let $h \in N(Q)\setminus\{0\}$. Then the space $\mathbf{R}^n$ can be represented as the direct sum of the straight line $\operatorname{span}\{h\}$ and some linear subspace $\tilde{X}$ in $\mathbf{R}^n$, $\dim \tilde{X} = n - 1$. This means that any vector $x \in X$ can be uniquely represented as $x = th + \tilde{x}$, where $t = t(x) \in \mathbf{R}$, $\tilde{x} = \tilde{x}(x) \in \tilde{X}$. Then

$$Q(x) = 2t(x)Q(h, \tilde{x}(x)) + Q(\tilde{x}(x)), \quad x \in \mathbf{R}^n. \tag{2}$$

Let $Y_1 = \operatorname{Im} Q_h$, Y_2 be an algebraic complement of Y_1 in $\mathbf{R}^n$, and P be the projector onto Y_2 parallel to Y_1 in $\mathbf{R}^n$. Note that $h \in \operatorname{Ker} Q_h \setminus \{0\}$, hence

$$\dim Y_2 = \operatorname{corank} Q_h \geq 1. \tag{3}$$

Note also that the case $n = 1$ is trivial as a quadratic mapping from $\mathbf{R}$ to $\mathbf{R}$ (and hence to $\mathbf{R}^m$ for any m) cannot be surjective.

Let $n = 2$. Then according to (2), (3), $PQ(\mathbf{R}^2) \neq Y_2$, as PQ is the quadratic mapping which actually depends on one scalar variable $\tilde{x} \in \tilde{X}$. But it means that Q cannot be surjective (the subspace Y_2 contains elements which do not belong to $Q(\mathbf{R}^2)$).

Let us now turn our attention to the case $n = 3$, and let us consider the possible values of $\operatorname{corank} Q_h$ separately (recall that the case $\operatorname{corank} Q_h = 0$ cannot occur because of (3)).

If $\operatorname{corank} Q_h = 3$ (i.e. $Q_h = 0$), then according to (2) we have

$$Q(x) = Q(\tilde{x}(x)), \quad x \in \mathbf{R}^3.$$

Then Q has to be surjective as a mapping from $\tilde{X}$ ($\dim \tilde{X} = 2$) to $\mathbf{R}^3$, but this is obviously impossible (for instance, because of the proven fact for $n = 2$, or because of some general facts of differential topology).

Let $\operatorname{corank} Q_h = 2$. Since Q is surjective and (2) holds, PQ is surjective as a mapping from $\tilde{X}$ to Y_2, and $\dim \tilde{X} = \dim Y_2 = 2$. It means that $N(PQ) \cap \tilde{X} = \{0\}$ (here again we take into account the proven fact for $n = 2$). But then according to (2), the nonempty set $Y_1 \setminus \{0\}$ has an empty intersection with $Q(\mathbf{R}^3)$, and this is in contradiction with the condition that Q is surjective.

Finally, let $\operatorname{corank} Q_h = 1$. Recall again that Q is surjective and (2) holds, hence the mapping PQ on $\tilde{X}$ can be considered as a quadratic form which is not semi-definite. Taking into account the equality $\dim \tilde{X} = 2$, we see now that the set $N(PQ) \cap \tilde{X}$ consists of two straight lines spanned by some elements h^1, $h^2 \in \tilde{X} \setminus \{0\}$ (as a matter of fact, the vectors h^1 and h^2 are linearly independent by necessity, but this is not important here).

Now in (2) we take $\tilde{x} = \tau h^i$, $\tau \in \mathbf{R}$. Taking into account that $Q(h^i) \in Y_1$, $i = 1, 2$, let us define the quadratic mappings

$$Q_i : \mathbf{R}^2 \to Y_1, \quad Q_i(\theta) =$$

$$= 2t\tau Q(h, h^i) + \tau^2 Q(h^i), \quad \theta = (t, \tau), \, i = 1, 2.$$

According to the argument above, the condition that Q is surjective implies the equality

$$Q_1(\mathbf{R}^2) \cup Q_2(\mathbf{R}^2) = Y_1, \tag{4}$$

because in the other case the subspace Y_1 contains some elements which do not belong to $Q(\mathbf{R}^3)$.

Now we have only to prove that the equality (4) is contradictory. There are several ways to do it, but the most simple is a straightforward analysis of the structure of $Q_i(\mathbf{R}^2)$, $i = 1, 2$.

For any $i = 1, 2$, the appearance of the mapping Q_i allows us to state the following:

- if the vectors $Q(h, h^i)$ and $Q(h^i)$ are linearly independent then $Q_i(\mathbf{R}^2)$ is an open half-plane with the additional point 0 in the plain Y_1;
- if the vectors $Q(h, h^i)$ and $Q(h^i)$ are linearly dependent then $Q_i(\mathbf{R}^2)$ is either a straight line, or a ray, or $Q_i(\mathbf{R}^2) = \{0\}$.

Clearly, in any case the sets $Q_1(\mathbf{R}^2)$ and $Q_2(\mathbf{R}^2)$ together do not cover the entire plane Y_1. □

According to Theorem 1 we have that, under the assumptions of Proposition 3, if Q is surjective then its right inverse is bounded. But again, we have to admit that we do not know whether or not a similar result holds for $n > 3$.

It is interesting that Propositions 2 and 3 hold true only for quadratic mappings. For polynomial mappings which are homogeneous of degree greater than two, similar results are not obtained. Moreover, we provide an example of a mapping of degree 5 which is onto, but its right inverse is not bounded.

EXAMPLE 2. Consider the mapping

$$Q : \mathbf{R}^2 \to \mathbf{R}^2, \quad Q(x) = (x_1^5 - x_1^2 x_2^3, \, x_1^3 x_2^2).$$

One can show that the image of a neighborhood of zero element in the original space is not a neighborhood of zero element in the image space here; at the same time, Q is onto. The easiest (but of course, not accurate) way to see this is to draw the image of the unit square (e.g., using a computer).

Clearly, these notes contain more questions than answers. A closely related question is: under what assumptions is it the case that not only is a given quadratic mapping surjective but any quadratic mapping close to it is surjective? In [**Ag**], a quadratic mapping with such a property was referred to as *substantially surjective*. There is a good reason to think that a complete answer to one of the questions in these notes will result in a complete answer to another, and perhaps, to numerous important questions about the topology and algebra of quadratic mappings.

The author would like to thank the organizers of QF99 for their hospitality and for financial aid they provided for his participation. The author is also grateful to two referees for their helpful comments.

References

[Av1] Y. R. Avakov, *Extremum conditions for smooth problems with equality-type constraints*, USSR Comput. Math. and Math. Phys **25** (1985), 24–32.

[Av2] Y. R. Avakov, *Theorems on estimates in the neighborhood of a singular point of a mapping*, Math. Notes **47** (1990), 425–432.

[Ag] A. A. Agrachyov, *Quadratic mappings topology and Hessians of smooth mappings* (in Russian), Itogi Nauki i Tekhniki. Algebra. Topologiya. Geometriya **26** (1983), 85–124.

[Ar1] A. V. Arutyunov, *Extremum conditions. Abnormal and degenerate cases* (in Russian), Factorial, Moscow, 1997.

[Ar2] A. V. Arutyunov, *On the theory of quadratic mappings in Banach spaces*, Soviet Math. Dokl. **42** (1991), 621–623.

[BT] K. N. Belash and A. A. Tretyakov, *Methods for solving singular problems*, USSR Comput. Math. and Math. Phys. **28** (1988), 1097–1102.

[D] S. Dineen, *Complex Analysis on Infinite Dimensional Spaces,* Springer Monographs in Mathematics, Springer-Verlag, London Berlin Heidelberg (1999).

[IT] A. F. Izmailov and A. A. Tretyakov, *Factoranalysis of nonlinear mappings* (in Russian), Nauka, Moscow, 1994.

[T] A. A. Tretyakov, *Necessary and sufficient conditions for optimality of p-th order*, USSR Comput. Math. and Math. Physics **24** (1984), 123–127.

Computing Center of the Russian Academy of Sciences
Vavilova Str. 40
117967, Moscow, Russia

Current address: Instituto de Matematica Pura e Applicada
Estrada Dona Castorina
110, Jardim Botanico
Rio de Janeiro, RJ 22460-320
Brazil

E-mail address: `izmaf@impa.br`, `izmaf@ccas.ru`

Contemporary Mathematics
Volume **272**, 2000

Quadratic Forms in Knot Theory

C. Kearton

Abstract. The purpose of this survey article is to show how quadratic and hermitian forms can give us geometric results in knot theory. In particular, we shall look at the knot cobordism groups, at questions of factorisation and cancellation of high-dimensional knots, and at branched cyclic covers.

Contents

1. The Seifert Matrix 135
2. Blanchfield Duality 142
3. Factorisation of Knots 144
4. Cancellation of Knots 146
5. Knot Cobordism 146
6. Branched Cyclic Covers 149
7. Spinning and Branched Cyclic Covers 151
8. Concluding Remarks 152
References 152

1. The Seifert Matrix

By an n-knot k we mean a smooth or locally-flat PL pair (S^{n+2}, S^n), both spheres being oriented. In the smooth case the embedded sphere S^n may

2000 *Mathematics Subject Classification.* 57Q45, 57M25, 11E12.
Key words and phrases. knot, knot module, Seifert matrix, quadratic form, hermitian form, Blanchfield duality.

have an exotic smooth structure. Two such pairs are to be regarded as equivalent if there is an orientation preserving (smooth or PL) homeomorphism between them.

As shown in [**33, 48**], a regular neighbourhood of S^n has the form $S^n \times B^2$; we set $K = S^{n+2} - \operatorname{int}\left(S^n \times B^2\right)$. Then K is the *exterior* of the knot k, and since a regular neighbourhood of S^n is unique up to ambient isotopy it follows that K is essentially unique.

PROPOSITION 1.1. *S^n is the boundary of an orientable $(n+1)$-manifold in S^{n+2}.*

This is proved in [**31, 33, 48**], but in the case $n = 1$ there is a construction due to Seifert [**41**] which we now give.

PROPOSITION 1.2. *Every classical knot k is the boundary of some compact orientable surface embedded in S^3.*

PROOF. Consider a diagram of k. Starting at any point of k, move along the knot in the positive direction. At each crossing point, jump to the other piece of the knot and follow that in the positive direction. Eventually

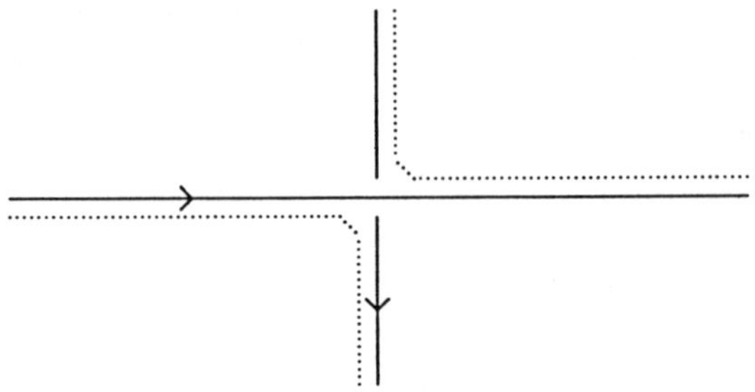

FIGURE 1.1

we return to the starting point, having traced out a *Seifert circuit*. Now start somewhere else, and continue until the knot is exhausted. The Seifert circuits are disjoint circles, which can be capped off by disjoint discs, and joined by half-twists at the crossing points. Hence we get a surface V with

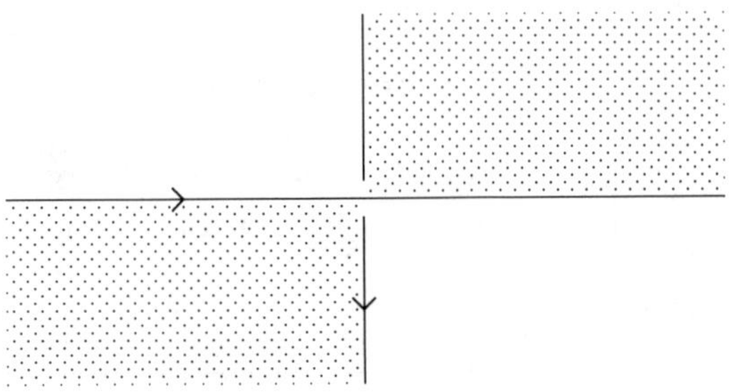

FIGURE 1.2

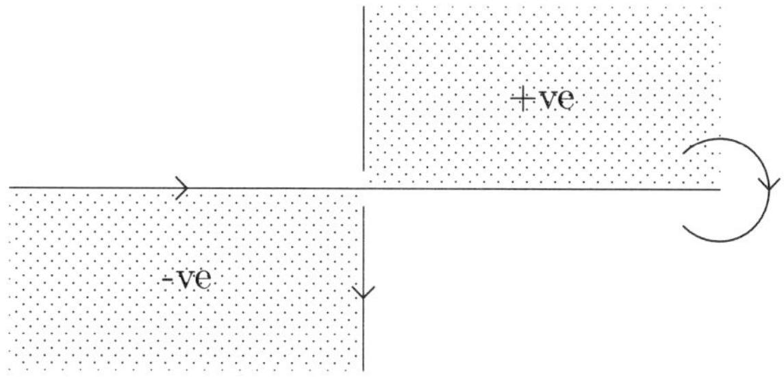

FIGURE 1.3

$\partial V = k$. To see that V is orientable, attach a normal to each disc using a right hand screw along the knot. Note that in passing from one disc to another the normal is preserved. Thus there are no closed paths on V which reverse the sense of the normal: hence V is orientable. □

COROLLARY 1.3. *If there are c crossing points and s Seifert circuits, then the genus of V is $\frac{1}{2}(c - s + 1)$.*

PROOF. The *genus* of V is g where $H_1(V) = \bigoplus_1^{2g} \mathbb{Z}$. We have a handle decomposition of V with s 0-handles and c 1-handles, which is equivalent to one with a single 0-handle and $c-(s-1)$ 1-handles (by cancelling 0-handles). Thus $2g = c - s + 1$. □

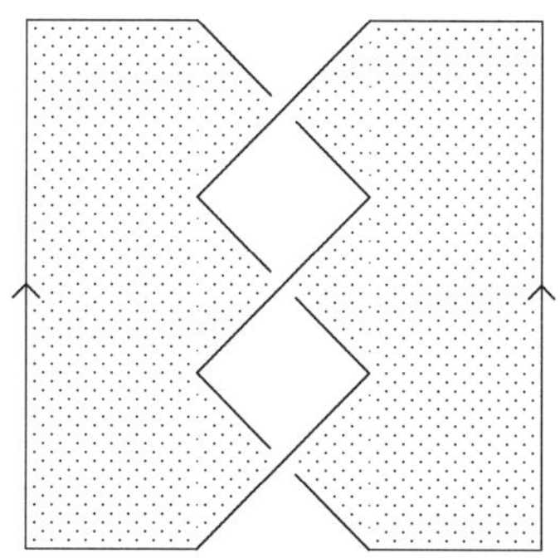

FIGURE 1.4. Trefoil knot

As an example we have a genus one Seifert surface of the trefoil knot in Figure 1.4.

DEFINITION 1.4. Let u, v be two oriented disjoint copies of S^1 in S^3 and assign a linking number as follows. Span v by a Seifert surface V and move u slightly so that it intersects V transversely. To each point of intersection we assign $+1$ or -1 according as u is crossing in the positive or negative direction, and taking the sum of these integers gives us $L(u, v)$. Two simple examples are indicated in Figure 1.5.

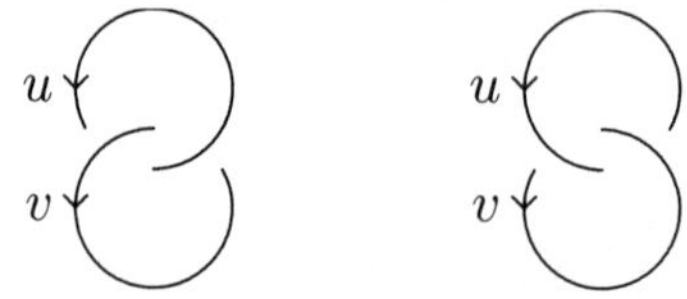

$L(u, v) = +1 \qquad L(u, v) = -1$

FIGURE 1.5. The linking number

The two copies of S^1 do not have to be embedded: the general definition is in terms of cycles and bounding chains.

DEFINITION 1.5. Let $u, v \in H_1(V)$ for some Seifert surface V. Let i_+u be the result of pushing u a small distance along the positive normal to V. Then set $\theta(u, v) = L(i_+u, v)$.

LEMMA 1.6. $\theta : H_1(V) \times H_1(V) \to \mathbb{Z}$ *is bilinear.*

DEFINITION 1.7. Let $x_1, \ldots, x_{2g}$ be a basis for $H_1(V) = \bigoplus_1^{2g} \mathbb{Z}$, and set $a_{ij} = \theta(x_i, x_j)$. The matrix $A = (a_{ij})$ is a *Seifert matrix* of k.

Note that if we choose another basis for $H_1(V)$, then A is replaced by PAP', where $\det P = \pm 1$, since P is a matrix over $\mathbb{Z}$ which is invertible over $\mathbb{Z}$.

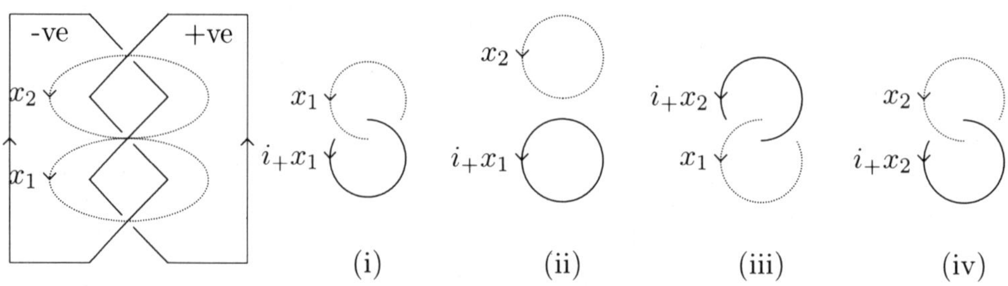

FIGURE 1.6

(i)	$a_{11} = \theta(x_1, x_1) = L(i_+x_1, x_1)$	-1
(ii)	$a_{12} = \theta(x_1, x_2) = L(i_+x_1, x_2)$	0
(iii)	$a_{21} = \theta(x_2, x_1) = L(i_+x_2, x_1)$	1
(iv)	$a_{22} = \theta(x_2, x_2) = L(i_+x_2, x_2)$	-1

TABLE 1.1

EXAMPLE 1.8. We see from Table 1.1 that the Seifert matrix of the trefoil knot in Figure 1.6 is

$$A = \begin{pmatrix} -1 & 0 \\ 1 & -1 \end{pmatrix}$$

Given a knot k, there will be infinitely many Seifert surfaces of k; for example, we can excise the interiors of two disjoint closed discs from any given Seifert surface V and glue a tube $S^1 \times B^1$ to what remains of V by the boundary circles, as illustrated in Figure 1.7

DEFINITION 1.9. A Seifert surface U is obtained from a Seifert surface V by *ambient surgery* if U and V are related as in Figure 1.7 or Figure 1.8. In the first case, the interiors of two disjoint closed discs in the interior of V are excised and a tube $S^1 \times B^1$ is attached, the two attachments being made on the same side of V. In symbols, $S^0 \times B^2$ is replaced by $S^1 \times B^1$. In the second case, the procedure is reversed.

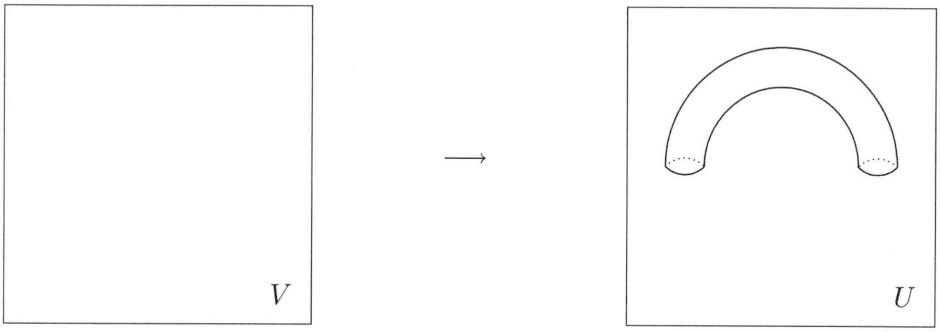

FIGURE 1.7. Ambient surgery (i)

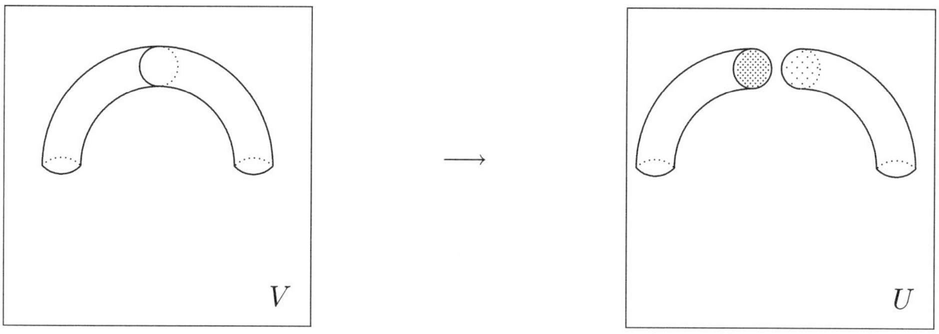

FIGURE 1.8. Ambient surgery (ii)

Note that the "hollow handle" may be knotted, and that these are inverse operations.

PROPOSITION 1.10. *For a given knot k, any two Seifert surfaces are related by a sequence of ambient surgeries.*

DEFINITION 1.11. Let A be a Seifert matrix. An *elementary S-equivalence* on A is one of the following, or its inverse.

(i) $A \mapsto PAP'$ for P a unimodular integer matrix.

(ii)

$$A \mapsto \begin{pmatrix} A & 0 & 0 \\ \alpha & 0 & 0 \\ 0 & 1 & 0 \end{pmatrix}$$

(iii)

$$A \mapsto \begin{pmatrix} A & \beta & 0 \\ 0 & 0 & 1 \\ 0 & 0 & 0 \end{pmatrix}$$

where α is a row vector of integers and β is a column vector of integers. Two matrices are *S-equivalent* if they are related by a finite sequence of such moves.

THEOREM 1.12. *Any two Seifert matrices of a given knot k are S-equivalent.*

PROOF. Let A be the matrix obtained from a Seifert surface U, B from V. After Proposition 1.10, it is enough to assume that V is obtained by an ambient surgery on U. Consider the diagram in Figure 1.9, and choose gen-

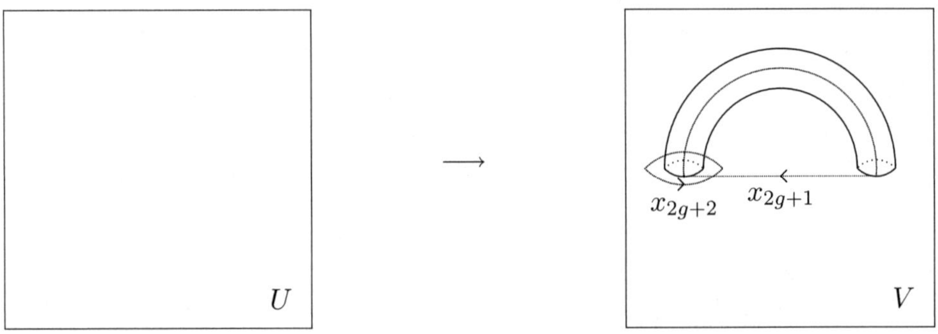

FIGURE 1.9

erators $x_1, \ldots, x_{2g}$ of $H_1(U)$, and $x_1, \ldots, x_{2g}, x_{2g+1}, x_{2g+2}$ of $H_1(V)$. Then $\theta\left(x_{2g+1}, x_{2g+1}\right) = 0$ if we choose the right number of twists around the handle (see Figure 1.10 for a different choice)

$\theta\left(x_{2g+1}, x_{2g+2}\right) = 0$
$\theta\left(x_{2g+2}, x_{2g+1}\right) = 1$
$\theta\left(x_{2g+2}, x_{2g+2}\right) = 0$
$\theta\left(x_i, x_{2g+2}\right) = 0$
$\theta\left(x_{2g+2}, x_i\right) = 0$ for $1 \leq i \leq 2g$.

Thus

$$B = \begin{pmatrix} A & \gamma & 0 \\ \alpha & 0 & 0 \\ 0 & 1 & 0 \end{pmatrix}$$

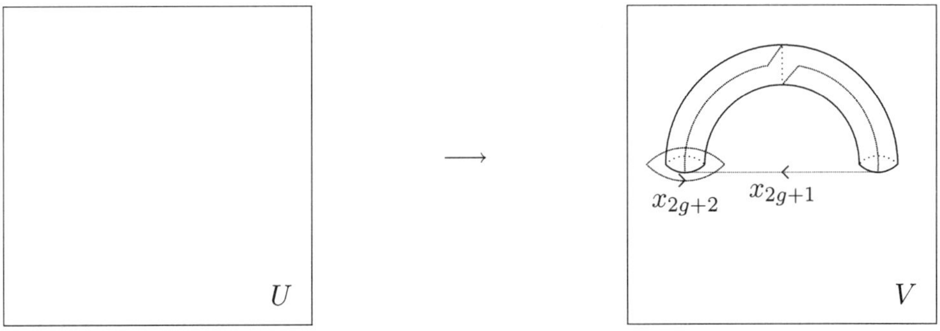

FIGURE 1.10

By a change of basis we can subtract multiples of the last row from the first $2g$ rows to eliminate γ; and the same multiples of the last column from the first $2g$ columns. Whence

$$P'BP = \begin{pmatrix} A & 0 & 0 \\ \alpha & 0 & 0 \\ 0 & 1 & 0 \end{pmatrix}$$

Adding the handle on the other side of U gives the other kind of S-equivalence. □

LEMMA 1.13. *If A is a Seifert matrix and A' is its transpose, then $A - A'$ is unimodular.*

PROOF. Recall that if $x_1, \ldots, x_{2g}$ is a basis for $H_1(V) = \bigoplus_1^{2g} \mathbb{Z}$, and $a_{ij} = \theta(x_i, x_j)$, then the matrix $A = (a_{ij})$ is a *Seifert matrix* of k. Let $i_{-}u$ be the result of pushing u a small distance along the negative normal to V. Then

$$\begin{aligned} a_{ij} &= \theta(x_i, x_j) \\ &= L\,(i_{+}x_i, x_j) \\ &= L\,(x_i, i_{-}x_j) \\ &= L\,(i_{-}x_j, x_i) \end{aligned}$$

and so

$$a_{ij} - a_{ji} = L\,(i_{+}x_i, x_j) - L\,(i_{-}x_i, x_j) = L\,(i_{+}x_i - i_{-}x_i, x_j)$$

Now $i_{+}x_i - i_{-}x_i$ is the boundary of a chain $S^1 \times I$ normal to V which meets V in x_i, and so $L\,(i_{+}x_i - i_{-}x_i, x_j)$ is the algebraic intersection of this chain with x_j, i.e. the algebraic intersection of x_i and x_j in V. Hence $A - A'$ represents the intersection pairing on $H_1(V)$, whence the result. □

Now let us state what happens in higher dimensions.

DEFINITION 1.14. A $(2q-1)$-knot k is *simple* if its exterior K satisfies $\pi_i(K) \cong \pi_i(S^1)$ for $1 \le i < q$.

The following result is proved in [**33**, Theorem 2].

THEOREM 1.15. *A $(2q-1)$-knot k is simple if and only if it bounds a $(q-1)$-connected Seifert submanifold.*

Now suppose that k is a simple $(2q-1)$-knot bounding an $(q-1)$-connected submanifold V. We can repeat the construction above, using $H_q(V)$, to obtain a Seifert matrix A of k satisfying the following.

THEOREM 1.16. *If A is a Seifert matrix of a simple $(2q-1)$-knot k and A' is its transpose, then $A+(-1)^q A'$ is unimodular. Moreover, if $q=2$, then the signature of $A+A'$ is a multiple of 16.*

Theorem 1.12 remains true for simple knots. The following result is proved in [**41**] for $q=1$, in [**36**, Theorem 2] for $q=2$, and in [**31**, Théorème II.3] for $q>2$.

THEOREM 1.17. *Let q be a positive integer and A a square integral matrix such that $A+(-1)^q A'$ is unimodular and, if $q=2$, $A+A'$ has signature a multiple of 16. If $q \neq 2$, there is a simple $(2q-1)$-knot k with Seifert matrix A. If $q=2$, there is a simple 3-knot k with Seifert matrix S-equivalent to A.*

In [**36**, Theorem 3] the classification of simple knots is completed.

THEOREM 1.18. *A simple $(2q-1)$-knot k, $q>1$, is determined up to ambient isotopy by the S-equivalence class of its Seifert matrix.*

2. Blanchfield Duality

Let us set $\Lambda = \mathbb{Z}\left[t, t^{-1}\right]$, the ring of Laurent polynomials in a variable t with integer coefficients.

THEOREM 2.1. *If A is a Seifert matrix of a simple $(2q-1)$-knot k, then the Λ-module M_A presented by the matrix $tA+(-1)^q A'$ depends only on the S-equivalence class of A, and so is an invariant of k. Moreover, there is a non-singular $(-1)^{q+1}$-hermitian pairing*

$$\langle\ ,\ \rangle_A : M_A \times M_A \to \Lambda_0/\Lambda$$

given by the matrix $(1-t)\left(tA+(-1)^q A'\right)^{-1}$ which is also an invariant of k. Conjugation is the linear extension of $t \mapsto t^{-1}$, Λ_0 is the field of fractions of Λ, and non-singular *means that the adjoint map $M_A \to \overline{\mathrm{Hom}}\left(M_A, \Lambda_0/\Lambda\right)$ is an isomorphism.*

DEFINITION 2.2. The Λ-module in Theorem 2.1 is called the *knot module* of k, and has a geometric significance which is explained in §6. The determinant of $tA + (-1)^q A'$ is the *Alexander polynomial* of k, and is defined up to multiplication by a unit of Λ. The hermitian pairing is due to R.C. Blanchfield [**9**]. The formula given here was discovered independently in [**22, 44**].

The following two results are proved in [**22, 23, 44, 45**].

THEOREM 2.3. *If A is a Seifert matrix of a simple $(2q-1)$-knot k, then the module and pairing $(M_A, \langle\ ,\ \rangle_A)$ satisfy:*

(i) M_A is a finitely-generated Λ-torsion-module;
(ii) $(t-1): M_A \to M_A$ is an isomorphism;
(iii) $\langle\ ,\ \rangle_A : M_A \times M_A \to \Lambda_0/\Lambda$ is a non-singular $(-1)^{q+1}$-hermitian pairing.

For $q = 2$ the signature is divisible by 16. Moreover, for $q > 1$, the module and pairing determine the knot k up to ambient isotopy.

THEOREM 2.4. *Suppose that $(M, \langle\ ,\ \rangle)$ satisfies*

(i) M is a finitely-generated Λ-torsion-module;
(ii) $(t-1): M \to M$ is an isomorphism;
(iii) $\langle\ ,\ \rangle : M \times M \to \Lambda_0/\Lambda$ is a non-singular $(-1)^{q+1}$-hermitian pairing.

and that, for $q = 2$, the signature is divisible by 16. Then for $q \geq 1$, $(M, \langle\ ,\ \rangle)$ arises from some simple $(2q-1)$-knot as $(M_A, \langle\ ,\ \rangle_A)$.

In [**43**, pp 485-489] Trotter proves the following.

PROPOSITION 2.5. *If A is a Seifert matrix, then A is S-equivalent to a matrix which is non-degenerate; that is, to a matrix with non-zero determinant.*

PROPOSITION 2.6. *If A and B are non-degenerate Seifert matrices of a simple $(2q-1)$-knot k, then A and B are congruent over any subring of $\mathbb{Q}$ in which $\det A$ is a unit. (Of course, $\det A$ is the leading coefficient of the Alexander polynomial of k.)*

DEFINITION 2.7. Let $\varepsilon = (-1)^q$ and let A be a non-degenerate Seifert matrix of a simple $(2q-1)$-knot k. Set $S = (A + \varepsilon A')^{-1}$ and $T = -\varepsilon A' A^{-1}$.

PROPOSITION 2.8. *The pair (S, T) have the following properties:*

(i) S is integral, unimodular, ε-symmetric;
(ii) $(I-T)^{-1}$ exists and is integral;
(iii) $T'ST = S$;
(iv) $A = (I-T)^{-1}S^{-1}$.

The following result is proved in [**44**, p179]

THEOREM 2.9. *The matrix S gives a $(-1)^q$-symmetric bilinear pairing $(\ ,\ )$ on M_A on which T (i.e. t) acts as an isometry. The pair $(M_A, (\ ,\ ))$ determines and is determined by the S-equivalence class of A.*

3. Factorisation of Knots

If we have two classical knots, there is a natural way to take their sum: just tie one after another in the same piece of string. Alternatively, we can think of each knot as a knotted ball-pair and identify the boundaries so that the orientations match up. The latter procedure generalises to higher dimensions.

DEFINITION 3.1. Let k_1, k_2 be two n-knots, say $k_i = \left(S_i^{n+2}, S_i^n\right)$. Choose a point on each S_i^n and excise a tubular neighbourhood, i.e. an unknotted ball-pair, leaving a knotted ball-pair $\left(B_i^{n+2}, B_i^n\right)$. Identify the boundaries so that the orientations match up, giving a sphere-pair $k_1 + k_2$.

If k_1, k_2 are simple knots with Seifert matrices A_1, A_2, then clearly $k_1 + k_2$ is also simple and has a Seifert matrix

$$\begin{pmatrix} A_1 & 0 \\ 0 & A_2 \end{pmatrix}$$

and the pairings in §2 are given by the orthogonal direct sum.

For the case $n = 1$, H. Schubert showed in [**40**] that every knot factorises uniquely as a sum of irreducible knots. In [**25**] it is shown that unique factorisation fails for $n = 3$, and in [**1**] E. Bayer showed that it fails for $n = 5$ and for $n \geq 7$. The following example is contained in [**1**], although presented here in a slightly different way.

Let $\Phi_{15}(t)$ denote the 15th cyclotomic polynomial, which we normalise so that $\Phi_{15}(t) = \Phi_{15}\left(t^{-1}\right)$, and let $\zeta = e^{\frac{2\pi i}{15}}$. Then $\mathbb{Z}[\zeta] \cong \mathbb{Z}\left[t, t^{-1}\right] / \left(\Phi_{15}(t)\right)$ is the ring of integers, and conjugation in $\mathbb{Z}\left[t, t^{-1}\right]$ corresponds to complex conjugation in $\mathbb{Z}[\zeta]$. Moreover, $\zeta - 1$ is a unit in $\mathbb{Z}[\zeta]$, and so we can think of $\mathbb{Z}[\zeta]$ as a Λ-module satisfying properties (i) and (ii) of Theorem 2.3. If $u \in U_0$, the set of units of $\mathbb{Z}[\zeta + \overline{\zeta}]$, then we can define a hermitian pairing on $\mathbb{Z}[\zeta]$ by $(x, y) = ux\overline{y}$. This corresponds to a hermitian pairing as in Theorem 2.3(iii) by

$$\langle x(t), y(t) \rangle = \frac{u(t)x(t)y\left(t^{-1}\right)}{\Phi_{15}(t)} \longleftrightarrow (x(\zeta), y(\zeta)) = ux(\zeta)\overline{y(\zeta)}.$$

The case of skew-hermitian pairings is dealt with by using $\left(\zeta - \overline{\zeta}\right) u$ in place of u. Note that $\zeta - \overline{\zeta}$ is a unit:

$$(3.1) \qquad \left(\zeta - \overline{\zeta}\right)^2 \left(\zeta + \overline{\zeta}\right) \left(1 - \zeta - \overline{\zeta}\right) = 1.$$

LEMMA 3.2. *Let* $u_r = \zeta^r + \zeta^{-r} - 1$ *for* $r = 0, 1, 2, 7$. *Then* $u_r \in U_0$ *and*

$$\begin{pmatrix} u_r & 0 \\ 0 & -u_r \end{pmatrix} \quad \text{and} \quad \begin{pmatrix} 1 & 0 \\ 0 & -1 \end{pmatrix}$$

represent isometric pairings on $\mathbb{Z}[\zeta] \times \mathbb{Z}[\zeta]$.

PROOF. Since ζ^2, ζ^7 are also primitive 15th roots of unity, (3.1) shows that $u_r \in U_0$ for $r = 1, 2, 7$. Of course, $u_0 = 1$. Now consider

$$\begin{pmatrix} \bar{a} & \bar{b} \\ b & a \end{pmatrix} \begin{pmatrix} 1 & 0 \\ 0 & -1 \end{pmatrix} \begin{pmatrix} a & \bar{b} \\ b & \bar{a} \end{pmatrix} = \begin{pmatrix} a\bar{a} - b\bar{b} & 0 \\ 0 & b\bar{b} - a\bar{a} \end{pmatrix} = \begin{pmatrix} u_r & 0 \\ 0 & -u_r \end{pmatrix}$$

For each r, we can write

$$u_r = \zeta^r + \zeta^{-r} - 1 = 1 - (\zeta^r - 1)(\zeta^{-r} - 1)$$

and so we can take $a = 1, b = (\zeta^r - 1)$. □

LEMMA 3.3. *The hermitian forms on $\mathbb{Z}[\zeta]$ given by $\pm u_r$ for $r = 0, 1, 2, 7$ are distinct.*

PROOF. Two such forms represented by $u, v \in U_0$ are equivalent if and only if $uv^{-1} \in N(U)$, where $N(c) = c\bar{c}$; write $u \sim v$ to denote this equivalence. Then $u_1 > 0$ but u_1 is conjugate to $u_7 < 0$, so $\pm u_1 \notin N(U)$ and hence the forms represented by $\pm 1, \pm u_1$ are distinct. Similarly $u_2/u_1 > 0$ but u_2/u_1 is conjugate to $u_4/u_2 < 0$, so $\pm u_2/u_1 \notin N(U)$. Similar arguments apply in the other cases. □

COROLLARY 3.4. *For each $q > 2$ there exist eight distinct irreducible simple $(2q-1)$-knots k_r, k_r^-, $r = 0, 1, 2, 7$, such that $k_r + k_r^- = k_s + k_s^-$ for all $r, s \in \{0, 1, 2, 7\}$.*

PROOF. By Theorem 2.3 there exist unique simple $(2q-1)$-knots k_r, k_r^- corresponding to $u_r, -u_r$ respectively. These are irreducible because in each case the Alexander polynomial is $\Phi_{15}(t)$. □

REMARK 3.5. It is shown in [**1**] that $U_0/N(U)$ has exactly eight elements, so that these are all the forms there are in this case.

There is another method, due to J.A. Hillman, depending upon the module structure, and this can be used to show that unique factorisation fails for $n \geq 3$ (see [**5**]). There are further results on this topic in [**18, 17, 19**]. It is known from [**10**] that every n-knot, $n \geq 3$, factorises into finitely many irreducibles, and that a large class of knots factorise into irreducibles in at most finitely many different ways (see [**6, 8**]).

I should mention that the method of [**25**] relies on the signature of a smooth 3-knot being divisible by 16, and hence does not generalise to higher dimensions. The work of Hillman in [**20**] shows that the classification theorems 1.16, 1.17, 1.18, and 2.3 hold for locally-flat topological 3-knots without any restriction on the signature.

4. Cancellation of Knots

In [**3**] Eva Bayer proves a stronger result, that cancellation fails for simple $(2q-1)$-knots, $q > 1$.

EXAMPLE 4.1. Let A be the ring of integers associated with Φ_{12}, and define Γ_{4n} to be the following lattice given in [**39**]. We take $\mathbb{R}^{4n}$ to denote the euclidean space, and let $e_1, \ldots, e_{4n}$ be an orthonormal basis with respect to the usual innerproduct. Then Γ_{4n} is the lattice spanned by the vectors $e_i + e_j$ and $\frac{1}{2}(e_1 + \cdots + e_{4n})$. Let $A\Gamma_{4n}$ be the corresponding hermitian lattice. It is shown in [**3**] that the hermitian form $A\Gamma_{4n}$ is irreducible if $n > 1$. Moreover

$$A\Gamma_8 \perp A\Gamma_8 \perp < -1 > \cong A\Gamma_{16} \perp < -1 >; \tag{4.1}$$

indeed, this holds already over $\mathbb{Z}$ by [**39**, Chap. II, Proposition 6.5]. But $A\Gamma_8 \perp A\Gamma_8 \ncong A\Gamma_{16}$ because the latter is irreducible. By Theorem 2.1 this shows that cancellation fails for q odd, $q \neq 1$. For q even, just multiply 4.1 by the unit $\tau - \tau^{-1}$ where τ is a root of Φ_{12}.

In [**4**], examples are given where the failure of cancellation depends on the structure of the knot module, and by the device known as spinning this result is extended to even dimensional knots. (See §7 for the definition of spinning.)

5. Knot Cobordism

There is an equivalence relation defined on the set of n-knots as follows.

DEFINITION 5.1. Two n-knots $k_i = \left(S_i^{n+2}, S_i^n\right)$ are *cobordant* if there is a manifold pair $\left(S^{n+2} \times I, V\right)$ such that

$$V \cap \left(S^{n+2} \times \{i\}\right) = \partial V \cap \left(S^{n+2} \times \{i\}\right) = S_i^n$$

(with the orientations reversed for $i = 0$) and $S_i^n \hookrightarrow V$ is a homotopy equivalence for $i = 0, 1$.

REMARK 5.2. For $n \geq 6$ the manifold V is a product, i.e. it is homeomorphic to $S^n \times I$, by the h-cobordism theorem.

This equivalence relation respects the operation of knot sum, and the equivalence classes form an abelian group C_n under this operation, with the trivial knot as the zero. In [**31**] M.A. Kervaire shows that $C_{2q} = 0$ for all q.

To tackle the odd-dimensional case, we begin by quoting a result of Levine: [**35**, Lemma 4].

LEMMA 5.3. *Every $(2q-1)$-knot is cobordant to a simple knot.*

DEFINITION 5.4. Let A_0, A_1 be two Seifert matrices of simple $(2q-1)$-knots. If $\begin{pmatrix} -A_0 & 0 \\ 0 & A_1 \end{pmatrix}$ is congruent to one of the form $\begin{pmatrix} 0 & N_1 \\ N_2 & N_3 \end{pmatrix}$ where each of the N_i are square of the same size then A_0, A_1 are said to be *cobordant.* This is an equivalence relation for Seifert matrices, and the set of equivalence classes forms a group under the operation induced by taking the block sum:

$$(A_0, A_1) \mapsto \begin{pmatrix} A_0 & 0 \\ 0 & A_1 \end{pmatrix}.$$

LEMMA 5.5. *Let k_0, k_1 be simple $(2q-1)$-knots which are cobordant, with Seifert matrices A_0, A_1 respectively. Then A_0, A_1 are cobordant.*

From Proposition 2.5 we deduce:

COROLLARY 5.6. *Every Seifert matrix is cobordant to a non-degenerate matrix.*

DEFINITION 5.7. Setting $\varepsilon_q = \operatorname{sign}(-1)^q$, the group obtained from the Seifert matrices of simple $(2q-1)$-knots is denoted by G_{ε_q}. The subgroup of G_+ given by matrices A such that the signature of $A + A'$ is divisible by 16 is denoted G_+^0. The map $\varphi_q : C_{2q-1} \to G_{\varepsilon_q}$ is induced by taking a simple knot to one of its Seifert matrices, and is a homomorphism.

The definition above appears in [**35**], where the following result is proved.

THEOREM 5.8. *The map φ_q is*

(a) an isomorphism onto G_{ε_q} for $q \geq 3$;
(b) an isomorphism onto G_+^0 for $q = 2$;
(c) an epimorphism onto G_- for $q = 1$.

The proof of [**34**, Lemma 8] shows that two Seifert matrices are cobordant if and only if they are cobordant over the rationals. This leads to the idea of Witt classes for the $(-1)^q$-symmetric forms and isometries in Theorem 2.9, and this is the strategy that Levine uses to investigate G_ε, and to prove Theorem 5.9 below (see [**35**, p 108]).

THEOREM 5.9. *G_ε is the direct sum of cyclic groups of orders 2, 4 and ∞, and there are an infinite number of summands of each of these orders.*

Both Levine's treatment and that of Kervaire in [**32**] rely on Milnor's classification of isometries of innerproduct spaces in terms of hermitian forms in [**38**].

I shall not attempt to prove any of these results, but it is easy to give Milnor's proof in [**37**] of infinitely many summands of infinite order for q odd, and at the same time to suggest an alternative way of looking at G_{ε_q}. First make the following definition, taken from [**24**].

DEFINITION 5.10. A module and pairing $(M_A, <\ ,\ >_A)$ as in Theorem 2.1 is *null-cobordant* if there is a submodule of half the dimension of M_A which is self-annihilating under $<\ ,\ >_A$. And $(M_A, <\ ,\ >_A)$, $(M_B, <\ ,\ >_B)$ are *cobordant* if the orthogonal direct sum $A \perp (-B)$ is null-cobordant. The *dimension* of M_A is the dimension over $\mathbb{Q}$ of M_A after passing to rational coefficients.

It is shown in [**24**] that $(M_A, <\ ,\ >_A)$ is null-cobordant if and only if A is null-cobordant. If we use rational coefficients, which we may as well do in light of [**34**, Lemma 8], the proof is even easier. We shall treat $(M_A, <\ ,\ >_A)$ in this way for the rest of this section, and set $\Gamma = \mathbb{Q}\left[t, t^{-1}\right]$.

DEFINITION 5.11. Given $p(t) \in \Lambda$, define $p^*(t) = t^{\deg(p(t))} p\left(t^{-1}\right)$. And for an irreducible $p(t) \in \Lambda$ define $M_A(p(t))$ to be the p-primary component of M_A, i.e. the submodule annihilated by powers of $p(t)$.

The following result is essentially Cases 1 and 3 of [**38**, p93, Theorem 3.2]: note that Case 2 does not arise here since we are dealing with knot modules, i.e. $p(t) \neq t \pm 1$.

PROPOSITION 5.12. *$(M_A, <\ ,\ >_A)$ splits as the orthogonal direct sum of $M_A(p(t))$ where $p(t) = p^*(t)$, and of $M_A(p(t)) \oplus M_A(p^*(t))$ where $p(t) \neq p^*(t)$. Furthermore, for $p(t) = p^*(t)$ the space $M_A(p(t))$ splits as an orthogonal direct sum $M^1 \oplus M^2 \oplus \dots$ where M^i is annihilated by $p(t)^i$ but is free over the quotient ring $\Gamma/p(t)^i\Gamma$.*

It is slightly easier to prove [**38**, Theorem 3.3] here.

THEOREM 5.13. *When $p(t) = p^*(t)$, for each i, the vector space*

$$H^i = M^i / p(t) M^i$$

over the field $E = \Gamma/p(t)\Gamma$ admits one and only one hermitian inner product $((x), (y))$ such that

$$\left\langle p(t)^{i-1} x, y \right\rangle = \frac{a(t)}{p(t)} \longleftrightarrow ((x), (y)) = a(\zeta) \tag{5.1}$$

where (x) denotes the image of $x \in M^i$ in H^i. The sequence of these hermitian inner product spaces determines $(M_A(p(t)), <\ ,\ >)$ up to isometry.

The following is easy to prove, where we think of the hermitian form on H^i as taking values in Γ_0/Γ.

LEMMA 5.14. *For i even, M^i is null-cobordant. For i odd, M^i is cobordant to H^i.*

This enables us to make the following definition of the Milnor signatures (compare [**38**]).

DEFINITION 5.15. For each $p(t) = p^*(t)$, define

$$\sigma_p = \sum_{i \text{ odd}} \sigma_i,$$

where σ_i is the signature of H^i.

LEMMA 5.16. *The signature σ_p is additive over knot composition and zero when k is null-cobordant, and so is a cobordism invariant.*

EXAMPLE 5.17 (Milnor, [**37**]). For each positive integer m, let $p_m(x) = mt + (1-2m) + mt^{-1}$. Then p_m is irreducible and is the Alexander polynomial of a simple $(4q+1)$-knot k_m for each $q \geq 1$ having $\begin{pmatrix} m & 1 \\ 0 & 1 \end{pmatrix}$ as a Seifert matrix. Then $\sigma_{p_m} = \pm 2$ for each m, and so for each $q \geq 1$ we have infinitely many independent knots of infinite order in C_{4q+1}.

6. Branched Cyclic Covers

Recall that if k is an n-knot (S^{n+2}, S^n), then a regular neighbourhood of S^n has the form $S^n \times B^2$, and the *exterior* of k is $K = S^{n+2} - \operatorname{int}(S^n \times B^2)$.

Choose a base-point $* \in K$. By Alexander-Poincaré duality, K has the homology of a circle, and so the Hurewicz theorem gives a map $\pi_1(K,*) \twoheadrightarrow H_1(K)$ whose kernel is the commutator subgroup $[\pi_1(K,*), \pi_1(K,*)]$ of the group $\pi_1(K,*)$. We write the infinite cyclic group $H_1(K)$ multiplicatively, as $(t:\,)$; the generator t is represented by $\{a\} \times S^1 \subset S^n \times S^1 = \partial K$ for some point $a \in S^n$, and is chosen so that the oriented circle has linking number $+1$ with S^n.

Let $\tilde{K} \to K$ be the infinite cyclic cover corresponding to the kernel of the Hurewicz map. A triangulation of K lifts to a triangulation of $\tilde{K}$ on which $(t:\,)$ acts as the group of covering transformations. This induces an action of $(t:\,)$ on the chain complex $C_*(\tilde{K})$, which extends by linearity to make $C_*(\tilde{K})$ a $\Lambda = \mathbb{Z}[t, t^{-1}]$-module. The Λ-module $C_*(\tilde{K})$ is finitely-generated because the original triangulation of K is finite. Passing to homology we obtain $H_*(\tilde{K})$ as a finitely-generated, indeed finitely-presented, Λ-module. For a simple $(2q-1)$-knot, the only non-trivial module is $H_q(\tilde{K})$, and this is in fact the same as the Λ-module M_A presented by the matrix $tA + (-1)^q A'$ in Theorem 2.1.

To recover K from $\tilde{K}$ all we do is identify x with tx, for each $x \in \tilde{K}$.

Compose the Hurewicz map with the map sending $(t:\,)$ onto the finite cyclic group of order r, and denote the r-fold cover of K corresponding to the kernel of this map by $\tilde{K}^r$. Since $\partial\tilde{K}^r \cong S^n \times S^1$, being an r-fold cover of $S^n \times S^1$, we may set $K^r = \tilde{K}^r \cup_\partial (S^n \times B^2)$ to obtain the r-fold cover of S^{n+2} branched over S^n. It may happen that $K^r \cong S^{n+2}$, in which case we have another n-knot k^r, which we refer to as the *r-fold branched cyclic* *cover* of k. In this case $H_*(\tilde{K})$ is a $\mathbb{Z}[t^r, t^{-r}]$-module when $\tilde{K}$ is regarded as the infinite cyclic cover of $\tilde{K}^r$. Note that k^r is the fixed point set of the $\mathbb{Z}_r$ action on $S^{n+2} = K^r$ given by the covering transformations.

Let k be a simple $(2q-1)$-knot giving rise to a pair of matrices (S, T) as in Proposition 2.8, and define U, V by

$$U = \begin{pmatrix} 0 & \dots & 0 & T \\ I & \ddots & & 0 \\ & \ddots & \ddots & \vdots \\ 0 & & I & 0 \end{pmatrix} \qquad V = \begin{pmatrix} S & 0 & \dots & 0 \\ 0 & \ddots & & \vdots \\ \vdots & & \ddots & 0 \\ 0 & \dots & 0 & S \end{pmatrix}$$

there being $r \times r$ blocks in each case. It is not hard to show that the pair (V, U) satisfies the conditions of Proposition 2.8, and so corresponds to a unique simple $(2q-1)$-knot k_r if $q \geq 2$ (see Theorem 2.9). Moreover,

$$U^r = \begin{pmatrix} T & 0 & \dots & 0 \\ 0 & \ddots & & \vdots \\ \vdots & & \ddots & 0 \\ 0 & \dots & 0 & T \end{pmatrix}$$

from which it follows without much difficulty that the r-fold branched cyclic cover of k_r is $\#_1^r k$, the sum of r copies of k.

The topological construction of k_r may be described as follows. Take a Seifert surface W of k which meets a tubular neighbourhood N of k in a collar neighbourhood of $k = \partial W$. Take r close parallel copies $W_1, \dots, W_r$ of W, so that the space between W_i and W_{i+1} is diffeomorphic to $W \times [0,1]$ for $i = 1, \dots, r-1$, and that between W_r and W_1 is diffeomorphic to the exterior of k split open along W. We can join ∂W_i to ∂W_{i+1} for $1 \leq i \leq r-1$ by bands within N to get the the boundary connected sum of the W_i. Then it is not hard to see that the Seifert surface we have constructed has V, U as above. Moreover, for $q > 1$, the resulting knot k_r is independent of the bands used, since the bands unknot in these dimensions.

In [**29**] examples are given of simple $(2q-1)$-knots k, l, $q \geq 2$, for which $\#_1^r k = \#_1^r l$ but $k_r \neq l_r$. It is known that there are examples for any odd r. The argument runs as follows.

Let $\Phi_m(t)$ denote the mth cyclotomic polynomial, where m is divisible by at least two distinct odd primes, and let ζ be a primitive m^{th} root of unity. We write $\mathbb{K} = \mathbb{Q}(\zeta)$ and $\mathbb{F} = \mathbb{Q}\left(\zeta + \zeta^{-1}\right)$. Let $h_{\mathbb{K}}$ denote the class number of $\mathbb{K}$, $h_{\mathbb{F}}$ that of $\mathbb{F}$, and $h_- = h_{\mathbb{K}}/h_{\mathbb{F}}$. According to the work of Eva Bayer in [**2**], the number of distinct simple $(2q-1)$-knots, $q \geq 3$, with Alexander polynomial $\Phi_m(t)$ is $h_- 2^{d-1}$ where $2d = \varphi(m) = [\mathbb{K} : \mathbb{Q}]$. The factor h_- represents the number of isomorphism classes of Λ-modules supporting a Blanchfield pairing [**2**, Corollary 1.3], and the factor 2^{d-1} represents the number of pairings (up to isometry) which a given module supports. The latter is in one-one correspondence with $U_0/N(U)$ where U is the group of units in (the ring of integers of) $\mathbb{K}$, U_0 the group of units of $\mathbb{F}$, and $N : \mathbb{K} \to \mathbb{F}$ is the norm.

If h_- has an odd factor $r > 1$ coprime to m, then there exists an ideal a of $\mathbb{Q}(\zeta)$ which has order r in the class group and supports a unimodular hermitian pairing h, i.e. as a Λ-module it supports a Blanchfield pairing. Then $\perp_1^r (a, h)$ has determinant (I, u) for some $u \in U_0/N(U)$, where I denotes a principal ideal. Since r is odd and $|U_0/N(U)| = 2^{d-1}$, there exists $v \in U_0/N(U)$ such that $v^r = u$.

Let k, l be the simple $(2q-1)$-knots, $q \geq 2$, corresponding to $\kappa = (a, h) \perp (a, -h)$, $\lambda = (I, v) \perp (I, -v)$ respectively. Then $\perp_1^r \kappa$, $\perp_1^r \lambda$ are indefinite and have the same rank, signatures and determinant. Hence by [**2**, Corollary 4.10] they are isometric, and so $\#_1^r k = \#_1^r l$. But κ is not isometric to λ, for the determinant of κ is $\left(a^2, \alpha\right)$ for some α, and a^2 is non-zero in the ideal class group since r is odd. Hence $k \neq l$. A similar, but more involved, argument shows that $k_r \neq l_r$.

Many examples may be obtained from the tables in [**46**] or [**47**].

7. Spinning and Branched Cyclic Covers

First we recall a definition of spinning. Let k be the n-knot $\left(S^{n+2}, S^n\right)$ and let B be a regular neighbourhood of a point on S^n such that $(B, B \cap S^n)$ is an unknotted ball pair. Then the closure of the complement of B in S^{n+2} is a knotted ball pair $\left(B^{n+2}, B^n\right)$, and $\sigma(k)$ is the pair $\partial\left[\left(B^{n+2}, B^n\right) \times B^2\right]$. The following is proved in [**28**, Theorem 3].

THEOREM 7.1. *Let k, l be simple $(2q-1)$-knots, $q \geq 3$; then $\sigma(k) = \sigma(l)$ if and only if $H_q(\tilde{K}) \cong H_q(\tilde{L})$.*

Note that [**28**] only covers the case $q \geq 5$; the theorem is extended to $q \geq 3$ by the results of [**21**].

In [**7**] it is shown that the following holds.

PROPOSITION 7.2. *If $q \geq 4$, then the map σ acting on simple $(2q-1)$-knots is finite-to-one.*

The following is easy to prove (see [**30**, Lemma 3.1]).

LEMMA 7.3. *Let k be an n-knot and r an integer such that the r-fold cyclic cover of S^{n+2} branched over k is a sphere. Then the r-fold cyclic cover of S^{n+3} branched over $\sigma(k)$ is also a sphere, and $\sigma\left(k^r\right) = \sigma(k)^r$.*

In [**42**], Strickland proves the following result.

THEOREM 7.4. *Let k be a simple $(2q-1)$-knot, $q \geq 2$. Then k is the r-fold branched cyclic cover of a knot if and only if there exists an isometry u of $\left(H_q(\tilde{K}), <\ ,\ >\right)$ such that $u^r = t$.*

A careful reading of [**42**] shows that the same proofs apply, almost verbatim, to yield the following result (see [**30**, Theorem 2.5]).

THEOREM 7.5. *Let k be a simple $(2q-1)$-knot, $q \geq 5$. Then $\sigma(k)$ is the r-fold branched cyclic cover of a knot if and only if there is a Λ-module isomorphism $u\colon H_q(\tilde{K}) \to H_q(\tilde{K})$ such that $u^r = t$.*

Thus if we can find a simple $(2q-1)$-knot k, $q \geq 5$, such that there is a Λ-module isomorphism $u\colon H_q(\tilde{K}) \to H_q(\tilde{K})$ with $u^r = t$, but no such isometry of $H_q(\tilde{K})$, then $\sigma(k)$ will be the r-fold branched cyclic cover of a knot but k will not. Examples of such knots are given in [**30**] for all $q \geq 5$ and all even r. Here is an example, due to S.M.J. Wilson, much simpler than the ones in [**30**] but not capable of generalising to the $2r$-fold case.

EXAMPLE 7.6. Let $f(t) = t^2 - 3t + 1$, which has roots $\frac{3\pm\sqrt{5}}{2}$. Set $\tau = \frac{3+\sqrt{5}}{2}$, $\xi = \frac{1+\sqrt{5}}{2}$, and note that $\xi^2 = \tau$. Put $R = \mathbb{Z}[\xi] = \mathbb{Z}\left[\tau, \tau^{-1}\right]$ and define conjugation in the obvious way by $\tilde{\xi} = \frac{1-\sqrt{5}}{2}$. Think of R as an R-module, and put a hermitian form on it by setting $(x,y) = x\tilde{y}$. Since $\xi\tilde{\xi} = -1$, ξ is an isomorphism on R but not an isometry. Since τ only has two square roots, $\pm\xi$, there are no isometries whose square is τ. In the usual way, $(R, (\ ,\ ))$ corresponds to a knot module and pairing for a simple $(4q+1)$-knot, $q > 1$.

8. Concluding Remarks

So far we have only dealt with odd dimensional knots, but much of what has been said can be extended to even dimensions. By Proposition 1.1, every $2q$-knot k bounds an orientable $(2q+1)$-manifold V in S^{2q+2}. A *simple* $2q$-knot k is one for which there is a $(q-1)$-connected V, so that $H_q(V)$ and $H_{q+1}(V)$ are the only non-trivial homology groups. Then $H_q\left(\widetilde{K}\right)$, $H_{q+1}\left(\widetilde{K}\right)$ are the only non-trivial homology modules. There is a sesquilinear duality pairing on $H_q\left(\widetilde{K}\right) \times H_{q+1}\left(\widetilde{K}\right)$, and this, together with some more algebraic structure connecting them, can be used to classify the simple $2q$-knots in high dimensions. See [**13, 14, 15, 26, 27**] for details.

Classification results have also been obtained for more general classes of knots; see [**11, 12, 16**].

References

[1] E. Bayer, *Factorisation in not unique for higher dimensional knots*, Comm. Math. Helv. **55** (1980), 583–592.

[2] ———, *Unimodular hermitian and skew-hermitian forms*, Jour. of Algebra **74** (1982), 341–373.

[3] ———, *Definite hermitian forms and the cancellation of simple knots*, Archiv der Math. **40** (1983), 182–185.

[4] ———, *Cancellation of hyperbolic ε-hermitian forms and of simple knots*, Math. Proc. Camb. Phil. Soc. **98** (1985), 111–115.

[5] E. Bayer, J.A. Hillman, and C. Kearton, *The factorization of simple knots*, Math. Proc. Camb. Phil. Soc. **90** (1981), 495–506.

[6] E. Bayer-Fluckiger, C. Kearton, and S.M.J. Wilson, *Decomposition of modules, forms and simple knots*, Journal für die reine und angewandte Mathematik **375/376** (1987), 167–183.

[7] ———, *Finiteness theorems for conjugacy classes and branched cyclic covers of knots*, Mathematische Zeitschrift **201** (1989), 485–493.

[8] ———, *Hermitian forms in additive categories: finiteness results*, Journal of Algebra **123** (1989), 336–350.

[9] R.C. Blanchfield, *Intersection theory of manifolds with operators with applications to knot theory*, Annals of Maths **(2) 65** (1957), 340–356.

[10] M.J. Dunwoody and R.R. Fenn, *On the finiteness of higher knot sums*, Topology **26** (1986), 337–343.

[11] M.Sh. Farber, *Duality in an infinite cyclic covering and even-dimensional knots*, Math. USSR Izvestija **11** (1977), no. 4, 749–781.

[12] ———, *Isotopy types of knots of codimension two*, Trans. Amer. Math. Soc. **261** (1980), 185–209.

[13] ———, *The classification of simple knots*, Russian Math. Surveys **38** (1983), 63–117.

[14] ———, *An algebraic classification of some even-dimensional spherical knots. I*, Trans. Amer. Math. Soc. **281** (1984), no. 2, 507–527.

[15] ———, *An algebraic classification of some even-dimensional spherical knots. II*, Trans. Amer. Math. Soc. **281** (1984), no. 2, 529–570.

[16] ———, *Minimal Seifert manifolds and the knot finiteness theorem*, Israel Journal of Mathematics **66** (1989), no. 1–3, 179–215.

[17] J.A. Hillman, *Blanchfield pairings with squarefree Alexander polynomials*, Math. Zeit. **176** (1981), 551–563.

[18] ———, *Finite knot modules and the factorization of certain simple knots*, Math. Annalen **257** (1981), 261–274.

[19] ———, *Factorization of Kojima knots and hyperbolic concordance of Levine pairings*, Houston Journal of Mathematics **10** (1984), 187–194.

[20] ———, *Simple locally flat 3-knots*, Bull. Lond. Math. Soc. **16** (1984), 599–602.

[21] J.A. Hillman and C. Kearton, *Seifert matrices and 6-knots*, Transactions of the American Mathematical Society **309** (1988), 843–855.

[22] C. Kearton, *Classification of simple knots by Blanchfield duality*, Bulletin of the American Mathematical Society **79** (1973), 952–955.

[23] ———, *Blanchfield duality and simple knots*, Transactions of the American Mathematical Society **202** (1975), 141–160.

[24] ———, *Cobordism of knots and Blanchfield duality*, Journal of the London Mathematical Society **10** (1975), 406–408.

[25] ———, *Factorisation is not unique for 3-knots*, Indiana University Mathematics Journal **28** (1979), 451–452.

[26] ———, *An algebraic classification of certain simple knots*, Archiv der Mathematik **35** (1980), 391–393.

[27] ———, *An algebraic classification of certain simple even-dimensional knots*, Transactions of the American Mathematical Society **276** (1983), 1–53.

[28] ———, *Simple spun knots*, Topology **23** (1984), 91–95.

[29] C. Kearton and S.M.J. Wilson, *Cyclic group actions on odd-dimensional spheres*, Comm. Math. Helv. **56** (1981), 615–626.

[30] ______, *Spinning and branched cyclic covers of knots*, Royal Society of London Proceedings A **455** (1999), 2235–2244.

[31] M.A. Kervaire, *Les noeuds de dimensions supérieures*, Bull. Soc. math. France **93** (1965), 225–271.

[32] ______, *Knot cobordism in codimension 2*, Manifolds Amsterdam 1970, Lecture Notes in Mathematics, vol. 197, Springer Verlag, Berlin-Heidelberg-New York, 1971, pp. 83–105.

[33] J. Levine, *Unknotting spheres in codimension two*, Topology **4** (1965), 9–16.

[34] ______, *Invariants of knot cobordism*, Invent. math. **8** (1969), 98–110.

[35] ______, *Knot cobordism groups in codimension two*, Comm. Math. Helv. **44** (1969), 229–244.

[36] ______, *An algebraic classification of some knots of codimension two*, Comm. Math. Helv. **45** (1970), 185–198.

[37] J.W. Milnor, *Infinite cyclic coverings*, Conference on the Topology of Manifolds (ed. J.G. Hocking), Complementary Series in Mathematics, vol. 13, Prindle, Weber & Schmidt, 1968, pp. 115–133.

[38] ______, *On isometries of inner product spaces*, Invent. math. **8** (1969), 83–97.

[39] J.W. Milnor and D. Husemoller, *Symmetric bilinear forms*, Springer-Verlag, Berlin-Heidelberg-New York, 1973.

[40] H. Schubert, *Die eindeutige Zerlegbarkeit eines Knotens in Primknoten*, S.-B. Heidelberger Akad. Wiss. Math.-Natur. Kl. 1 **3** (1949), 57–104.

[41] H. Seifert, *Über das Geschlecht von Knoten*, Math. Ann. **110** (1934), 571–592.

[42] Paul Strickland, *Branched cyclic covers of simple knots*, Proc. Amer. Math. Soc. **90** (1984), 440–444.

[43] H.F. Trotter, *Homology of group systems with applications to knot theory*, Annals of Math. **76** (1962), 464–498.

[44] ______, *On S-equivalence of Seifert matrices*, Invent. math. **20** (1973), 173–207.

[45] ______, *Knot modules and Seifert matrices*, Knot Theory, Proceedings, Plans-sur-Bex 1977 (Berlin-Heidelberg-New York) (J.-C. Hausmann, ed.), Lecture Notes in Mathematics, vol. 685, Springer-Verlag, 1978, pp. 291–299.

[46] G. Schrutka v. Rechtenstamm, *Tabelle der (relativ-) Klassenzahlen von Kreiskörpern*, Abh. Deutsch Akad. Wiss. Berlin **2** (1964), 1–64.

[47] L.C. Washington, *Introduction to cyclotomic fields*, Graduate Texts in Mathematics, vol. 83, Springer-Verlag, New York-Heidelberg-Berlin, 1982.

[48] E.C. Zeeman, *Twisting spun knots*, Trans. Amer. Math. Soc. **115** (1965), 471–495.

MATHEMATICS DEPARTMENT
UNIVERSITY OF DURHAM
SOUTH ROAD
DURHAM DH1 3LE, ENGLAND

E-mail address: Cherry.Kearton@durham.ac.uk

Contemporary Mathematics
Volume **272**, 2000

Biography of Ernst Witt (1911-1991)

Ina Kersten

ABSTRACT. Ernst Witt (1911–1991) was one of the most important influences on the development of quadratic forms in the 20th century. His collected papers [**CP**] were published in 1998, but for copyright reasons without a biography. This article repairs the omission.

1. Childhood and youth

The grandfather of Ernst Witt, Heinrich Witt (1830–1893) worked as school teacher, author and lay preacher. He made his name as the author of three volumes of biblical commentary[1]. Many years later, his grandchildren were educated from these books. Heinrich Witt became widely known for his engagement in religious education, including seminars for adults and Bible classes, held in the spirit of the widespread religious revivalist movement of the 19th century. He sacrificed everything for his faith, expecting the same from his family. His children had to take second place to his great task.

[1] *Die biblischen Geschichten Alten und Neuen Testaments* mit Bibelwort und freier Zwischenrede anschaulich dargestellt (Biblical Histories From the Old and the New Testament, Illustrated by Quotations from the Bible and Devotional Commentaries), published by Bertelsmann Gütersloh, 8 Marks.

This was the spiritual atmosphere into which Ernst Witt's father Heinrich (1871–1959) was born, the seventh of thirteen children. Like nearly all of his brothers and sisters, he decided to join the Christian Movement. He studied theology in Halle (Saale) with the intention of becoming a missionary. In 1893 a personal meeting with Hudson Taylor (1832–1905), the founder of the China Home Mission, became a crucial aspect of his future life. In 1896, Heinrich Witt was appointed as the travelling secretary of the German Christian Student Association, of which he was a co-founder.

In 1900, the mission of Liebenzell, a German branch of the China Home Mission, sent him to China. There he adopted Chinese customs wearing Chinese clothes and plaiting his hair in order to gain the confidence of the local people. He totally submitted himself to the principles of Taylor and always did as he was instructed, distributing tracts, selling books and brochures and holding street meetings. In 1906, he married Charlotte Jepsen from Sonderburg (which is now in Denmark) where he had done his voluntary military service eleven years before. The couple moved to a small, moderately furnished rice farmer's hut in Yüanchow. In spite of the primitive conditions, they were content. As the mission of Liebenzell was very poor, the missionaries did not receive a fixed salary and had to get by on bare necessities.

In 1911, they were granted their first home leave. With their three-year-old daughter they went back to Germany, where Heinrich Witt worked again as a travelling secretary until 1913. Their son Ernst was born on June 26, 1911 on Alsen, a Baltic Sea island, which was German at that time.

As a child of missionaries, an unusual fate awaited him. At the age of two he came to China, where his father became superintendent and head of the Liebenzell Mission in Changsha. A high wall closed off the mission station, Ernst's home for the following years, from the hostile outside world. Heinrich Witt often went for long journeys and did not have much time for his children – there were to be six in all. They were educated strictly; above all, the father attached great importance to honesty. His eldest son Ernst, being without any playmates of the same age, was mostly left to himself. He had a passion for any kind of machine, and learnt Chinese from his Chinese nannies. His received his first lessons from his father, who realized and promoted his son's great talent for arithmetic. The father neglected almost all other subjects, for lack of time. However, he studied with his sons, Ernst and Otto, all of the biblical histories by Heinrich Witt mentioned at the beginning of this article.

When Ernst Witt was nearly nine years old, he and his younger brother Otto were sent to Germany for their schooling. A brother of their father was to take care of their further education. They were sent to their uncle's house at Müllheim, in the south of Baden, where their elder sister had already been living for some time. The uncle was a preacher with eight children of his own. In due course he became the housemaster of a children's home which had at times from 20–25 children of missionaries. His pedagogical work overwhelmed him and he ran a strict regimen. But he could not break the individualistic spirit of the young Ernst. In later years Ernst Witt still used to stand by his convictions determinedly and consistently. Besides, he had a dry sense of humour which he retained his whole life. He always took a keen pleasure in making other people laugh.

Ernst Witt visited the *Realschule* (grammar school) at Müllheim, where he was mainly interested in mathematics and chemistry. After his *mittlere Reife* he went to the *Oberrealschule* (high school) leading to the *Abitur,* the university entrance qualification – in Freiburg. There his class teacher was Karl Öttinger, an excellent mathematician, who discerned the extraordinary mathematical talent of his pupil and advanced him wherever he could. Ernst Witt maintained contact with this teacher after his schooldays, and decades later he remembered him with gratitude.

During his school years in Freiburg, Ernst's parents came to Germany for a second leave from 1927–29. When they went back to China, they left all of their children in Germany. Again they could communicate only by letter. In the course of the following years, Heinrich Witt endeavoured to stay in touch with his son. He tried to convince his son of his way of religious thinking, without success. Ernst rejected the biblical faith represented by his father. But his early education had shaped his character. He detested lying – honesty and straightforwardness were his characteristic feature. He always said what he thought, without any diplomacy, often causing misunderstandings and even hostility, which later worried him a lot.

2. University studies

After having passed his final exams, the *Abitur,* in 1929, Ernst Witt started to study mathematics and physics. For the first two terms he stayed in Freiburg, where he attended lectures by Loewy, Bolza and Mie. From the summer term of 1930 to the winter term of 1933/34 he studied in Göttingen, mainly under Herglotz, Emmy Noether, Weyl, and Franck and astronomy under Meyermann. He particularly liked the lectures of Gustav Herglotz (1881–1953), who had an extremely broad knowledge and whose talks stood out because of their special clarity. Herglotz, for his part, always followed the mathematical development of his student with interest and sympathy, and they maintained contact until Herglotz's death.

Seminar Excursion, Göttingen: Bernays, Scherk, Schilling, Schwerdtfeger, Taussky, Bannow, Noether, –, W. Weber. *Sitting:* Witt, Ulm, –, –, Wichmann, Tsen, –.

At the age of nineteen, Ernst Witt published his first paper including a new proof of Wedderburn's theorem that every finite skew field is commutative [**1931**]. Herglotz later remembered this ([**6**], 1946):

After one lecture, a young lad showed me a torn off piece of paper and said: 'This is a proof of Dickson's theorem.' In fact, it contained a proof of an extraordinary simplicity, not achieved up to then by any of the well-known mathematicians. When I got to know him closer, he showed an eminent mathematical gift, with which he, after a first approach, observed a thing in his own way and independently followed it to its furthest development.

In the summer of 1932, Emil Artin (1898–1962) came to Göttingen and held his famous lectures on class field theory, which strongly influenced Witt's further scientific development. In the same year, Artin invited him to Hamburg, where Witt intensively studied class field theory of number fields. In the course of the following years, he carried the theory over to function fields [1934–36].

Witt had to study under extremely harsh economic conditions. He tried hard to keep his expenses to a minimum, taking special examinations for grants or the remission of tuition fees. He completed his Ph.D. after only four years, in the summer of 1933. The title of his thesis was: *Riemann-Rochscher Satz und Zeta-Funktion im Hyperkomplexen.* The idea for it came to him through a problem set by Emmy Noether. She asked him whether the thesis by Artin's student Käte Hey *Analytische Zahlentheorie in Systemen hyperkomplexer Zahlen* could be transferred to algebraic function fields to find the hypercomplex analogue of the Riemann-Roch theorem. Witt could answer Noether's question in the affirmative by first proving, however, a Riemann-Roch theorem for central simple algebras over algebraic function fields with perfect field of constants and then using this theorem to transfer Hey's thesis.

As Witt liked to relate in later years, he wrote down his thesis within one week. In a letter of August 1933, his sister told her parents [**2**]:

He did not have a thesis topic until July 1, and the thesis was to be submitted by July 7. He did not want to have a topic assigned to him, and when he finally had the idea, [...] he started working day and night, and eventually managed to finish in time. [...] Artin also liked Ernst's paper and has told him to go over it again, so that others might understand it as well.

The examiner of his thesis was Herglotz since Emmy Noether had been suspended from her duties by the Nazi regime. The oral examination held by Herglotz, Weyl, and Pohl was at the end of July 1933.

3. The time between 1933 and 1945

On 1st May, 1933 Ernst Witt joined the Nazi party and the SA, a step which provoked disappointment and bewilderment in the Göttingen mathematics institute. By that time Emmy Noether was already suspended. Her seminar now took place at her home, where Witt turned up one day dressed in his SA uniform. Apparently, she did not take it amiss.

Herglotz later wrote in a report [**6**] of 22nd November, 1946:

And then one day he had joined the S.A., urged by the simple wish – as I was convinced in those days, and as I am still today – not to stand apart, where others carried their burden. I asked him about his impression of his comrades, whose ideas, as I suspected, would often come into conflict with his devotion to science. His answer was: 'I don't know much about them. During our night marches I never talk to them, and in the morning I go home immediately, to continue my studies where I left them the evening before.' In those days we had much trouble with certain 'activists', also among the younger lecturers. I would like to particularly emphasize the fact that Witt always stood away from this group and its troublemaking. He was completely absorbed in his mathematical work, which he only interrupted for night and pack marches. The way he looked at the time caused quite a bit of worry.

One of the Nazi students at the mathematical institute in Göttingen was Oswald Teichmüller (1913–1943). He took part in vicious acts leading the boycott against Landau's first year course in November 1933. Decades later, when Witt was asked by his students in Hamburg on his relationship to Teichmüller, he emphasized that Teichmüller was his friend. He related, that Teichmüller had asked him whether one could refer to Albert, a Jewish mathematician who had obtained similar results on p-algebras. Because of Witt's answer, Teichmüller (cf. his Collected Papers, pp. 121, 122, 138) made the reference. There is a handwritten note by Witt on a conversation with Teichmüller who had investigated algebraic function fields in the beginning of the fortieth (cf. Collected Papers, pp. 611–621):

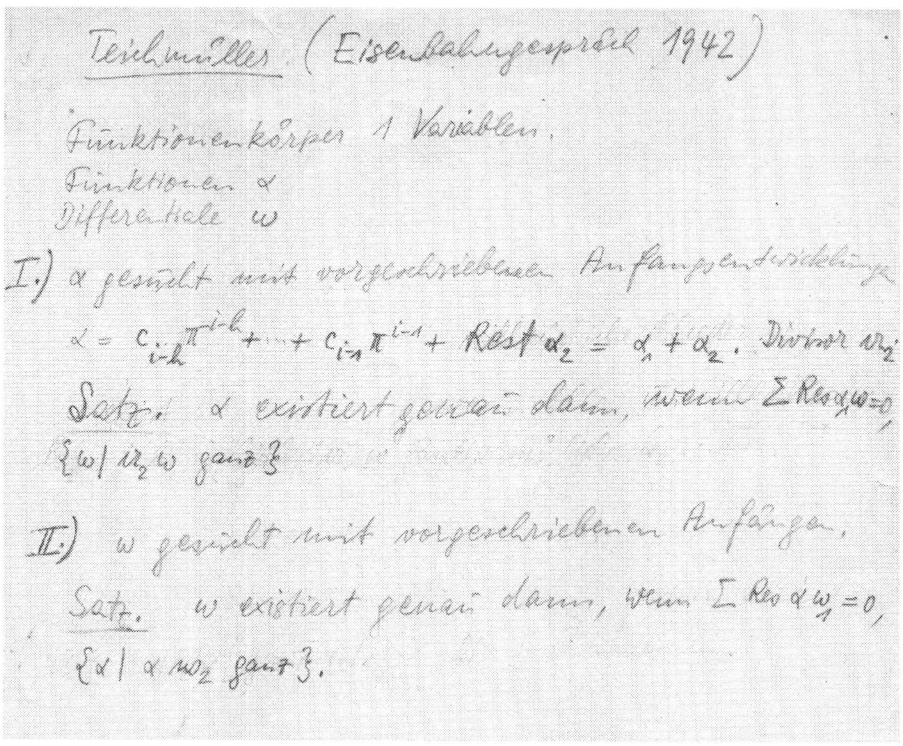
Teichmüller (Eisenbahngespräch 1942)

Funktionenkörper 1 Variablen,
Funktionen α
Differentiale ω

I.) α gesucht mit vorgeschriebenen Anfangsentwicklungen
$\alpha = c_{i-k}\pi^{i-k} + \dots + c_{i-1}\pi^{i-1} + \text{Rest}\ \alpha_2 = \alpha_1 + \alpha_2$. Divisor w_2
Satz: α existiert genau dann, wenn $\sum \operatorname{Res} \alpha_1 \omega = 0$,
$\{\omega \mid w_2 \omega \text{ ganz}\}$

II.) ω gesucht mit vorgeschriebenen Anfängen,
Satz. ω existiert genau dann, wenn $\sum \operatorname{Res} \alpha \omega_1 = 0$,
$\{\alpha \mid \alpha w_2 \text{ ganz}\}$.

Note by Witt

Teichmüller. (Railway conversation 1942)

function field, 1 variable,
functions α
differentials ω

I.) *To find α with prescribed initial expansions.*

$$\alpha = c_{i-h}\pi^{i-h} + \cdots + c_{i-1}\pi^{i-1} + \textit{remainder}\ \alpha_2 = \alpha_1 + \alpha_2. \quad \textit{Divisor}\ \mathfrak{a}_2\,.$$

Proposition. α *exists if and only if* $\sum \operatorname{res} \alpha_1\,\omega = 0$, $\{\omega \mid \mathfrak{a}_2\,\omega\ \textit{integral}\}$.

II.) *To find ω with prescribed initial parts.*

Proposition. ω *exists if and only if* $\sum \operatorname{res} \alpha\,\omega_1 = 0$, $\{\alpha \mid \alpha\,\mathfrak{w}_2\ \textit{integral}\}$.

Witt did not take part in the political discussions during the years of 1933/34. But when in 1934 the university of Göttingen looked for a new head of the mathematical institute and the students suggested a scientifically insignificant Nazi, he openly expressed his indignation, despite threats by the SA.

In 1934, Witt became an assistant of Helmut Hasse (1898–1979) in Göttingen, where he qualified as a university lecturer in 1936. The oral exam took place in February, his 'habilitation lecture' on convex bodies in June 1936. His 'habilitation' on the theory of quadratic forms in arbitrary fields ranks as one of his most famous works. In it he introduced what was later named the *Witt ring* of quadratic forms. Shortly after that, Witt introduced the ring of *Witt vectors,* which had a great influence on the development of modern algebraic geometry (cf. G. Harder: "An essay on Witt vectors" in [CP]).

Since Witt wanted to work as a lecturer, he had to attend and pass a compulsory National Socialist course for lecturers, which took place from August 2 to 28, 1937 in Thüringen. At its end Witt received the following assessment [**6**]:

National Socialist thinking: Mediocre
Independent propagandist in any situation: No
National Socialist disposition: Limited
Physical capabilities: weakly-built, cannot be established because of a sporting injury.
General enthusiasm for his duty: shirker
Behaviour towards people around him: quiet, restrained, his manners are somewhat insecure.

Description of his character: W. has shown himself to be quiet, modest and restrained, with a tendency to keep to himself; characteristic features are a certain naivety and eccentricity. He is honest and straightforward. He dedicates himself to his work with dogged tenacity, continuously brooding and thinking and thus represents the typical, politically indifferent researcher and scientist, who will probably be successful in his subject, but who, at least for the time being, is lacking any of the qualities of a leader or educator.

This certificate just about allowed him to become a lecturer. It is striking how accurate the 'description of his character' is. Some years later, Herglotz stressed (in his report of November 22, 1946), apart from Witt's significant mathematical talent his striking unworldliness.

From 1933 until 1938 when he left Göttingen, Witt founded and organized a study group on higher algebra and number theory. It produced seven important papers, all published in the celebrated volume 176 of the Crelle Journal.

In 1938, Ernst Witt became a lecturer at the mathematics department in Hamburg, where, in 1939, he was appointed as an associate professor. He got the downgraded position of Emil Artin, who was forced to emigrate to the United States in 1937. Witt's move to Hamburg enabled him to break off his service in the SA.

In Hamburg, Witt met a fellow student, Erna Bannow, from Göttingen, who had earlier gone to Hamburg to work with Artin. She continued her studies with Witt and finished her doctorate in 1939. A brief report on her thesis "Die Automorphismengruppen der Cayleyzahlen" (cf. Hamb. Abh. 13) was given by Witt in Crelles Journal. They married in 1940 and had two daughters.

In February 1940, Witt was called up by the Wehrmacht. He submitted his paper on modular forms of degree two at the end of January 1940, thereby cutting short some of his investigations. In February 1940, with Blaschke's help, he succeeded in deferring his military service for one year. This enabled him to finish his paper on reflection groups and the enumeration of semisimple rings, cf. [1941].

In February 1941, Witt came to Lübeck to be trained as a radio operator. From there, in June 1941, he was sent to the Russian front. Later he said, that he had the worst time of his life there, and that he was surprised to survive it. In November 1941, he fell ill and was sent back in several stages, until finally he arrived at a hospital in Lübeck. This illness probably saved his life, because shortly afterwards his company was almost completely wiped out.

After his recovery, Witt did not have to go back to the front, because the headquarters of the Wehrmacht in Berlin needed him for decoding work. Some information on Witt's work as a mathematician at the army is given in the book [**3**], pp. 340 and 357, for example, that he built special equipment which functioned on an optical basis. In a two-page letter of 31st January 1943 Witt wrote to Herglotz:

Witt Berlin SW 61
Wartenburgstr. 11^{II}*r* *Berlin, 31st January 43*

Dear Professor Herglotz!

I would like to cordially congratulate you on your birthday!

Many thanks for the invitation to give a talk in Göttingen. My civilian institution would immediately grant the corresponding leave, but unfortunately, my company is only interested in military things. According to some rumour circulating since half a year I will "soon" be dismissed from military service in order to continue my present occupation as a civilian. But one can hardly figure out to which (presumably complicated) chronology the word "soon" refers. In this sense, I am looking forward to seeing Göttingen again "soon".[2]

My wife and I often think of the beautiful days we had in Göttingen. The time of mathematical expeditions is over for the time being, and I can just make some short walks in an already familiar mathematical territory. Occasionally, however, I succeed in finding an unknown plant there as well. So I realized that the Klein bottle can always be coloured with six colours, whereas the torus needs seven colours, as

[2]Witt gave the talk in Göttingen on 17th June 1944. He talked about his investigation of subrings of free Lie rings which is described on the second page of his letter to Herglotz.

is well-known, although the topological "connectivity number" [3] *is the same in both cases and only the possibility of orientation is responsible for the difference.*

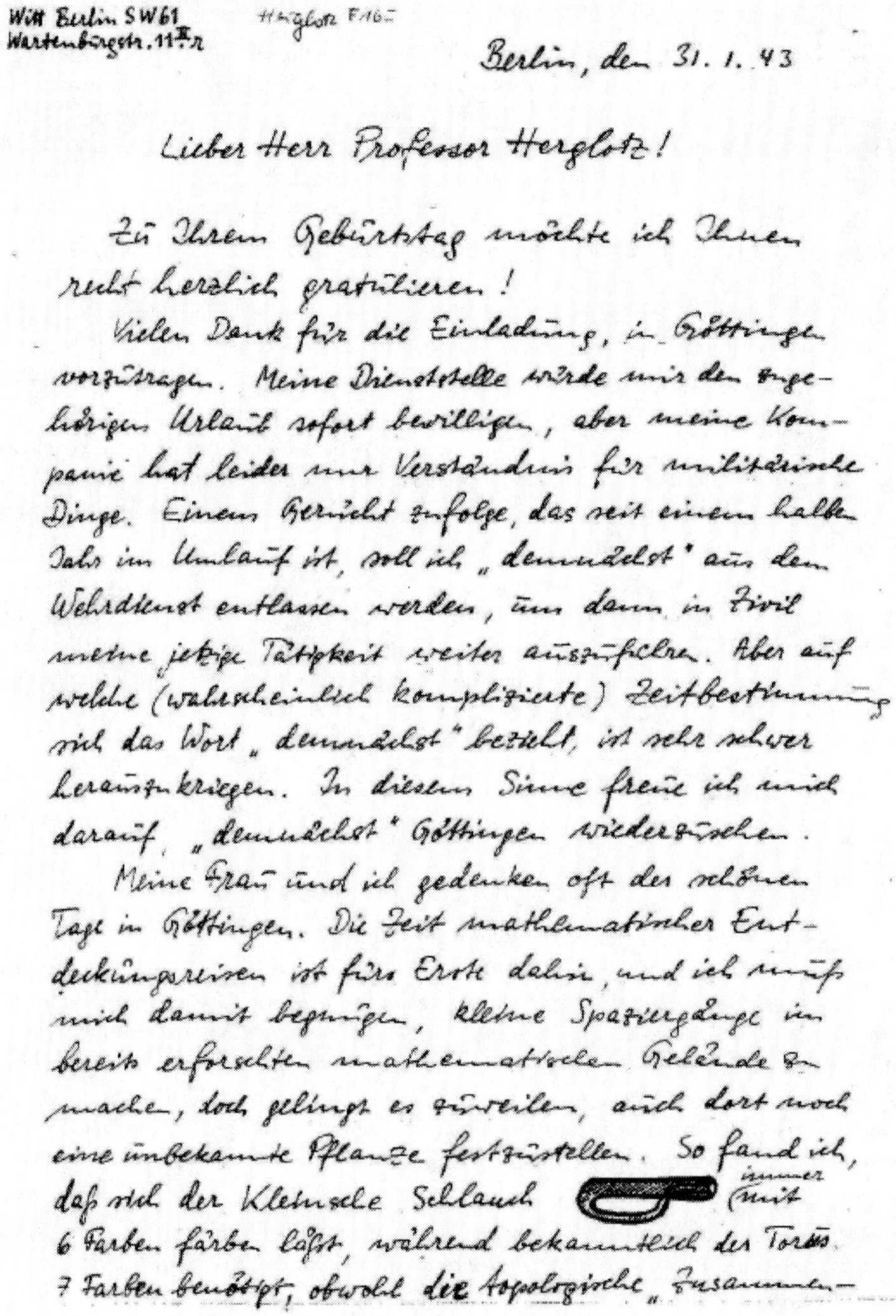

Witt Berlin SW61
Wartenburgstr. 11 III r

Herglotz F16

Berlin, den 31. 1. 43

Lieber Herr Professor Herglotz!

Zu Ihrem Geburtstag möchte ich Ihnen recht herzlich gratulieren!

Vielen Dank für die Einladung, in Göttingen vorzutragen. Meine Dienststelle würde mir den zugehörigen Urlaub sofort bewilligen, aber meine Kompanie hat leider nur Verständnis für militärische Dinge. Einem Gerücht zufolge, das seit einem halben Jahr im Umlauf ist, soll ich „demnächst" aus dem Wehrdienst entlassen werden, um dann in Zivil meine jetzige Tätigkeit weiter auszuführen. Aber auf welche (wahrscheinlich komplizierte) Zeitbestimmung sich das Wort „demnächst" bezieht, ist sehr schwer herauszukriegen. In diesem Sinne freue ich mich darauf, „demnächst" Göttingen wiederzusehen.

Meine Frau und ich gedenken oft der schönen Tage in Göttingen. Die Zeit mathematischer Entdeckungsreisen ist fürs Erste dahin, und ich muß mich damit begnügen, kleine Spaziergänge im bereits erforschten mathematischen Gelände zu machen, doch gelingt es zuweilen, auch dort noch eine unbekannte Pflanze festzustellen. So fand ich, daß sich der Kleinsche Schlauch (immer mit 6 Farben färben läßt, während bekanntlich der Torus 7 Farben benötigt, obwohl die topologische „Zusammen-

Two-page letter by Witt to Herglotz [**4**]

[3] i.e. the maximal number of linearly independent non-separating closed curves, which is 2 for both the Klein bottle and the torus, this being the rank of mod 2 homology in dimension 1

Rangzahl* in beiden Fällen gleich ist und nur die Orientierbarkeit ~~bewirkt~~ den Unterschied bewirkt.

Aus der Zeit vor meiner Einziehung habe ich noch eine Untersuchung über „Unterringe freier Liescher Ringe" liegen. (Ein freier Liescher Ring hat Erzeugende x_ν ohne definierende Relationen — Analogon zur freien Gruppe). Sämtliche Unterringe sind wieder frei (bei geeigneten Erzeugenden). Aus diesem „Satz von Schreier, auf Liesche Ringe übertragen" ergeben sich wieder Konsequenzen für freie Gruppen. Zum Aufschreiben bin ich leider noch nicht gekommen.

Für kommutative und assoziative Ringe ist der entsprechende Satz falsch!

Obwohl zur Zeit für mich die Mathematik sehr stark rationiert ist, bin ich mit meinem Schicksal zufrieden, daß ich in Deutschland bleiben darf. Voriges Jahr war ich ja in Rußland, und dahin habe ich keinerlei Sehnsucht. Ich darf hier privat wohnen, bekomme in einigen Tagen Zivilerlaubnis (nur bezüglich der Bekleidung), darf zwei Sonntage im Monat nach Hamburg fahren.

[A private remark about Witt's family has been erased here]

und der Papper freut sich auf jeden Sonntagsurlaub!

Viele herzliche Grüße

von Ihrem Ernst Witt

I have some results on "subrings of free Lie rings", dating from the time before my conscription. (A free Lie ring has generators x_ν <u>without</u> defining relations – an analogue of a free group). All subrings are again free (for suitable generators). The corresponding theorem for commutative or associative rings is false! The programme "to carry over the theorem of Schreier to Lie rings" yields results on free groups as well. Unfortunately, I have not yet found time to write this down.[4]

Though at present there is very little time for me to do mathematics I am content with my fate for allowing me to stay in Germany. Last year I was in Russia and I have no desire to go there again. I am allowed to live in a private

[4] This was done in 1955, cf. [**1956**].

house, shall get permission, in a few days, to wear civilian clothes, and am allowed to travel to Hamburg on two Sundays a month. [...]

And Papa is looking forward to each holiday Sunday.

Many cordial greetings

Yours Ernst Witt

4. Dismissal and rehabilitation

At the end of the war, Ernst Witt was in the south of Germany, where his section had fled to during the last days of the war. After a short time as a prisoner of war, he returned to Hamburg in 1945. Three months later, the British military government dismissed him from his position as a professor at the mathematics department of Hamburg University. Witt immediately appealed this measure, writing on September 15, 1945 among other things [**6**]:

As an expatriate German – my father was a missionary in China – when I was young, I felt particularly obliged to the concepts of 'Heimat' and 'Vaterland'. Thus, when in May 1933, following the general mood of my fellow students, I joined the SA and the National Socialist Party, I thought I served a good cause. However, I never went along with the attitude of the party towards Jews, as I highly regarded my teachers, many of whom were either Jewish themselves or had Jewish wives. I continued to work under them, finishing my doctorate. Realizing very soon that science was no issue of importance in the party, – at the time, in an address to the students, Rust underlined that marching was more important than studying – my interest in the party quickly decreased. I refused to join the Studentenbund (National Socialist Student Organization) and later the Dozentenbund (National Socialist Lecturer Organization) and avoided all other factions of the party. [...]. *The best proof that I did not bother about party matters is the fact that nobody in our house knew that I was a party member.*

Along with his dismissal, Witt's accounts were blocked, he was forbidden to enter the university and his food ration-cards were withdrawn.

In the summer of 1946, Witt was requested to testify that he had never been more than a nominal member of the National Socialist Party and not a convinced militarist. He could easily furnish proof of this, as he had never held any position within the party, and his military rank had only been that of a lance-corporal.

The requested testimonies were furnished by Herglotz, Magnus, Rellich and F. K. Schmidt. In their reports of winter 1946/47, each of Witt's colleagues independently stressed his total commitment to science and his lack of political experience [**6**]. F. K. Schmidt added: "when I was in trouble, Witt stayed in contact with me despite pressure on him to do otherwise."

Speiser wrote in a letter July 5, 1946, to Blaschke [**6**]:

I hope that Witt will be rehabilitated. He really is completely innocent. At that time, I think in 1936, I very much liked him as a mathematician, who had no other interests, who did not know anything of politics and other public matters, a character, which probably can be found only in Germany. Why then are they making troubles for him now.

His former student Ho-Jui Chang gave the following statement [**6**] on 8th February 1946:

Dr. Ernst Witt was my teacher and I have had personal contacts with him as well. During all these years I never had the impression that he was a national socialist, though as a Chinese I usually noticed this without difficulties. On the contrary, after the outbreak of the war I often heard him criticising the German government. As a scientist Dr. Witt is very qualified.

Witt also wanted a statement from Richard Courant (1888-1972), because in Göttingen the latter had always been very friendly to him. So Blaschke asked him, on Witt's behalf, for a report. Courant, who had been forced to leave Göttingen in 1934, wrote in a letter of 23rd December 1946 to Blaschke, obviously under the erroneous assumption that it was only a question of an educational task [**6**]:

A few days ago I received your letter of November 16 asking for a statement on behalf of Dr. Witt, who, as I understand, has been reappointed to a research position at Hamburg University but is seeking reinstatement as a teacher. I am answering in English for easier use, but I am afraid that my statement may not be exactly what you desire.

When I was Director of the Mathematics Institute in Göttingen, I discovered Mr. Witt, then a young student, in poor material circumstances, inarticulate, but obviously a budding mathematical talent. I immediately took steps to secure financial help for him, so that he could leisurely finish his studies, which he continued under the scientific and personal guidance of his Jewish teacher, Emmy Noether. It was for all of us a painful disappointment, when in spring 1933 Mr. Witt, still a student, revealed himself firmly entrenched in the Nazi camp as an old party member. A somewhat mitigating fact was that he did not seem to be one of those opportunists of the Nazi era, whose motives were to remain on top or to improve their chances for a career; he probably was governed by a sort of romantic confusion. To my recollection, he did not take a personal part in vicious acts of violence at the university, as some of his Nazi fellow-students did. To what extent he actually dissociated himself from the spirit of deconstruction is not known to me, and I suppose that as to further development after 1933, unbiased observers such as Artin and Hecke will be most competent witnesses.

Courant also send a copy directly to the military government adding the following remark:

I personally would very much hesitate to entrust a man like Witt with an educational task unless during the Nazi regime he has proved himself basically opposed to his masters – this is a matter of which I have no knowledge.

In contrast, Erich Hecke (1887–1947), who enjoyed considerable respect for his courageous and upright behaviour in the Nazi years, stood up for Witt and testified that Witt was "very soon convinced that methods and aims of the Nazi-Party were incompatible with his attitude as a scientifically minded person. Accordingly he was in opposition and did not join either the NSDSTB or the Dozentenbund and in 1938 he resigned from the SA. As a lecturer he avoided all political topics in his official relations to his students." He recommended Witt's reinstatement as a Professor as early as in October 1945 ([**6**]).

In April 1947, Witt was taken on again, and later, after a second screening, he was fully rehabilitated.

5. After 1947

His first lectures after his reinstatement were on Lie groups. The reason for these lectures was that F. K. Schmidt had suggested him to write a book on Lie groups for the yellow Springer series. But as Witt could not receive the latest American publications, and still the lack of food was reducing his capacity for work, the book project drew out longer and longer. After several attempts and in spite of Schmidt's repeated appeals, he eventually gave it up in 1961.

In 1947, a ray of hope appeared with an invitation from his former student Ho-Jui Chang, to stay with his family as a guest professor at the National University of Peking for at least one year. (In the *Hamburger Abhandlungen* 14, Chang had introduced the phrase: "Witt's Lie ring", which today is generally used.) Because of the uncertain political situation in China, however, Witt declined the invitation after all.

At the end of the forties, Witt started to deal with the foundations of mathematics and especially intuitionism. He wanted to know how far he could carry this and even gave 'finitistic' lectures on differential and integral calculus. He was invited to give talks about it and was temporarily considered to be an intuitionist, which he declined saying that a person who works on non-euclidean geometry is not necessarily a "non-euclidean".

At that time, there were a number of available chairs in the zone occupied by the Soviets, and some people tried to bring Witt over there. As early as in July 1948, the university of Jena was interested in him. In 1949, he was offered a chair at the Humboldt University in East Berlin. He was invited to give three talks and set out for the adventurous journey from Hamburg to Berlin, across two borders with their controls, which was a whole day's journey at the time. Though they had better working conditions he decided after mature reflection to decline the offer.

Foreign mathematical institutes now also opened up for him. Wilhelm Blaschke (1885–1962) had reestablished his international contacts, inviting speakers to Hamburg and making sure that mathematicians from Hamburg were invited to foreign countries. In fact, Witt was invited to Spain, where he gave twelve talks on intuitionism, complex analysis and algebra. To learn Spanish, he hired a waiter in Madrid, whom he paid by the hour. He wrote two papers in Spanish, cf. [1950] and [1951].

In the early fifties, he studied the books of Bourbaki in seminars with his students, out of which in 1954 emerged the thesis of his student Banaschewski on filter spaces.

From December 1952 to April 1953, Severi, at the request of Blaschke, invited Witt to Rome. There he lectured on the algebraic theory of quadratic forms in Italian. In the fall of 1953, Witt travelled again, this time to Barcelona at the invitation of his friend Teixidor. He then lectured on Lie rings, cf. [**1956**], footnote 6.

In 1954, Witt's position in Hamburg was changed into a personal chair (without financial consequences). At that time his colleagues in Halle would have liked him to become the successor of Brandt. Witt declined this offer. However, from September to the middle of November 1955, he gave a four-hour lecture on differential geometry. In addition, he presented the general result on quadratic forms and inner product spaces there, which reached its final form in a joint paper with Lenz, cf. [**1957**].

In the following spring in Jena, he held two two-hour lectures on Lie groups and Algebra III for two months.

In 1957 Witt's personal chair was changed into a regular full professorship. "That was about time", one colleague congratulated him.

In April 1957, Emil Artin came to visit Hamburg, his former university. When he entered Witt's office, Witt spontaneously said "Mr. Artin, this is your place." Witt wanted to give back the chair he had got because of Artin's emigration. Artin was touched by this unworldly gesture. In spring 1958, Hamburg University created a new chair for Artin, which he held until his sudden death in 1962.

In spring 1958, Čahit Arf, Witt's Turkish friend since his time in Göttingen, invited him and his family to his house in Istanbul. Witt was to lecture on "non-commutative algebra, in particular Galois theory". He insisted on travelling in his small Ford, though travelling right across four countries, with lots of snow, a crammed car and his family was a real adventure in those days. When he came back to Turkey later, in 1962, 1963, and 1964, being invited to Ankara by Ulucay, he chose the more comfortable way by plane.

In 1959, Blaschke suggested to Witt to apply for a Fulbright travel grant, to get to know leading mathematicians in the United States. André Weil proposed him to the Institute for Advanced Study at Princeton for the academic year of 1960/61, and Witt received an official invitation from the director Robert Oppenheimer. He was very pleased to be invited to the "hub of mathematics", and travelled to Princeton with high hopes. There, one day, during a discussion about a member of the National Socialist Party, he felt obliged to declare that he had also been a member of that party. To behave otherwise would have seemed insincere to him. He found, to his utter astonishment, that his contacts with his colleagues were suddenly severed.

In the course of the following years, Witt accepted many invitations to universities in the United States and in Canada, e.g. in the spring of 1963 to St. John's University in Newfoundland. At the winter term of 1963/64, Banaschewski invited him to McMaster University in Hamilton in Ontario, from where he visited several Canadian universities and received offers for temporary as well as permanent positions. He did not accept since this would mean less financial security concerning retirement and surviving dependents' pension. The following winter terms of 1964/65 and 1965/66, he spent at Stony Brook, where again people would have liked him to stay. He declined the offer for similar reasons. Thus he remained faithful to Hamburg University until he retired in 1979.

As early as in 1959, Ernst Witt realized the advantages of the use of computers in almost all branches of mathematics, and thus was – as so often before – ahead of his time. At that time, programs still had to be punched into punch cards at the computer center and one had to wait for them to be processed. For more than twenty years, Witt spent most of his time programming, until in 1982 the center got a new computer, and he did not want to learn how to use. Among other things, he calculated twin primes and coefficients of cyclotomic polynomials and drew up programs for the classification of Steiner systems. He customized the programming language Algol into what his students called "Wittgol".

In Hamburg, Witt and Deuring organized a seminar which in 1951, after Deuring had left became the Hasse-Witt seminar. In 1955, Witt founded his seminar on topology, in which, until 1981, many subsequently famous mathematicians and often also physicists gave talks as students and assistants. During his long years of teaching at Hamburg, his students always enjoyed a lot of freedom in their work, but on the other hand Witt also expected them to think for themselves. Thus he did not propose the topics for diplomas or Ph.D.'s, but wanted the students to find their own topics. Topics were allowed from any field of mathematics. Among others, his students were: Sigrid Böge, née Becken, Walter Borho, Günter Harder, Manfred Knebusch, Horst Leptin and Jürgen Rohlfs. The following assessment of the teacher Ernst Witt, from a report written as early as 1937 always held good for him [**5**]:

Witt's lectures are models of clarity and concision. He has a beautiful gift for creating in his audience, with the help of comparatively brief hints, the exact image fundamental for a deeper understanding, which goes beyond the formal thread. However, he will have to learn to adapt his assumptions to the level of an audience which does not consist of scientists but of students. Witt's personality is straightforward and sincere. He can be relied on in every respect.

In the sixties Witt gave several colloquium talks in which he significantly simplified the algebraic theory of quadratic forms by means of his notion of a *round form.*

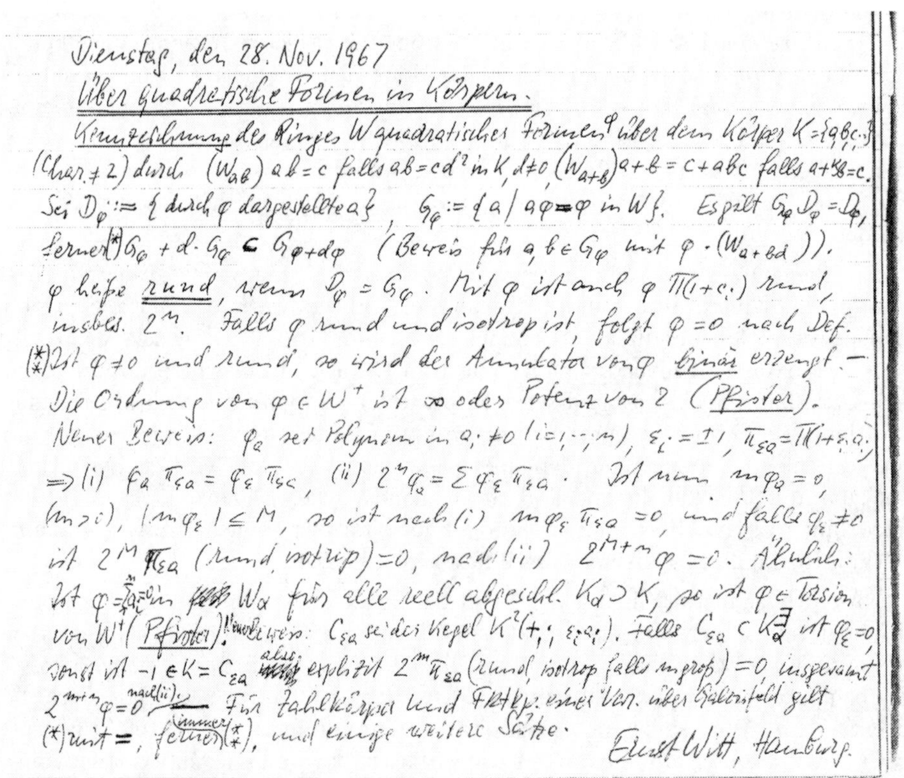

Dienstag, den 28. Nov. 1967

Über quadratische Formen in Körpern.

Kennzeichnung des Ringes W quadratischer Formen φ über dem Körper $K = \{a, b, c, \ldots\}$ (Char. $\neq 2$) durch (W_{ab}) $ab = c$ falls $ab = cd^2$ in K, $d \neq 0$, (W_{a+b}) $a + b = c + abc$ falls $a + b = c$.

Sei $D_\varphi := \{$durch φ dargestellte $a\}$, $G_\varphi := \{a \mid a\varphi = \varphi$ in $W\}$. Es gilt $G_\varphi D_\varphi = D_\varphi$, ferner (*) $G_\varphi + d \cdot G_\varphi \subset G_{\varphi + d\varphi}$ (Beweis für $a, b \in G_\varphi$ mit $\varphi \cdot (W_{a+bd})$).

φ heiße rund, wenn $D_\varphi = G_\varphi$. Mit φ ist auch $\varphi \prod(1 + c_i)$ rund, insbes. 2^n. Falls φ rund und isotrop ist, folgt $\varphi = 0$ nach Def.

(**) Ist $\varphi \neq 0$ und rund, so wird der Annulator von φ binär erzeugt. — Die Ordnung von $\varphi \in W^+$ ist ∞ oder Potenz von 2 (Pfister).

Neuer Beweis: φ_a sei Polynom in $a_i \neq 0$ $(i = 1, \ldots, n)$, $\varepsilon_i = \pm 1$, $\pi_{\varepsilon a} = \prod(1 + \varepsilon_i a_i)$ $\Rightarrow$ (i) $\varphi_a \pi_{\varepsilon a} = \varphi_\varepsilon \pi_{\varepsilon a}$, (ii) $2^n \varphi_a = \sum \varphi_\varepsilon \pi_{\varepsilon a}$. Ist nun $m\varphi_a = 0$, $(m > 0)$, $|m\varphi_\varepsilon| \leq M$, so ist nach (i) $m\varphi_\varepsilon \pi_{\varepsilon a} = 0$, und falls $\varphi_\varepsilon \neq 0$ ist $2^M \pi_{\varepsilon a}$ (rund, isotrop) $= 0$, nach (ii) $2^{M+n} \varphi = 0$. Ähnlich: Ist $\varphi = \sum_1^m a_i = 0$ in W_α für alle reell abgeschl. $K_\alpha \supset K$, so ist $\varphi \in$ Torsion von W^+ (Pfister)! Neuer Beweis: $C_{\varepsilon a}$ sei der Kegel $K^2(+_i\, \varepsilon_i a_i)$. Falls $C_{\varepsilon a} \subset K_\alpha^{\exists}$ ist $\varphi_\varepsilon = 0$, sonst ist $-1 \in K = C_{\varepsilon a}$ also explizit $2^m \pi_{\varepsilon a}$ (rund, isotrop falls m groß) $= 0$, insgesamt $2^{m+n} \varphi = 0$ nach (ii). — Für Zahlkörper und Fktkp. einer Var. über Galoisfeld gilt (*) mit $=$, ferner immer (**), und einige weitere Sätze.

Ernst Witt, Hamburg.

Colloquium talk in Hamburg. *Tuesday, 28th Nov. 1967*
On quadratic forms in fields.[5]

Characterization of the ring W of quadratic forms φ over the field $K = \{a, b, c, \dots\}$ (char. $\neq 2$) by (W_{ab}) $ab = c$ if $ab = cd^2$ in K, $d \neq 0$, and (W_{a+b}) $a + b = c + abc$ if $a +^K b = c$. Let $D_\varphi := \{a \text{ represented by } \varphi\}$, $G_\varphi := \{a \mid a\varphi = \varphi \text{ in } W\}$. We have $G_\varphi D_\varphi = D_\varphi$, in addition $(*)$ $G_\varphi + d \cdot G_\varphi \subset G_{\varphi + d\varphi}$ (proof by $\varphi \cdot (W_{a+bd})$ for $a, b \in G_\varphi$). φ is said to be round if $D_\varphi = G_\varphi$. If φ is round then $\varphi \prod(1 + c_i)$ is round, in particular 2^n. If φ is round and isotropic then $\varphi = 0$ by definition.
$\binom{*}{*}$ If $\varphi \neq 0$ and round then the annihilator of φ is binary generated. –
The order of $\varphi \in W^+$ is ∞ or a power of 2 (Pfister). New proof: let φ_a be a polynomial in $a_i \neq 0$ $(i = 1, \dots, n)$, $\varepsilon_i = \pm 1$, $\pi_{\varepsilon a} = \prod(1 + \varepsilon_i a_i) \implies$
(i) $\varphi_a \pi_{\varepsilon a} = \varphi_\varepsilon \pi_{\varepsilon a}$, (ii) $2^n \varphi_a = \sum \varphi_\varepsilon \pi_{\varepsilon a}$. Assume that $m\varphi_a = 0$, $(m > 0)$, $|m\varphi_\varepsilon| \leq M$ then $m\varphi_\varepsilon \pi_{\varepsilon a} = 0$ by (i), and if $\varphi_\varepsilon \neq 0$ then $2^M \pi_{\varepsilon a}$ (round and isotropic) $= 0$, thus (ii) yields $2^{M+n}\varphi = 0$. Similarly: If $\varphi = \sum_1^n a_i = 0$ in W_α for each real closed $K_\alpha \supset K$ then $\varphi \in$ torsion part of W^+ (Pfister). New proof: let $C_{\varepsilon a}$ be the cone $K^2(+, \cdot, \varepsilon_i \cdot a_i)$. If $C_{\varepsilon a} \subset K_\alpha^\exists$ then $\varphi_\varepsilon = 0$, otherwise $-1 \in K = C_{\varepsilon a}$, hence, explicitly, $2^m \pi_{\varepsilon a}$ (round and isotropic if m large) $= 0$; summarizing, we get $2^{m+n}\varphi = 0$ by (ii). – For number fields and function fields in one variable over a Galois field, $(*)$ holds with $=$, in addition, $\binom{*}{*}$ always holds as well as some further theorems.

Also in his written work, Witt always points out exactly the things which are fundamental for a true understanding. His lectures were unconventional both for the choice of topics as for their style. From 1954 on, his lectures on Calculus I and II for students in their first and second semester were based on the notion of filters. Even the students who had difficulties in understanding his lectures did not get bored, as he always wove in humorous remarks and witty comparisons. Local coordinates, for example, he illustrated as follows: "If the mayor of a little town in the middle-of-nowhere has to draw a map, his little town will become the center of it". He also taught a lot of things from his own works, as, for example, the theory of quadratic forms, the Mathieu groups and Steiner systems or the Riemann-Roch theorem for skew fields. The reader is referred to P. M. Neumann: "An essay on Witt's work on the Mathieu groups and on Steiner systems" in [CP].

In 1978, Witt became an ordinary member of the *Akademie der Wissenschaften zu Göttingen.*

Though he had always been interested in the latest findings, Witt no longer actively followed the development of modern algebraic geometry by Grothendieck and others. One reason for this might have been that he was not very healthy. Since 1969, he suffered from allergies to various detergents and adhesives for wallpapers and carpets. He complained about dizziness and a diminishing ability to concentrate. Because of these troubles, he had to decline several invitations for talks at home and abroad, and in 1975, he experienced a slight stroke. Moreover, because of his allergies, he could not join the move of the mathematics department to a modern highrise building, in which, because of some kind of air conditioning, the windows could not be opened. Thus he got more and more isolated, and his general reputation to be an eccentric was strengthened. On his account the colloquia of the

[5]Published with kind permission of the coordinator of the colloquium on pure mathematics at Hamburg University. See [**CP**] for a detailed version.

mathematics department were held in another building, and he could attend them until shortly before his death. He died after a short illness on July 3, 1991 at an age of 80 years.

Acknowledgement. I would like to thank Bärbel Deninger, Samuel Patterson and Andrew Ranicki very much for help with the English version.

Author's remark. I was a student of Ernst Witt at the University of Hamburg. After my examination in Mathematics, Witt asked me if I would like to become his assistant. Due to rationalization, Hamburg University offered me a contract for only five months. Since Witt, however, had promised me a three year contract, his reaction to the university administration was the serious threat to resign, and in consequence of that, the contract was extended. So I became his assistant until his retirement in 1979. The above descriptions refer to [1] and [2], some are from my personal recollection.

References

[CP] Ernst Witt: *Collected Papers.* Springer-Verlag 1998.

[1] Ina Kersten: *Ernst Witt 1911–1991.* Jber. d. Dt. Math.-Verein. **95** (1993), 166–180.

[2] Ina Kersten: *Ernst Witt – ein großer Mathematiker.* DMV–Mitteilungen 2–1995, 28–30.

[3] Friedrich L. Bauer: *Entzifferte Geheimnisse, Methoden und Maximen der Kryptologie.* Springer-Verlag 1997.

[4] SUB Göttingen, Handschriftenabteilung. Signatur: Cod. Ms. Herglotz F 160.

[5] Staatsarchiv Hamburg. Bestandsnr.: 113–5, Signatur: 92d UA 19.

[6] Staatsarchiv Hamburg. Bestandsnr.: 221–11, Signatur: Ed 852.

[Papers by Ernst Witt] The numbers following the dots refer to page numbers in **[CP]**. Papers with no source mentioned are published in **[CP]** for the first time.
Hamb. Abh.:= Abh. Math. Sem. Univ. Hamburg; Crelle := J. reine angew. Math.

[1931] Über die Kommutativität endlicher Schiefkörper, Hamb. Abh. **8** 395

[1934] Riemann-Rochscher Satz und Z-Funktion im Hyperkomplexen, Math. Ann. **110** .. 43–59

[1934] Zerlegung reeller algebraischer Funktionen in Quadrate. Schiefkörper über reellem Funktionenkörper, Crelle **171** 81–88

[1934] Algebra (a, b, ζ), Facsimile 117

[1935] Über ein Gegenbeispiel zum Normensatz, Math. Z. **39** 63–68

[1935] Über die Invarianz des Geschlechts eines algebraischen Funktionenkörpers, Crelle **172** 90–91

[1935] Der Existenzsatz für abelsche Funktionenkörper, Crelle **173** 69–77

[1935] Zwei Regeln über verschränkte Produkte, Crelle **173** 118–119

[1936] Konstruktion von galoisschen Körpern der Charakteristik p zu vorgegebener Gruppe der Ordnung p^f, Crelle **174** 120–128

[1936] Bemerkungen zum Beweis des Hauptidealsatzes von S. Iyanaga, Hamb. Abh. **11** 114

[1936] (mit Hasse) Zyklische unverzweigte Erweiterungskörper vom Primzahlgrade p über einem algebraischen Funktionenkörper der Charakteristik p, Monatsh. Math. Phys. **43** ... 92–107

[1936] Zur Funktionalgleichung der L-Reihen im Funktionenkörperfall. Handwritten Notes Summarized by Rainer Schulze-Pillot 78–80

[1936] Der Satz von v. Staudt-Clausen 396

[- - - -] Primzahlsatz 397

[1937] Theorie der quadratischen Formen in beliebigen Körpern, Crelle **176** 2–15

[1937] Zyklische Körper und Algebren der Charakteristik p vom Grad p^n. Struktur diskret bewerteter perfekter Körper mit vollkommenem Restklassenkörper der Charakteristik p, Crelle **176** 142–156

[1937] Schiefkörper über diskret bewerteten Körpern, Crelle **176** 129–132

[1937] (mit Schmid) Unverzweigte abelsche Körper vom Exponenten p^n über einem algebraischen Funktionenkörper der Charakteristik p, Crelle **176** 108–113

[1937] Treue Darstellung Liescher Ringe, Crelle **177** 195–203

[1937] Über Normalteiler, deren Ordnung prim ist zum Index 277–278
[1938] Die 5-fach transitiven Gruppen von Mathieu, Hamb. Abh. **12** 279–287
[1938] Über Steinersche Systeme, Hamb. Abh. **12** 288–298
[1938] Zum Problem der 36 Offiziere, Jber. DMV **48**, ... 305
[1940] Ein Identitätssatz für Polynome, Mitt. Math. Ges. Hamburg VIII 402–403
[1940] Die Automorphismengruppen der Cayleyzahlen, Crelle **182** 253
[1941] Eine Identität zwischen Modulformen zweiten Grades Hamb. Abh. **14** 313–327
[1941] Spiegelungsgruppen und Aufzählung halbeinfacher Liescher Ringe, Hamb. Abh. **14** 213–246
[1949] Rekursionsformel für Volumina sphärischer Polyeder, Arch. Math. **1** 371–372
[1950] Sobre el Teorema de Zorn, Rev. Mat. Hisp.-Amer. (4) **10** 373–376
[1951] Beweisstudien zum Satz von M. Zorn, Math. Nachr. **4** 377–381
[1951] Algunas cuestiones de matemática intuicionista, Conferencias de Matemática II, Publ. del Instituto de Matemáticas "Jorge Juan", Madrid. 382–387
[1952] Die algebraische Struktur des Gruppenringes einer endlichen Gruppe über einem Zahlkörper, Crelle **190** 330–344
[1952] Ein kombinatorischer Satz der Elementargeometrie, Math. Nachr. **6** 349–350
[1952] Über einen Satz von Ostrowski, Arch. Math. **3** 404
[1953] Treue Darstellungen beliebiger Liescher Ringe, Collect. Math. **6** 204–212
[1953] Über freie Ringe und ihre Unterringe, Math. Z. **58** 351–352
[1953] (mit Jordan) Zur Theorie der Schrägverbände, Akad. Wiss. Lit. Mainz, Abh. math.-naturw. Kl. 1953, Nr. 5 353–360
[1954] Über eine Invariante quadratischer Formen mod 2, Crelle **193** 17–18
[1954] (mit Klingenberg) Über die Arfsche Invariante quadratischer Formen mod 2, Crelle **193** 19–20
[1954] Konstruktion von Fundamentalbereichen, Ann. Mat. Pura Appl. **36** 388–394
[1954] Verlagerung von Gruppen und Hauptidealsatz, Proc. ICM 1954 115–116
[1955] Über die Kommutatorgruppe kompakter Gruppen Rendiconti di Matematica e delle suo Applicazioni **14** 308–312
[1955] Über den Auswahlsatz von Blaschke, Hamb. Abh. **19** 405
[1956] Die Unterringe der freien Lieschen Ringe, Math. Z. **64** 254–275
[1957] Verschiedene Bemerkungen zur Theorie der quadratischen Formen über einem Körper, Centre Belge Rech. Math., Coll. d'Algèbre sup. Bruxelles 21–26
[1957] (mit Lenz) Euklidische Kongruenzsätze in metrischen Vektorräumen 27–32
[1958] p-Algebren und Pfaffsche Formen, Hamb. Abh. **22** 133–140
[1961] Beziehungen zwischen Klassengruppen von Zahlkörpern 361–367
[1962] Konstruktion endlicher Ebenen, Intern. Nachr. **72** 307
[1964] Vektorkalkül und Endomorphismen von 1-Potenzreihen, Facsimile 164
[1967] Über quadratische Formen in Körpern 35–39
[1969] Vektorkalkül und Endomorphismen der Einspotenzreihengruppe 157–163
[1969] Über die Divergenz gewisser ζ-Reihen 369–370
[1973] Über die Sätze von Artin-Springer und Knebusch 41
[1975] Homöomorphie einiger Tychonoffprodukte, in: Topology and its Applications, Lecture Notes in Pure and Appl. Math. **12**, Dekker, New York 406–407
[1976] Über eine Klasse von Algebren mit lauter endlich erzeugbaren Unteralgebren, Mitt. Math. Ges. Hamburg X 276
[1976] Über die Ordnungen endlicher Neokörper 408–409
[1983] Vorstellungsbericht, Jahrbuch Akad. Wiss. Göttingen 1
[1989] Erinnerungen an Gabriel Dirac in Hamburg, Ann. Discr. Math. **41** 410–411

MATHEMATISCHES INSTITUT
BUNSENSTR. 3–5
D-37073 GÖTTINGEN
GERMANY

E-mail address: `kersten@mathematik.uni-bielefeld.de`

Contemporary Mathematics
Volume **272**, 2000

Generic Splitting Towers and Generic Splitting Preparation of Quadratic Forms

Manfred Knebusch and Ulf Rehmann

Abstract. This manuscript describes how a generic splitting tower of a regular anisotropic quadratic form digests the form down to a form which is totally split.

Introduction

We work with quadratic forms on finite dimensional vector spaces over an arbitrary field k. We call such a form $q: V \to k$ *regular* if the radical $V^{\perp}$ of the associated bilinear form B_q has dimension ≤ 1 and the quasilinear part $q|V^{\perp}$ of q is anisotropic. If k has characteristic char $k \neq 2$ this means that $V^{\perp} = \{0\}$. If char $k = 2$ it means that either $V^{\perp} = \{0\}$ or $V^{\perp} = kv$ with $q(v) \neq 0$.

In the present article a "form" always means a regular quadratic form. Our first goal is to develop a generic splitting theory of forms. Such a theory has been given in [K$_2$] for the case of char $k \neq 2$. Without any restriction on the characteristic, a generic splitting theory for complete quotients of reductive groups was given in [KR], which is closely related to our topic.

In §1 we present a generic splitting theory of forms in a somewhat different manner than in [K$_2$]. We start with a key result from [KR] (cf. Theorem 1.3 below), then develop the notion of a *generic splitting tower* of a given form q over k with associated higher indices and kernel forms, and finally explain how such a tower $(K_r \mid 0 \leq r \leq h)$ together with the sequence of higher kernel forms $(q_r \mid 0 \leq r \leq h)$ of q controls the splitting of $q \otimes L$ into a sum of hyperbolic planes and an anisotropic form (called the anisotropic part or kernel form of $q \otimes L$), cf. 1.19 below. More generally we explain how the generic splitting tower $(K_r \mid 0 \leq r \leq h)$ together with $(q_r \mid 0 \leq r \leq h)$ controls the splitting of the specialization $\gamma_*(q)$ of q by a place

2000 *Mathematics Subject Classification.* Primary 11E04, 11E81.

Key words and phrases. quadratic forms, generic splitting.

This research was supported by the European Union TMR Network FMRX CT-97-0107 "Algebraic K-Theory, Linear Algebraic Groups and Related Structures", and by the Sonderforschungsbereich "Diskrete Mathematik" at the University of Bielefeld.

$\gamma\colon k \to L \cup \infty$, if q has good reduction under q, cf. Theorem 1.18 below. Then in §2 we study how a generic splitting tower of $q \otimes L$ can be constructed from a generic splitting tower of q for any field extension L/k. All these results are an expansion of the corresponding results in [K$_2$] to fields of any characteristic.

We mention that a reasonable generic splitting theory holds more generally for a quadratic form $q\colon V \to k$ such that the quasilinear part $q|V^{\perp}$ is anisotropic, without the additional assumption $\dim V^{\perp} \leq 1$. This needs more work. It will be contained in the forthcoming book [K$_5$].

In §3 we prove that for any form q over k there exists a generic splitting tower $(K_r \mid 0 \leq r \leq h)$ of q which contains a subtower $(K'_r \mid 0 \leq r \leq h)$ of field extensions of k such that K'_r/K'_{r-1} is purely transcendental, and such that the anisotropic part of $q \otimes K_r$ can be defined over K'_r for every $r \in [1, h]$. {We have $K'_0 = K_0 = k$.} This result, which may be surprising at first glance, leads us in §4 to the second theme of this article, namely *generic splitting preparations* (Def. 4.3) and the closely related *generic splitting decompositions* (Def. 4.8) of a form q. We focus now on the second notion, since its meaning can slightly more easily be grasped than that of the first (more general) notion.

A generic splitting decomposition of a form q over k consists of a purely transcendental field extension K'/k and an orthogonal decomposition

$$q \otimes K' \cong \eta_0 \perp \eta_1 \perp \cdots \perp \eta_h \perp \varphi_h \qquad (*)$$

with certain properties. In particular, $\dim \varphi_h \leq 1$, all η_i have even dimension, and η_0 is the hyperbolic part of $q \otimes K'$ (which comes from the hyperbolic part of q by going up from k to K'). The generic splitting decomposition in a certain sense controls the splitting behavior of $q \otimes L$ for any field extension L of k, more generally of $\gamma_*(q)$, for any place $\gamma\colon k \to L \cup \infty$ such that q has good reduction under γ. This control can be made explicit in much the same way as the control by generic splitting towers, using "*quadratic places*" (or "Q-places" for short) instead of ordinary places, cf. §6.

Quadratic places have been introduced in the recent article [K$_4$] and used there for another purpose. We recapitulate here what is necessary in §5. We are sorry to say that our theory in §6 demands that the occurring fields have characteristic $\neq 2$. This is forced by the article [K$_4$], where the specialization theory of forms under quadratic places is only done in the case of characteristics $\neq 2$. It seems that major new work and probably also new concepts are needed to establish a specialization theory of forms under quadratic places in all characteristics.

An overall idea behind generic splitting decompositions is the following. If we allow for the form q over k a suitable linear change of coordinates with coefficients in a purely transcendental field extension $K' \supset k$, then the form – now called $q \otimes K'$ – decomposes orthogonally into subforms $\eta_0, \eta_1, \ldots, \eta_h, \varphi_h$ such that the forms $\eta_k \perp \cdots \perp \eta_h \perp \varphi_h$ with $1 \leq k \leq h$ give the higher kernel forms of q, when we go up further from K' to suitable field extensions of K'. Thus, after the change of coordinates, the form q is "well prepared" for an investigation of its splitting behavior. This reminds a little of the Weierstrass preparation theorem, where an analytic function germ becomes well prepared after a linear change of coordinates. In contrast to Weierstrass preparation we allow a purely transcendental field extension for the coefficients of the linear change of coordinates. But no essential information about the form q is lost by passing from q to $q \otimes K'$, since q is the specialization $\lambda_*(q \otimes K')$ of $q \otimes K'$ under any place $\lambda\colon K' \to k \cup \infty$ over k.

The idea behind generic splitting preparations is similar. {Generic splitting decompositions form a special class of generic splitting preparations.} Now the forms η_i are defined over fields K'_i such that $K'_0 = k$ and every K'_i is a purely transcendental extension of K'_{i-1}.
Generic splitting decompositions and, more generally, generic splitting preparations give new possibilities for manipulations with forms. For example, if $q \otimes K' \cong \eta_0 \perp \cdots \perp \eta_h \perp \varphi_h$ is a generic splitting decomposition of q, then we may look how many hyperbolic planes split off in $q \otimes E_r$ for E_r the total generic splitting field of one of the summands η_r. We do not enter these matters here, leaving all experiments to the future and to the interested reader.

§1. Generic splitting in all characteristics

1.0. NOTATIONS. For a, b elements of a field k we denote the form $a\xi^2 + \xi\eta + b\eta^2$ over k by $[a, b]$. Since we only allow regular forms, we demand $1 - 4ab \neq 0$. By $H\colon= [0, 0]$ we denote the hyperbolic plane.
If q is a (regular quadratic) form over k then we have the Witt decomposition ([W], [A]) $q \cong r \times H \perp \varphi$ with an anisotropic form φ and $r \in \mathbb{N}_0$. We call r the *index* of q and write $r = \text{ind}\,(q)$. We further call φ the *kernel form* [1)] or *anisotropic part* of q and use both notations $\varphi = \ker(q)$, $\varphi = q_{\text{an}}$.
If $L \supset k$ is a field extension then $q \otimes L$ or q_L denotes the form over L obtained from q by extension of the base field k to L. Thus, if q lives on the k-vector space V, then $q \otimes L$ lives on the L-vector space $L \otimes_k V$. A major theme of this article is the study of ind $(q \otimes L)$ and $\ker(q \otimes L)$ for varying extensions L/k.
$\dim q$ denotes the dimension of the vector space V on which q lives, i.e., the number of variables occurring in the form q. We have $\dim q = \dim(q \otimes L)$. The zero form $q = 0$ is not excluded. Then $V = \{0\}$ and $\dim q = 0$.
We say that q *splits totally* if $\dim(q_{\text{an}}) \leq 1$. This is equivalent to ind $(q) = [\dim q/2]$.
For another form φ over k we write $\varphi < q$ if φ is isometric to a subform of q (including the case $\varphi \cong q$).

1.1. DEFINITION/FURTHER NOTATIONS. If $q \neq 0$ and $\dim q$ is even, let

$$\delta q\colon= \begin{cases} (\text{discriminant of } q) \in k^*/k^{*2} & \text{if char } k \neq 2 \\ (\text{Arf invariant of } q) \in k^+/\wp k & \text{if char } k = 2, \end{cases}$$

$$p_{\delta q}(X)\colon= \begin{cases} X^2 - \delta q & \text{if char } k \neq 2 \\ X^2 + X + \delta q & \text{if char } k = 2. \end{cases}$$

The separable polynomial $p_{\delta q}(X)$ splits over k if and only if δq is trivial. If $\dim q$ is odd or if δq is trivial we say that q is of *inner type*, otherwise we say that q is of *outer type*.
These notions are adjusted to the corresponding notions in the theory of reductive groups. q is inner (resp. outer) if and only if $SO(q)$ is inner (resp. outer). {N.B. $SO(q)$ is almost simple for $\dim q \geq 3$ since the form q is regular.}
We define

$$k_{\delta q}\colon= \begin{cases} k & \text{if } q \text{ is of inner type} \\ k[X]/p_{\delta q}(X) & \text{if } q \text{ is of outer type.} \end{cases}$$

For $i = 1, \ldots, [\dim q/2]$, we denote by $V_i(q)$ the projective variety of totally isotropic subspaces of dimension i in the underlying space of q, and by $k_i(q)$ we denote the

---

1) = “Kernform” in [W]

function field of $V_i(q)$, unless $\dim q = 2$. In the latter case V_1 consists of two irreducible components defined over $k_{\delta q}$. We then set $k_1(q) = k_{\delta q}$.
In general, we will also write $k(q) = k_1(q)$, which is, with the above interpretation, the function field of the quadric $V_1(q)$ associated to q by the equation $q = 0$. □

1.2. LEMMA. *Let q be a (regular quadratic) form over k.*

i) *If K/k is any field extension such that q_K is of inner type, then K contains a subfield isomorphic to $k_{\delta q}$.*
ii) *Let $i \in \{1, \dots, [\dim q/2]\}$. If q is of inner type or if $i \le \dim q/2 - 1$, then $V_i(q)$ is defined over k. If q is of outer type, then $V_{\dim q/2}(q)$ is defined over $k_{\delta q}$.*
iii) *$V_i(q)$ is geometrically irreducible unless q is of outer type and $i = \dim q/2 - 1$, in which case it decomposes, over $k_{\delta q}$, into two geometrically irreducible components isomorphic to $V_{\dim q/2}(q)$.*

PROOF. i): Let $\dim q$ be even. Clearly q_K is inner if and only if the polynomial $p_{\delta q}$ has a zero in K, hence i) follows.
ii), iii): Since the stabilizer of an i-dimensional totally isotropic subspace of the underlying space of q is a parabolic subgroup of $SO(q)$, the statements follow from [KR, 3.7, p. 44f]. □

A key observation for the generic splitting theory of quadratic forms is the following theorem, which has a generalization for arbitrary homogeneous projective varieties [KR, 3.16, p. 47]:

1.3. THEOREM. *Let q be a form over k, let F_i denote the function field of $V_i(q)$ as a regular extension of k resp. $k_{\delta q}$ according to 1.2.ii, and let L/k be a field extension. The following statements are equivalent.*

i) $\operatorname{ind}(q \otimes L) \ge i$.
ii) *The projective variety $V_i(q)$ has an L-rational point.*
iii) *There is a k-place $F_i \to L \cup \infty$.*
iv) *L contains a subfield isomorphic to the algebraic closure F_i^0 of k in F_i, and the free composite LF_i over F_i^0 is a purely transcendental extension of L.*

REMARK. Only in the case of an outer q and $i = \dim q/2$, we have $F_i^0 = k_{\delta q} \neq k$; in all other cases, i.e., if q is inner or $i \le \dim q/2 - 1$, we have $F_i^0 = k$ in iv).

PROOF of 1.3. The equivalence of i) and ii) is obvious. The other equivalences follow from [KR, 3.16, p.47], again after observing that the stabilizer of an i-dimensional totally isotropic subspace is a parabolic subgroup of $SO(q)$. □

1.4. DEFINITION. We call two field extensions $K \supset k$ and $L \supset k$ *specialization equivalent* over k, and we write $K \sim_k L$, if there exists a place from K to L over k and also a place from L to K over k. □

1.5. COROLLARY. *If $q' = l \times H \perp q$, and if F_i' is the function field of $V_i(q')$, then F_{l+i}' and F_i are specialization equivalent over k.* [2] .

---

[2] *We will denote the hyperbolic plane $[0, 0]$ over* **any** *field by H*

PROOF. This is obvious by the equivalence of i) and iii) in 1.3. □

In the following q is a (regular quadratic) form over k. We want to associate to q partial generic splitting fields and partial generic splitting towers as has been done in [K₂] for char $k \neq 2$. We will proceed in a different way than in [K₂], starting with a formal consequence of Theorem 1.3.

1.6. COROLLARY. *Let L and L' be field extensions of k. Assume there exists a place $\lambda\colon L \to L' \cup \infty$ over k. Then* $\operatorname{ind}(q \otimes L') \geq \operatorname{ind}(q \otimes L)$.

PROOF. Let $i := \operatorname{ind}(q \otimes L)$. By the theorem there exists a place $\rho\colon F_i \to L \cup \infty$ over k. Then $\lambda \circ \rho$ is a place from F_i to $L' \cup \infty$. Again by the theorem, $\operatorname{ind}(q \otimes L') \geq i$.□

1.7. DEFINITION. (Cf. [HR₁]). The *splitting pattern* $\mathrm{SP}(q)$ is the (naturally ordered) sequence of Witt indices $\operatorname{ind}(q \otimes L)$ with L running through all field extensions of k. □

Notice that the sequence $\mathrm{SP}(q)$ is finite, consisting of at most $[\dim q/2]+1$ elements $j_0 < j_1 < \cdots < j_h$. Of course, $j_0 = \operatorname{ind}(q)$ and $j_h = [\dim q/2]$. We call h the *height* of q, and we write $h = h(q)$. Notice also that $\mathrm{SP}(q)$ is the sequence of all numbers $i \leq [\dim q/2]$ with $\operatorname{ind}(q \otimes F_i) = i$.

1.8. DEFINITION. Let $r \in \{0, 1, \ldots, h\} = [0, h]$. A *generic splitting field of* q of level r is a field extension F/k with the following properties:

a) $\operatorname{ind}(q \otimes F) = j_r$.
b) For every field $L \supset k$ with $\operatorname{ind}(q \otimes L) \geq j_r$ there exists a place $\lambda\colon F \to L \cup \infty$ over k.

Such a field extension F/k, for any level r, is also called a *partial generic splitting field of* q, and, in the case $r = h$, a *total generic splitting* field of k. □

It is evident from the definitions and from Corollary 1.6 that, if K is a generic splitting field of q of some level r and L is a field extension of k, then L is a generic splitting field of q of level r if and only if K and L are specialization equivalent over k.

1.9. PROPOSITION. *Let $r \in [0, h]$. All the fields F_i from Theorem* 1.3 *with $j_{r-1} < i \leq j_r$ are generic splitting fields of q of level r. $\{$Read $j_{-1} = -1.\}$ In particular, the fields $k(q_{\mathrm{an}}), F_{j_0+1}, \ldots, F_{j_1}$ are generic splitting fields of q of level* 1.

PROOF. By Theorem 1.3, we certainly have $\operatorname{ind}(q \otimes F_i) \geq i$, hence $\operatorname{ind}(q \otimes F_i) \geq j_r$. If L/k is any field extension with $\operatorname{ind}(q \otimes L) \geq j_r$, then, again by Theorem 1.3, there exists a place $\lambda\colon F_i \to L \cup \infty$ over k. Thus condition b) in Definition 1.8 is fulfilled. We can choose L as an extension of k with $\operatorname{ind}(q \otimes L) = j_r$. By Corollary 1.6 we have $\operatorname{ind}(q \otimes F_i) \leq \operatorname{ind}(q \otimes L)$. Thus $\operatorname{ind}(q \otimes F_i) = j_r$.
Moreover, since $k(q_{\mathrm{an}})$ is the function field of $V_1(q_{\mathrm{an}})$, it follows from 1.5 that this field is specialization equivalent over k to F_{j_0+1}. □

1.10. COROLLARY. *If F is a generic splitting field of q (of some level r), then the algebraic closure of k in F is always k, except if q is outer and* $\operatorname{ind}(q_F) = \dim q/2$, *in which case it is $k_{\delta q}$.*

PROOF. Clearly $F \sim_k F_i$ for $i = \operatorname{ind} q_F$, hence we have k-places from F to F_i and vice versa, which are of course injective on the algebraic closure of k in F resp. F_i. Our claim now follows from 1.3 and the remark after 1.3. □

1.11. SCHOLIUM. *Let $(K_r \mid 0 \le r \le h)$ be a sequence of field extensions of k such that for each $r \in [0,h]$ the field K_r is a generic splitting field of q of level r. Let L/k be a field extension of k. We choose $s \in [0,h]$ maximal such that there exists a place from K_s to L over k. Then* $\operatorname{ind}(q \otimes L) = j_s$.

PROOF. By Corollary 1.6 we have $\operatorname{ind}(q \otimes L) \ge j_s$. Suppose that $\operatorname{ind}(q \otimes L) > j_s$. Then $\operatorname{ind}(q \otimes L) = j_r$ for some $r \in [0,h]$ with $r > s$. Thus there exists a place from K_r to L over k. This contradicts the maximality of s. We conclude that $\operatorname{ind}(q \otimes L) = j_s$. □

If q is anisotropic and $\dim q \ge 2$, then a generic splitting field of q of level 1 is called a *generic zero field* of q. Proposition 1.9 tells us that, in general, F_{j_0+1} and $k(q_{\mathrm{an}})$ are generic zero fields of q_{an}. {N.B. The notion of generic zero field has also been established if q is isotropic, cf. [K_2, p. 69]. Then it still is true that $k(q)$ is a generic zero field of q.}

1.12. DEFINITION. A *generic splitting tower* of q is a sequence of field extensions $K_0 \subset \cdots \subset K_h$ of k such that K_0 is specialization equivalent over k with k, and such that K_{r+1} is specialization equivalent over K_r with $K_r(q_{K_r,\mathrm{an}})$. [3)] In particular, the inductively defined sequence $K_0 = k, K_{r+1} = K_r(q_{K_r,\mathrm{an}})$ is the *standard generic splitting tower of* q (cf. [K_2, p. 78]). We call $q_r := (q_{K_r})_{\mathrm{an}}$ the *r-th higher kernel form* of q (with respect to the tower). We define $i_0 := \operatorname{ind} q$ and $i_r := \operatorname{ind} q_{r-1} \otimes K_r$ for $1 \le r \le h$, and we call i_r the *r-th higher index* of q $\quad (0 \le r \le h)$.

1.13. THEOREM. *If $K_0 \subset \cdots \subset K_h$ is a generic splitting tower of q, then, for every $r \in [0,h]$, the field K_r is a generic splitting field of q of level r.*

PROOF. We denote the function fields of the varieties $V_i(q)$ by F_i, as in 1.3. By 1.9, it suffices to show $K_r \sim_k F_{j_r}$, for every $r \ge 0$. For $r = 0$ this is obvious, since F_{j_0} is a purely transcendental extension of k. For $r = 1$ we have, by 1.5, applied to $q_{K_0} = j_0 \times H \perp q_{K_0,\mathrm{an}}$,

$$K_1 \sim_{K_0} K_0(q_{K_0,\mathrm{an}}) \sim_{K_0} F_{j_0+1}K_0 \sim_k F_{j_0+1},$$

hence $K_1 \sim_k F_{j_0+1} \sim_k F_{j_1}$, by 1.9.
We proceed by induction on $\dim q_{\mathrm{an}}$. By induction assumption, our claim is true for $q_1 := q_{K_1,\mathrm{an}}$ over K_1, and hence for $q_{K_1} = (j_0 + j_1) \times H \perp q_1$ by 1.5. That is, for $r \ge 1$, the field K_r is a generic splitting field of q_{K_1} of level $r-1$, and, as such, specialization equivalent over K_1 with the function field of $V_{j_r}(q_{K_1}) \cong V_{j_r}(q) \times_k K_1$ resp. $\cong V_{j_r}(q) \times_{k_{\delta q}} K_1 k_{\delta q}$, which is $F_{j_r} \cdot K_1$ (free product over k resp. $k_{\delta q}$). Hence it remains to show that $F_{j_r} \cdot K_1$ is specialization equivalent to F_{j_r} over k. We have a trivial k-place $F_{j_r} \to F_{j_r} \cdot K_1 \cup \infty$. On the other hand, since $r \ge 1$, we also have a k-place $K_1 \to F_{j_r} \cup \infty$, which gives us a k-place $F_{j_r} \cdot K_1 \to F_{j_r} \cup \infty$. □

The rest of this paragraph will be used in paragraphs 5 and 6 only. For the next statements we need the notion of "good reduction" of a quadratic form.

[3)] This definition of generic splitting towers is slightly broader than the definition in [K_2, p.78]. There it is demanded that $K_0 = k$.

1.14. DEFINITION/REMARK. Let $q\colon K^n \to K$ be a quadratic form over a field K and $\lambda\colon K \to L \cup \infty$ be a place to a second field L. Let $\mathfrak{o} = \mathfrak{o}_\lambda$ denote the valuation ring of λ.

a) We say that q has *good reduction* (abbreviated: GR) under the place λ, if there exists a linear change of coordinates $T \in \mathrm{GL}(n, K)$ such that ($x := (x_1, \ldots, x_n)$)

$$q(Tx) = \sum_{i \le j} a_{ij} x_i x_j$$

with coefficients $a_{ij} \in \mathfrak{o}$, and such that the form $\sum_{i \le j} \lambda(a_{ij}) x_i x_j$ over L is regular.

b) In this situation it can be proved that, up to isometry, the form

$$\sum_{i \le j} \lambda(a_{ij}) x_i x_j$$

does not depend on the choice of T (cf. [K_1, Lemma 2.8], [K_5, §8]). Abusively we denote this form by $\lambda_*(q)$, and we call $\lambda_*(q)$ *"the" specialization of q under λ*.

c) Let q be a regular form over k, let K, L be field extensions of k and let $\lambda : K \to L \cup \infty$ be a k-place with valuation ring $\mathfrak{o}$. Then q_K has GR under λ and $\lambda_*(q_K) = q_L$. By Lemma 1.14.b below it follows that also $q_{K,\mathrm{an}}$ has GR under λ. □

If q has GR under λ then certainly q itself is regular. Moreover it can be proved that

$$(*) \qquad q \cong [a_1, b_1] \perp \cdots \perp [a_m, b_m] \quad (\perp [\varepsilon])$$

with elements $a_i, b_i \in \mathfrak{o}$ and $\varepsilon \in \mathfrak{o}^*$ (cf. [K_1], [K_5, §6]). Here the last summand $[\varepsilon]$ denotes the form εX^2 in one variable X. It appears if and only if $n = \dim q$ is odd. Of course, $(*)$ implies

$$\lambda_*(q) \cong [\lambda(a_1), \lambda(b_1)] \perp \cdots \perp [\lambda(a_m), \lambda(b_m)] \quad (\perp [\lambda(\varepsilon)]).$$

1.15. LEMMA. *Let q and q' be forms over K, and assume that $\dim q$ is even.*

a) *If q and q' have GR under λ then $q \perp q'$ has GR under λ, and*

$$\lambda_*(q \perp q') \cong \lambda_*(q) \perp \lambda_*(q').$$

b) *If q and $q \perp q'$ have GR under λ, then q' has GR under λ.*

PROOF. Part a) of this lemma is trivial, but b) needs a proof. A proof can be found in [K_1, §2] in the case that also q' has even dimension, and in [K_5, §8] in general. □

Part b) will be crucial for the arguments to follow.

1.16. PROPOSITION. *Let $\lambda\colon K \to L \cup \infty$ be a place and φ a form over K with GR under λ. Then $\varphi_0 := \ker(\varphi)$ has again GR under λ and $\lambda_*(\varphi) \sim \lambda_*(\varphi_0)$, $\mathrm{ind}\,(\lambda_*(\varphi)) \ge \mathrm{ind}\,(\varphi)$. If $\mathrm{ind}\,(\lambda_*(\varphi)) = \mathrm{ind}\,(\varphi)$, then $\ker \lambda_*(\varphi) = \lambda_*(\varphi_0)$.*

PROOF. Let $\varphi_0 := \ker \varphi$. We have $\varphi \cong j \times H \perp \varphi_0$ with $j = \operatorname{ind}(\varphi)$. The form $j \times H$ has GR under λ. By Lemma 1.15 it follows that φ_0 has GR under λ and $\lambda_*(\varphi) \cong j \times H \perp \lambda_*(\varphi_0)$. Now all the claims are evident. □

1.17. PROPOSITION. *Let φ be an anisotropic form over K of dimension ≥ 2 which has GR under a place $\lambda: K \to L \cup \infty$. Let $K_1 \supset K$ be a generic zero field of φ. Then $\lambda_*(\varphi)$ is isotropic if and only if λ extends to a place $\mu: K_1 \to L \cup \infty$.*

SKETCH OF PROOF. a) If λ extends to a place $\mu: K_1 \to L \cup \infty$ then it is obvious that $\varphi \otimes K_1$ has GR under μ and $\mu_*(\varphi \otimes K_1) = \lambda_*(\varphi)$. Since $\varphi \otimes K_1$ is isotropic we conclude by Proposition 1.16 that $\lambda_*(\varphi)$ is isotropic.
b) Assume now that $\lambda_*(\varphi)$ is isotropic. We denote this form by $\overline{\varphi}$ for short. By use of elementary valuation theory it is rather easy to extend λ to a place $\tilde{\lambda}: K(\varphi) \to L(\overline{\varphi}) \cup \infty$ (cf. [K$_5$, §9]; here we do not need that $\overline{\varphi}$ is isotropic). Since $\overline{\varphi}$ is isotropic the field extension $L(\overline{\varphi})/L$ is purely transcendental (cf. Th. 1.3). Thus there exists a place $\rho: L(\overline{\varphi}) \to L \cup \infty$ over L. Now $\rho \circ \tilde{\lambda}: K(\varphi) \to L \cup \infty$ is a place extending λ. Since $K(\varphi) \sim_K K_1$, there exists also a place $\mu: K_1 \to L \cup \infty$ extending λ. □

We return to our form q over k.

1.18. THEOREM. *Let $(K_r \mid 0 \leq r \leq h)$ be a generic splitting tower of q with higher indices i_r and higher kernel forms q_r $(0 \leq r \leq h)$. Let $\gamma: k \to L \cup \infty$ be a place such that q has GR under γ. Moreover let $m \in [0, h]$ and $\lambda: K_m \to L \cup \infty$ be a place extending γ. Assume in the case $m < h$ that λ does not extend to a place from K_{m+1} to L. Then* $\operatorname{ind}(\gamma_*(q)) = i_0 + \cdots + i_m = j_m$. *The form q_m has GR under λ and* $\ker(\gamma_*(q)) \cong \lambda_*(q_m)$.

PROOF. We have $i_0 + \cdots + i_m = j_m$ and $q \otimes K_m \cong j_m \times H \perp q_m$. This implies, that q_m has GR under λ and

$$\gamma_*(q) = \lambda_*(q \otimes K_m) \cong j_m \times H \perp \lambda_*(q_m)$$

(cf. Proof of Prop. 1.16.) It remains to prove that $\lambda_*(q_m)$ is anisotropic. This is trivial if $m = h$. Assume now that $m < h$. If $\lambda_*(q_m)$ would be isotropic then Proposition 1.17 would imply that λ extends to a place from K_{m+1} to L, contradicting our assumptions in the theorem. Thus $\lambda_*(q_m)$ is anisotropic. □

Applying the theorem to the special case that L is a field extension of k and γ is the trivial place $k \hookrightarrow L$, we obtain a result on the Witt decomposition of $q \otimes L$ which is much stronger than 1.11.

1.19. COROLLARY. *Let $(K_r \mid 0 \leq r \leq h)$ be a generic splitting tower of q. If L/k is a field extension and* $\operatorname{ind}(q \otimes L) = j_m$, *and if $\rho: K_r \to L \cup \infty$ is a place over k for some $r \in [0, h]$, then $r \leq m$ and ρ extends to a place $\lambda: K_m \to L \cup \infty$. For every such place λ the kernel form q_m of $q \otimes K_m$ has GR under λ and $\lambda_*(q_m) \cong$* $\ker(q \otimes L)$.□

An easy consequence is the following statement.

1.20. SCHOLIUM. *("Uniqueness" of generic splitting towers and higher kernel forms). Let $(K_r \mid 0 \leq r \leq h)$ and $(K'_r \mid 0 \leq r \leq h)$ be generic splitting towers of q with associated sequences of higher indices $(i_r \mid 0 \leq r \leq h)$, $(i'_r \mid 0 \leq r \leq h)$ and sequences of kernel forms $(q_r \mid 0 \leq r \leq h)$, $(q'_r \mid 0 \leq r \leq h)$. Then $i_r = i'_r$ for every $r \in [0, h]$. There exists a place $\lambda: K_h \to K'_h \cup \infty$ over k which restricts to a* *place* $\lambda_r: K_r \to K_r' \cup \infty$ *for every* $r \in [0, h]$. *If there is given a place* $\mu: K_r \to K_r' \cup \infty$ *over* k *for some* $r \in [0, h]$ *then* q_r *has good reduction under* μ *and* $\mu_*(q_r) \cong q_r'$. □

§2. Behavior of generic splitting fields and generic splitting towers under base field extension

2.1. DEFINITION/REMARK. If K/k is a partial generic splitting field of q of some level r, then we denote the algebraic closure of k in K by K°. The extension K°/k is k or $k_{\delta q}$ (cf. 1.10).

For systematic reasons we retain the notation K° for later use, although most often $K^\circ = k$.

2.2. DEFINITION. We call a generic splitting field K of q of some level $r \in [0, h]$ *regular*, if K is regular over the algebraic closure K° of k in K. We then denote by $L \cdot K$, or more precisely by $L \cdot_k K$, the free composite of $L \cdot K^\circ$ and K over K°.
Explanation. Here we have to read $K^\circ = k$, $L \cdot K^\circ = L$ if $r < h$ or if $r = h$ and q is inner. If $r = h$ and q is outer we have two cases. Either L splits the discriminant of q. In this case $K^\circ = k_{\delta q}$ embeds into L and we read $L \cdot K^\circ = L$. Or L does not split δq. In this case $L \cdot K^\circ = L \otimes_k K^\circ = L_{\delta(q \otimes L)}$. □

2.3. DEFINITION. We call a generic splitting tower $(K_r \mid 0 \le r \le h)$ of q *regular* if K_r/K_{r-1} is a regular field extension for every r with $1 \le r < h$, and also for $r = h$, if the form q is inner. If $r = h$ and q is outer, we demand that K_h is regular over the composite $K_{h-1} \cdot K_h^\circ = K_{h-1} \cdot k_{\delta q} = K_{h-1} \otimes_k k_{\delta q}$ over k. □

Let L/k be any field extension. We want to construct partial generic splitting fields and generic splitting towers for $q \otimes L$ from corresponding data for q.
Assume that $(K_r \mid 0 \le r \le h)$ is a *regular* generic splitting tower of q. For every $r \in [0, h]$ we have the free composite $L \cdot K_r = L \cdot_k K_r$ as explained in 2.2. (The existence of the free products $L \cdot K_r$ is the only assumption needed for the following theorem. This is generally true if either L or K_r is regular over k resp. K^0. Thus, instead of the regularity of the generic splitting tower, we could also assume that the field L is separable over k.)

2.4. THEOREM. *Let* $J = (r_0, \dots r_e)$ *denote the sequence of increasing numbers* $r \in \{0, \dots, h\}$ *such that* $\operatorname{ind}(q \otimes K_r) = \operatorname{ind}(q \otimes L \cdot K_r)$.

a) *Then the sequence*

$$L \cdot K_{r_0} \subset L \cdot K_{r_1} \subset \cdots \subset L \cdot K_{r_e}$$

is a regular generic splitting tower of $q \otimes L \cdot K_0$, *and hence of* $q \otimes L$.

b) *For every* $r \in [0, h] \setminus J$ *we have an* $L \cdot K_r$*-place* $L \cdot K_{r+1} \to L \cdot K_r \cup \infty$.

PROOF. The claim is obvious if $\dim q_{\text{an}} \le 1$. We proceed by induction on $\dim q_{\text{an}}$. Let $r' := \min J \setminus \{0\}$. The induction hypothesis, applied to q_{K_1}, gives a regular generic splitting tower $L \cdot K_{r'} \subset \cdots \subset L \cdot K_{r_e}$ for $q_{L \cdot K_1}$, as well as b) for $r \ge 1$. In particular, the latter implies that $L \cdot K_{r'} \sim_{L \cdot K_1} L \cdot K_1 \sim_{L \cdot K_0} L \cdot K_0(q_{L \cdot K_0, \text{an}})$, and this proves a).
It remains to show b) for $r = 0$. But $0 \notin J$ means $\operatorname{ind} q_L > \operatorname{ind} q$, hence $\operatorname{ind} q_{L \cdot K_0} > \operatorname{ind} q_{K_0}$. Therefore there is a K_0-place $K_1 \to L \cdot K_0 \cup \infty$, which yields an $L \cdot K_0$-place $L \cdot K_1 \to L \cdot K_0 \cup \infty$. □

In [K_2, p. 85] another proof of Theorem 2.4 and its corollary has been given, which clearly remains valid if char $k = 2$. We believe that the present proof albeit shorter gives more insight than the proof in [K_2].

The sequence $\mathrm{SP}(q \otimes L)$ is a subsequence of $\mathrm{SP}(q) = (j_0, \dots, j_h)$, say $\mathrm{SP}(q \otimes L) = (j_{t(0)}, \dots, j_{t(e)})$, with

$$0 \leq t(0) < t(1) < \cdots < t(e) = h.$$

{ It is evident, that $j_{t(e)} = j_h = [\dim q/2] \in \mathrm{SP}(q \otimes L.$}

It follows from Theorem 2.4 that the $t(i)$ coincide with the numbers r_i there. Thus we have the following corollary.

2.5. COROLLARY. *a) For every $s \in [0, e]$ the anisotropic part of $q \otimes L \cdot K_{t(s)}$ is $q_{t(s)} \otimes L \cdot K_{t(s)}$.*
b) $\mathrm{SP}(q \otimes L)$ is the sequence of all $r \in [0, h]$ such that the form $q_r \otimes L \cdot K_r$ is anisotropic. □

2.6. PROPOSITION. *Let K be a regular generic splitting field of q of some level $r \in [0, h]$. Then $L \cdot K$ is a generic splitting field of $q \otimes L$ of level s, where $s \in [0, e]$ is the number with $t(s-1) < r \leq t(s)$. {Read $t(-1) = -1$.}*

PROOF. We return to the fields F_i in Theorem 1.3. Let $i := j_r$. We have $K \sim_k F_i$ by Proposition 1.9. This implies $K \cdot L \sim_L F_i \cdot L$. Thus it suffices to prove the claim for F_i instead of K. Now $L \cdot F_i$ is the function field of the variety $V_i(q \otimes L)$. Proposition 1.9 gives the claim. □

For later use, it is convenient to insert a digression about "inessential" field extensions.

2.7. DEFINITION. We call a field extension E/k *inessential*, if there exists a place $\alpha : E \to k \cup \infty$ over k, i.e., $E \sim_k k$. □

The idea behind this definition is that, if E/k is inessential, then $q \otimes E$ has essentially the same splitting behavior as q. This will now be verified.
We know already from Corollary 1.6 that ind $(q \otimes E) =$ ind (q), hence $\ker(q \otimes E) = \ker(q) \otimes E$.

2.8. COROLLARY. *Assume again that E/k is an inessential field extension.*

i) *If $(E_r \mid 0 \leq r \leq h')$ is a generic splitting tower of $q \otimes E$, then it is also a generic splitting tower of q. In particular $h' = h$, i.e., $h(q \otimes E) = h(q)$. Moreover $\mathrm{SP}(q \otimes E) = \mathrm{SP}(q)$.*
ii) *If K/E is a generic splitting field of $q \otimes E$ of some level $r \in [0, h]$, then K/k is a generic splitting field of q of the same level r.*

PROOF. i): This follows from the definition 1.12 of the notion of a generic splitting tower, together with 2.4.
ii): We have $K \sim_E E_r$. This implies $K \sim_k E_r$, and we are done. □

2.9. REMARK. Theorem 2.4 tells us that the splitting pattern of a quadratic form becomes coarser under base field extension. This may even happen with anisotropic k-forms, which stay anisotropic over the extension field L. The classical example is a quadratic form ψ of dimension 4 and with a non trivial discriminant (or Arf invariant) $\delta\psi$. The form ψ remains anisotropic over the quadratic discriminant extension $L = k_{\delta\psi}$. Of course the height of ψ_L is one, which means, that, over L, the form ψ is 'simpler' than over k. Such a transition $\psi \mapsto \psi_L$ is called an *anisotropic splitting*, since it reduces the complexity of the quadratic form ψ without disturbing its anisotropy.

This phenomenon can be more subtle than in the example just given. For the rest of this remark we assume that char $k \neq 2$.

i) For an an anisotropic r-Pfister form φ with pure part φ', and ψ as above, we study the form $q := \varphi' \otimes \psi$. As mentioned above, ψ_L is anisotropic for $L = k(\sqrt{\delta q})$. We also assume that φ and hence φ' as well as q stay anisotropic over L: For example, we can start with some ground field k_0, and let $k = k_0(X_1, X_2, Y_1, \dots Y_r)$ be the function field of $r+2$ indeterminates over k_0, and then take

$$\psi = \langle 1, X_1, X_2, \delta X_1 X_2 \rangle \text{ with } \delta \in k_0^* \setminus k_0^{*2},\ \varphi = \langle\langle Y_1, \dots, Y_r \rangle\rangle.$$

Then $(\varphi \otimes \psi)_L$ is an anisotropic $r+2$-Pfister form, and q_L is a Pfister neighbor of that form with complement ψ_L.

Hence q_L is an excellent form of height 2 with splitting pattern

$$\mathrm{SP}(q_L) = (2^{r+1} - 4,\ 2^{r+1} - 2).$$

On the other hand, if $E = k(\sqrt{-X_1})$, then $\psi_E = H \perp \psi_{E,\mathrm{an}}$, and hence $q_E = (\varphi' \otimes \psi)_E = \varphi'_E \otimes H \perp \varphi'_E \otimes \psi_{E,\mathrm{an}}$. Using, e.g., [HR$_2$, 1.2, p. 165] one sees easily that $\varphi'_E \otimes \psi_{E,\mathrm{an}}$ is anisotropic. It is similar to a Pfister neighbor with complement $\psi_{E,\mathrm{an}}$, hence of height two with splitting pattern $(2^r - 1, 2^{r+1} - 3)$.

The splitting pattern of q therefore contains the numbers

$$2^r - 1,\ 2^{r+1} - 4,\ 2^{r+1} - 3,\ 2^{r+1} - 2,$$

and possibly more, but only two of them survive for q_L.

ii) The following example may be even more instructive. We refer to [HR$_2$, 2.5–2.10, p. 167ff.]. Assume $n \geq r > 0$, and let $k = k_0(X_1, \dots, X_n, Y_1, \dots, Y_r, Z$ be the function field over some field k_0 in $n + r + 1$ indeterminates. We let k' denote the subfield $k_0(X_1, \dots, X_n, Y_1, \dots, Y_r)$ of k, hence $k = k'(Z)$ is an inessential extension of k'.

We consider the anisotropic forms

$$q := \langle\langle X_1, \dots, X_n \rangle\rangle \perp Z\langle\langle Y_1, \dots, Y_r \rangle\rangle \qquad \text{over } k$$

and

$$\psi := \langle\langle X_1, \dots, X_n \rangle\rangle \perp \langle\langle Y_1, \dots, Y_r \rangle\rangle \qquad \text{over } k'.$$

According to [l.c., 2.6], their splitting pattern is given by

$$(*) \qquad \begin{array}{ll} (0,\ 2^0,\ 2^1, \dots, 2^r,\ 2^{n-1},\ 2^{n-1} + 2^{r-1}) & \text{if } 1 \leq r \leq n-2 \\ (0,\ 2^0,\ 2^1, \dots, 2^{n-1},\ 2^{n-1} + 2^{n-2}) & \text{if } r = n-1 \\ (0,\ 2^0,\ 2^1, \dots, 2^n) & \text{if } r = n \end{array}$$

(The proof is given in [l.c.] for q, but works, mutatis mutandis, for ψ as well.) We consider the standard generic splitting tower $K_0 = k, K_1, \dots, K_h$ of ψ_k, which is a generic splitting tower of ψ as well, since k is inessential over k', and note that $h = r+3, r+2, r+1$ respectively in the three cases distinguished above.

The forms $\sigma := \langle\langle X_1, \dots, X_n\rangle\rangle_{K_i}$ and $\tau := \langle\langle Y_1, \dots, Y_r\rangle\rangle_{K_i}$ are anisotropic for $r = 0, \dots, r$. Hence, using [l.c., Thm. 1.2, p. 165], we conclude that q_{K_i} is anisotropic for $i = 0, \dots, r$.

In [l.c., 2.5, p. 167] the following well known linkage result is stated: If a linear combination of two anisotropic Pfister forms σ, τ over a given field is isotropic, then its index is the dimension of a Pfister form of maximal dimension dividing both σ and τ.

Since ψ_{K_i} is isotropic, it follows that its index is a power of two, since it is the dimension of the common maximal Pfister divisor of σ and τ. Hence, by the same result, the first higher index of q_{K_i} is exactly the dimension of this Pfister divisor. Therefore, for $i \le r$, the splitting pattern of q_{K_i} consists of 0, followed by the suffix starting with 2^i of the appropriate sequence $(*)$.

This shows that the gaps occurring in a splitting pattern by base field extension can be arbitrarily large, even for a form which stays anisotropic over the extension.

§3. Defining higher kernel forms over purely transcendental field extensions

3.1. Definition. Let L/K be a field extension and φ a form over L. If we have $\varphi \cong \varphi'_L = \varphi' \otimes L$ with some form φ' over K, then we say that φ is *definable over K (by φ')*. We say that φ is *defined over K (by φ')* if φ' is unique up to isometry over K.

It is known for a field k of characteristic $\neq 2$, that all higher kernels of a form q over k are defined over k if and only the form q is excellent [K_3, 7.14,p. 6], which is a very strong condition on that form: q is excellent if either $\dim q \le 3$, or if q is a Pfister neighbor with excellent complement. E.g., $\sum_{i=1}^n X_i^2$ is excellent over any field of characteristic $\neq 2$.

In this section, as before, q is a (regular quadratic) form over a field k. We want to prove the surprising fact, that, for a suitable generic splitting tower of q, every higher kernel form of q is definable over some finitely generated purely transcendental extension of k.

The following lemma is well known, but we will need its precise statement as given here later on.

3.2. Lemma. *Assume that q is anisotropic. Let L be a separable quadratic field extension of k, such that q_L is isotropic. Then*

$$q \cong \alpha \perp \beta$$

for some regular quadratic forms α, β over k, such that α_L is hyperbolic and β_L is anisotropic. Let $L = k[X]/(aX^2 + X + b)$ with $a \neq 0$, $b \neq 0$ (which can always be achieved). Then α is divisible by $[a, b]$. More precisely, if $i = \operatorname{ind}(q_L)$, then there are pairwise orthogonal vectors $y_1, \dots, y_i$ over k such that

$$\alpha = [a, b] \otimes \langle q(y_1), \dots, q(y_i)\rangle.$$

In case $\operatorname{char} k \neq 2$, we may assume that $L = k(\sqrt{\delta})$. Then, α is divisible by $\langle 1, -\delta\rangle$.

Proof. Let $e \neq 0$ be an isotropic vector for q_L. We denote the image of X in L by θ. Then, for $e = x + y\theta$, where x, y have coordinates in k, we obtain [4] $0 = aq_L(e) =$

[4] We briefly write $(x, y) := B_q(x, y)$ with B_q the bilinear form associated to q

$aq(x) + aq(y)\theta^2 + a(x,y)\theta = aq(x) - bq(y) + (a(x,y) - q(y))\theta$, hence $aq(x) = bq(y)$ and $a(x,y) = q(y)$. Therefore $(x,y) \neq 0$ and $q(\xi y + \eta x) = (a\xi^2 + \xi\eta + b\eta^2)(x,y)$ for arbitrary $\xi, \eta \in k$, which gives a binary orthogonal summand of q of the requested type. If its complement is anisotropic over L we are done. Otherwise, an induction on $\dim q$ gives the general result, and the special result for char $k \neq 2$ is obtained as usual by the substitution $X = X' - 1/(2a)$, $\delta = b/a - 1/(4a^2)$. □

3.3. COROLLARY. *Let $k(q)$ denote the function field of the quadric given by $q = 0$, let $k' \subset k(q)$ be a subfield containing k such that $k(q)/k'$ is separable quadratic and k'/k is purely transcendental. Then there is a decomposition*

$$q_{k'} = \alpha \perp q',$$

over k', such that $\alpha \neq 0$, $\alpha_{k(q)}$ is hyperbolic and $q'_{k(q)}$ is anisotropic. Hence the first higher kernel form of q is definable by q' over the purely transcendental extension k' of k.

PROOF. This follows immediately from 3.2. □

As before $h = h(q)$ denotes the height of q.

3.4. THEOREM. *There exists a regular (cf. 2.3) generic splitting tower $(K_r \mid 0 \leq r \leq h)$ of q, a tower of fields $(K'_r \mid 0 \leq r \leq h)$ with $k \subset K'_r \subset K_r$ for every r, $k = K'_0 = K_0$, and a sequence $(\varphi_r \mid 0 \leq r \leq h)$ of forms φ_r over K'_r, such that the following holds. {N.B. All the fields K_r, K'_r are subfields of K_h.}*

(1) *$\varphi_r \otimes K_r = \ker(q \otimes K_r)$ for every $r \in [0,h]$.*
(2) *$\varphi_{r+1} < \varphi_r \otimes K'_{r+1}$ $(0 \leq r < h)$.*[5)]
(3) *K'_{r+1}/K'_r is purely transcendental of finite transcendence degree $(0 \leq r < h)$.*
(4) *K_r/K'_r is a finite multiquadratic extension $(0 \leq r \leq h)$.*

PROOF. We proceed by induction on $\dim q$. We may assume that q is anisotropic. If $\dim q \leq 1$ nothing has to be done. Assume now that $\dim q > 1$. Let $K_1 = k(q)$, the function field of the projective quadric $q = 0$. We choose for K'_1 a subfield of K_1 containing k such that K'_1/k is purely transcendental and K_1/K'_1 is quadratic, which is possible. By 3.3 we have a (not unique) decomposition $q \otimes K'_1 = \eta_1 \perp \varphi_1$ with $\dim \varphi_1 < \dim q$, $\varphi_1 \otimes K_1$ anisotropic, $\eta_1 \otimes K_1$ hyperbolic. If the height $h = 1$, we have finished with K_1, K'_1, φ_1.
Assume now that $h > 1$. We apply the induction hypothesis to φ_1. Let $h(\varphi_1) = e$. There exists a regular generic splitting tower $(L_j \mid 0 \leq j \leq e)$ of φ_1, a tower of fields $(L'_j \mid 0 \leq j \leq e)$, and forms ψ_j over L'_j $(0 \leq j \leq e)$, such that $K'_1 \subset L'_j \subset L_j$, $\psi_{j+1} < \psi_j \otimes L'_{j+1}$, $\psi_j \otimes L_j = \ker(\varphi_1 \otimes L_j)$, L'_{j+1}/L'_j is purely transcendental of finite degree, and L_j/L'_j is finite multiquadratic. Certainly $e \geq h - 1 \geq 1$ since $h(\varphi_1 \otimes K_1) = h - 1$.
We form the field composites $K_1 \cdot L_j = K_1 \cdot_{K'_1} L_j$ as explained in 2.2. Let J denote the set of indices $j \in [0,e]$ with $\psi_j \otimes K_1 \cdot L_j$ anisotropic, and let $\mu(0) < \mu(1) < \cdots < \mu(t)$ be a list of these indices. (N.B. $\mu(0) = 0, \mu(t) = e$.) By 2.4 and 2.9 the sequence of fields $(K_1 \cdot L_{\mu(r)} \mid 0 \leq r \leq t)$ is a regular generic splitting tower of $\varphi_1 \otimes K_1$, hence of $\varphi \otimes K_1$, and $\varphi \otimes (K_1 \cdot L_{\mu(r)})$ has the kernel form $\psi_{\mu(r)} \otimes (K_1 \cdot L_{\mu(r)})$. Clearly the tower $(K_1 \cdot L_{\mu(r)} \mid 0 \leq r \leq t)$ is regular, and $t = h - 1$.

---

5) *cf. Notations 1.0*

For $2 \le i \le h$ we put $K_i' := L'_{\mu(i-1)}$, $K_i = K_1 \cdot L_{\mu(i-1)}$, $\varphi_i := \psi_{\mu(i-1)}$. Adding to these fields and forms the fields K_1', K_1, $K_0 = K_0' = k$, and the forms $\varphi_1, \varphi_0 := q$, we have towers $(K_r \mid 0 \le r \le h)$, $(K_r' \mid 0 \le r \le h)$ and a sequence $(\varphi_r \mid 0 \le r \le h)$ of anisotropic forms with all the properties listed in the theorem. □

We add to this theorem a further observation.

3.5. PROPOSITION. *We stay in the situation of Theorem 3.4. Assume that $h \ge 1$, i.e., q is not split. By property (2) we have a sequence $(\eta_r \mid 1 \le r \le h)$ of forms η_r over K_r' such that*

$$\varphi_{r-1} \otimes K_r' \cong \eta_r \perp \varphi_r \qquad (1 \le r \le h).$$

We choose for each $r \in [1, h]$ a total generic splitting field E_r of η_r. Let q_r denote the kernel form of $q \otimes K_r$, $0 \le r \le h$.

CLAIM. *The field composite $K_{r-1} \cdot_{K'_{r-1}} E_r =: L_r$ is specialization equivalent to K_r over K_{r-1}. Thus L_r is a generic zero field of q_{r-1} and a generic splitting field of q of level r.*

PROOF. $q_{r-1} \otimes L_r$ is isotropic. Thus there exists a place $\lambda: K_r \to L_r \cup \infty$ over K_{r-1}. On the other hand $\eta_r \otimes K_r \sim 0$. Thus there exists a place $\rho: E_r \to K_r \cup \infty$ over K_r'. The field extension K_{r-1}/K'_{r-1} is finite multiquadratic. By standard valuation theoretic arguments ρ extends to a place μ: $K_{r-1} \cdot_{K'_{r-1}} E_r \to K_r \cup \infty$ over $K_{r-1} \cdot K_r'$, hence over K_{r-1}. □

EXPLANATIONS OF THE DIAGRAM. The tower on the left consists of purely transcendental extensions (labeled by "p.t."), the tower on the right is a generic splitting tower of q. The "horizontal" extensions labeled by "m.q." are multiquadratic, splitting the direct sums $\varepsilon_r = \perp_{i=1}^{r-1} \eta_{i,K_i'} \perp \eta_r$ (for which we simply have written $\eta_1 \perp \cdots \perp \eta_r$) totally and leaving φ_r anisotropic, making it isometric to the r-th higher kernel form q_r of q. The field $E_r = K_r'(\eta_r \to 0)$ is a generic total splitting field of the form η_r over K_r'.

A GENERIC SPLITTING TOWER ACCORDING TO 3.4 AND 3.5

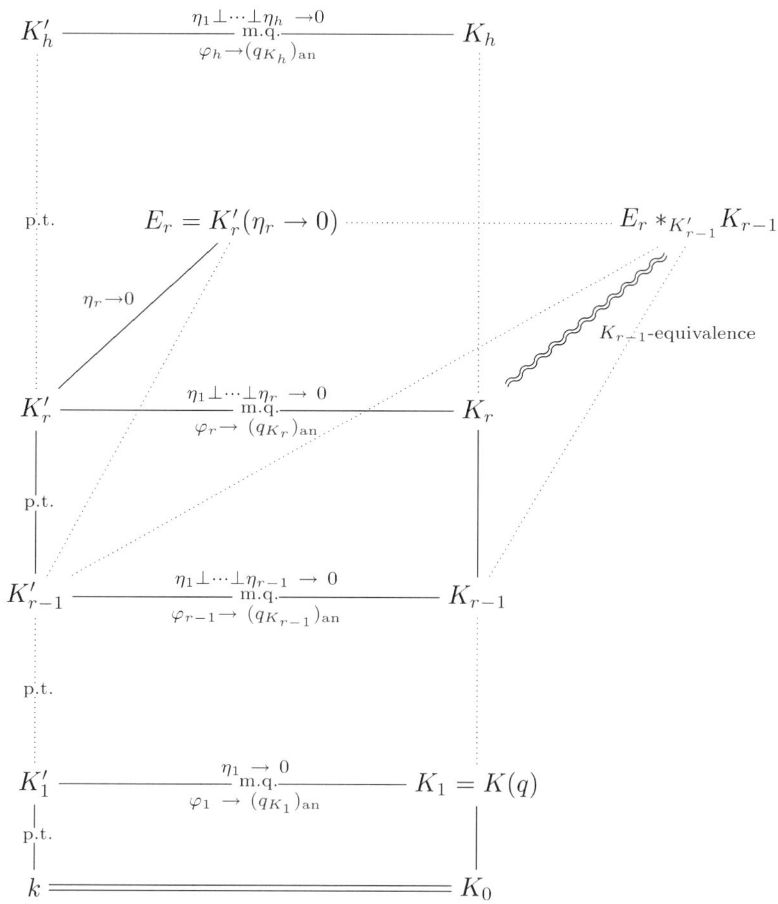

§4. Generic splitting preparation

In 3.4 and 3.5 we have associated to the form q over k – among other things – a tower $(K'_r \mid 0 \le r \le h)$ of purely transcendental field extensions of k together with a sequence $(\eta_r \mid 1 \le r \le h)$ of anisotropic subforms η_r of $q \otimes K'_r$. We want to understand in which way these data control the splitting behavior of q under field extensions, forgetting the generic splitting tower $(K_r \mid 0 \le r \le h)$ in 3.4. (We will have only partial success, cf. §6 below.) In addition we strive for an abstraction of the situation established in 3.4 and 3.5.

As before q is any regular quadratic form over a field k and h denotes the height $h(q)$.

4.1. DEFINITION. Let $r \in [0, h]$ and let K/k be an inessential field extension (cf. 2.7). A form η over K is called a *generic splitting form of q of* level r, if $\dim \eta$ is even and there exists an orthogonal decomposition $q \otimes K \cong \eta \perp \psi$ such that the following holds: If E/K is a total generic splitting field of η, then E/k is a partial generic splitting field of q of level r, while $\psi \otimes E$ is anisotropic. We further call ψ the *complement* (or complementary form) *of* η. □

N.B. This property does not depend on the choice of the total generic splitting field of E.
Generic splitting forms occur whenever a higher kernel form of q is definable over an inessential field extension of k.

4.2. PROPOSITION. *Let K/k be a partial generic splitting tower of q of level r. Assume there is given an inessential subextension K'/k of K/k and a subform ψ of $q \otimes K'$ such that $\psi \otimes K = \ker(q \otimes K)$. Let η denote the complement of ψ in $q \otimes K'$, i.e., $q \otimes K' \cong \eta \perp \psi$. (N.B. η is uniquely determined up to isometry.) Then η is a generic splitting form of q of level r.*

PROOF. $\dim q \equiv \dim \psi \mod 2$. Thus $\dim \eta$ is even. Let E/K' be a total generic splitting field of η. Then $q \otimes E \sim \psi \otimes E$. Thus there exists a place $\lambda\colon K \to E \cup \infty$ over k. On the other hand, $\eta \otimes K \sim 0$. Thus there exists a place $\mu\colon E \to K \cup \infty$ over K'. Since μ is also a place over k, the fields E and K are specialization equivalent over k. We conclude that also E is a generic splitting field of q of level r. Now it is evident that $\dim \ker(q \otimes E) = \dim \psi$. Thus $\psi \otimes E$ is anisotropic. □

4.3. DEFINITION. A *generic splitting preparation* of q is a tower of fields $(K'_r \mid 0 \leq r \leq h)$ together with a sequence $(\eta_r \mid 0 \leq r \leq h)$ of forms η_r over K'_r such that the following holds:

(1) $K'_0 = k$, and η_0 is the hyperbolic part of φ.
(2) K'_{r+1}/K'_r is purely transcendental for every r, $0 \leq r < h$. [6)]
(3) There exist orthogonal decompositions

$$q \cong \eta_0 \perp \varphi_0, \quad \varphi_r \otimes K'_{r+1} \cong \eta_{r+1} \perp \varphi_{r+1}, \quad (0 \leq r < h).$$

(4) For every $r \in [0, h]$ the form $\mathop{\perp}\limits_{j=0}^{r} \eta_j \otimes K'_r$ is a generic splitting form of q of level r. □

The forms φ_r $(0 \leq r \leq h)$ are uniquely determined (up to isometry, as always) by condition (3). We call φ_r *the r-th residual form* and η_r the r-th *splitting form* of the given generic splitting preparation. Clearly $i_r := \dim \eta_r/2$ is the r-th higher index (cf. 1.1) of q, and $\dim \varphi_h \leq 1$. If q is anisotropic then we sometimes denote the generic splitting preparation by $(K'_r \mid 0 \leq r \leq h)$, $(\eta_r \mid 1 \leq r \leq h)$, omitting the trivial form $\eta_0 = 0$. □

4.4. SCHOLIUM. *Generic splitting preparations of q exist in abundance. Indeed, in the situation described in 3.4 and 3.5, the tower of fields $(K'_r \mid 0 \leq r \leq h)$ together with the sequence $(\eta_r \mid 0 \leq r \leq h)$ of forms η_r over K'_r, where η_r for $r \geq 1$ has been introduced in 3.5 and η_0 denotes the hyperbolic part of q, is a generic splitting preparation of q.*

[6)] Things below would not change much if we merely demanded that the extensions K'_{r+1}/K'_r are inessential.

PROOF. Let $r \in [1,h]$ be fixed and $\varepsilon_r := \perp_{j=0}^{r} \eta_j \otimes K'_r$. Then $q \otimes K'_r \cong \varepsilon_r \perp \varphi_r$ and $\ker(\varphi \otimes K_r) = \varphi_r \otimes K_r$. Proposition 4.2 tells us that ε_r is a generic splitting form of q of level r. □

Notice that in this argument property (4) of Theorem 3.4 has not been used.

In the following, we study a fixed generic splitting preparation $(K'_r \mid 0 \le r \le h)$, $(0 \le \eta_r \le h)$ *of* q *with associated residual forms* φ_r $(0 \le r \le h)$. *For every* $r \in \{1, \dots, h\}$ *let* ε_r *denote the form* $\perp_{j=0}^{r} \eta_j \otimes K'_r$. We do *not* assume that the preparation arises in the way described in the scholium.

Our next goal is to derive a generic splitting preparation of $q \otimes L$ from these data for a given field extension L/k.

4.5. LEMMA. *Assume that* K/k *is an inessential field extension. Let* η *be a form over* K *which is a generic splitting form of* q, *and let* ψ *be the complementary form,* $\varphi \otimes K \cong \eta \perp \psi$. *Let* E/K *be a regular total generic splitting field of* η *(cf. 2.2). Finally, let* $K \cdot L = K \cdot_k L$ *denote the free field composite of* K *and* L *over* k, *and* $E \cdot L = E \cdot_k L$ *the composite of* E *and* L *as explained in 2.2. Assume that* $\psi \otimes E \cdot L$ *remains anisotropic. Then* $K \cdot L/L$ *is again inessential and* $\eta \otimes K \cdot L$ *is a generic splitting form of* $q \otimes L$ *with complementary form* $\psi \otimes K \cdot L$.

PROOF. Any place $\alpha: K \to L \cup \infty$ over k extends to a place from $L \cdot K$ to L over L. Thus $K \cdot L/L$ is inessential. By Proposition 2.6 the field $E \cdot_k L = E \cdot_K (K \cdot_k L)$ is a total generic splitting field of $\eta \otimes K \cdot L$ and also a partial generic splitting field of $\varphi \otimes L$. Now the claim is obvious from Definition 4.1. □

As in §1 and §2 we enumerate the splitting pattern $\mathrm{SP}(q) = \{j_r \mid 0 \le r \le h\}$ by

$$0 \le j_0 < j_1 < \cdots < j_h = [\frac{\dim q}{2}]$$

and write $\mathrm{SP}(q \otimes L) = \{j_r \mid r \in J\}$ with $J = \{t(s) \mid 0 \le s \le e\}$,

$$0 \le t(0) < t(1) < \cdots < t(e) = h.$$

We have $e = h(q \otimes L)$.

4.6. PROPOSITION. *For* $0 < s \le e$ *we define* $L'_s = L \cdot K'_{t(s)}$ *as the free composite of the fields* L *and* $K'_{t(s)}$ *over* k, *and we put*

$$\zeta_s := \eta_{t(s-1)+1} \otimes L'_s \perp \cdots \perp \eta_{t(s)-1} \otimes L'_s \perp \eta_{t(s)}.$$

We further define $L'_0 = L$ *and* ζ_0 *as the hyperbolic part of* $\varphi \otimes L$. *Then* $(L'_s \mid 0 \le s \le e)$, $(\zeta_s \mid 0 \le s \le e)$ *is a generic splitting preparation of* $q \otimes L$.

PROOF. For every r with $0 < r \le h$ we choose a regular total generic splitting field F_r/K'_r of the form ε_r.*) [7] Then F_r/k is a partial generic splitting field of q of level r. Let $L \cdot F_r = L \cdot_k F_r$ denote the composite of L with F_r over k as explained in

[7] The fields F_i from 1.3 will not be used in the following.

2.2. If $0 < s \le e$, then the field $L \cdot F_{t(s)}$ is a generic splitting field of $\varphi \otimes L$ and $\ker(\varphi \otimes L \cdot F_{t(s)}) = \varphi_{t(s)} \otimes L \cdot F_{t(s)}$ by Proposition 2.6. We have

$$\varphi \otimes L'_s \cong \varepsilon_{t(s)} \otimes L'_s \perp \varphi_{t(s)} \otimes L'_s.$$

Again by 2.6, the field

$$L \cdot_k F_{t(s)} = (L \cdot_k K'_{t(s)}) \cdot_{K'_{t(s)}} F_{t(s)} = L'_s \cdot_{K'_{t(s)}} F_{t(s)}$$

is a total generic splitting field of $\varepsilon_{t(s)} \otimes L'_s$. The form $\varphi_{t(s)} \otimes L'_s$ remains anisotropic over $L \cdot_k F_{t(s)}$. Now Proposition 4.2 tells us that $\varepsilon_{t(s)} \otimes L'_s$ is a generic splitting form of $\varphi \otimes L'_s$, and we are done. □

4.7. COROLLARY. *Let $m \in [1, h]$ be fixed, and let F be a regular total generic splitting field of ε_m. For every r with $m < r \le h$ let $F \cdot K'_r$ denote the free composite of F and K'_r over k. Then the tower $F \subset F \cdot K'_{m+1} \subset \cdots \subset F \cdot K'_h$ together with the sequence $(\eta_r \otimes F \cdot K'_r \mid m < r \le h)$ is a generic splitting preparation of the anisotropic form $\varphi_m \otimes F$.*

PROOF. We apply Proposition 4.6 with $L = F$. Now $J = \{m, m+1, \dots h\}$. Thus $e = h - m$ and $t(s) = s + m \quad (0 \le s \le h - m)$. The form $q \otimes F$ has the kernel form $\varphi_m \otimes F$. □

We now look for generic splitting preparations with $K'_1 = \cdots = K'_h$. These are the "generic decompositions" according to the following definition.

4.8. DEFINITION. Let $K' \supset k$ be a purely transcendental field extension. A *generic splitting decomposition* of q over K' is a sequence $(\alpha_0, \alpha_1, \dots, \alpha_h)$ of forms over K' of even dimensions such that the following holds.

i) $q \otimes K' \cong \alpha_0 \perp \alpha_1 \perp \cdots \perp \alpha_h \perp \varphi_h$ with $\dim \varphi_h \le 1$.

ii) α_0 is the hyperbolic part of $q \otimes K'$. For every r with $1 \le r \le h$ the form $\alpha_0 \perp \cdots \perp \alpha_r$ is a generic splitting form of q of level r. □

The next proposition tells us that we can always pass from a generic splitting preparation of q to a generic splitting decomposition of q.

4.9. PROPOSITION. *As before, let $(K'_r \mid 0 \le r \le h)$, $(\eta_r \mid 0 \le r \le h)$ be a generic splitting preparation of q. Then $(\eta_r \otimes K'_h \mid 0 \le r \le h)$ is a generic splitting decomposition of q over K'_h.*

PROOF. Let $K' := K'_h$ and $\alpha_r := \eta_r \otimes K'_h \quad (0 \le r \le h)$. Since η_0 is the hyperbolic part of q and K'/k is purely transcendental, the form α'_0 is the hyperbolic part of $q \otimes K'$. For $1 \le r \le h$ we have – with the notations from above – $\varepsilon_r \otimes K' = \alpha_0 \perp \cdots \perp \alpha_r$. Let F_r be a regular total generic splitting field of ε_r (over K'_r), hence a partial generic splitting field of q (over k) of level r. The free composite $F_r \cdot K' = F_r \cdot_{K'_r} K'$ is a total generic splitting field of $\varepsilon_r \otimes K' = \alpha_r$ by Proposition 2.6. Now $F_r \cdot K'$ is purely transcendental over F_r. Thus $F_r \cdot K'$ is also a partial generic splitting field of q of level r. We have $q \otimes K' \cong \alpha_r \perp (\varphi_r \otimes K')$. The form $\varphi_r \otimes F_r$ is anisotropic. Since $F_r \cdot K'/F_r$ is purely transcendental, it follows that $(\varphi_r \otimes K') \otimes F_r \cdot K' = (\varphi_r \otimes F_r) \otimes F_r \cdot K'$ is anisotropic. Thus α_r is a generic splitting form of q of level r. □

Although generic splitting decompositions look simpler than generic splitting preparations, it is up to now not clear to us which of the two concepts is better to work with. See also our discussion below at the end of §6.

§5. A brief look at quadratic places

We need some more terminology.
If K is a field, we denote the group of square classes K^*/K^{*2} by $Q(K)$ and a single square class aK^{*2} by $\langle a\rangle$, identifying this class with the bilinear form $\langle a\rangle$ over K. If $\lambda: K \to L \cup \infty$ is a place with associated valuation ring $\mathfrak{o} = \mathfrak{o}_\lambda$, we denote the image of the unit group $\mathfrak{o}^*$ in $Q(K)$ by $Q(\mathfrak{o})$. Notice that $Q(\mathfrak{o}) \cong \mathfrak{o}^*/\mathfrak{o}^{*2}$, and that λ gives us a homomorphism λ_*: $Q(\mathfrak{o}) \to Q(L)$, $\lambda_*(\langle\varepsilon\rangle) = \langle\lambda(\varepsilon)\rangle$. (The bilinear form $\langle\varepsilon\rangle$ has good reduction under λ and $\lambda_*(\langle\varepsilon\rangle)$ is the specialization of this form under λ.)

5.1. DEFINITION. A *quadratic place*, or *Q-place* for short, from a field K to a field L is a triple (λ, H, χ) consisting of a place $\lambda: K \to L \cup \infty$, a subgroup H of $Q(K)$ containing $Q(\mathfrak{o}_\lambda)$, and a homomorphism $\chi: H \to Q(L)$ (called a "character" in the following) extending the homomorphism λ_*: $Q(\mathfrak{o}_\lambda) \to Q(L)$. □

We often denote such a triple (λ, H, χ) by a capital Greek letter Λ and symbolically write $\Lambda: K \to L \cup \infty$ for the Q-place Λ.
Every place $\lambda: K \to L \cup \infty$ gives us a Q-place $\hat{\lambda} = (\lambda, Q(\mathfrak{o}_\lambda), \lambda_*)$: $K \to L \cup \infty$, where $\lambda_*: Q(\mathfrak{o}_\lambda) \to Q(L)$ is defined as above. We regard λ and $\hat{\lambda}$ essentially as the same object. In this sense Q-places are a generalization of the usual places.

5.2. DEFINITION. If $\Lambda = (\lambda, H, \chi)$: $K \to L \cup \infty$ is a Q-place then an *expansion of* Λ is a Q-place $\Lambda' = (\lambda, H', \chi'): K \to L \cup \infty$ with the same first component λ as Λ, a subgroup H' of $Q(K)$ containing H, and a character $\chi': H' \to Q(L)$ extending χ. □

Usually Λ allows many expansions, and Λ itself is an expansion of $\hat{\lambda}$.
In the following $\Lambda = (\lambda, H, \chi)$: $K \to L \cup \infty$ a Q-place and $\mathfrak{o} := \mathfrak{o}_\lambda$.

5.3. DEFINITIONS. Let k be a subfield of K.

a) The *restriction* $\Lambda|k$ of the Q-place Λ to k is the Q-place (ρ, E, σ) where $\rho = \lambda|k: k \to L\cup\infty$ denotes the restriction of the place $\lambda: K \to L\cup\infty$ in the usual sense, E denotes the preimage of H in $Q(k)$ under the natural map $j: Q(k) \to Q(K)$, and σ denotes the character $\chi \circ (j|D)$ from D to $Q(L)$.
b) If $\Gamma: k \to L \cup \infty$ is any Q-place from k to L then we say that Λ *extends* Γ (or, that Λ is an extension of Γ), if $\Lambda|k$ is an expansion of Γ. □

At the first glance one might think that this notion of extension is not strong enough. One should demand $\Lambda|k = \Gamma$, in which case we say that Λ is a *strict extension* of Γ. But it will become clear below (cf. 5.8.ii, 5.12, 6.1) that the weaker notion of extension as above is the one needed most often.
As before, we stay with a Q-place $\Lambda = (\lambda, H, \chi)$: $K \to L \cup \infty$.

5.4. DEFINITION. Let φ be a regular quadratic form over K. We say that φ has *good reduction* (abbreviated: GR) *under* Λ, if there exists an orthogonal decomposition

$$(*) \qquad \varphi = \perp_{h\in H} h\psi_h$$

with forms ψ_h over K which all have GR under λ. Here $h\psi_h$ denotes the product $\langle h\rangle \otimes \psi_h$ of the bilinear form $\langle h\rangle$ with ψ_h. {This amounts to scaling ψ_h by a representative of the square class $\langle h\rangle$.} Of course, $\psi_h \neq 0$ for only finitely many $h \in H$.

Alternatively we say in this situation that φ is Λ-*unimodular*. In harmony with this speaking we call a form ψ over K, which has GR under λ, also a λ-*unimodular form.* □

5.5. REMARK. We may choose a subgroup U of H such that $H = U \times Q(\mathfrak{o})$. If φ has GR under Λ then we can simplify the decomposition $(*)$ to a decomposition

$$\varphi \cong \perp_{u \in U} u\varphi_u$$

with λ-unimodular forms φ_u. □

5.6. PROPOSITION. *If φ has GR under Λ, and a decomposition $(*)$ is given, then the form $\perp_{h\in H} \chi(h)\lambda_*(\psi_h)$ is up to isometry independent of the choice of the decomposition $(*)$.*

This has been proved in [K$_4$] if char$L \neq 2$. A proof in general (which is rather different) will be contained in $[K_5]$.

5.7. DEFINITION. If φ has GR under Λ then we denote the form $\perp_{h\in H} \chi(h)\lambda_*(\psi_h)$ (cf. 5.6) by $\Lambda_*(\varphi)$, and we call $\Lambda_*(\varphi)$ the *specialization of φ under λ.* □

5.8. REMARKS.

i) *If ψ is a second form over K such that $\varphi \perp \psi$ is regular, and if both φ and ψ have GR under Λ then $\varphi \perp \psi$ has GR under Λ and*

$$\Lambda_*(\varphi \perp \psi) \cong \Lambda_*(\varphi) \perp \Lambda_*(\psi).$$

ii) *Assume that k is a subfield of K and $\Gamma: k \to L \cup \infty$ is a Q-place such that Λ extends Γ (cf.5.3.b). Let q be a regular form over k which has GR under Γ. Then $q \otimes K$ has GR under Λ and $\Lambda_*(q \otimes K) \cong \Gamma_*(q)$.*

We omit the easy proofs. □

It seems that quadratic places come up in connection with generic splitting forms (cf. 4.1) in a natural way. We illustrate this by a little proposition, which will also serve us to indicate some of the difficulties we have to face if we want to make good use of quadratic places in generic splitting business.

5.9. PROPOSITION. *Let q be an anisotropic regular form over a field k, $\dim q > 2$. Let $k(q)$ denote the function field of the projective quadric $q = 0$. Let $L \supset k$ be a field extension such that $q_L = q \otimes L$ is isotropic. Then there is a purely transcendental subextension k'/k of $k(q)/k$, such that $k(q)/k'$ is separable quadratic, and a quadratic place $\Lambda: k' \to L \cup \infty$ over k, such that the form α described in 3.3 $\{q_{k'} \cong \alpha \perp q',\ \alpha_{k(q)} \sim 0,\ q'_{k(q)}$ anisotropic$\}$ has GR under Λ and $\Lambda_*(\alpha) \sim 0$.*

PROOF. Since q_L is isotropic we have a place $\lambda: k(q) \to L \cup \infty$ over k. Let $\mathfrak{o}$ denote the valuation ring of λ, let V denote the underlying k-vector space of q. We take a decomposition $V \cong \perp_{i=1}^{r} (ke_i \oplus kf_i) \perp V'$ with $ke_i \oplus kf_i = [a_i, b_i]$, and $V' = \langle a_0 \rangle$ or $V' = 0$, and $a_i, b_i, a_0 \in k^*$.

We choose a primitive isotropic vector $x \in V_{\mathfrak{o}} = \mathfrak{o} \otimes_k V$ with $(x, V_{\mathfrak{o}}) = \mathfrak{o}$. By rearranging coordinates over k we may assume that $x = Xe_1 + Yf_1 + z$, and $X \in \mathfrak{o} \setminus 0$, $Y \in \mathfrak{o}^*$, $z \in \perp_{i=2}^{r} (\mathfrak{o}e_i \oplus k\mathfrak{o}f_i) \perp \mathfrak{o}V'$, and after dividing by Y, we

may assume that $Y = 1$. We take the coordinates of $\perp_{i=2}^{r} (\mathfrak{o}e_i \oplus \mathfrak{o}f_i) \perp \mathfrak{o}V'$ as independent variables over k, which generate a purely transcendental subfield k' of $k(q)$. The equation $0 = q_{\mathfrak{o}}(x) = a_1X^2 + X + b$ with $b = b_1 + q_{k(q)}(z) \in \mathfrak{o} \cap k'$ defines the field $k(q)$ as a separable quadratic extension of k'.
By construction we have $\lambda(b) \neq \infty$. Hence the quadratic form $[a_1, b]$ has good reduction under this place. We write $\alpha = [a_1, b] \otimes \langle c_1, \ldots, c_i \rangle$ with $c_\nu \in \Im(q_{k'}) \setminus 0$ according to 3.2. We denote the restriction of λ to k' by λ', and the associated valuation ring by $\mathfrak{o}'(= \mathfrak{o} \cap k')$.
Let H denote the subgroup of $Q(k')$ which is generated by $Q(\mathfrak{o}')$ and the classes $\langle c_1 \rangle, \ldots, \langle c_i \rangle$. We choose some extension $\chi: H \to Q(L)$ of the character $\lambda'_*: Q(\mathfrak{o}') \to Q(L)$. The quadratic place $\Lambda = (\lambda', H, \chi)$ has the desired properties. Alternatively we may choose for Λ the restriction of $\hat{\lambda}$ to k'. □

This proposition leaves at least two things to be desired. Firstly, it would be nice and much more useful, to have the subextension k'/k to be chosen in advance, independently of the place λ in the proof. Secondly it would be pleasant if also the form q' has GR under Λ. This is by no means guaranteed by our proof. We see no reason, why the analogue of Lemma 1.15.b for quadratic places instead of ordinary places should be true. The main crux here is the case char $k = 2$. Then usually many forms over k' do not admit λ'-modular decompositions for a given place $\lambda': k' \to L \cup \infty$ over k. {It is easy to give counterexamples in a sufficiently general situation, cf. [K5]}. On the other hand, if we can achieve in Proposition 5.9 in addition that q' has GR under Λ, then the equation

$$q \otimes L = \Lambda_*(\alpha) \perp \Lambda_*(q'),$$

which follows from the remarks 5.8 above, together with $\Lambda_*(\alpha) \sim 0$ would "explain" that $q \otimes L$ is isotropic, and how the anisotropic part of $q \otimes L$ is connected with the generic splitting form α of q of level 1.
If char $k \neq 2$ it is still not evident that the analogue of 1.15.b holds for the quadratic place Λ' but now at least every form over k' has a λ'-modular decomposition. Indeed, this trivially holds for forms of dimension 1, hence for all forms. In [K4] a way has been found, to force an analogue of 1.15.b for quadratic places to be true, by relaxing the notion "good reduction" to a slightly weaker – but still useful – notion "almost good reduction".
We can define "almost good reduction" without restriction to characteristic $\neq 2$ as follows.

5.10. Definition. Let $\Lambda = (\lambda, H, \chi)$: $K \to L \cup \infty$ be a Q-place, $\mathfrak{o} := \mathfrak{o}_\lambda$, and let S be a subgroup of $Q(K)$ such that $Q(K) = S \times H$. A form φ over K has *almost good reduction* (abbreviated AGR) under Λ if φ has a decomposition

$$(**) \qquad \varphi \cong \perp_{s \in S} s\varphi_s$$

with Λ-unimodular forms φ_s and $\Lambda_*(\varphi_s) \sim 0$ for every $s \in S$, $s \neq 1$. In this case we call the form

$$\Lambda_*(\varphi) := \Lambda_*(\varphi_1) \perp (\dim \varphi - \dim \varphi_1)/2 \times H$$

the *specialization of φ under Λ*. □

The point here is that $\Lambda_*(\varphi)$ is independent of the choice of the decomposition $(**)$ and also of the choice of S. This has been proved in [K$_4$] in the case char $L \neq 2$. It also holds if char $L = 2$, cf. [K$_5$].
It now is almost trivial that the analogue of 1.15. holds for Q-places with AGR instead of GR, provided char $L \neq 2$, cf. [K$_4$, §2].

5.11. PROPOSITION. *Let $\Lambda: K \to L \cup \infty$ be a Q-place and let φ and ψ be regular forms over K. Assume that char $L \neq 2$.*

a) *If φ and ψ have AGR under Λ, then $\varphi \perp \psi$ has AGR under Λ and*

$$\Lambda_*(\varphi \perp \psi) \cong \Lambda_*(\varphi) \perp \Lambda_*(\psi).$$

b) *If φ and $\varphi \perp \psi$ have AGR under Λ, then ψ has AGR under Λ.* □

5.12. PROPOSITION. *Let $k \subset K$ be a field extension. Let $\Gamma: k \to L \cup \infty$ and $\Lambda: K \to L \cup \infty$ be Q-places with Λ extending Γ. Assume finally that q is a form over k with AGR under Γ. Then $q \otimes K$ has AGR under Λ and $\Lambda_*(q \otimes K) \cong \Gamma_*(q)$.*

The proof has been given in [K$_4$, §3] for char $L \neq 2$. The arguments are merely book keeping. They remain true if char $L = 2$. □
Now we can repeat the arguments in the proof of Proposition 1.16 for quadratic places and AGR instead of usual places and GR, provided char $L \neq 2$. We obtain the following.

5.13. PROPOSITION. *Let $\Lambda: K \to L \cup \infty$ be a Q-place and φ a form over K with AGR under Λ. Assume that char $L \neq 2$. Then $\varphi_0 := \ker(\varphi)$ has again AGR under Λ and $\Lambda_*(\varphi) \sim \Lambda_*(\varphi_0)$, $\text{ind}\ \Lambda_*(\varphi) \geq \text{ind}\ (\varphi)$. If $\text{ind}\ \Lambda_*(\varphi) = \text{ind}\ (\varphi)$, then $\ker \Lambda_*(\varphi) = \Lambda_*(\varphi_0)$.* □

We finally state an important fact, proved in [K$_4$, §3], which has no counterpart on the level of ordinary places.

5.14. PROPOSITION. *Let again $\Lambda: K \to L \cup \infty$ be a Q-place with char $L \neq 2$. Let k be a subfield of K and $\Gamma := \Lambda|k$. Let q be a regular form over k. Then $q \otimes L$ has AGR under Λ if* and only if *q has AGR under Γ, and in this case $\Lambda_*(q \otimes L) \cong \Gamma_*(q)$.* □

§6. Control of the splitting behavior by use of quadratic places

If we stay with fields of characteristic $\neq 2$ then the propositions 5.11 – 5.13 indicate that it should be possible to obtain a complete analogue of the generic splitting theory displayed in §1 using quadratic places instead of ordinary places. Indeed such a theory has been developed in [K$_4$, §3]. We quote here the main result obtained there.
Let q be a form over a field k. We return to some notations from §1: $(K_r \mid 0 \leq r \leq h)$ is a generic splitting tower of q with higher indices $(i_r \mid 0 \leq r \leq h)$ and higher kernel forms $(\varphi_r \mid 0 \leq r \leq h)$. Further $(j_r \mid 0 \leq r \leq h)$, with $j_r = i_0 + \cdots + i_r$, is the splitting pattern $\text{SP}(q)$ of q.

6.1. THEOREM. *[K$_4$, Th. 3.7]. Let $\Gamma: k \to L \cup \infty$ be a Q-place into a field L of characteristic $\neq 2$. Assume that q has AGR under Γ. We choose a Q-place $\Lambda: K_m \to L \cup \infty$ extending Γ such that either $m = h$ or $m < h$ and Λ does not extend to a Q-place from K_{m+1} to L. Then $\text{ind}\ (\Gamma_*(q)) = j_m$, the form φ_m has GR under Λ and $\ker(\Gamma_*(q)) \cong \Lambda_*(\varphi_m)$.* □

A small point here – which we will not really need below – is that φ_m has GR under Λ, not just AGR.

6.2. REMARK.
The theorem shows that the generic splitting tower $(K_r \mid 0 \le r \le h)$ "controls" the splitting behavior of $\Gamma_*(q)$. Indeed, if L'/L is any field extension then we obtain from Γ a Q-place $j \circ \Gamma \colon k \to L' \cup \infty$ in a rather obvious way (cf. 6.4.iii below). The form q has also AGR under $j \circ \Gamma$, and $(j \circ \Gamma)_*(q) = \Gamma_*(q) \otimes L'$. We can apply the theorem to $j \circ \Gamma$ and q instead of Γ and q. In particular we see that ind $(\Gamma_*(q) \otimes L')$ is one of the numbers j_r. Thus the splitting pattern $\mathrm{SP}(\Gamma_*(q))$ is a subset of $\mathrm{SP}(q)$.□

We now aim at a result similar to Theorem 6.1, where the field K_m is replaced by an arbitrary partial generic splitting field for q. (This is not covered by [K$_4$].) For that reason we briefly discuss the "composition" of Q-places.

6.3. DEFINITION. Let $\Lambda = (\lambda, H, \chi) \colon K \to L \cup \infty$ and $M = (\mu, D, \psi) \colon L \to F \cup \infty$ be Q-places. The *composition* $M \circ \Lambda$ of M and Λ is the Q-place

$$(M \circ \Lambda) = (\mu \circ \lambda, H_0, \psi \circ (\chi | H_0))$$

with $H_0 := \{\alpha \in H \mid \chi(\alpha) \in D\}$. □

6.4. REMARKS.

i) If $N \colon F \to E \cup \infty$ is a third Q-place then $N \circ (M \circ \Lambda) = (N \circ M) \circ \Lambda$, as is easily checked.
ii) Let $i \colon k \hookrightarrow K$ be a field extension, regarded as a trivial place. This gives us a "trivial" Q-place $\hat{\imath} = (i, Q(k), i_* \colon Q(k) \to Q(K))$ from k to K. If $\Lambda = (K, H, \chi) \colon K \to L \cup \infty$ is any Q-places starting at K, then $\Lambda \circ \hat{\imath}$ is the restriction $\Lambda | k \colon k \to L \cup \infty$.
iii) The Q-place $j \circ \Gamma$ alluded to in 6.2 is $\hat{\jmath} \circ \Gamma$. □

In all the following $\Gamma \colon k \to L \cup \infty$ *is a Q-place into a field of characteristic* $\neq 2$ *such that q has AGR under* Γ.

6.5. PROPOSITION. *Let F/k be a generic splitting field of the form q of some level $r \in [0, h]$. Assume that* ind $\Gamma_*(q) \ge j_r$. *Then there exists a Q-place $\Lambda \colon F \to L \cup \infty$ extending Γ. For any such Q-place Λ the anisotropic part $\varphi = \ker(q \otimes F)$ of $q \otimes F$ has AGR under Λ and $\Lambda_*(\varphi) \sim \ker \Gamma_*(q)$. if* ind $\Gamma_*(q) = j_r$, *then $\Lambda_*(\varphi) = \ker \Gamma_*(q)$.*

PROOF. We only need to prove the existence of a Q-place $\Lambda \colon F \to L \cup \infty$ extending Γ. The other statements are clear from §5 (cf. 5.12, 5.13). By Theorem 6.1 we have a Q-place $\Lambda' \colon K_r \to L \cup \infty$ extending Γ. We further have a place $\rho \colon F \to K_r$ over k. It now can be checked in a straightforward way that the Q-place $\Lambda = \Lambda' \circ \hat{\rho}$ from F to L extends Γ. □

6.6. THEOREM. *Let K/k be an inessential field extension and $q \otimes K \cong \eta \perp \varphi$ with η a generic splitting form of q of some level $r \in [0, h]$. Assume that* ind $\Gamma_*(q) \ge j_r$. *Then there exists a Q-place $\Lambda \colon K \to L \cup \infty$ extending Γ such that η has AGR under Λ and $\Lambda_*(\eta) \sim 0$. For every such Q-place Λ the form φ has AGR under Λ and $\Lambda_*(\varphi) \sim \Gamma_*(q)$. If* ind $\Gamma_*(q) = j_r$, *then $\Lambda_*(\varphi) = \ker \Gamma_*(q)$.*

PROOF. Again it suffices to prove the existence of a Q-place Λ extending Γ such that η has AGR under Λ and $\Lambda_*(\eta) \sim 0$, the other statements being covered by §5 (cf. 5.11, 5.12).
Let E be a total generic splitting field of η. Then E is a partial generic splitting field of q (cf. 4.1). By the preceding proposition 6.5 there exists a Q-place $M \colon E \to L \cup \infty$

extending Γ. Let $\Lambda\colon K \to L \cup \infty$ denote the restriction of M to K (cf. 5.3). Of course, also Λ extends Γ. The form $\eta \otimes E$ is hyperbolic, hence certainly has AGR under M and $M_*(\eta \otimes E) \sim 0$. Now Proposition 5.14 tells us that η has AGR under Λ and $\Lambda_*(\eta) \sim 0$. □

6.7. REMARK. Suppose that $\Gamma = \hat{\gamma}$ with $\gamma\colon k \to L \cup \infty$ a place and that q has GR under γ. Then Theorem 6.1 and Proposition 6.5 give us nothing more than we know from the generic splitting theory in §1. Indeed, if $\Lambda = (\lambda, H, \chi)$ is a Q-place as stated there, then the form φ_m in 6.1, resp. φ in 6.5, automatically has GR under λ.
This is different with Theorem 6.6. Even in the case $\Gamma = \hat{j}$ for $j\colon k \hookrightarrow L$ the inclusion map into an overfield L of k (actually the case which is perhaps the most urgent at present), the Q-place Λ will be different from $\hat{\lambda}$. Thus, in 6.6, Q-places instead of usual places are needed even in the case $\Gamma = \hat{j}$. □

We now choose again a generic splitting preparation $(K'_r \mid 0 \le r \le h)$, $(\eta_r \mid 0 \le r \le h)$ of q with residual forms φ_r and generic splitting forms $\varepsilon_r = \perp_{j=0}^{r} \eta_j \otimes K'_j$ $(0 \le r \le h)$, cf. 4.3.

6.8. DISCUSSION. Assume that ind $\Gamma_*(q) \ge j_m$ for some $m \in [0, h]$. Theorem 6.6 tells us that there exists a Q-place $\Lambda\colon K'_m \to L \cup \infty$ extending Γ such that ε_m has AGR under Λ and $\Lambda_*(\varepsilon_m) \sim 0$. Moreover, for every such Q-place λ the form φ_m has AGR under Λ and $\Lambda_*(\varphi_m) \sim \Gamma_*(q)$. If ind $\Gamma_*(q) = j_m$, it follows that $\Lambda_*(\varphi_m)$ is the anisotropic part of $\Gamma_*(q)$.
Assume now that ind $\Gamma_*(q) > j_m$ and $\Lambda\colon K'_m \to L \cup \infty$ is a Q-place as just described. Then by analogy with 1.18 one might suspect at first glance that Λ extends to a place $M\colon K'_{m+1} \to L \cup \infty$ such that ε_{m+1} has AGR under M and $M_*(\varepsilon_{m+1}) \sim 0$. {N.B. Since $\Lambda_*(\varepsilon_m) \sim 0$, this is equivalent to the property that η_{m+1} has AGR under M and $M_*(\eta_{m+1}) \sim 0$.} But this is too much to be hoped for. Indeed, let us consider the special case that $K'_1 = \cdots = K'_h =: K'$, i.e., we are given a generic splitting preparation $(\eta_0, \eta_1, \ldots, \eta_h)$ over K'. If Λ: $K' = K'_m \to L \cup \infty$ is as above, then we can expand $\Lambda = (\lambda, H, \chi)$ to a Q-place $\Lambda' = (\lambda, Q(K'), \psi')\colon K' \to L \cup \infty$. Also Λ' extends Γ, further ε_m has AGR under Λ' and $\Lambda'_*(\varepsilon_m) = \Lambda_*(\varepsilon_m) \sim 0$ (cf. 5.12). But since $K'_{m+1} = K'$, the only extension of Λ' to K'_{m+1} is Λ' itself, and there is no reason why η_{m+1} should have AGR under Λ' and $\Lambda'_*(\eta_{m+1}) \sim 0$. □

Nevertheless, if ind $\Gamma_*(q) > j_{m+1}$, there exists a somewhat natural procedure to obtain from a Q-place $\Lambda\colon K'_m \to L \cup \infty$ as above a Q-place $M\colon K'_{m+1} \to L \cup \infty$ extending Γ such that both $\varepsilon_m \otimes K'_{m+1}$ and ε_{m+1} have AGR under M and we have $M_*(\varepsilon_m \otimes K'_{m+1}) \sim 0$ as well as $M_*(\varepsilon_{m+1}) \sim 0$. This runs as follows.

6.9. PROCEDURE. Assume that ind $\Gamma_*(q) > j_m$. Let $\Lambda\colon K'_m \to L \cup \infty$ be a Q-place extending Γ such that ε_m has AGR under Λ and $\Lambda_*(\varepsilon_m) \sim 0$. We choose a regular total generic splitting field F of ε_m. By Proposition 6.5 the Q-place Λ extends to a Q-place $\tilde{\Lambda}\colon F \to L \cup \infty$. The form $\varphi_m \otimes F$ is anisotropic. We now invoke Corollary 4.7, which tells us that the tower $F \subset F \cdot K'_{m+1} \subset \cdots \subset F \cdot K'_h$ together with the sequence of forms $(\eta_r \otimes F \cdot K'_r \mid m < r \le h)$ is a generic splitting preparation of $\varphi_m \otimes F$. Here $F \cdot K'_r$ denotes the free composite of the fields F and K'_r over k. In particular $\eta_{m+1} \otimes F \cdot K'_r$ is a generic splitting form of $\varphi_m \otimes F$ over $F \cdot K'_r$ of level 1. We have $\tilde{\Lambda}_*(\varphi_m \otimes F) \sim \tilde{\Lambda}_*(q \otimes F) = \Gamma_*(q)$. Since ind $\Gamma_*(q) > j_m$, we conclude that $\tilde{\Lambda}_*(\varphi_m \otimes F)$ is isotropic. Thus, again by Proposition 6.5, $\tilde{\Lambda}$ extends to a Q-place $\tilde{M}: F \cdot K'_{m+1} \to L \cup \infty$ such that $\eta_{m+1} \otimes F \cdot K'_{m+1}$ has AGR under $\tilde{M}$ and $\tilde{M}_*(\eta_{m+1} \otimes F \cdot K'_{m+1}) \sim 0$. Let $M: K'_{m+1} \to L \cup \infty$ denote the restriction of $\tilde{M}$ to K'_{m+1}. By 5.14, the form η_{m+1} has AGR under M and $M_*(\eta_{m+1}) \sim 0$. Now we need a delicate argument to prove that also $\varepsilon_m \otimes K'_{m+1}$ has AGR under M and $M_*(\varepsilon_m \otimes K'_{m+1}) \sim 0$. The problem is that the diagram of field embeddings

$$\begin{array}{ccc} F & \longrightarrow & F \cdot K'_{m+1} \\ \uparrow & & \uparrow \\ K'_m & \longrightarrow & K'_{m+1} \end{array}$$

does not commute, since the field composite $F \cdot K'_{m+1}$ is built over k instead of K'_m. Thus M probably does not extend the Q-place Λ.
The argument runs as follows. Also $\varphi_{m+1} \otimes F \cdot K'_{m+1}$ has AGR under $\tilde{M}$, hence φ_{m+1} has AGR under M, and

$$M_*(\varphi_{m+1}) \cong \tilde{M}_*(\varphi_{m+1} \otimes F \cdot K'_{m+1}) \sim \tilde{\Lambda}_*(\varphi_m \otimes F) \sim \tilde{\Lambda}_*(q \otimes F) \cong \Gamma_*(q).$$

Since $q \otimes K'_{m+1} \cong \varepsilon_m \otimes K'_{m+1} \perp \eta_{m+1} \perp \varphi_{m+1}$ and both, η_{m+1} and φ_{m+1}, have AGR under M, also $\varepsilon_m \otimes K'_{m+1}$ has AGR under M, cf. 5.11, and

$$\Gamma_*(q) \cong M_*(q \otimes K'_{m+1}) \cong M_*(\varepsilon_m \otimes K'_{m+1}) \perp M_*(\eta_{m+1}) \perp M_*(\varphi_{m+1}).$$

Since $M_*(\eta_{m+1}) \sim 0$ and $M_*(\varphi_{m+1}) \sim \Gamma_*(q)$, we conclude that $M_*(\varepsilon_m \otimes K'_{m+1}) \sim 0$. □

We have ind $\Gamma_*(q) = j_r$ for some $r \in [0, h]$ (cf. 6.2). Iterating the procedure with $m = 0, 1, \ldots, r-1$, we obtain the following theorem.

6.10. THEOREM. *Let* ind $\Gamma_*(q) = j_r$. *Then there exists a Q-place $\Lambda: K'_r \to L \cup \infty$ extending Γ such that $\eta_m \otimes K'_r$ has AGR under Λ and $\Lambda_*(\eta_m \otimes K'_r) \sim 0$ for every $m \in [0, r]$. If Λ is any such Q-place then φ_r has AGR under Λ and $\Lambda_*(\varphi_r) =$* ker $\Gamma_*(q)$.

We briefly discuss the case that L is a field extension of k and $\Gamma = \hat{j}$ with $j: k \hookrightarrow L$ the inclusion map.

6.11. DEFINITION/REMARK. Let K and L be field extensions of k. A Q-place from K to L *over* k is a Q-place $\Lambda = (\lambda, H, \chi): K \to L \cup \infty$ such that the first component λ is a place over k. It is evident from Definitions 5.3 that this condition just means that Λ extends the quadratic place $\hat{j}: k \to L \cup \infty$, and also that $\Lambda | k = \hat{j}$. □

6.12. SCHOLIUM. *Let L/k be a field extension,* ind $(q \otimes L) = j_r$. *Then there exists a Q-place $\Lambda: K'_r \to L \cup \infty$ over k such that $\eta_m \otimes K'_r$ has AGR under Λ and $\Lambda_*(\eta_m \otimes K'_r) \sim 0$ for every $m \in [0, r]$. For any such Q-place Λ the form φ_r has AGR under Λ and $\Lambda_*(\varphi_r) =$* ker$(q \otimes L)$. □

We return to an arbitrary Q-place $\Gamma: k \to L \cup \infty$ such that q has AGR under Γ.

6.13. DEFINITION. We call the generic splitting preparation $(K'_r \mid 0 \le r \le h)$, $(\eta_r \mid 0 \le r \le h)$ of q *tame*, if there exists a generic splitting tower $(K_r \mid 0 \le r \le h)$ of q such that K'_r is a subfield of K_r and $\eta_r \otimes K_r \sim 0$ for every $r \in [0, h]$. □

Notice that a generic splitting preparation $(K'_r \mid 0 \le r \le h)$, $(\eta_r \mid 0 \le r \le h)$ as described by 3.4 and 3.5 is tame. Thus every form admits tame generic splitting preparations. On the other hand we suspect that there exist generic splitting preparations which are wild (= not tame), although we did not look for examples. If our given generic splitting preparation of q is tame then there exists a much simpler procedure than the one described above, to obtain a Q-place $\Lambda\colon K'_r \to L\cup\infty$ with the properties stated in Theorem 6.10.

6.14. PROCEDURE. Assume that ind $\Gamma_*(q) = j_r$, and that $(K_i \mid 0 \le i \le h)$ is a generic splitting tower of q such that K'_i is a subfield of K_i and $\varepsilon_i \otimes K_i \sim 0$ for every $i \in [0,h]$. By 6.1 there exists a Q-place $\tilde{\Lambda}\colon K_r \to L \cup \infty$ extending Γ. Let Λ denote the restriction of $\tilde{\Lambda}$ to K'_r. Then Λ again extends Γ. For every $i \in [0,r]$ we have $(\varepsilon_i \otimes K'_r) \otimes K_r = \varepsilon_i \otimes K_r \sim 0$. Certainly $\varepsilon_i \otimes K_r$ has AGR under $\tilde{\Lambda}$ and $\tilde{\Lambda}_*(\varepsilon_i \otimes K_r) \sim 0$. By 5.14 the form $\varepsilon_i \otimes K'_r$ has AGR under Λ and $\Lambda_*(\varepsilon_i \otimes K'_r) \sim 0$. Since this holds for every $i \in [0,r]$ we conclude (using 5.11) that $\eta_i \otimes K'_r$ has AGR under Λ and $\Lambda_*(\eta_i \otimes K'_r) \sim 0$ for every $i \in [0,r]$.

Thus it seems that life is easier if we have a tame generic splitting preparation at our disposal than an arbitrary one. Up to now this is an argument in favor of working with generic splitting preparations instead of the more special generic splitting decompositions (cf. 4.8), in spite of Proposition 4.9, for we do not know whether every form admits a *tame* generic splitting decomposition.

References

[A] C. Arf, Untersuchungen über quadratische Formen in Körpern der Charakteristik 2 (Teil I). *J. reine angew. Math.* **183** (1941), 148–167.

[HR$_1$] J. Hurrelbrink and U. Rehmann, Splitting patterns of excellent quadratic forms, *J. reine angew. Math.* **444** (1993), 183–192.

[HR$_2$] J. Hurrelbrink and U. Rehmann, Splitting patterns and linear combinations of two Pfister forms. *J. reine angew. Math.* **495** (1998), 163–174

[J] N. Jacobson, "Lectures in abstract algebra" III, Theory of fields and Galois theory, New York, Heidelberg, Berlin: Springer 1964.

[KR] I. Kersten and U. Rehmann, Generic splitting of reductive groups, *Tôhoku Math. J.* **46** (1994), 35–70.

[K$_1$] M. Knebusch, Specialization of quadratic forms and symmetric bilinear forms, and a norm theorem, *Acta Arithmetica*, **24** (1973), 279–299.

[K$_2$] M. Knebusch, Generic splitting of quadratic forms, I, *Proc. London Math. Soc.* **33** (1976), 65–93.

[K$_3$] M. Knebusch, Generic splitting of quadratic forms, II, *Proc. London Math. Soc.* **34** (1977), 1–31.

[K$_4$] M. Knebusch, Generic splitting of quadratic forms in the presence of bad reduction, *J. reine angew. Math.* **517** (1999), 117–130.

[K$_5$] M. Knebusch, "Spezialisierung von quadratischen und symmetrischen bilinearen Formen", Book in preparation.

[Lg] S. Lang, "Introduction to algebraic geometry". New York: Interscience 1958.

[W] E. Witt, Theorie der quadratischen Formen in beliebigen Koerpern. *J. reine angew. Math.* **176** (1937), 31–44.

Fachbereich Mathematik
Universität Regensburg
D93040 Regensburg
Germany
E-mail address: manfred.knebusch@mathematik.uni-regensburg.de

Fakultät für Mathematik
Universität Bielefeld
D33501 Bielefeld
Germany
E-mail address: rehmann@mathematik.uni-bielefeld.de

Contemporary Mathematics
Volume **272**, 2000

Local densities of hermitian forms

Maurice Mischler

ABSTRACT. Local densities are used in the calculation of mass formulae. The aim of this paper is to describe how to compute the local density of a given hermitian form, knowing its usual invariants such as the Jordan decomposition or invariant factors. Note that explicit formulae have been already obtained by Y. Hironaka in [**5**] and [**6**] but only in the case of inert primes.

1. Introduction

Let E be a totally complex number field with complex multiplication, and K the fixed field under complex conjugation. Let B be the ring of integers of E, and A be the ring of integers of K.

We say that (M,h) is a *hermitian B-lattice* if M is a projective module of finite rank over B equipped with a nondegenerate hermitian form h, the involution being defined by complex conjugation. Two hermitian B-lattices (M,h) and (N,k) are said to be in the same *genus* if

$$M_{\mathfrak{p}} := (M,h) \otimes_B (B \otimes_A A_{\mathfrak{p}}) \text{ is isometric to } (N,k) \otimes_B (B \otimes_A A_{\mathfrak{p}}) =: N_{\mathfrak{p}}$$

for every spot $\mathfrak{p}$ of K and where $A_{\mathfrak{p}}$ is the completion of A at $\mathfrak{p}$. We suppose that (M,h) is totally definite. This means that h is positive at every infinite spot. The following theorem gives an example of the application of local densities when E is a cyclotomic field.

THEOREM 1.1. *Suppose that $E = \mathbb{Q}(\zeta_d)$ is a cyclotomic field. Let (M,h) be a totally definite hermitian B-lattice of rank n. Denote by $\mathcal{G}_M$ the genus of M (up to isometry). The mass of $\mathcal{G}_M$ is the following sum:*

$$\omega(M,h) := \sum_{L \in \mathcal{G}_M} \frac{1}{|U(L)|}$$

Key words and phrases. Local densities, mass of hermitian forms.

The author thanks the referee for interesting remarks, Prof.J. Boéchat for fruitful discussions and the organizers of the conference for the opportunity to give a talk.

2000 *Mathematics Subject Classification*: 11E39.

where $|U(L)|$ *is the cardinality of the unitary group of any representative of the class* L. *Then we have*

$$\omega(M,h) = 2\cdot|d(E)|^{\frac{n(n+1)}{4}}\cdot|d(K)|^{-\frac{n}{2}}\cdot\left(\prod_{j=1}^{n}\frac{(j-1)!}{(2\pi)^j}\right)^{\frac{\varphi(d)}{2}}\cdot\det(M)^{\frac{n}{2}}\cdot\prod_{\mathfrak{p}\in\mathcal{P}}\mathfrak{B}_{\mathfrak{p}}(M_{\mathfrak{p}})^{-1}$$

where $\det(M) = [M^{\#} : M]$ *with* $M^{\#} = \{x \in M\otimes E \mid h(x,M)\subset B\}$, $d(E)$ *and* $d(K)$ *are the discriminants of the fields* E *and* K *respectively,* $\mathcal{P}$ *is the set of all finite spots of* K *and* $\mathfrak{B}_{\mathfrak{p}}(M_{\mathfrak{p}})$ *is the local density of* $M_{\mathfrak{p}}$.

We will give the precise definition of local density in the next part.

The proof of this theorem is very long and technical. See ([**9**], formula (4.5), p. 20), ([**2**], p. 112]), and ([**3**], Satz VI). For more general formulae, see the wonderful book of Shimura [**11**]. The usefulness of this theorem is evident (for classifications for instance), so we have to be able to precisely compute local densities.

2. Definitions

Let E, K, B, A be as in the introduction and $\mathfrak{p}$ a finite spot of K. Suppose that $2\notin\mathfrak{p}$. Three cases are possible:

$$\mathfrak{p}B = \begin{cases} \mathfrak{P} & \text{inert case} \\ \mathfrak{P}^2 & \text{ramified case} \\ \mathfrak{P}_1\cdot\mathfrak{P}_2 & \text{decomposed case} \end{cases}$$

for some prime ideals $\mathfrak{P},\mathfrak{P}_1,\mathfrak{P}_2$ of B. Write $R = B\otimes_A A_{\mathfrak{p}}$. The classical theory of localisation shows that

$$R = \begin{cases} B_{\mathfrak{P}} & \text{in the inert and ramified case} \\ B_{\mathfrak{P}_1}\times B_{\mathfrak{P}_2}\simeq A_{\mathfrak{p}}\times A_{\mathfrak{p}} & \text{in the decomposed case.} \end{cases}$$

Finally, if π is a uniformizing parameter of $A_{\mathfrak{p}}$ such that $\mathfrak{p}A_{\mathfrak{p}} = \pi A_{\mathfrak{p}}$, we can suppose that

$$\mathfrak{p}R = \begin{cases} \mathrm{p}\cdot R & \text{in the inert case, with } \mathrm{p} = \pi \\ \mathrm{p}^2\cdot R & \text{in the ramified case, with } \mathrm{p}^2 = \pi \\ \mathrm{p}\cdot R\simeq \pi A_{\mathfrak{p}}\times\pi A_{\mathfrak{p}} & \text{in the decomposed case, with } \mathrm{p} = (\pi,\pi). \end{cases}$$

In any case, let $\mathfrak{P}$ denote the ideal $\mathrm{p}\cdot R$.

Definition 2.1. Let (M,h) be a hermitian R-lattice. Suppose that $\mathfrak{p}$ does not decompose. A vector x of M is said to be *primitive* if there exists an R-basis $(x_1,\dots,x_n)$ of M such that $x = x_1$. Let $\mathfrak{a}$ be an ideal of R. We say that (M,h) is $\mathfrak{a}$-*modular* if $h(x,M) = \mathfrak{a}$ for every primitive vector of M. If there exists $m_1 < \cdots < m_l$ such that the following orthogonal sum holds:

$$M = M_1 \boxplus\cdots\boxplus M_l$$

where M_i is $\mathfrak{P}^{m_i}$-modular for every i, we say that $M_1\boxplus\cdots\boxplus M_l$ is a *Jordan decomposition of* (M,h). It is possible to prove that Jordan decomposition exists and that l, $\dim(M_i)$, and m_i do not depend on the decomposition. In this case, $\mathfrak{P}^{m_1}\supset\cdots\supset\mathfrak{P}^{m_l}$ are called the *invariant factors of* (M,h). (See ([**4**], p. 33) and ([**7**], p. 449) for proofs.)

DEFINITION 2.2. Let (M, h) be a hermitian R-lattice. Suppose that $\mathfrak{p}$ decomposes. There exist $x_1, \ldots, x_n \in M$ such that

$$M = (A_\mathfrak{p} \times \pi^{m_1} A_\mathfrak{p})x_1 \oplus \cdots \oplus (A_\mathfrak{p} \times \pi^{m_n} A_\mathfrak{p})x_n$$

and such that $h(x_i, x_j) = \delta_{ij} \cdot (1,1)$ for all $1 \leq i, j \leq n$ and $m_1 \leq \cdots \leq m_n$. In this case

$$\pi^{m_1} A_\mathfrak{p} \supset \cdots \supset \pi^{m_n} A_\mathfrak{p}$$

are called the *invariant factors of* (M, h), because

$$M^{\#} = (\pi^{-m_1} A_\mathfrak{p} \times A_\mathfrak{p})x_1 \oplus \cdots \oplus (\pi^{-m_n} A_\mathfrak{p} \times A_\mathfrak{p})x_n.$$

(See ([**4**], p. 25) or ([**10**], Proposition 3.2) for proofs.) In addition, (M, h) is said to be $\mathfrak{P}^m$*-modular* if

$$m_1 = \cdots = m_n = m.$$

DEFINITION 2.3. Let (M, h) and (N, k) be two hermitian R-lattices of rank $r \geq s$ respectively. Let $m \in \mathbb{N}$. We define $\mathcal{A}_{\mathfrak{p}^m}(M, N)$ to be the set of R-linear maps $u : N \to M$ which are distinct modulo $\mathfrak{p}^m RM$ and satisfy

$$h(u(x), u(y)) \equiv k(x, y) \pmod{\mathfrak{p}^m R} \quad \text{for all } x, y \in N.$$

We write $A_{\mathfrak{p}^m}(M, N)$ the cardinality of $\mathcal{A}_{\mathfrak{p}^m}(M, N)$.

PROPOSITION 2.4. *Let (M, h) and (N, k) be two hermitian R-lattices of rank $r \geq s$ respectively. Suppose that $\mathfrak{p} \cap \mathbb{Z} = p\mathbb{Z}$ and write $q = p^f = |A/\mathfrak{p}|$. Then we have:*

$$A_{\mathfrak{p}^{m+1}}(M, N) = q^{s(2r-s)} A_{\mathfrak{p}^m}(M, N)$$

for any $m \geq 2 \cdot v_\mathfrak{p}(\mathfrak{a}_l) + 1$. Here $\mathfrak{a}_1 \supset \cdots \supset \mathfrak{a}_l$ are the invariant factors of (N, k), and $v_\mathfrak{p}$ is the discrete valuation over $A_\mathfrak{p}$.

PROOF. See ([**9**], Hilfssatz 5.3) for the proof. □

DEFINITION 2.5 (DEFINITION OF LOCAL DENSITIES). Let (M, h) be a hermitian R-lattice of rank r. We write

$$\mathfrak{B}_\mathfrak{p}(M) = \lim_{m \to \infty} q^{-r^2 m} A_{\mathfrak{p}^m}(M, M).$$

The last proposition shows that this limit exists. This limit is called the *local density of* (M, h).

Now, the natural question is the following: if we know the invariant factors of a given hermitian R-lattice, is it possible to calculate its local density ? We will show that the answer is yes!

3. Results

Let $E, K, B, A, \mathfrak{p}, R$, p, and q be as in the last part.

PROPOSITION 3.1. *Let (M,h) be a hermitian R-lattice of rank n.*

(a) Suppose that $\mathfrak{p}$ does not decompose. If $h = \pi^m h'$ with $M \subset M^{\#}_{h'}$ and $m \geq 1$, write M' for (M,h'). Let l be a strictly positive integer . Then we have $A_{\mathfrak{p}^{l+m}}(M,M) = q^{2mn^2} A_{\mathfrak{p}^l}(M',M')$. This implies that

$$\mathfrak{B}_{\mathfrak{p}}(M) = q^{mn^2}\mathfrak{B}_{\mathfrak{p}}(M').$$

(b) Suppose that $\mathfrak{p}$ decomposes. We have seen that we can suppose that $M = (A_{\mathfrak{p}} \times \pi^{m_1} A_{\mathfrak{p}})x_1 \oplus \cdots \oplus (A_{\mathfrak{p}} \times \pi^{m_n} A_{\mathfrak{p}})x_n$ with $h(x_i,x_j) = \delta_{ij}\cdot(1,1)$. Now define h' such that $h'(x_i,x_j) = (1,\pi^{-m_1})h(x_i,x_j)$ for all $1 \leq i,j \leq n$. Then we have the same results with m_1 instead of m.

PROOF. In each case, let

$$\begin{aligned} \Psi \,:\, \mathcal{A}_{\mathfrak{p}^{m+l}}(M,M) &\longrightarrow \mathcal{A}_{\mathfrak{p}^l}(M',M') \\ u \pmod{\mathfrak{p}^{l+m}R} &\longmapsto u \pmod{\mathfrak{p}^l R} \end{aligned}$$

The function Ψ is surjective. More precisely, if $u \in \mathcal{A}_{\mathfrak{p}^l}(M',M')$, it is easy to verify that the set of $u+\pi^l v$, where v is any endomorphism modulo $\mathfrak{p}^m R$, is the set of u' such that $\Psi(u') = u$. Since $|R/\mathfrak{p}^m R| = |A_{\mathfrak{p}}/\mathfrak{p}^m|^2 = q^{2m}$, the number of v which is the number of u' is equal to q^{2mn^2}. Hence $A_{\mathfrak{p}^{l+m}}(M,M) = q^{2mn^2} A_{\mathfrak{p}^l}(M',M')$. So we have proved the proposition. Indeed, for l big enough, we have:

$$\mathfrak{B}_{\mathfrak{p}}(M) = \frac{A_{\mathfrak{p}^{l+m}}(M,M)}{q^{(l+m)n^2}} = \frac{q^{2mn^2}A_{\mathfrak{p}^l}(M',M')}{q^{mn^2}q^{ln^2}} = q^{mn^2}\mathfrak{B}_{\mathfrak{p}}(M'). \quad \square$$

PROPOSITION 3.2. *Let (M,h) be a hermitian R-lattice of rank n. Suppose that (M,h) is $\mathfrak{P}^m$ modular. Let (N,k) be another hermitian R-lattice of rank n.*

Suppose there exist R-bases $(e_1,\ldots,e_n)$ and $(f_1,\ldots,f_n)$ of M and N respectively such that

$$h(e_i,e_j) \equiv k(f_i,f_j) \pmod{\mathfrak{p}^{m+1}R} \quad \textit{in the non-ramified cases}$$

and such that

$$h(e_i,e_j) \equiv k(f_i,f_j) \pmod{\mathfrak{p}^{[\frac{m}{2}]+1}R} \quad \textit{in the ramified case}$$

where $[x]$ is the integer part of x. Then (M,h) is isometric to (N,k) over R.

PROOF. In each case, it is easy to prove that (N,k) is $\mathfrak{P}^m$-modular. If $\mathfrak{p}$ ramifies and m is odd, or if $\mathfrak{p}$ does not ramify, all $\mathfrak{P}^m$-modular forms are isometric. See [**4,5**] for proofs. The proposition is hence proved in these cases. Suppose now that $\mathfrak{p}$ ramifies and that m is even. We only have to prove that $\det(M,h) = \det(N,k)$. We need Hensel's lemma. It is easy to prove that $\det(M,h) = \lambda\pi^{\frac{mn}{2}}$ for a given unit λ of $A_{\mathfrak{p}}$. Without loss of generality, we can suppose that for all $i = 1,\ldots,n-1$, we have $k(f_i,f_i) = \pi^{\frac{m}{2}} + \pi^{\frac{m}{2}+1}\alpha_i$, and $k(f_n,f_n) = \lambda\pi^{\frac{m}{2}} + \pi^{\frac{m}{2}+1}\alpha_n$, with $\alpha_i, \alpha_n \in A_{\mathfrak{p}}$. If $i \neq j$, then $k(f_i,f_j)$ is a multiple of $\pi^{\frac{m}{2}+1}$. Hence $\det(N,k) = \lambda\pi^{\frac{mn}{2}} + \pi^{\frac{mn}{2}+1}\alpha = \pi^{\frac{mn}{2}}(\lambda+\pi\alpha)$ for some α in $A_{\mathfrak{p}}$. Let $\eta = \frac{\lambda+\pi\alpha}{\lambda} = 1 + \pi\frac{\alpha}{\lambda} \equiv 1 \pmod{\mathfrak{p}}$. Then Hensel's lemma shows that η is a square of a unit of $A_{\mathfrak{p}}$. Hence, $\det(M,h) = \det(N,k)$. $\square$

PROPOSITION 3.3. *Let (M,h) and (N,k) be hermitian R-lattices of rank n_1 and n_2 respectively such that $n_1 \geq n_2$. Suppose that (N,k) and $m \geq 1$ have the property that $(N',k') \simeq (N,k) \pmod{\mathfrak{p}^m R}$ implies that (N,k) is isometric to (N',k') over R. Then we have:*

$$A_{\mathfrak{p}^{m+1}}(M,N) = q^{n_2(2n_1-n_2)} A_{\mathfrak{p}^m}(M,N).$$

PROOF. This is the Proposition 5.4 of [**1**], in the ramified case. The proof in the other cases is similar. □

PROPOSITION 3.4. *Let (M,h) be a hermitian R-lattice of rank n. Suppose that $M = M_1 \boxplus \cdots \boxplus M_l$ is a Jordan decomposition of M, that is, every M_i is $\mathfrak{P}^{m_i}$-modular for all i, and $m_1 < \cdots < m_l$. Assume that the rank of M_1 is n_1 and that $m_1 = 0$ if $\mathfrak{p}$ does not ramify and that $m_1 = 0$ or 1 in the ramified case. Then we have*

$$A_{\mathfrak{p}^m}(M,M) = q^{m_1 n_1(n-n_1)} A_{\mathfrak{p}^m}(M,M_1)\, A_{\mathfrak{p}^m}(N,N) \text{ for all } m \geq 1$$

where $N = M_2 \boxplus \cdots \boxplus M_l$.

PROOF. Let $v \in \mathcal{A}_{\mathfrak{p}^m}(M,M_1)$. By Proposition 3.2, M_1 is isometric to $v(M_1)$. Then, $v(M_1)$ splits M (see ([**1**], proposition 5.3) or ([**4**], p. 32) for proof in the inert and ramified cases, in the decomposed case use the same arguments). It follows that there exists $u \in \mathcal{A}_{\mathfrak{p}^m}(M,M)$ such that $u|_{M_1} \equiv v \pmod{\mathfrak{p}^m R}$. Let u' be another element of $\mathcal{A}_{\mathfrak{p}^m}(M,M)$ such that $u'|_{M_1} \equiv v \pmod{\mathfrak{p}^m R}$. Then $(u^{-1}\circ u')|_{M_1} \equiv Id_{M_1} \pmod{\mathfrak{p}^m R}$ and $(u^{-1}\circ u')|_N \in \mathcal{A}_{\mathfrak{p}^m}((u^{-1}\circ u')(N),N) \simeq \mathcal{A}_{\mathfrak{p}^m}(N,N)$. Let $x_1,\dots,x_{n_1}$ be a basis of M_1 and $x_{n_1+1},\dots,x_n$ a basis of N. Let $l \geq n_1+1$. Write $(u^{-1}\circ u')(x_l) = \sum_{i=1}^{n} \lambda_{i,l} x_i$.

(a) Suppose that $\mathfrak{p}$ does not ramify. Using a classification given in [**4,7,10**], we can suppose that $h(x_i,x_i) \equiv 1_R \pmod{\mathfrak{p}^m R}$ for all $i \leq n_1$ and that $h(x_i,x_j) \equiv 0 \pmod{\mathfrak{p}^m R}$ for all $i \neq j$. Then we have

$$0 \equiv h((u^{-1}\circ u')(x_l),x_i) \equiv \lambda_{i,l} h(x_i,x_i) \equiv \lambda_{i,l} \pmod{\mathfrak{p}^m R}.$$

So we have no choice for the $\lambda_{i,l}$. This completes the proof in the non-ramified case.

(b) The ramified case is Proposition 5.5 of [**1**]. □

THEOREM 3.5. *Let (M,h) be a hermitian R-lattice of rank n. Consider $M = M_1 \boxplus \cdots \boxplus M_l$, a Jordan decomposition of (M,h), such that M_i is $\mathfrak{P}^{m_i}$-modular, $m_1 < \cdots < m_l$, and the rank of M_i is n_i for all i. Thanks to Proposition 3.1, we can suppose that $m_1 = 0$ or 1 in the ramified case, and that $m_1 = 0$ in the inert and decomposed cases.*

(a) Suppose that $\mathfrak{p}$ is ramified. Let $m_2 = 2m_2' + \epsilon(m_2)$, with $\epsilon(m_2) = 1$ if m_2 is odd and $\epsilon(m_2) = 0$ otherwise. If we define $M_2' \boxplus \cdots \boxplus M_l' = (M_2,h_2') \boxplus \cdots \boxplus (M_l,h_l')$, with $h_i = \pi^{m_2'} h_i'$, we have $M_i \subset (M_i)^{\#}_{h_i'}$ for all $i = 2,\dots,l$, and M_2' is $\mathfrak{P}^{\epsilon(m_2')}$-modular. Under these hypothesis, we have:

(i) if $(m_1,m_2) \neq (0,1)$, then

$$\mathfrak{B}_{\mathfrak{p}}(M) = q^{m_2'(n-n_1)^2 + m_1 n_1 (n-n_1)}\, \mathfrak{B}_{\mathfrak{p}}(M_1)\, \mathfrak{B}_{\mathfrak{p}}(M_2' \boxplus \cdots \boxplus M_l').$$

(ii) *if* $(m_1, m_2) = (0,1)$, *then*

$$\mathfrak{B}_\mathfrak{p}(M) = \mathfrak{B}_\mathfrak{p}(M_2)^{-1}\, \mathfrak{B}_\mathfrak{p}(M_1 \boxplus M_2)\, \mathfrak{B}_\mathfrak{p}(M_2 \boxplus \cdots \boxplus M_l).$$

(b) *Suppose that* $\mathfrak{p}$ *is inert. Define* $M_2' \boxplus \cdots \boxplus M_l' := (M_2, h_2') \boxplus \cdots \boxplus (M_l, h_l')$, *with* $h_i = \pi^{m_2} h_i'$. *In this case,* M_2' *is unimodular, and* $M_i \subset (M_i)^{\#}_{h_i'}$ *for all* $i = 2, \ldots, l$. *Then we have:*

$$\mathfrak{B}_\mathfrak{p}(M) = q^{m_2(n-n_1)^2}\, \mathfrak{B}_\mathfrak{p}(M_1)\, \mathfrak{B}_\mathfrak{p}(M_2' \boxplus \cdots \boxplus M_l').$$

(c) *Suppose that* $\mathfrak{p}$ *decomposes. Define* $M_2' \boxplus \cdots \boxplus M_l' := (M_2, h_2') \boxplus \cdots \boxplus (M_l, h_l')$, *with* $h_i = (1, \pi^{m_2}) h_i'$. *In this case,* M_2' *is unimodular, and* $M_i \subset (M_i)^{\#}_{h_i'}$ *for all* $i = 2, \ldots, l$. *Then again:*

$$\mathfrak{B}_\mathfrak{p}(M) = q^{m_2(n-n_1)^2}\, \mathfrak{B}_\mathfrak{p}(M_1)\, \mathfrak{B}_\mathfrak{p}(M_2' \boxplus \cdots \boxplus M_l').$$

PROOF. Let $\mu = 2m_l$. In any case, thanks to Proposition 2.4, we have

$$\mathfrak{B}_\mathfrak{p}(M) = \frac{A_{\mathfrak{p}^{\mu+1}}(M, M)}{q^{(\mu+1)n^2}}.$$

On the other hand, Proposition 3.4 says that

$$A_{\mathfrak{p}^m}(M, M) = q^{m_1 n_1 (n-n_1)}\, A_{\mathfrak{p}^m}(M, M_1)\, A_{\mathfrak{p}^m}(N, N) \text{ for all } m \geq 1,$$

where $N = M_2 \boxplus \cdots \boxplus M_l$. Finally, Propositions 3.2 and 3.3 show that

$$A_{\mathfrak{p}^{\mu+1}}(M, M_1) = q^{\mu n_1 (2n-n_1)} A_\mathfrak{p}(M, M_1).$$

Proof of part (i) of (a): in this case, we know that $m_2 \geq 2$. We claim that $A_\mathfrak{p}(M, M_1) = q^{2n_1(n-n_1)} A_\mathfrak{p}(M_1, M_1)$. Indeed, let $u \in \mathcal{A}_\mathfrak{p}(M, M_1)$. We represent u by a matrix $U = \overbrace{\begin{pmatrix} U_1 \\ U_2 \end{pmatrix}}^{n_1} \begin{matrix} \}n_1 \\ \}n-n_1 \end{matrix} \in \mathrm{M}_{n \times n_1}(R)$ such that $\overline{U}^t H_M U \equiv H_{M_1} \pmod{\mathfrak{p}R}$, where H_M and H_{M_1} are the matrices of h and $h|_{M_1}$ respectively. In this case, we know that $\mathfrak{p}R = \mathfrak{P}^2$. Hence, $H_M = H_{M_1} \oplus H_{M_2} \oplus \cdots \oplus H_{M_l} \equiv \begin{pmatrix} H_{M_1} & 0 \\ 0 & 0 \end{pmatrix}$ $\pmod{\mathfrak{P}^2}$, since $m_2 \geq 2$. It follows that $\overline{U}^t H_M U \equiv \overline{U_1}^t H_{M_1} U_1 \equiv H_{M_1} \pmod{\mathfrak{P}^2}$, and hence $U_1 \in \mathcal{A}_\mathfrak{p}(M_1, M_1)$. That is, $A_\mathfrak{p}(M, M_1) = q^{2n_1(n-n_1)} A_\mathfrak{p}(M_1, M_1)$.

We also verify, thanks to Proposition 3.1, that

$$A_{\mathfrak{p}^{k+m_2'}}(N, N) = q^{2m_2'(n-n_1)^2} A_{\mathfrak{p}^k}(N', N') \text{ for all positive } k$$

where $N = M_2 \boxplus \cdots \boxplus M_l$ and $N' = M_2' \boxplus \cdots \boxplus M_l'$.

Now, we are able to compute $\mathfrak{B}_\mathfrak{p}(M)$.

$$\begin{aligned} \mathfrak{B}_\mathfrak{p}(M) &= q^{-(\mu+1)n^2 + m_1 n_1 (n-n_1) + \mu n_1 (2n-n_1) + 2n_1(n-n_1) + 2m_2'(n-n_1)^2} \\ &\quad \cdot q^{n_1^2 + (n-n_1)^2(\mu+1-m_2')} \cdot \frac{A_\mathfrak{p}(M_1, M_1)}{q^{n_1^2}} \cdot \frac{A_{\mathfrak{p}^{\mu+1-m_2'}}(N', N')}{q^{(n-n_1)^2(\mu+1-m_2')}} \\ &= q^{m_2'(n-n_1)^2 + m_1 n_1 (n-n_1)}\, \mathfrak{B}_\mathfrak{p}(M_1)\, \mathfrak{B}_\mathfrak{p}(M_2' \boxplus \cdots \boxplus M_l'). \end{aligned}$$

Proof of part (ii) of (a):

Proposition 3.4 tells us that $A_{\mathfrak{p}^3}(M_1 \boxplus M_2, M_1) = \frac{A_{\mathfrak{p}^3}(M_1 \boxplus M_2, M_1 \boxplus M_2)}{A_{\mathfrak{p}^3}(M_2, M_2)}$. On the other hand, by Proposition 3.3 we have $A_{\mathfrak{p}^3}(M_1 \boxplus M_2, M_1) = q^{2n_1(2(n_1+n_2)-n_1)} \cdot A_{\mathfrak{p}}(M_1 \boxplus M_2, M_1)$. It follows that,

$$A_{\mathfrak{p}}(M_1 \boxplus M_2, M_1) = p^{-2n_1(2(n_1+n_2)-n_1)} \frac{A_{\mathfrak{p}^3}(M_1 \boxplus M_2, M_1 \boxplus M_2)}{A_{\mathfrak{p}^3}(M_2, M_2)}.$$

Using the same arguments as for part (i), we can easily see that

$$A_{\mathfrak{p}}(M, M_1) = p^{2(n-(n_1+n_2))n_1} A_{\mathfrak{p}}(M_1 \boxplus M_2, M_1).$$

Again, we are able to compute:

$$\begin{aligned}\mathfrak{B}_{\mathfrak{p}}(M) &= q^{\mu n_1(2n-n_1)-(\mu+1)n^2+2n_1(n-(n_1+n_2))-2n_1(2n_2+n_1)} \\ &\quad \cdot q^{3(n_1+n_2)^2-3n_2^2+(\mu+1)(n-n_1)^2} \cdot \mathfrak{B}_{\mathfrak{p}}(M_2)^{-1}\, \mathfrak{B}_{\mathfrak{p}}(M_1 \boxplus M_2)\, \mathfrak{B}_{\mathfrak{p}}(N) \\ &= \mathfrak{B}_{\mathfrak{p}}(M_2)^{-1}\, \mathfrak{B}_{\mathfrak{p}}(M_1 \boxplus M_2)\, \mathfrak{B}_{\mathfrak{p}}(M_2 \boxplus \cdots \boxplus M_l).\end{aligned}$$

The proofs of parts (b) and (c) are similar to that of part (i) of (a), except that in these cases, $m_1 = 0$. □

With these results, we have to compute local densities only in four cases:
(a) when M is unimodular and $\mathfrak{p}$ ramifies,
(b) when M is $\mathfrak{P}$-modular and $\mathfrak{p}$ ramifies,
(c) when $M = M_1 \boxplus M_2$ with M_1 unimodular, M_2 $\mathfrak{P}$-modular, and $\mathfrak{p}$ ramifies,
(d) when M is unimodular and $\mathfrak{p}$ does not ramify.

But these local densities are already calculated in the literature. We recall these computations in the next theorem.

THEOREM 3.6. *Let (M, h) be a hermitian R-lattice of rank n.*
(a) Suppose that $\mathfrak{p}$ is ramified.
(i) Suppose that M is unimodular, that $U(A_{\mathfrak{p}})/N_{R/A_{\mathfrak{p}}}(U(R)) = \{1, \epsilon\}$ and that the determinant of (M, h) is $\lambda \in \{1, \epsilon\}$. If n is even, then

$$\mathfrak{B}_{\mathfrak{p}}(M) = 2(1 - \left(\frac{(-1)^{\frac{n}{2}}\lambda}{\mathfrak{p}}\right) q^{-\frac{n}{2}}) \prod_{i=1}^{\frac{n-2}{2}} (1 - q^{-2i})$$

where $\left(\frac{x}{\mathfrak{p}}\right) = \begin{cases} 1 & \textit{if } x \textit{ is a square} \pmod{\mathfrak{p}} \\ -1 & \textit{otherwise.} \end{cases}$

(ii) If M is unimodular and n is odd, then

$$\mathfrak{B}_{\mathfrak{p}}(M) = 2 \prod_{i=1}^{\frac{n-1}{2}} (1 - q^{-2i}).$$

(iii) If M is $\mathfrak{P}$-modular. Then n must be even, and we have:

$$\mathfrak{B}_{\mathfrak{p}}(M) = q^{\frac{n(n+1)}{2}} \prod_{i=1}^{\frac{n}{2}} (1 - q^{-2i}).$$

(iv) *Suppose that $M_1 \boxplus M_2$ is a Jordan decomposition of M with M_1 unimodular of dimension n_1 even and of determinant $\lambda \in \{1, \epsilon\}$, and with M_2 $\mathfrak{P}$-modular, of dimension $n_2 = n - n_1$ necessarily even. Then we have:*

$$\mathfrak{B}_\mathfrak{p}(M) = \frac{2q^{\frac{n_2(n_2+1)}{2}} \prod_{i=1}^{\frac{n_1}{2}} (1 - q^{-2i}) \prod_{i=1}^{\frac{n_2}{2}} (1 - q^{-2i})}{1 + \left(\frac{(-1)^{\frac{n_1}{2}} \lambda}{\mathfrak{p}} \right) q^{-\frac{n_1}{2}}}.$$

(v) *Suppose that we are in the same situation as (iv), but the dimension of M_1 is odd. Then we have:*

$$\mathfrak{B}_\mathfrak{p}(M) = 2\, q^{\frac{n_2(n_2+1)}{2}} \prod_{i=1}^{\frac{n_1-1}{2}} (1 - q^{-2i}) \prod_{i=1}^{\frac{n_2}{2}} (1 - q^{-2i}).$$

(b) *Suppose that $\mathfrak{p}$ is inert and M is unimodular. Then:*

$$\mathfrak{B}_\mathfrak{p}(M) = \prod_{i=1}^{n} (1 - (-1)^i\, q^{-i}).$$

(c) *If $\mathfrak{p}$ decomposes, and M is unimodular. Then:*

$$\mathfrak{B}_\mathfrak{p}(M) = \prod_{i=1}^{n} (1 - q^{-i}).$$

PROOF. See ([**1**], p. 24-28) for part (a) and ([**9**], Hilfssatz 5.3) for parts (b) and (c). □

References

1. E. Bannai, *Positive definite unimodular lattices with trivial automorphism groups*, Memoirs of the AMS **429** (1990).
2. S. Boge, *Schiefhermitesche Formen über Zahlkörpern und Quaternionenschiefkörpern*, J. Reine Angew. Math **221** (1966), 85–112.
3. H. Braun, *Zur Theorie der hermitischen Formen*, Abh. Math. Sem. Hamburg **14** (1941), 61–150.
4. P. Calame, *Genre et facteurs invariants de formes hermitiennes*, Théorie des nombres, Années 1996/97–1997/98, Publ. Math. UFR Sci. Tech, Univ. Franche-Comté, Besançon, 1999.
5. Y. Hironaka, *Local zeta functions on hermitian forms and its application to local densities*, J. Number Theory **71** (1998), 40–64.
6. Y. Hironaka, *Local densities of hermitian forms*, Contemporary Mathematics **249** (1999), 135–148.
7. R. Jacobowitz, *Hermitian forms over local fields*, Am. J. Math. **84** (1962), 441–465.
8. M. Mischler, *Un lien entre les $\mathbb{Z}$-réseaux et les formes hermitiennes: les F-réseaux*, Publ. Fac. Sci. Besançon (1997), Univ. Franche-Comté, Besançon.
9. U. Rehmann, *Klassenzahlen einiger totaldefiniter klassischer Gruppen über Zahlkörpern*, (Diss.) (1971), Göttingen.
10. G. Shimura, *Arithmetic of unitary groups*, Ann. of Math. **79 no 2** (1964), 369–409.
11. G. Shimura, *Euler Products and Eisenstein Series*, CMBS Reg. Conf. Series in Math. **93** (1997), 369–409.

INSTITUT DE MATHÉMATIQUES
FACULTÉ DES SCIENCES, UNIVERSITÉ DE LAUSANNE
CH-1015 LAUSANNE, SWITZERLAND
E-mail address: maurice.mischler@ima.unil.ch

Contemporary Mathematics
Volume **272**, 2000

Notes towards a constructive proof of Hilbert's Theorem on ternary quartics

Victoria Powers and Bruce Reznick

1. Introduction

In 1888, Hilbert [**5**] proved that a real ternary quartic which is positive semi-definite (psd) must have a representation as a sum of three squares of quadratic forms. Hilbert's proof is short, but difficult; a high point of 19th century algebraic geometry. There have been two modern expositions of the proof – one by Cassels in the 1993 book [**6**] by Rajwade, and one by Swan [**8**] in these Proceedings – but there are apparently no other proofs of this theorem in the literature. In 1977, Choi and Lam [**2**] gave a short elementary proof that a psd ternary quartic must be a sum of (five) squares of quadratic forms, but as we shall see, the number "three" is critical.

Hilbert's approach does not address two interesting computational issues:

1. Given a psd ternary quartic, how can one find three such quadratics?
2. How many "fundamentally different" ways can this be done?

In this paper, we describe some methods for finding and counting representations of a psd ternary quartic as a sum of three squares. In certain special cases, we can answer these questions completely, describing all representations in detail. For example, if $p(x,y,z) = x^4 + F(y,z)$, where F is a psd quartic, then we give an algorithm for constructing all representations of p as a sum of three squares. We show that if F is not the fourth power of a linear form, then there are at most 8 such representations. The key idea to our work is the simple observation that if $p = f^2 + g^2 + h^2$, then $p - f^2$ is a sum of two squares. We also give an equivalent form of Hilbert's Theorem which involves only binary forms.

2. Preliminaries

Suppose

$$p(x,y,z) = \sum_{i+j+k=4} \alpha_{i,j,k} x^i y^j z^k \tag{1}$$

is a ternary quartic. How can we tell whether p is psd? The general answer, by the theory of quantifier elimination (see, e.g., [**1**]) tells us that this is the case if and only if the coefficients of p belong to a particular semi-algebraic set. This general

2000 *Mathematics Subject Classification.* 11E20, 11E76, 14N15, 12Y05.

set is likely to be rather unedifying to look at in detail, so it will be convenient to make a few harmless assumptions about p.

Suppose that $p(x_1,\dots,x_n)$ is a homogeneous polynomial. By an *invertible change* taking p to p', we will mean a formal identity:

$$p(x'_1\dots,x'_n)=p'(x_1,\dots,x_n),\qquad \begin{bmatrix} x'_1\\ \vdots\\ x'_n\end{bmatrix}=\begin{bmatrix} m_{11} & \cdots & m_{1n}\\ \vdots & \ddots & \vdots\\ m_{n1} & \cdots & m_{nn}\end{bmatrix}\begin{bmatrix} x_1\\ \vdots\\ x_n\end{bmatrix},$$

where the matrix $M=[m_{ij}]$ above is in $GL(n,\mathbf{R})$. Note that p is psd if and only if p' is psd, and representations of p as a sum of m squares are immediately transformed into similar representations of p' and vice versa.

For example, if $p(x_1,\dots,x_n)$ is a psd quadratic form of rank r, then after an invertible change, $p=x_1^2+\cdots+x_r^2$. If *deg* $p=d$ and $M=cI$, then $p'=c^dp$. Thus, multiplying p by a positive constant is an invertible change. A non-trivial application of invertible changes is given in Theorem 6 below. When making invertible changes, we will customarily drop the primes as soon as no confusion would result.

Suppose now that p is a non-zero psd ternary quartic. Then there exists a point (a,b,c) for which $p(a,b,c)>0$. By an (invertible) rotation, we may assume that $p(t,0,0)=t^4p(1,0,0)=u>0$, and so we may assume that $p(1,0,0)=1$; hence $\alpha_{4,0,0}=1$. Writing p in decreasing powers of x, we have

$$p(x,y,z)=x^4+\alpha_{3,1,0}x^3y+\alpha_{3,0,1}x^3z+\dots.$$

If we now let $x'=x+\frac{1}{4}(\alpha_{3,1,0}y+\alpha_{3,0,1}z)$, $y'=y$, $z'=z$, then $x=x'-\frac{1}{4}(\alpha_{3,1,0}y'+\alpha_{3,0,1}z')$, $y=y'$, $z=z'$, and it's easy to see that $p'(x,y,z)=x^4+0\cdot x^3y+0\cdot x^3z+\cdots$.

We may thus assume without loss of generality that

$$p(x,y,z)=x^4+2F_2(y,z)x^2+2F_3(y,z)x+F_4(y,z), \tag{2}$$

where F_j is a binary form of degree j in (y,z). Henceforth, we shall restrict our attention to ternary quartics of this shape.

We present a condition for p to be psd. No novelty is claimed for this result, which has surely been known in various guises for centuries. Note that p is psd if and only if, for all $(y,z)\in\mathbf{R}^2$ and all real t,

$$\Phi_{(y,z)}(t):=t^4+2F_2(y,z)t^2+2F_3(y,z)t+F_4(y,z)\ge 0.$$

THEOREM 1. *The quartic* $\Phi(t)=t^4+2at^2+2bt+c$ *satisfies* $\Phi(t)\ge 0$ *for all* t *if and only if* $c\ge 0$ *and*

$$|b|\le\frac{2}{3\sqrt{3}}\left(-a+\sqrt{a^2+3c}\right)^{1/2}\left(2a+\sqrt{a^2+3c}\right):=K(a,c). \tag{3}$$

PROOF. A necessary condition for $\Phi(t)\ge 0$ for all t is that $\Phi(0)=c\ge 0$. If $\Phi(0)=0$, then $\Phi'(0)=2b=0$ as well, and clearly $t^4+2at^2\ge 0$ if and only if $a\ge 0$. Thus, one possibility is that $c=0$, $b=0$, and $a\ge 0$.

We may henceforth assume that $\Phi(0)=c>0$, and so, dividing by $|t|$, $\Phi(t)\ge 0$ for all $|t|$ if and only if $|t|^3+2a|t|+2b\cdot\mathrm{Sign}(t)+c|t|^{-1}\ge 0$, which holds if and only if

$$\min_{u>0}\left(u^3+2au+\frac{c}{u}\right)\ge 2|b|.$$

The minimum occurs when $3u_0^4 + 2au_0^2 - c = 0$. The only positive solution to this equation is

$$u_0 = \left(\frac{-a + \sqrt{a^2 + 3c}}{3}\right)^{1/2}.$$

Thus, using $c = 3u_0^4 + 2au_0^2$ to simplify the computation, we see that $\Phi(t) \geq 0$ if and only if

$$2|b| \leq u_0^3 + 2au_0 + cu_0^{-1} = 4u_0^3 + 4au_0 = 4u_0(u_0^2 + a) \tag{4}$$

$$= 4\left(\frac{1}{3}\left(-a + \sqrt{a^2 + 3c}\right)\right)^{1/2}\left(\frac{1}{3}\left(-a + \sqrt{a^2 + 3c}\right) + a\right) = 2K(a, c). \tag{5}$$

We are nearly done, because this case assumes that $c > 0$. But note that if $c = 0$, we have $K(a, 0) = 0$ if $a \geq 0$ and $K(a, 0) < 0$ if $a < 0$, so $|b| \leq K(a, 0)$ implies $b = 0$ and $a \geq 0$ when $c = 0$, subsuming the first case. □

Note that if (a, b, c) satisfies (3), then $K(a, c) \geq 0$, and it's easy to check that this implies that $a \geq -\sqrt{c}$. However, it is not necessary to write this as a separate condition.

COROLLARY 2. *Suppose p is given by (2). Then p is psd if and only if F_4 is psd, and for all $(r, s) \in \mathbf{R}^2$,*

$$|F_3(r, s)| \leq K(F_2(r, s), F_4(r, s)). \tag{6}$$

We remark, that, even after squaring, (6) is not a "true" illustration of quantifier elimination, because there will still be square roots on the right-hand side.

3. The Gram matrix method

Observe that for polynomials in $f, g \in \mathbf{R}[X] := \mathbf{R}[x_1, \ldots, x_n]$ and for all θ,

$$f^2 + g^2 = (\cos\theta f + \sin\theta g)^2 + (\pm\sin\theta f \mp \cos\theta g)^2. \tag{7}$$

More generally, if $M = [m_{ij}]$ is a real $t \times t$ orthogonal matrix, then

$$\sum_{i=1}^{t}\left(\sum_{j=1}^{t} m_{ij} f_j\right)^2 = \sum_{j=1}^{t}\sum_{k=1}^{t}\left(\sum_{i=1}^{t} m_{ij} m_{ik}\right) f_j f_k = \sum_{j=1}^{t} f_j^2. \tag{8}$$

(Note that (7) includes all real 2×2 orthogonal matrices.) Thus, any attempt to count the number of representations of a form as a sum of squares must mod out the action of the orthogonal group.

Choi, Lam and Reznick [4] have developed a method for studying representations of a form $p \in \mathbf{R}[X]$ as a sum of squares, called the *Gram matrix method.* For $\alpha = (\alpha_1, \ldots, \alpha_n) \in \mathbf{N}^n$, we write $|\alpha|$ to denote $\sum \alpha_i$ and X^α to denote $x_1^{\alpha_1} \cdot \cdots \cdot x_n^{\alpha_n}$. Suppose p is a form in $\mathbf{R}[X]$ which is a sum of squares of forms. Then p must have even degree $2d$ and thus can be written

$$p = \sum_{|\alpha| = 2d} a_\alpha X^\alpha.$$

Suppose now that p has a representation

$$p = h_1^2 + \cdots + h_t^2 \tag{9}$$

where $h_i = \sum_{|\beta|=d} b_\beta^{(i)} X^\beta$. For each $\beta \in \mathbf{N}^n$ of degree d, set $U_\beta = (b_\beta^{(1)}, \ldots, b_\beta^{(t)})$. Then (9) becomes $p = \sum_{\beta,\beta'} U_\beta \cdot U_{\beta'} X^{\beta+\beta'}$. Hence, for each α,

$$a_\alpha = \sum_{\beta+\beta'=\alpha} U_\beta \cdot U_{\beta'}. \tag{10}$$

The matrix $V := [U_\beta \cdot U_{\beta'}]$ (indexed by $\beta \in \mathbf{N}^n$ with $|\beta| = d$) is the *Gram matrix* of p associated to (9). Note that $V = (v_{\beta,\beta'})$ is symmetric, positive semidefinite, and the entries satisfy the equations

$$a_\alpha = \sum_{\beta+\beta'=\alpha} v_{\beta,\beta'}. \tag{11}$$

The following result is proven in [**4**, Thm. 2.4, Prop. 2.10]:

THEOREM 3. *Suppose $p = \sum_{|\alpha|=2d} a_\alpha X^\alpha$ and $V = [v_{\beta,\beta'}]$ is a real symmetric matrix indexed by all $\beta \in \mathbf{N}^n$ such that $|\beta| = d$.*

1. *The following are equivalent: (a) p is a sum of squares of forms and V is the Gram matrix associated to a representation $p = \sum h_i^2$, (b) V is positive semidefinite and the entries of V satisfy the equations* (11).
2. *If V is the Gram matrix of a representation of p as a sum of squares, then the minimum number of squares needed in a representation corresponding to V is the rank of V.*
3. *Two representations of p as a sum of t squares are orthogonally equivalent, as in* (8), *if and only if they have the same Gram matrix.*

We now form the (general) Gram matrix of p by solving the linear system corresponding to the equations (11), where the $v_{\beta,\beta'}$ are variables, with $v_{\beta,\beta'} = v_{\beta',\beta}$. This gives the $v_{\beta,\beta'}$'s as linear polynomials in some parameters. Then $V = [v_{\beta,\beta'}]$ is the Gram matrix of p. By Theorem 3, values of the parameters for which V is psd correspond to representations of p as a sum of squares, with the minimum number of squares needed equal to the rank of V.

If we consider the two sets of vectors of coefficients from the two representations given in (8), we see that one set is the image of the other upon by the action of M, and since M is orthogonal, the dot products of the vectors are unaltered. If p happens to be a quadratic form, then upon arranging the monomials in the usual order, it's easy to see that the (unique) Gram matrix for p is simply the usual matrix representation for p. It follows that a psd quadratic form has, in effect, only one representation as a sum of squares.

Henceforth, when we say that $p \in \mathbf{R}[X]$ is *a sum of t real squares in m ways*, we shall mean that the sums of t squares comprise m distinct orbits under the action of the orthogonal group, or, equivalently, that there are exactly m different psd matrices of rank t which satisfy (11).

Finally, we remark that a real Gram matrix for p of rank t which is *not* psd corresponds to a representation of p as a sum or difference of t squares over $\mathbf{R}$ and that a complex Gram matrix of rank t corresponds to a sum of t squares over $\mathbf{C}$. These facts require relatively simple proofs, but we defer these to a future publication.

4. Hilbert's Theorem and Gram matrices – an introduction

We describe how the Gram matrix method works for ternary quartics. There are 6 monomials in a quadratic form in three variables, and 15 coefficients in the ternary quartic. This means that there are 21 distinct entries in the Gram matrix and 15 equations in (11), and hence the solution to the linear system will have $6 = 21 - 15$ parameters. Thus the Gram matrix of a ternary quartic is 6×6 with entries linear in 6 parameters. If we recall (1), denote the parameters by $\{a, b, c, d, e, f\}$, and write the monomials of degree 2 in the order $x^2, y^2, z^2, xy, xz, yz$, then we find the general form of a Gram matrix of a ternary quartic p:

$$\begin{bmatrix} \alpha_{4,0,0} & a & b & \frac{1}{2}\alpha_{3,1,0} & \frac{1}{2}\alpha_{3,0,1} & d \\ a & \alpha_{0,4,0} & c & \frac{1}{2}\alpha_{1,3,0} & e & \frac{1}{2}\alpha_{0,3,1} \\ b & c & \alpha_{0,0,4} & f & \frac{1}{2}\alpha_{1,0,3} & \frac{1}{2}\alpha_{0,1,3} \\ \frac{1}{2}\alpha_{3,1,0} & \frac{1}{2}\alpha_{1,3,0} & f & \alpha_{2,2,0} - 2a & \frac{1}{2}\alpha_{2,1,1} - d & \frac{1}{2}\alpha_{1,2,1} - e \\ \frac{1}{2}\alpha_{3,0,1} & e & \frac{1}{2}\alpha_{1,0,3} & \frac{1}{2}\alpha_{2,1,1} - d & \alpha_{2,0,2} - 2b & \frac{1}{2}\alpha_{1,1,2} - f \\ d & \frac{1}{2}\alpha_{0,3,1} & \frac{1}{2}\alpha_{0,1,3} & \frac{1}{2}\alpha_{1,2,1} - e & \frac{1}{2}\alpha_{1,1,2} - f & \alpha_{0,2,2} - 2c \end{bmatrix}$$

Hilbert's Theorem together with Theorem 3 says that if p is psd, then for some choice of the parameters $\{a, b, c, d, e, f\}$, this matrix will be psd and have rank 3.

We ignore the psd requirement for the moment and consider the problem of finding choices of parameter for which this Gram matrix has rank 3. For any such matrix, all 4×4 minors will equal zero. There are 225 such minors, although by symmetry there are at most 120 different minors. Each minor is the determinant of a 4×4 matrix with entries linear in the parameters, and hence its vanishing is an equation of degree at most 4 in the 6 parameters.

Thus for a specific ternary quartic p we can form a system of 120 equations of degree at most 4 in 6 variables so that the solutions correspond to rank 3 Gram matrices for p. We can attempt to solve this system, however in almost all cases, the system is much too complicated to solve "by hand". We have made use of a computational tool called RealSolving, which can count the number of solutions, both complex and real, in the case where there are only finitely many complex solutions. For details on RealSolving, see **[7]** and the RealSolving webpage

`www.loria.fr/~rouillie`

EXAMPLE. We consider $p(x, y, z) = x^4 + y^4 + z^4$. The Gram matrix of p is

$$V = V(a, b, c, d, e, f) := \begin{bmatrix} 1 & a & b & 0 & 0 & d \\ a & 1 & c & 0 & e & 0 \\ b & c & 1 & f & 0 & 0 \\ 0 & 0 & f & -2a & -d & -e \\ 0 & e & 0 & -d & -2b & -f \\ d & 0 & 0 & -e & -f & -2c \end{bmatrix}.$$

Since p is psd, Hilbert's Theorem states that it is a sum of three squares; indeed, one such representation is evident. In terms of the Gram matrix, this means that there is a choice of values for the parameters so that $V(a,b,c,d,e,f)$ is psd with rank 3. The obvious representation

$$x^4+y^4+z^4=(x^2)^2+(y^2)^2+(z^2)^2$$

corresponds to $V(0,0,0,0,0,0)$. But p has other representations. In fact, it's easy to see that $V(-1,0,0,0,0,0)$ is also psd with rank 3. If we seek vectors whose dot products are given by this matrix, we are easily led to the following representation:

$$x^4+y^4+z^4=(x^2-y^2)^2+2(xy)^2+(z^2)^2.$$

Clearly two other such representations can be found by cycling the variables: $V(0,-1,0,0,0,0)$ and $V(0,0,-1,0,0,0)$. It turns out that there are four others. One of them is $V(r,r,r,s,s,s)$, with $r=1-\sqrt{2}$ and $s=\sqrt{2}-2$; the three others correspond to the symmetry of p under the sign changes $y\to -y$ and $z\to -z$. (See (15), (16) below.) We will later show how these representations can be derived without using a Gram matrix.

Using RealSolving, for $p=x^4+y^4+z^4$ we have found that there are 15 choices of parameter in which V is a real matrix of rank 3, and 63 choices of parameter in which V is a complex matrix of rank 3. As noted above, the non-psd cases correspond to the representations of p as a sum or *difference* of three real squares or as a sum of three *complex* squares. Thus we know that there are exactly 63 (orthogonally inequivalent) ways to write p as a sum of three squares of forms over $\mathbf{C}$, of which 15 are a sum or difference of three squares over $\mathbf{R}$.

In this case, after "by hand" manipulation of the 120 equations, we can find the following 15 representations of p as a sum or difference of three squares of real quadratic forms:

$$(x^2)^2+(y^2)^2+(z^2)^2 \tag{12}$$

$$(x^2-y^2)^2+2(xy)^2+(z^2)^2 \tag{13}$$

$$(x^2+y^2)^2-2(xy)^2+(z^2)^2 \tag{14}$$

$$\begin{aligned}(x^2+(1-\sqrt{2})(y^2+\sqrt{2}yz+z^2))^2+(\sqrt{2}-1)(x(\sqrt{2}y+z)+yz-z^2)^2\\ +(\sqrt{2}-1)(xz-(y-z)(\sqrt{2}y+z))^2\end{aligned} \tag{15}$$

$$\begin{aligned}(x^2+(1+\sqrt{2})(y^2-\sqrt{2}yz+z^2))^2-(\sqrt{2}+1)(x(-\sqrt{2}y+z)+yz-z^2)^2\\ -(\sqrt{2}+1)(xz-(y-z)(-\sqrt{2}y+z))^2.\end{aligned} \tag{16}$$

These five equations correspond to 15 different representations, because p is both symmetric under permutation of the variables and even in each of the variables. Thus, $p=\sum f_i(x,y,z)^2$ implies that $p=\sum f_i(x,\pm y,\pm z)^2=\sum f_i(x,\pm z,\pm y)^2=$ etc. The "obvious" representation (12) is unaffected by these symmetries. The equations (13) and (14) correspond to three psd and three non-psd representations each, after the cyclic permutation of the variables. It is not obvious, but (15) and (16) are already symmetric in the variables (this shows up in their Gram matrices); however, the substitutions $(y,z)\to(\pm_1 y,\pm_2 z)$ make them correspond to four psd and four non-psd representations respectively.

If we consider p as a sum of three complex quadratic forms, we need to allow the entries of the Gram matrix to be complex. There are 48 non-real Gram matrices of rank 3. We find, for example, that $V(1, i, i, 0, 0, 2i)$ has rank 3, and this gives us a representation of p as a sum of three squares:

$$(x^2 + y^2 + iz^2)^2 + 2(ixy + z^2)^2 - 2i(xz + yz)^2. \tag{17}$$

Since $p(x, y, z) = p(x, i^m y, i^n z)$, a cyclic permutation of the variables gives potentially $3 \times 4^2 = 48$ different sums of squares. However, (17) is symmetric under $z \to -z$, so that it corresponds to only 24 non-real representations. We turn to the real representations of the previous paragraph, and note that (13) and (14) are now equivalent under $y \to iy$. There are also $4^2 - 2^2 = 12$ ways to take $(y, z) \to (i^m y, i^n z)$, with $0 \le m, n \le 3$, where at least one of (m, n) is odd, and 12 non-real representations which correspond to such a substitution into each of (15) and (16), completing the inventory.

Finally, we note that by [**4**, Cor. 2.12], given a psd Gram matrix for p with rank 3, we may assume that x^2 appears only in the first square and xy appears only in the first two squares. Thus, we can view the totality of sums of three squares as inducing a polynomial map from $\mathbf{R}^{15} \to \mathbf{R}^{15}$:

$$\begin{aligned}(b_1x^2 + b_2xy + b_3xz + b_4y^2 + b_5yz + b_6z^2)^2 + \\ (b_7xy + b_8xz + b_9y^2 + b_{10}yz + b_{11}z^2)^2 + (b_{12}xz + b_{13}y^2 + b_{14}yz + b_{15}z^2)^2 \\ = \textstyle\sum_{i+j+k=4} \alpha_{i,j,k}(b_1, \dots, b_{15})x^i y^j z^k.\end{aligned}$$

Hilbert's Theorem, in these terms, is that $\{\alpha_{i,j,k}(\mathbf{R}^{15})\}$ is precisely the set of coefficients of psd ternary quartics. It is not unreasonable to expect that the degree of this mapping would (usually) be finite, but we have not seen this issue discussed in detail in the other proofs of Hilbert's Theorem. We know of no studies of Hilbert's Theorem over $\mathbf{C}$.

We have applied the method of the example to a number of different real ternary quartics. In all cases, we have obtained the values $(63, 15)$ for the number of complex and real solutions, apart from a couple of "degenerate" cases where the numbers are less. Our experiments suggest that the values $(63, 15)$ are generic. We hope to have much more to say about this in a future publication.

5. Some preparatory results on binary forms

We now show how the representations of certain psd ternary quartics as a sum of three squares can be analyzed without using Gram matrices explicitly. This is done by reducing the analysis to certain questions about binary forms.

Suppose $p(t, u)$ is a psd binary form of degree $2d$. An invertible change is now defined by

$$p'(t, u) = p(at + bu, ct + du), \qquad ad \ne bc.$$

By the same reasoning applied to ternary quartics, we may assume that, after an invertible change, $p(1, 0) = 1$, so $p(t, u) = t^{2d} + \cdots$. In any given representation $p = f_1^2 + f_2^2$, we have $f_1(t, u) = at^d + \dots$ and $f_2(t, u) = bt^d + \dots$. Then $a^2 + b^2 = 1$, hence there exists α such that $a = \cos\alpha$ and $b = \sin\alpha$, and we have from (7),

$$\begin{aligned}p(t, u) &= (\cos\theta f_1 + \sin\theta f_2)^2 + (\pm\sin\theta f_1 \mp \cos\theta f_2)^2 \\ &= (\cos(\theta - \alpha)t^d + \dots)^2 + (\pm(\sin(\theta - \alpha)t^d + \dots))^2 \\ &:= f_{1,\theta,\pm}^2(t, u) + f_{2,\theta,\pm}^2(t, u).\end{aligned}$$

We see that, for exactly one value of θ (namely α) and one choice of sign in $\pm$, the coefficients of t^d in $f_{1,\theta,\pm}$ and $f_{2,\theta,\pm}$ are 1 and 0 respectively, and the highest power of t in $f_{2,\theta,\pm}$ has a non-negative coefficient. We will call this a *standard form* for writing p as a sum of two squares; in our terminology, p is a sum of two squares in m ways means that there are exactly m standard forms for p.

Sums of two squares always factor over $\mathbf{C}$: $p = f_1^2 + f_2^2 \implies p = (f_1 + if_2)(f_1 - if_2)$, so the expression of p in standard form as a sum of squares is equivalent to a factorization $p = G_+G_-$ over $\mathbf{C}[t,u]$ as a product of conjugate factors so that $G_\pm(1,0) = f_1(1,0) \pm if_2(1,0) = 1$. Note also that if $p = G_+G_-$, where $G_\pm = f_1 \pm if_2$, then for all θ, $p = (e^{-i\theta}G_+)(e^{i\theta}G_-)$, where

$$e^{\mp i\theta}G_\pm = (\cos\theta f_1 + \sin\theta f_2) \mp i(\sin\theta f_1 - \cos\theta f_2).$$

The linear factors of $p(t,u)$ over $\mathbf{C}[t,u]$ are either real or appear as conjugate pairs, and since the coefficient of t^{2d} in p is 1, we may arrange that the coefficient of t is 1 in each of these factors:

$$p(t,u) = \prod_{j=1}^{q}(t+\lambda_j u)^{m_j}\prod_{k=1}^{r}(t+(\mu_k+i\nu_k)u)^{n_k}\prod_{k=1}^{r}(t+(\mu_k-i\nu_k)u)^{n_k}. \tag{18}$$

Furthermore, since $p \geq 0$, the exponents of the real factors, m_j, must be even.

THEOREM 4. *Suppose $p(t,u)$ is a psd binary form of degree $2d$ with $p(1,0) = 1$, and suppose that p factors over $\mathbf{C}$ as in* (18). *Then p is a sum of two squares in $\left\lceil \frac{1}{2}\prod_{k=1}^{r}(n_k+1)\right\rceil$ ways.*

PROOF. Suppose $p = f_1^2 + f_2^2$ is given in standard form, with $f_1(1,0) = 1$, $f_2(1,0) = 0$. Suppose first that p has the real linear factor $\ell(t,u) = t + \lambda u$. Then $p(\lambda,-1) = 0$ for $j = 1,2$, hence $f_j(\lambda,-1) = 0$ as well, and so ℓ divides both f_1 and f_2. In this way, we can "peel off" all the real linear factors of p, and we may assume without loss of generality that p has only the complex conjugate factors.

As noted above, we consider the possible factorizations of $p = G_+G_-$. Since $G_+ \mid p$, there exist $0 \leq a_k, b_k \leq n_k$ such that

$$G_+(t,u) = \prod_{k=1}^{r}(t+(\mu_k+i\nu_k)u)^{a_k}\prod_{k=1}^{r}(t+(\mu_k-i\nu_k)u)^{b_k}.$$

Taking conjugates, we see that

$$G_-(t,u) = \prod_{k=1}^{r}(t+(\mu_k+i\nu_k)u)^{b_k}\prod_{k=1}^{r}(t+(\mu_k-i\nu_k)u)^{a_k}.$$

Comparison with the factorization of p shows that $a_k + b_k = n_k$, hence $b_k = n_k - a_k$ for all k. There are $N = \prod_{k=1}^{r}(n_k+1)$ ways to choose the a_k's, giving N pairs (G_+, G_-) of complex conjugate factors of p, which in turn define N pairs $(f_1, f_2) = (\frac{1}{2}(G_+ + G_-), \frac{1}{2i}(G_+ - G_-))$. If $G_+ \neq G_-$, then exactly one of the pairs $\{(G_+, G_-), (G_-, G_+)\}$ will leave f_2 in standard form. There is one exceptional case: if all n_k's are even, then taking $a_k = \frac{1}{2}n_k$ gives $G_+ = G_-$, and $f_2 = 0$. This occurs in the case that p is already a square. □

We shall need the following result, though not in its full generality for n variables.

THEOREM 5. *Suppose $p \in \mathbf{R}[X]$ is quartic and can be written as a sum of two squares. If p has no linear factors over $\mathbf{R}[X]$, but factors as a product of linear* *forms over* $\mathbf{C}[X]$, *then* p *is a sum of two squares in 2 ways. Otherwise,* p *is a sum of two squares in 1 way.*

PROOF. Since p is psd, if ℓ is a real linear factor and $\ell \mid p$, then $\ell^2 \mid p$, and if $p = f_1^2 + f_2^2$, then $\ell \mid f_j$. Writing $p = \ell^2 \bar{p}$, $f_j = \ell \bar{f}_j$, we'd have $\bar{p} = \bar{f}_1^2 + \bar{f}_2^2$. Since $\bar{p}$ is quadratic, this means it has rank two, and there is only one way to write it as a sum of two squares (up to (8), as always.)

We now assume that p has no linear factors, $p(x_1, \dots, x_n) = x_1^4 + \dots$ and that a representation $p = f_1^2 + f_2^2$ has $f_1(1, 0, \dots, 0) = 1$ and $f_2(1, 0, \dots, 0) = 0$. Then $p = (f_1 + if_2)(f_1 - if_2)$ factors over $\mathbf{C}[X]$ as a product of conjugate quadratics, and conversely, any such factorization gives p as a sum of two squares. If p has a different standard form representation $p = g_1^2 + g_2^2$, then p has a different factorization $p = (g_1 + ig_2)(g_1 - ig_2)$, with $g_1 \pm ig_2 \neq c(f_1 \pm if_2)$. Let $\ell_1 = \gcd(f_1 + if_2, g_1 + ig_2)$. Then ℓ_1 has to be linear, and we can normalize so that $\ell_1(1, 0, \dots, 0) = 1$. It is now easy to show by unique factorization in $\mathbf{C}[X]$ that there are linear factors ℓ_j so that $\ell_j(1, 0, \dots, 0) = 1$ and

$$f_1 + if_2 = \ell_1 \ell_2, \quad f_1 - if_2 = \ell_3 \ell_4; \qquad g_1 + ig_2 = \ell_1 \ell_3, \quad g_1 - ig_2 = \ell_2 \ell_4$$

It follows that $\ell_4 = \bar{\ell}_1$ and $\ell_3 = \bar{\ell}_2$, so that $p = \ell_1 \bar{\ell}_1 \ell_2 \bar{\ell}_2$, and this implies that the two representations are all that are possible. (It does not matter whether ℓ_1 and ℓ_2 are distinct in this case; in the notation of the last theorem, $2 = \lceil \frac{2+1}{2} \rceil = \lceil \frac{(1+1)(1+1)}{2} \rceil$.) □

Finally, we shall need the following result. It is similar to the classical canonical form for the binary quartic, which is in the literature. However, the classical theorem allows invertible changes in $GL(2, \mathbf{C})$; it is unclear whether our analysis of the real case is in the literature.

THEOREM 6. *If* $p(t, u)$ *is a psd quartic form, then using an invertible change,* $p(t, u)$ *can be put into one of the following shapes:* t^4, t^2u^2, $t^2(t^2 + u^2)$, $(t^2 + u^2)^2$, *or* $t^4 + \lambda t^2 u^2 + u^4$ *with* $|\lambda| < 2$. *The particular shape of* p *depends only on the factorization of* p *over* $\mathbf{C}[t, u]$.

PROOF. Factor p as in (18). If $\sum m_j = 4$, then since the m_j's are even, either $p = \ell_1^4$ or $p = \ell_1^2 \ell_2^2$, where ℓ_1 and ℓ_2 are non-proportional linear forms. Make the invertible change $t' = \ell_1(t, u)$ and $u' = \ell_2(t, u)$ to get the first two cases.

If $\sum m_j = 2$, then $p(t, u) = \ell^2 q(t, u)$, where ℓ is linear and q is a positive definite quadratic. Make a preliminary invertible change so that $\ell = t'$, drop the prime and note that $q(t, u) = at^2 + 2btu + cu^2$, where $c > 0, ac > b^2$. Thus,

$$q(t, u) = (a - \tfrac{b^2}{c})t^2 + c(\tfrac{b}{c}t + u)^2.$$

Writing $d = a - \frac{b^2}{c} > 0$ and $\ell'(t, u) = \frac{b}{c}t + u$, we can make another invertible change so that $\ell' = u'$. This shows that p can be turned into $t^2(dt^2 + u^2)$. By taking $u = \sqrt{d}u'$ and dividing by d, we obtain the third case.

In the last two cases, p has only complex conjugate roots. If they are repeated, then p is the square of a positive definite binary quadratic form, which after an invertible change is $t^2 + u^2$.

Otherwise, we may assume that $p(t, u) = (t^2 + u^2)(at^2 + 2btu + cu^2)$, where the second factor is positive definite. Under an orthogonal change of variables $t = ct' + su', u = -st' + cu'$, where $s = \sin \alpha$, $c = \cos \alpha$, the first factor becomes $(t')^2 + (u')^2$ and the coefficient of $t'u'$ in the second becomes $(a - c) \sin 2\alpha + 2b \cos 2\alpha$.

Thus, we may choose α so that the second factor is also even in t' and u'. (In fact, any two positive definite quadratic forms can be simultaneously diagonalized.) In other words, after an invertible change, we may assume that p is a product of two even positive definite quadratic forms, and after rescaling t and u if necessary, we have $p(t,u) = (t^2+ru^2)(t^2+\frac{1}{r}u^2) = t^4+Rt^2u^2+u^4$, with $R = r+\frac{1}{r} > 2$. A final invertible change gives

$$p(t+u,t-u) = (2+R)\left(t^4+\left(\frac{12-2R}{2+R}\right)t^2u^2+u^4\right),$$

and since $\lambda = \frac{12-2R}{2+R} = -2+\frac{16}{2+R} = 2-\frac{4R-8}{2+R}$, we have $|\lambda| < 2$. $\square$

6. The direct approach to Hilbert's Theorem

Let us now assume Hilbert's Theorem, and write

$$p(x,y,z) = x^4+2x^2F_2(y,z)+2xF_3(y,z)+F_4(y,z) = \sum_{j=1}^{3} f_j^2(x,y,z). \tag{19}$$

As noted earlier, we may assume that the term x^2 appears only in f_1, and up to sign, we may assume that its coefficient is 1. Thus,

$$p(x,y,z) = \Big(x^2+g_{1,1}(y,z)x+g_{2,1}(y,z)\Big)^2+\sum_{j=2}^{3}\Big(g_{1,j}(y,z)x+g_{2,j}(y,z)\Big)^2. \tag{20}$$

Comparing the coefficients of x^3 in (19) and (20), we see that $0 = 2g_{1,1}(y,z)$, hence we may assume that $f_1(x,y,z) = x^2+Q(y,z)$ for a binary quadratic Q and

$$p(x,y,z) = \Big(x^2+Q(y,z)\Big)^2+\sum_{j=2}^{3}\Big(g_{1,j}(y,z)x+g_{2,j}(y,z)\Big)^2. \tag{21}$$

We now exploit the algebraic properties of sums of two squares, in a lemma which will be applied to $p-(x^2+Q)^2$. The basic idea is similar to [**3**, Lemma 7.5].

LEMMA 7. *Suppose*

$$\phi(x,y,z) = h_2(y,z)x^2+2h_3(y,z)x+h_4(y,z)$$

is a quartic form (so that h_k is a form of degree k). Then there exist forms $\psi_{(j)}$ so that $\phi = \psi_{(1)}^2+\psi_{(2)}^2$ if and only if ϕ is psd and the discriminant of ϕ as a quadratic in x,

$$\Delta(y,z) := h_2(y,z)h_4(y,z)-h_3^2(y,z),$$

is the square of a real cubic form.

PROOF. First, if $\phi = \psi_{(1)}^2+\psi_{(2)}^2$, then it is psd and we have

$$\psi_{(j)}(x,y,z) = \lambda_{(j)}(y,z)x+\mu_{(j)}(y,z),$$

hence $h_2 = \lambda_{(1)}^2+\lambda_{(2)}^2$, $h_3 = \lambda_{(1)}\mu_{(1)}+\lambda_{(2)}\mu_{(2)}$, $h_4 = \mu_{(1)}^2+\mu_{(2)}^2$. It follows that

$$\Delta = h_2h_4-h_3^2 = (\lambda_{(1)}\mu_{(2)}-\lambda_{(2)}\mu_{(1)})^2.$$

Conversely, suppose ϕ is psd and Δ is a square. Then $h_2(y,z)$ is a psd quadratic form, so after an invertible change in (y,z), which will affect neither the hypothesis nor the conclusion, we may consider one of three cases: $h_2(y,z) = 0$, $h_2(y,z) = y^2$, $h_2(y,z) = y^2+z^2$.

In the first case, $\Delta = -h_3^2$, so $h_3 = 0$ as well and $\phi(x, y, z) = h_4(y, z)$ is a psd binary quartic. By Theorem 4, $\phi = h_4$ is a sum of two squares.

In the second case, $\Delta(y, z) = y^2 h_4(y, z) - h_3^2(y, z) \geq 0$, hence $\Delta(0, z) = -h_3^2(0, z) \geq 0$, so $h_3(0, z) = 0$. Thus $h_3(y, z) = yk_2(y, z)$ for some quadratic k_2. Further, there exists a cubic form $c_3(y, z)$ so that

$$\Delta(y, z) = y^2(h_4(y, z) - k_2^2(y, z)) = c_3^2(y, z).$$

Thus, $c_3(y, z) = ys_2(y, z)$ for some quadratic s_2. But this means that $h_4 - k_2^2 = s_2^2$, hence

$$\phi(y, z) = x^2y^2 + 2xyk_2(y, z) + k_2^2(y, z) + s_2^2(y, z) = (xy + k_2(y, z))^2 + s_2^2(y, z)$$

is a sum of two squares.

Finally, in the third case, since Δ is a square, there exists real c_3 so that

$$\Delta(y, z) = (y^2 + z^2)h_4(y, z) - h_3^2(y, z) = c_3^2(y, z).$$

It follows that, over $\mathbf{C}[y, z]$,

$$\begin{aligned}(y + iz)(y - iz)h_4(y, z) &= (y^2 + z^2)h_4(y, z) = h_3^2(y, z) + c_3^2(y, z) \\ &= (h_3(y, z) + ic_3(y, z))(h_3(y, z) - ic_3(y, z)).\end{aligned} \tag{22}$$

Thus, up to choice of sign of c_3, $y + iz$ is a factor of $h_3(y, z) + ic_3(y, z)$. Write

$$h_3(y, z) + ic_3(y, z) = (y + iz)(k_2(y, z) + is_2(y, z)). \tag{23}$$

so that

$$h_3(y, z) = yk_2(y, z) - zs_2(y, z), \qquad c_3(y, z) = ys_2(y, z) + zk_2(y, z).$$

Taking conjugates in (23) and substituting into (22), we get

$$h_4(y, z) = (k_2(y, z) + is_2(y, z))(k_2(y, z) - is_2(y, z)) = k_2^2(y, z) + s_2^2(y, z).$$

Thus,

$$\begin{aligned}\phi(y, z) &= x^2(y^2 + z^2) + 2x(yk_2(y, z) - zs_2(y, z)) + k_2^2(y, z) + s_2^2(y, z) \\ &= (xy + k_2(y, z))^2 + (xz - s_2(y, z))^2.\end{aligned}$$

□

This lemma leads to the fundamental constructive theorem of this paper.

THEOREM 8. *If p is a quartic satisfying* (19), *then p can be written as in* (21) *if and only if*

$$p(x, y, z) - (x^2 + Q(y, z))^2 = 2(F_2(y, z) - Q(y, z))x^2 + 2F_3(y, z)x + F_4(y, z) - Q^2(y, z)$$

is psd and

$$\Delta(y, z) = 2(F_2(y, z) - Q(y, z))(F_4(y, z) - Q^2(y, z)) - F_3^2(y, z)$$

is the square of a real cubic form.

Note that for every Q which satisfies the above conditions, $p(x, y, z) - (x^2 + Q(y, z))^2$ is quadratic in x and is a sum of two squares, and hence by Theorem 5 can be written as a sum of two squares in at most two ways. That is, the number of representations of p as a sum of three squares is bounded by twice the number of suitable Q.

Whereas the Gram matrix approach involves a system of polynomial equations in the six parameters $\{a, b, c, d, e, f\}$, the method of Theorem 8 involves three parameters, the coefficients of Q. It is not difficult to set up necessary conditions for a binary sextic to be the square of a cubic form, and when applied to $\Delta = 2(F_2 - Q)(F_4 - Q^2) - F_3^2$, these give a non-trivial system of three equations, although the degree is much higher than that which arises in the Gram matrix approach.

Finally, by comparing Corollary 2 and Theorem 8, we see that Hilbert's Theorem can be reduced entirely to a theorem in binary forms.

COROLLARY 9. *Suppose F_2, F_3, F_4 are binary forms of degree 2, 3, 4 respectively, such that F_4 is psd and*

$$27F_3^2 \leq 4\left(-F_2 + \sqrt{F_2^2 + 3F_4}\right)\left(2F_2 + \sqrt{F_2^2 + 3F_4}\right)^2.$$

Then there exists a binary quadratic Q such that $2(F_2 - Q)(F_4 - Q^2) - F_3^2$ is a perfect square and $F_2 - Q$ and $F_4 - Q^2$ are psd.

We believe that it should be possible to prove Corollary 9 directly. This would provide a purely constructive proof of Hilbert's Theorem. We hope to validate this belief in a future publication.

7. Some constructions

The simplest applications of Theorem 8 occur when $F_3(y,z) = 0$; that is, when p is an even polynomial in x. (Unfortunately, a constant-counting argument which we omit shows that not every real ternary quartic can be put in this form after an invertible change.) We revisit Theorem 8 in this special case:

COROLLARY 10. *There is a representation*

$$x^4 + 2F_2(y,z)x^2 + F_4(y,z) = (x^2 + Q(y,z))^2 + \sum_{j=2}^{3} f_j^2(x,y,z) \tag{24}$$

if and only if one of the following four cases holds:

- **(a):** *$F_4 - F_2^2$ is psd and $Q = F_2$.*
- **(b):** *$F_4 = k_2^2$ is a square, $Q = \pm k_2$ and $F_2 \mp k_2$ is psd.*
- **(c):** *There is a linear form ℓ so that $Q = F_2 - \ell^2$, and $F_4 - (F_2 - \ell^2)^2$ is a square.*
- **(d):** *There is a linear form ℓ so that $F_4 - Q^2 = \ell^2(F_2 - Q)$ and $F_2 - Q$ is psd. (In this case, $F_2 - Q$ is a factor of $F_4 - F_2^2$.)*

PROOF. By Theorem 8, the necessary and sufficient conditions are that

$$2(F_2(y,z) - Q(y,z))x^2 + F_4(y,z) - Q^2(y,z)$$

be psd, and that

$$\Delta(y,z) = (F_2(y,z) - Q(y,z))(F_4(y,z) - Q^2(y,z)) \tag{25}$$

is the square of a real cubic form. The first condition is equivalent to $F_2 - Q$ and $F_4 - Q^2$ both being psd. We now turn to the second condition.

If the first factor in (25) is 0, then $\Delta = 0$ is trivially a square, and $Q = F_2$. Thus, the remaining condition is that $F_4 - F_2^2$ be psd. This is (a).

If the second factor in (25) is 0, then again Δ is trivially a square and $Q^2 = F_4$. Suppose $F_4 = k_2^2$, then $Q = \pm k_2$, and the remaining condition is that $F_2 - Q = F_2 \mp k_2$ be psd and we obtain case (b).

In the remaining two cases, we have a quadratic $q_2 = F_2 - Q$ and a quartic $q_4 = F_4 - Q^2$ whose product is a square. If q_2 and q_4 are relatively prime, then each must be a square. Thus, $F_2 - Q = \ell^2$ for some linear form ℓ, and $F_4 - Q^2 = s_2^2$ is a square. This is (c).

Finally, if $\gcd(q_2, q_4) = g$, then $q_2 = gu$ and $q_4 = gv$, with u and u relatively prime, so that $q_2 q_4 = g^2 uv$ is a square. This implies that u and v are squares, so that g has even degree. This last case is that g is quadratic, so we may take $g = q_2$ and write $v = \ell^2$ for a linear form ℓ; that is, $F_4 - Q^2 = (F_2 - Q)\ell^2$. Note that this implies that $(F_2 - Q)(\ell^2 - F_2 - Q) = F_4 - F_2^2$. Thus any Q which satisfies this condition will have the additional property that $F_2 - Q$ is a psd factor of $F_4 - F_2^2$. □

REMARK. We can use Corollary 10 to count the number of possible representations as a sum of three squares of $x^4 + 2F_2(y,z)x^2 + F_4(y,z)$. If (a) holds, then

$$p(x,y,z) = (x^2 + F_2(y,z))^2 + (F_4(y,z) - F_2^2(y,z)),$$

and the second summand above is a sum of two squares by Theorem 4, in one or two ways, depending on whether $F_4 - F_2^2$ has linear factors. (It may also happen to be a square: $q^2 + 0^2$ can be viewed as a sum of two squares.)

In case (b), the condition that $F_2 - Q = F_2 \mp k_2$ is psd may be true for zero, one or two choices of sign. If it is true, we have

$$p(x,y,z) = (x^2 + Q(y,z))^2 + 2x^2(F_2(y,z) - Q(y,z)),$$

If $F_2 - Q$ is psd, it is a sum of two squares (in exactly one way) by Theorem 4.

If (c) holds, then

$$p(x,y,z) = (x^2 + F_2(y,z) - \ell^2(y,z))^2 + 2\ell^2(y,z)x^2 + s_2(y,z)^2$$

is, as written, a sum of three squares. Furthermore, although $2\ell^2(y,z)x^2 + s_2(y,z)^2$ factors into quadratic forms over $\mathbf{C}[y,z]$, it does not factor into linear forms unless $\ell \mid s_2$, and so the sum of three squares is unique except in this case. It is not *a priori* clear how many different linear forms ℓ satisfy these conditions for a given pair (F_2, F_4).

Finally, in case (d),

$$p(x,y,z) = (x^2 + Q(y,z))^2 + 2(F_2(y,z) - Q(y,z))(x^2 + \ell^2(y,z)).$$

Since $F_2 - Q$ is a psd binary form, it splits into linear factors over $\mathbf{C}[y,z]$, and so any suitable Q leads to two representations of p as a sum of three squares. Again, it is not *a priori* clear how many such forms Q exist for given (F_2, F_4).

We conclude this section with some simple examples.

EXAMPLE. The psd quartic

$$p(x,y,z) = (x^2 + y^2)(x^2 + z^2) = x^4 + x^2(y^2 + z^2) + y^2 z^2$$

is a product of two sums of two squares and hence is a sum of two squares in two different ways. Are there other ways to write p as a sum of three squares? Using Theorem 8, if one of the squares is $x^2 + Q(y,z)$, then $F_2 - Q$ and $F_4 - Q^2$ must be psd. If $y^2 z^2 - Q^2(y,z)$ is psd, then $Q(y,z) = \alpha yz$ with $|\alpha| \le 1$ and $F_2 - Q = \frac{1}{2}(y^2 - 2\alpha yz + z^2)$ is psd. But

$$\Delta(y,z) = \tfrac{1-\alpha^2}{2}(y^2 - 2\alpha yz + z^2)y^2 z^2$$

will be a perfect square only when $\alpha = \pm 1$. This re-derives the familiar representations from the two-square identity:

$$p(x, y, z) = (x^2 - yz)^2 + x^2(y + z)^2 = (x^2 + yz)^2 + x^2(y - z)^2$$

EXAMPLE. The similar-looking psd quartic

$$p(x, y, z) = x^4 + x^2y^2 + y^2z^2 + z^4$$

is irreducible, and so is not a sum of two squares. It is not trivial to write p as a sum of three squares, so we apply the algorithm.

Here, $F_2(y, z) = \frac{1}{2}y^2$ and $F_4(y, z) = z^2(y^2 + z^2)$. If $F_4 - Q^2$ is psd then $z \mid Q$, so $Q(y, z) = ayz + bz^2$ for some (a, b). It is easily checked that $F_4 - Q^2$ is psd if and only if $a^2 + b^2 \le 1$ and it's a square, $z^2(by - az)^2$, if and only if $a^2 + b^2 = 1$. And $F_2 - Q$ is psd if and only if $a^2 + 2b \le 0$, and it's a square, $(y - \frac{1}{2}az)^2$, if and only if $b = -\frac{1}{2}a^2$.

Running through the cases, we see that (a) and (b) are not possible, because Q cannot equal F_2 and F_4 is not a square. For (c), $F_2 - Q$ and $F_4 - Q^2$ are both squares when $b = -\frac{1}{2}a^2$ and $a^2 + b^2 = 1$, which implies that

$$a = \pm\tau := \pm\sqrt{2\sqrt{2} - 2}, \qquad b = 1 - \sqrt{2}.$$

This gives the representation

$$p(x, y, z) = (x^2 \pm \tau yz + (1 - \sqrt{2})z^2)^2 + x^2(y \mp \tau z)^2 + z^2((\sqrt{2} - 1)y \pm \tau z)^2.$$

The sum of the last two squares does not split over $\mathbf{C}[y, z]$, so there are no additional representations in this case. In (d), $\frac{1}{2}y^2 - Q(y, z) = \frac{1}{2}(y^2 - 2ayz - 2bz^2)$ must be a psd factor of

$$F_4 - F_2^2 = z^4 + z^2y^2 - \tfrac{1}{4}y^4 = \left(z^2 + \tfrac{1-\sqrt{2}}{2}y^2\right)\left(z^2 + \tfrac{1+\sqrt{2}}{2}y^2\right).$$

Thus, it is a multiple of $z^2 + \frac{1+\sqrt{2}}{2}y^2$, and $a = 0$, $b = 1 - \sqrt{2}$. This leads to

$$p(x, y, z) = (x^2 + (1 - \sqrt{2})z^2)^2 + (x^2 + z^2)(y^2 + (2\sqrt{2} - 2)z^2);$$

since the last sum of two squares splits into linear factors over $\mathbf{C}$, there are two more representations of p as a sum of two squares, making four in all.

EXAMPLE. We consider the class of quartics: $p(x, y, z) = (x^2 + G(y, z))^2$, so that $F_2(y, z) = 2G(y, z)$ and $F_4(y, z) = G^2(y, z)$. By Corollary 10, p is a sum of three squares as in (24) if and only if $2(G - Q)$ and $G^2 - Q^2$ are both psd and $(G - Q)(G^2 - Q^2) = (G - Q)^2(G + Q)$ is a square. If $G = Q$, then these conditions are satisfied immediately, and of course, we recover the representation of p as a single square. If $G = -Q$, then we get another representation, provided G is psd:

$$(x^2 + G(y, z))^2 = (x^2 - G(y, z))^2 + 4x^2G(y, z).$$

Since G is a quadratic form, this gives p as a sum of two squares if $G = \ell^2$ and a sum of three squares if G is positive definite. Otherwise, we must have that $G - Q$ is psd and $G + Q$ is a square. This means that $G(y, z) \ge |Q(y, z)|$ for all (y, z), and hence G is psd. Thus $Q(y, z)$ can be $-(G(y, z) - (ay + bz)^2)$ for any (a, b) for which $2G(y, z) - (ay + bz)^2$ is psd.

If G has rank 1, then after an invertible change, $G(y,z) = y^2$, and $Q(y,z) = (1-a^2)y^2$, so that $(G+Q)(y,z) = (2-a^2)y^2 \geq 0$; that is, $Q(y,z) = -\lambda y^2$, with $-1 \leq \lambda \leq 1$. This gives an infinite family of representations:

$$(x^2+y^2)^2 = (x^2-\lambda y^2)^2 + (2+2\lambda)x^2y^2 + (1-\lambda^2)y^4.$$

If G has rank 2, then after an invertible change, $G(y,z) = y^2 + z^2$, and $G \geq |Q|$ if and only if $a^2 + b^2 \leq 2$. This gives a doubly infinite family of representations:

$$(x^2+y^2+z^2)^2 = (x^2-(y^2+z^2-(ay+bz)^2))^2+(2(y^2+z^2)-(ay+bz)^2)(2x^2+(ay+bz)^2).$$

If G is not psd, then p has only the trivial representation. This also can be seen directly: since $x^2 + G(y,z)$ is indefinite, in any representation $p = \sum f_j^2$, f_j must be a multiple of $x^2 + G(y,z)$; by degrees, it must be a scalar multiple. Thus any representation of p as a sum of squares is orthogonally equivalent to the trivial one.

8. A complete answer in a special case

We now simplify further still, by supposing that $F_2(y,z) = 0$ as well, so that

$$p(x,y,z) = x^4 + F_4(y,z),$$

where F_4 is a psd quartic form. Hilbert's Theorem is no mystery in this special case, because we already know that F_4 can be written as a sum of two squares, and this gives one way to write p as a sum of three squares. Are there any other representations? Note that necessary conditions on Q include that $-Q$ and $F_4 - Q^2$ are both psd.

There are five cases, based on the factorization of F_4; we shall need two lemmas about real binary forms.

LEMMA 11. *Suppose $F(y,z)$ is a positive definite quartic form, and consider the equation*

$$F(y,z) - (ay+bz)^4 = q^2(y,z) \tag{26}$$

for linear forms $ay+bz$ and quadratic forms q. If F is a square, then (26) has only the trivial solution $(a,b) = (0,0)$. If F is not a square, then there are two different q's for which (26) holds.

PROOF. By Theorem 6, we may assume that $F(y,z) = y^4 + \lambda y^2z^2 + z^4$ and that $-2 < \lambda \leq 2$. There are two trivial solutions to (26):

$$\begin{aligned} y^4 + \lambda y^2z^2 + z^4 - (1-\tfrac{\lambda^2}{4})z^4 &= (y^2 + \tfrac{\lambda}{2}z^2)^2, \\ y^4 + \lambda y^2z^2 + z^4 - (1-\tfrac{\lambda^2}{4})y^4 &= (\tfrac{\lambda}{2}y^2 + z^2)^2. \end{aligned} \tag{27}$$

If $\lambda = 2$, these are truly trivial! It is easy to see that these are the only possible expressions in which $a = 0$ or $b = 0$. For other solutions, assume $ab \neq 0$, and set up the five equations for the coefficients of $F(y,z) - (ay+bz)^4 = (ry^2+syz+tz^2)^2$:

$$1-a^4 = r^2, \quad -4a^3b = 2rs, \quad \lambda - 6a^2b^2 = 2rt + s^2, \quad -4ab^3 = 2st, \quad 1-b^4 = t^2.$$

Since $4r^2s^2t^2 = r^2(2st)^2 = t^2(2rs)^2$, we have

$$(1-a^4)(-4ab^3)^2 = (1-b^4)(-4a^3b)^2 \implies a^2b^6 - a^6b^6 = a^6b^2 - a^6b^6.$$

Since $ab \neq 0$ it follows that $a^4 = b^4$, so $a^2 = b^2$ and so $rs = st$. If $s = 0$, then $ab = 0$, which is impossible, so we conclude that $r = t$. But then

$$s^2 = s^2 + 2rt - 2r^2 = (\lambda - 6a^2b^2) - 2(1-a^4) = \lambda - 2 - 4a^4 < 0,$$

which is a contradiction. Thus, (27) gives the only solutions to (26). □

LEMMA 12. *If $F(y,z)$ and $G(y,z)$ are non-proportional positive definite quadratic forms, then there is a unique positive number μ_0 such that $F - \mu_0 G$ is the non-zero square of a linear form.*

PROOF. Since F and G are both positive definite, the following minimum is well-defined; it is positive, and achieved for $\theta = \theta_0$:

$$\mu_0 = \min_{0\le\theta\le 2\pi} \frac{F(\cos\theta, \sin\theta)}{G(\cos\theta, \sin\theta)}.$$

Let $H_\mu(y,z) = F(y,z) - \mu G(y,z)$. Then H_{μ_0} is psd and $H_{\mu_0}(\cos\theta_0, \sin\theta_0) = 0$, and as F and G are not proportional, H_{μ_0} is not identically zero. Thus H_{μ_0} is the non-zero square of a linear form. If $\mu < \mu_0$, then H_μ is positive definite, and so is not a square; if $\mu > \mu_0$, then $H_\mu(\cos\theta_0, \sin\theta_0) < 0$, so H_μ is not even psd. □

If $|\lambda| < 2$, then $\lambda = 2 - \nu^2$ with $0 < \nu < 2$, so

$$y^4 + \lambda y^2 z^2 + z^4 = (y^2 + \nu yz + z^2)(y^2 - \nu yz + z^2)$$

is a product of two positive definite quadratics. In this case, the computation of μ_0 is extremely easy: the minimum occurs at the extreme value of $\cos\theta\sin\theta$, namely, $\pm\frac{1}{2}$ and

$$\mu_0 = \min_{0\le\theta\le 2\pi} \frac{1 + \nu\cos\theta\sin\theta}{1 - \nu\cos\theta\sin\theta} = \frac{1 - \frac{\nu}{2}}{1 + \frac{\nu}{2}}.$$

In this case, note that

$$(y^2 \pm \nu yz + z^2) - \left(\frac{2-\nu}{2+\nu}\right)(y^2 \mp \nu yz + z^2) = \left(\frac{2\nu}{2+\nu}\right)(y \pm z)^2.$$

COROLLARY 13. *Suppose $p(x,y,z) = x^4 + F_4(y,z)$ is psd. The one of the following holds:*

1. *$F_4 = \ell^4$ for some linear form ℓ, and p is a sum of three squares in infinitely many ways.*
2. *$F_4 = \ell_1^2\ell_2^2$ for non-proportional linear forms ℓ_1 and ℓ_2, and p is a sum of three squares in exactly one way.*
3. *$F_4 = \ell^2 k_2$, where k_2 is positive definite, and p is a sum of three squares in exactly two ways.*
4. *$F_4 = k_2^2$, where k_2 is positive definite, and p is a sum of three squares in exactly three ways.*
5. *$F_4 = k_2 q_2$, where k_2 and q_2 are positive definite and not proportional, and p is a sum of three squares in exactly eight ways.*

PROOF. Throughout, we shall use the classification of Theorem 6 as the first step in the proof.
1. We assume that $\ell(y,z) = y$. We must have that $-Q(y,z)$ and $y^4 - Q^2(y,z)$ are psd. The second condition implies that $Q(y,z) = \alpha y^4$ with $1 \ge \alpha^2$, and the first implies that $\alpha < 0$. In this case $\Delta = -\alpha(1-\alpha^2)y^6$ is always a square and, writing $\alpha = -\beta^2$, $0 \le \beta \le 1$ we have

$$x^4 + y^4 = (x^2 - \beta^2 y^2)^2 + 2\beta^2 x^2 y^2 + (1-\beta^4)y^4.$$

The distinct values of β give orthogonally distinct different representations of p as a sum of three squares. This can't be too surprising, because p is obviously a sum of two squares. However, the next case gives another sum of two squares which has no additional representations as a sum of three squares.

2. In this case, $\ell_1(y,z) = y$ and $\ell_2(y,z) = z$. We must have that $-Q(y,z)$ and $y^2z^2 - Q^2(y,z)$ are psd. The second condition implies that $yz \mid Q$, hence $Q(y,z) = \alpha yz$. But the first condition then implies that $\alpha = 0$, so $Q = 0$ and we have

$$x^4 + y^2z^2 = (x^2)^2 + \sum_{j=2}^{3} f_j^2(x,y,z).$$

But this implies that $f_j(y,z) = \alpha_j yz$ and $1 = \alpha_2^2 + \alpha_3^2$; these are all orthogonally equivalent to $(yz)^2 + 0^2$. So the only representations of p as a sum of three squares are orthogonally equivalent to those as a sum of two squares. In fact, the psd Gram matrices for p have no parameters, and p has, up to orthogonal equivalence, a unique representation as a sum of squares.

3. In this case, we assume that $F_4(y,z) = y^2(y^2+z^2)$. The condition that $F_4 - Q^2$ is psd implies that $y \mid Q$, and the condition that $-Q$ is psd implies that $Q(y,z) = -\alpha y^2$, with $\alpha \geq 0$, so now $F_4(y,z) - Q^2(y,z) = y^2((1-\alpha^2)y^2 + z^2)$, hence $0 \leq \alpha \leq 1$. Finally, the condition that $\Delta = \alpha y^4((1-\alpha^2)y^2+z^2)$ be a square implies that $\alpha = 0$ or $\alpha = 1$. In the first case, we have

$$x^4 + y^2(y^2+z^2) = (x^2)^2 + \sum_{j=2}^{3} f_j^2(x,y,z).$$

There is by Theorem 4 exactly one way to write $y^2(y^2+z^2)$ as a sum of two squares, $(y^2)^2 + (yz)^2$. In the second case, we have

$$x^4 + y^2(y^2+z^2) = (x^2-y^2)^2 + \sum_{j=2}^{3} f_j^2(x,y,z) \implies 2x^2y^2 + y^2z^2 = \sum_{j=2}^{3} f_j^2(x,y,z).$$

By Theorem 5, there is also just one way to write $y^2(2x^2+z^2)$ as a sum of two squares, $2(xy)^2 + (yz)^2$, so altogether there are two ways to write p as a sum of three squares.

4. We assume that $k_2(y,z) = y^2 + z^2$. We now run through the four cases in Corollary 10. In case (a), we have $Q = 0$, and

$$x^4 + (y^2+z^2)^2 = (x^2)^2 + \sum_{j=2}^{3} f_j^2(x,y,z).$$

We know from Theorem 4 that there are two inequivalent choices for (f_2, f_3). These are easy to compute by hand and give

$$x^4 + (y^2+z^2)^2 = (x^2)^2 + (y^2+z^2)^2 + 0^2 = (x^2)^2 + (y^2-z^2)^2 + (2yz)^2.$$

In case (b), $Q(y,z) = \pm(y^2+z^2)$ and $-Q$ is psd, so $Q(y,z) = -(y^2+z^2)$ and

$$x^4 + (y^2+z^2)^2 = (x^2 - (y^2+z^2))^2 + \sum_{j=2}^{3} f_j^2(x,y,z).$$

This implies that $2x^2(y^2+z^2) = \sum_{j=2}^{3} f_j^2(x,y)$, and, as before, Theorem 5 implies that there is a unique representation:

$$x^4 + (y^2+z^2)^2 = (x^2 - (y^2+z^2))^2 + 2(xy)^2 + 2(xz)^2.$$

In case (c), we have that $Q(y,z) = -(ay+bz)^2$ and $(y^2+z^2)^2 - (ay+bz)^4$ is a square. We have seen in Lemma 11 that this is impossible. Finally, in case (d), $-Q$ is a psd factor of $F_4 - F_2^2 = (y^2+z^2)^2$, hence $Q(y,z) = -\alpha(y^2+z^2)$ for some $\alpha > 0$. This implies that $\Delta(y,z) = \alpha(1-\alpha^2)(y^2+z^2)^3$, which is only a square for $\alpha = 0, 1$, which have been already discussed. Altogether, there are only three representations.

5. We write $F_4(y,z) = y^4 + \lambda y^2 z^2 + z^4$, with $|\lambda| < 2$ and, as before, write $\lambda = 2 - \nu^2$, with $0 < \nu < 2$. In case (a), $Q = 0$, and as in the last case, F_4 is a sum of two squares in two ways:

$$y^4 + \lambda y^2 z^2 + z^4 = (y^2 + \tfrac{\lambda}{2} z^2)^2 + (1 - \tfrac{\lambda^2}{4}) z^4 = (y^2 - z^2)^2 + (2+\lambda)(yz)^2.$$

This gives two ways to write $x^4 + F_4(y,z)$ as a sum of three squares.

Case (b) does not apply, since F_4 is not a square. In case (c), $Q = -\ell^2$, and $F_4 - \ell^4 = s_2^2$ is a square. By Lemma 11, there are two different choices of (ℓ^2, s^2) for which this is the case. For simplicity, let $\rho = \sqrt{1 - \frac{\lambda^2}{4}}$. These give the representations

$$x^4 + y^4 + \lambda y^2 z^2 + z^4 = (x^2 - \rho y^2)^2 + 2\rho x^2 y^2 + (\tfrac{\lambda}{2} y^2 + z^2)^2,$$

and a similar one, with y and z permuted. Note that the factors of the two summands are $\sqrt{2\rho}xy \pm i(\frac{\lambda}{2}y^2 + z^2)$ which are irreducible over $\mathbf{C}$. Thus there is only one representation of p as a sum of three squares for each $Q = -\ell^2$, and so two in all.

Finally, in case (d), we have that $-Q$ is a psd factor of

$$F_4(y,z) = (y^2 + \nu yz + z^2)(y^2 - \nu yz + z^2)$$

Thus $Q = \kappa(y^2 \pm \nu yz + z^2)$ for some choice of sign. In this case

$$F_4(y,z) - Q^2(y,z) = \tfrac{1}{\kappa} Q(y,z)\big((y^2 - \mp\nu yz + z^2) - \kappa^2((y^2 \pm \nu yz + z^2)\big).$$

By Lemma 12, the last factor is a square if and only if $\kappa^2 = \frac{2-\nu}{2+\nu}$. In this case, we have

$$x^4+y^4+\lambda y^2z^2+z^4 = (x^2-\kappa(y^2\pm\nu yz+z^2))^2+(y^2\pm\nu yz+z^2)(2\kappa x^2+(1-\kappa^2)(y\mp z)^2)$$

Since the sum of these last two squares splits over $\mathbf{C}$, we get four different representations of p as a sum of three squares altogether, so there are four from case (d) and eight in all. □

EXAMPLE. We illustrate the eight representations of $p(x,y,z) = x^4 + y^4 + z^4$ as a sum of three squares of real quadratic forms. In this case, $\lambda = 0$, $\rho = 1$, $\nu = \sqrt{2}$ and $\kappa = \sqrt{\frac{2-\sqrt{2}}{2+\sqrt{2}}} = \sqrt{2} - 1$, so that $1 - \kappa^2 = 2\kappa$. The two from case (a) are

$$p(x,y,z) = (x^2)^2 + (y^2)^2 + (z^2)^2 = (x^2)^2 + (y^2 - z^2)^2 + 2(yz)^2.$$

That is, (12) and one of (13). The two cases from (c) become the other two from (13).

$$p(x,y,z) = (x^2 - y^2)^2 + 2(xy)^2 + (z^2)^2 = (x^2 - z^2)^2 + 2(xz)^2 + (y^2)^2.$$

Finally, from case (d), we get four representations from

$$(x^2 - (\sqrt{2}-1)(y^2 \pm \sqrt{2}yz + z^2))^2 + 2(\sqrt{2}-1)(y^2 \pm \sqrt{2}yz + z^2)(x^2 + (y \mp z)^2).$$

The two-square identity then gives (15).

References

[1] J. Bochnak, M. Coste, and M.-F. Roy, *Real Algebraic Geometry*, Springer-Verlag, Berlin, 1998.

[2] M.D. Choi and T.Y. Lam, *Extremal positive semidefinite forms*, Math. Ann. **231** (1977), 1-26.

[3] M.D. Choi, T.Y. Lam, and B. Reznick, *Real zeros of positive semidefinite forms I*, Math. Z. **171** (1980), 1-25.

[4] M.D. Choi, T.Y. Lam, and B. Reznick, *Sums of squares of real polynomials*, Symp. in Pure Math. **58** (1995), 103–126, Amer. Math. Soc., Providence, R.I.

[5] D. Hilbert, *Über die Darstellung definiter Formen als Summe von Formenquadraten*, Math. Ann. **32** (1888), 342-350.

[6] A.R. Rajwade, *Squares*, London Mathematical Society Lecture Notes **171**, Cambridge University Press, Cambridge, 1993.

[7] F. Rouillier, *Algorithmes efficaces pour l'étude des zéros réels des systèmes polynomiaux*, PhD. thesis, Université Rennes, France, 1997.

[8] R.G. Swan, *Hilbert's Theorem on positive ternary quartics*, these proceedings.

Department of Mathematics and Computer Science
Emory University
Atlanta, GA 30322
E-mail address: vicki@mathcs.emory.edu

Department of Mathematics
University of Illinois
Urbana, IL 61801
E-mail address: reznick@math.uiuc.edu

Contemporary Mathematics
Volume **272**, 2000

On the History of the Algebraic Theory of Quadratic Forms

Winfried Scharlau

The purpose of these remarks is not to give a complete and systematic history of the algebraic theory of quadratic forms. Rather, I shall concentrate on some aspects neglected in the existing literature, and others on which I can report from personal experience. References to the subject are Pfister's talk (1984) at the Hamilton conference, his paper (1990), and his survey (2000). A thorough discussion of Milnor's algebraic work can be found in Bass (1993); for the history of algebraic K-theory I refer to Weibel (1999). I am grateful to many colleagues for valuable information and useful discussions of special topics; I particularly want to mention M. Kneser, A. Pfister and M. Knebusch. I want to thank the referee for a critical reading of the paper and many corrections and helpful comments.

This paper is dedicated to the founder of the algebraic theory of quadratic forms, Albrecht Pfister, at the occasion of his 65th birthday.

1. A quick summary

In this section I try to give an overview of some of the most significant developments in the theory of quadratic forms since Witt's paper (1937). It is not possible to describe all important work and I apologize to all colleagues whose contributions will not be mentioned appropriately. (A number of names come to my mind immediately.) In the following sections I shall discuss selected topics in some detail.

1.1. Functorial properties. After the appearance of Witt's paper it was a natural task to study the functorial properties of the Witt ring and its relation to other functorial constructions like the Brauer group, Galois cohomology or K-theory. Already Witt himself did the first steps in this direction: he defines for every quadratic form its Clifford algebra and studies the functorial properties of this construction. He mentions (p. 35) the

Key words and phrases. quadratic form.

analogies between quadratic forms and simple algebras.[1] Springer (1959) and Serre (1964) pointed out the relation to the Galois cohomology of the orthogonal and related groups. Delzant (1962), Scharlau (1967), Belskii (1968), Milnor (1970), Arason (1975) and later many others investigated quadratic forms and (abelian) Galois cohomology. This development culminated in the proof of the Milnor conjecture (see Pfister (2000) for a more thorough discussion).

Durfee (1948) and Springer (1955) developed the theory for quadratic forms over a discrete valuation ring.[2] In particular, they classified quadratic forms over complete discrete valuation rings and their quotient field if the residue characteristic is not 2. Knebusch (1976, 1977) extended this to arbitrary places. Several authors investigated fields with an arbitrary henselian valuation (and residue characteristic not 2).

The behaviour of quadratic forms and Witt rings under ground field extensions was studied by Springer (1952), Scharlau (1969a, 1970), Rosenberg-Ware (1970), Knebusch-Scharlau (1971), Elman-Lam (1976), where in particular the case of quadratic extensions is clarified, which is crucial for many applications. As an application of these techniques, norm principles were proved by Scharlau (1969) and Knebusch (1971). Dress (1971) pointed out close relations between the Burnside ring of monomial representations and the Witt ring, the trace defining a natural functor. Milnor (1970) invented Milnor K-theory to formalize similar descriptions of the functors W and K_2 by generators and relations.

1.2. Structure theorems. The basic structure theorems for the Witt ring were proved by Pfister (1966) and Arason-Pfister (1971). Some complements were contributed by Witt, Harrison (1970) and Lorenz-Leicht (1970). Several authors developed alternative approaches to the structure theorems, e.g. Scharlau (1970), Lorenz-Leicht (1970), Dress (1971), Knebusch-Rosenberg-Ware (1972), and Lewis (1989). The structure of the graded Witt ring remained the most important problem of the whole theory for some time. After very important initial results of Arason (1975) (and some others) the first significant breakthrough is due to Merkurjev (1981). This was followed by work of Merkurjev-Suslin, Rost, Jacob-Rost and Szyjewski. The complete solution, i.e. the proof of the Milnor conjecture, was announced by Voevodsky and Orlov-Vishik-Voevodsky.

Since the Milnor conjecture is perhaps the most important single result in the theory of quadratic forms we briefly state what it claims: for every field F of characteristic not 2 and all n, three abelian groups are canonically isomorphic, namely

$$I^n(F)/I^{n+1}(F),\ \ H^n(F,\mathbb{Z}/2\mathbb{Z}),\ \ k_n(F) = K_n(F)/2K_n(F).$$

The groups are defined via quadratic form theory, Galois cohomology, and algebraic K-theory, respectively. $I(F)$ denotes the "fundamental ideal" of

[1]The footnotes can be found at the end of the paper.

even dimensional forms in the Witt ring $W(F)$, and $K_n(F)$ is the n-the Milnor K-group of F. - Presently, many details of the proof have not yet been published, not even in preliminary form. We refer again to Pfister (2000) for more details.

1.3. The Hasse principle. The Hasse principle (or local-global principle), originally discovered by Minkowski (1890) and Hasse (1923, 1924) in the classification of quadratic forms over algebraic number fields but extending far beyond this theory (see Colliot-Thélène (1992)), is a powerful tool which allows to reduce problems from algebraic number fields to the corresponding questions over the p-adic completions. It was systematically employed by Hasse to prove central results of algebraic number theory and class field theory. Landherr (1938) extended Hasse's results on quadratic forms to various kinds of hermitian forms. In the fifties it was recognized (Weil, Serre) that hermitian forms and similar objects can be described by Galois 1-cocycles of their automorphism groups. This led to the problem of the validity of the Hasse principle for arbitrary algebraic groups. It was also recognized that this depends essentially on the cohomological dimension of the ground field (see Serre (1964)). Kneser (1969), Harder and others obtained important results for semisimple simply connected linear algebraic groups. It is impossible to quote all the relevant work on this problem. However, after the proof of the Merkurjev-Suslin theorem further progress was possible concerning the classical groups (i.e. hermitian forms of various kinds; see Bayer-Fluckiger, Parimala (1995, 1998)). Also Scheiderer (1996) completely settled the case of virtual cohomological dimension 1 (where one has cohomological dimension 1 "except for real places"). Concerning in particular the Witt group, we want to mention also Sujatha (1995) and Parimala-Sujatha (1996) for results on the Hasse principle for algebraic curves over global fields. Concerning higher Galois cohomology groups, Hasse principles have been proved by Kato (1986) and Jannsen (1989/90 and unpublished). This has important applications to the Pythagoras numbers of algebraic function fields over algebraic number fields (Colliot-Thélène, Jannsen (1991)).

1.4. Generalizations. Arf (1941) extended Witt's results to fields of characteristic 2. Later, Baeza published several papers which extended known results to the characteristic 2 case. A systematic treatment was given by Milnor (1971) and Sah (1972). Surprisingly, the proof of the Milnor conjecture in characteristic 2 turned out to be easier than in the general case (Sah (1972), Kato (1982)). In principle, some aspects of the theory in characteristic 2 are more interesting and richer because one has to distinguish between symmetric bilinear and quadratic forms and inseparable extensions have to be considered. Nevertheless, fields of characteristic 2 have remained the pariahs of the theory.

Long before the emergence of the algebraic theory, quadratic forms were studied over rings, e.g. over $\mathbb{Z}$ or other rings of number theory. After the basic ideas of K-theory were formulated in the late fifties, it was clear that a systematic theory of quadratic forms over rings (and schemes) could and should be developed. Serre (1961/62) suggested to do this; Kneser (unpublished (1962)) worked out essential parts of this theory over Dedekind rings. Several authors considered the case of local rings and valuation rings; a comprehensive reference is Baeza (1978) where additional references are given. A systematic development of the theory over arbitrary schemes was given for the first time in Knebusch (1969/70) and later extended in Knebusch (1977a). A significant, if so far somewhat isolated result, is Arason's computation (1980) of the Witt group of projective spaces. Starting in the seventies quadratic forms over polynomial rings, affine algebras and related rings were systematically studied by Karoubi, Knus, Ojanguren, Parimala, Sridharan, Quebbemann and many others. This developed into a full fledged theory, systematically presented in Knus (1991). We refer to this book for further references.

It was also clear from the beginning that a similar theory could be developed in noncommutative settings, e.g. over division algebras with involution. The basic existence and classification theorems for involutions were found by Albert (1939) and simplified by Riehm (1970) and Scharlau (1975). After these first beginnings the theory of involutions developed into a vast research field of its own. An exhaustive treatment of the present state of the art is given in Knus-Merkurjev-Rost-Tignol (1998). The theory of hermitian forms was worked out in number theoretic situations by Landherr (1938) and Kneser (1969). Concerning the basic notions – e.g. of the Witt group - several variations are possible and lead to somewhat different but related theories. Foundational work in this direction is due to Bass (1967), Bak (1969, 1981), Fröhlich-McEvett (1969), C.T.C. Wall (1970, 1973) and others.

Since the time of Weierstraß and Kronecker the classification of arbitrary sesquilinear forms, of isometries and selfadjoint transformations had been a somewhat separated but related field. Dozens of papers by many authors are devoted to this topic. Initially, the ground field was $\mathbb{C}$ or $\mathbb{R}$ and the approach purely matrix theoretic. More or less all results lead to a unique Jordan decomposition. Later arbitrary ground fields were considered, e.g. in several papers by Williamson. A systematic, more conceptual treatment was given for isometries by G.E. Wall (1963) and Milnor (1969). Eventually one realized that all these different problems could be handled by the consideration of abelian k-categories with involution. This general concept was worked out in detail by Quebbemann-Scharlau-Scharlau-Schulte (1976) and Quebbemann-Scharlau-Schulte (1979) with further contributions and applications by several authors, among others Knus-Ojanguren-Parimala (1982), Bayer-Fluckiger, Kearton, Wilson (1989) and Knus (1991, Chap. II).

Perhaps a generalization in a completely different direction should be mentioned also in this section, namely quaternionic structures, abstract (reduced) Witt rings, and abstract ordering spaces. What first looked like a rather formal (and perhaps useless) generalization of the notion of Witt rings under the hands of Marshall (1980) (and other papers quoted there) turned into a most useful tool providing elementary and purely combinatorial proofs of deep and difficult structure theorems in real algebra and real algebraic geometry. (See also section 1.7.)

1.5. Pfister forms. The most important notion in the algebraic theory of quadratic forms is probably the notion of Pfister forms (Pfister (1965)). Pfister forms are 2^n-dimensional forms that can be written as n-fold tensor products of 2-dimensional forms $\langle 1, -a_i\rangle$, that is, forms of type

$$\langle\langle a_1, \dots, a_n\rangle\rangle := \bigotimes_{i=1}^{n} \langle 1, -a_1\rangle.$$

Hence 2-fold Pfister forms are of type $\langle 1, -a, -b, ab\rangle$, etc. They are an indispensable tool for almost every significant problem. They have many important properties, but the most significant seems to be that for Pfister forms the "difficult" problem of isotropy coincides with the "easy" problem of hyperbolicity. This makes it possible to apply functorial Witt ring methods to the study of individual forms. The existence of Pfister forms seems to be specific to the theory of quadratic forms. Decomposable elements $a_1 \cup \dots \cup a_n$ in Galois cohomology, respectively symbols $\{a_1, \dots, a_n\}$ in Milnor K-theory are analogues but do not seem to be of quite the same importance as Pfister forms.

1.6. Generic splitting. First examples of generic splitting fields occur already in H. Kneser (1934) and later in Pfister's work on the level and in the proof of the Arason-Pfister Hauptsatz. If $\varphi_0 = \varphi = \langle a_1, \dots, a_n\rangle$ is an anisotropic form over a field $K_0 = K$ its generic zero field is defined as the algebraic function field

$$K_1 = K(X_1, \dots, X_n)/(a_1X_1^2 + \dots + a_nX_n^2).$$

In a sense, which easily can be made precise, this is the most general field in which φ becomes isotropic. If φ_1 is the anisotropic part of $\varphi_0 \otimes K_1$ one can repeat this construction inductively and obtains finally the generic splitting field. A systematic study of generic splitting fields was initiated by Knebusch (1976, 1977). I remember that Knebusch and I discussed the possibility of developing such a theory already some years earlier; a paper of Roquette (1966) suggested such an approach. What first seemed to be a rather restricted circle of ideas, soon turned out to be a crucial tool for a number of important questions. Merkurjev's construction (1992) of fields with given even u-invariant uses generic splitting fields in an essential way and the same is true for all work on the Milnor conjecture. This also leads naturally to the investigation of quadratic forms over quadrics and similar varieties, a field of active research. All this work is based essentially also on Quillen's (1973) computation of the K-theory of projective spaces and Brauer-Severi varieties, on Swan's (1985) computation of the K-theory of quadrics, and, more recently, on the computation of motivic cohomology of some "motives" associated to Pfister quadrics (in particular by M. Rost).

1.7. Formally real fields. The first result relating quadratic forms and real fields is of course Sylvester's law of inertia. The relation between the two concepts comes from the trivial fact that sums of squares are always positive. This leads immediately to Hilbert's 17th problem: can positive definite functions be represented as sums of squares? Artin and Schreier (1927) solved this problem and developed for this purpose the theory of ordered and formally real fields. It is well known that Hilbert's 17th problem led to similar questions in real algebraic geometry (real Nullstellensatz, Positivstellensatz, Krivine (1964), Dubois (1969), Stengle (1974), Risler (1976)).

However the "real" starting point was again Pfister's work, namely his local-global principle, that is, the exact sequence

$$0 \longrightarrow W_t(K) \longrightarrow W(K) \xrightarrow{sign} \prod_P \mathbb{Z}$$

where P runs over all orderings of K, $sign$ is the total signature and $W_t(K)$ is the torsion subgroup. This leads immediately to a number of questions: What can be said about the structure of the ordering space $X(K)$, what is the functorial behavior of $X(K)$? This question led to Marshall's (1980) theory of abstract ordering spaces and over arbitrary ground rings to the notion and the theory of the real spectrum (see Knebusch (1984)). Closely related is the determination of the image and of the cokernel of the total signature map. This led to the notions of fans, reduced Witt rings, stability indices, etc. The first essential results in this direction are due to Bröcker (1974, 1977), Becker-Bröcker (1978), Marshall and others. Moreover, special classes of ground fields were considered in detail, e.g. pythagorean fields, hereditary pythagorean fields, euclidean fields, SAP- and WAP-property, etc.

The years 1968 – 1980 saw a "real" explosion in this area. Deep connections between quadratic forms, real and ordered fields, valuation theory, model theory and real algebraic geometry were discovered. Probably more papers were published in this area than in the rest of the algebraic theory of quadratic forms. We therefore refrain from mentioning more details and refer instead to the following books, lecture notes and expository articles containing also extensive references: Colliot-Thélène et al. (1982), Lam (1983), Marshall (1980), Prestel (1984), Bochnak-Coste-Roy (1987), Knebusch-Scheiderer (1989), Bröcker (1991), Coste et al. (1991).

1.8. Field invariants. So far we have mentioned mainly structural and functorial aspects of the algebraic theory. We come now to a somewhat different circle of ideas, concerned with the so called field invariants. The most important ones are *level, u-invariant,* and *Pythagoras number.* The level of a field K is the minimal number s such that -1 is a sum of s squares in K. If such a representation of -1 does not exist, that is, if K is formally real, the level is ∞. Pfister (1995) is an excellent reference. Again, Pfister's results on the level of fields were the first breakthrough: if s is finite it is a 2-power; all 2-powers occur as levels of suitable fields (see also section 3 for some details).

The definition and the first results on the *u-invariant* (the maximal dimension of anisotropic torsion forms) are due to Kaplansky (1953) and were consequences of the classification theorems. Since iterated power series and rational function fields (e.g. over algebraically closed fields) have 2-power u-invariants (Springer (1955), Tsen (1936)), it has been conjectured that the u-invariant is always a power of 2. A sensational result of Merkurjev's (1992) disproves this: there exist fields of given even u-invariant. (The problem of odd u-invariants is open for $u > 7$.)[4]. Merkurjev's proof uses essentially so called index-reduction formulas on the behaviour of the index of a central simple algebra under a base field extension given by the function field of a suitable variety. Their proof, in turn, depends on the computation of the K-theory of certain homogeneous spaces.

Before Merkurjev's result, Elman and Lam (1973), in a series of papers, studied the u-invariant for fields satisfying additional hypotheses (concerning e.g. I^3 or the quaternion algebras). Also, Leep (1984) proved a nice result on the zeros of systems of quadratic forms, with applications to the u-invariant of a finite field extension L/K. After Merkurjev's result a considerable number of papers appeared containing simplifications, variations and applications of his methods and results. Nevertheless, it had not hitherto been possible to compute the u-invariant of a given field, such as a rational function field. Therefore it was a remarkable success when Hoffmann-van Geel (1998) proved recently that $\mathbb{Q}_p(X)$ has finite u-invariant (in fact $u \leq 22$) if $p \neq 2$. The proof is based on an important theorem of Saltman (1997) concerning common splitting fields of quaternion algebras. The bound was later sharpened to $u \leq 10$ by Parimala and Suresh (1998).

The *Pythagoras number* of a field (or ring) originated from the quantitative version of Hilbert's 17th problem: It is the smallest number p such that every sum of squares is already a sum of p squares. For non-real fields $p = s$ or $p = s + 1$ ($s =$ level). The determination of $p(K)$ for algebraic number fields follows from the Hasse-Minkowski theorem. Besides this the first results were obtained for rational function fields $F = K(X_1, \dots, X_n)$. If K is real closed, Cassels (1964) proved $1+X_1^2+\dots+X_n^2$ is not a sum of n squares, that is $p(F) \geq n + 1$. (This result is one of the starting points of Pfister's work, see section 3.) Ax proved $p(F) \leq 8$ for $n = 3$, Pfister (1967) proved $p(F) \leq 2^n$ for arbitrary n, and Cassels-Ellison-Pfister (1971) proved for $n = 2$ that the (positive) Motzkin polynomial $1 - 3X_1^2X_2^2 + X_1^4X_2^2 + X_1^2X_2^4$ is not a sum of 3 squares, hence $p(F) = 4$ for $n = 2$. Already Landau proved $p(\mathbb{Q}(X)) \leq 8$ but Pourchet (1971) and Hsia-Johnson (1974) sharpened this to $p(K(X)) \leq 5$ for every algebraic number field. As mentioned in 1.3 work on the Kato conjecture and cohomological Hasse principles yields $p(F) \leq 2^{n+1}$ for K an algebraic number field and $n = 2, 3$, as well as a number of related results (Colliot-Thélène, Jannsen (1991)). See Hoffmann (1999) for a recent result on the Pythagoras numbers of fields.

The invariants level, u-invariant, and Pythagoras number can also be defined for commutative rings. An number of results are known, perhaps the most spectacular is due to Dai-Lam-Peng (1980): the integral domain $\mathbb{R}[X_1, \ldots, X_n]/(1 + X_1^2 + \ldots + X_n^2)$ has level n. The proof uses topological methods, namely the Borsuk-Ulam theorem. This connection between topology and algebra was in principle known since the fifties, certainly since Swan (1962). However, this application came as a surprise; it led on the one hand to algebraic proofs of the Borsuk-Ulam theorem and the Brouwer fixed point theorem (Knebusch (1982), Arason-Pfister (1982)), on the other hand to the notion of the level of a topological space with involution and the computation of this invariant for projective spaces (Dai-Lam (1984), Pfister-Stolz (1987), Stolz (1989)). Mahé proved that the level of a real affine algebra without a real point is finite. We refer to Pfister (1995) for some results on the Pythagoras number of rings.

2. Quadratic forms over arbitrary fields: The work of Dickson

Perhaps the most basic problem in the algebraic theory of quadratic forms is the classification problem: how can one decide whether two quadratic forms over a given field or ring are isometric? It is not clear (to me) who explicitly formulated this question for the first time. Certainly, in 1890 Minkowski (1864 – 1909) asked and solved this question for the field of rational numbers. Since the fundamental concepts of algebra and linear algebra were clarified at about the same time one can assume that the problem was on the agenda since about this time. In 1899 L. E. Dickson (1874 – 1954) solved the classification problem for finite fields (a rather trivial result) and he seems to have been aware of the general question. The details are certainly involved. It should be kept in mind that already in 1890 Kronecker (1823 – 1891) had discussed the more difficult problem of the classification of pairs of quadratic forms.

In 1907 Dickson published an article "On quadratic forms in a general field", a title very reminiscent of Witt's famous paper (1937). The paper begins with the words: *We investigate the equivalence, under linear transformation in a general field F, of two quadratic forms ...*

$$q \equiv \sum_{i=1}^{n} a_i x_i^2; \quad Q \equiv \sum_{i=1}^{n} \alpha_i X_i^2.$$

Though he does not say so, Dickson probably believed that he had given in some sense a complete solution of the classification problem, and in some sense he was right! He continued: *An obvious necessary condition is that α_1 shall be representable by q , viz., that there shall exist elements b_i in F such that*

$$\alpha_1 = \sum_{i=1}^{n} a_i b_i^2.$$

Using this representation he writes down an explicit transformation of q in a form

$$q' \equiv \alpha_1 x_1'^2 + \sum_{i=2}^{n} a_i' x_i'^2.$$

Then, he states and proves the cancellation law (yes!) and is thus reduced to the discussion of $(n-1)$-dimensional forms. It must have been a very natural idea for him to believe that in order to solve the classification problem, one must at least be able to decide the "obvious" necessary condition, whether q represents a given element. Today we have realized that the representation problem is in fact the more difficult problem.

In more detail and in modern terminology, Dickson's paper (1907) contains the following results:

1) Every quadratic form over a field of characteristic $\neq 2$ can be diagonalized. (Footnote on p. 108)
2) If $q = \langle a_1, \dots, a_n \rangle$ represents $b \neq 0$, then $q \cong \langle b, \dots \rangle$. (Theorem in §2 ; there is some inessential restriction on the characteristic.)
3) If $q = \langle a_1, \dots a_n \rangle$ and v is a vector with $q(v) = a_1$, then there exists an automorphism σ of q sending v to the first basis vector. (Theorem in §5 on p. 114. The statement is marred by the fact that Dickson considers simultaneously q and $q' = \langle a_1^{-1}, \dots, a_n^{-1} \rangle$. Then σ is an isometry of q if and only if σ^t is an isometry of q' . One has to take this trivial fact into account.)
4) Every isometry between regular subspaces can be extended to an isometry of the whole quadratic space. (Theorem in §6 on p. 115)
5) The cancellation law holds: $\langle a, a_2, \dots, a_n \rangle \cong \langle a, b_2, \dots, b_n \rangle$ implies $\langle a_2, \dots, a_n \rangle \cong \langle b_2, \dots, b_n \rangle$. (Theorem in §4 on p. 114.)
6) In section 1 Dickson discusses necessary and sufficient conditions for $\langle a_1, \dots, a_n \rangle \cong \langle \alpha_1, \dots, \alpha_n \rangle$. Writing out these conditions, one sees that he constructs this isometry stepwise. It is easy to see that his equation (1) means that one has to change only two diagonal coefficients every step (Witt's "Satz 7"). A remark in an earlier paper (Dickson (1906)), however, suggests that he was not aware of this fact.

It seems that Dickson's paper went by completely unnoticed; I could not find a single reference to it in the literature. However, one must admit, that this paper – like most of Dickson's work – is not very pleasant to read. It is entirely algebraic. Dickson is not aware of the geometric interpretation of quadratic forms that Witt (1937) uses so elegantly. Where a simple geometric argument could be used he performs lengthy calculations. He does not know the importance of reflections. When he has to construct a suitable isometry he uses the Cayley transform, a somewhat awkward tool. Nevertheless, I believe that, as far as "Witt's theorem" and related questions is concerned, some credit should be given to Dickson.

One may also note that Hasse in his classical papers (1923a, b; 1924 a, b) was not aware of the cancellation law which could have saved him a lot of work. Hasse proves in his first paper the strong local-global principle. As we know today, this, together with the cancellation law, immediately implies the weak local-global principle for isometry – but he does the whole theory again – and that in a somewhat complicated way.

Concerning the classification problem itself, it seems that before Voevodsky there had been not much real progress beyond the work of Minkowski and Hasse. All known results (real function fields, local fields, and a few others) rely on ideas that can be found in those papers; Witt (1937) gives a very clear discussion of the classification problem. See also Elman and Lam (1974).

3. The work of A. Pfister

The modern algebraic theory of quadratic forms begins with Witt's (1911 – 1991) fundamental paper (1937). The importance of this paper has been pointed out and discussed so often that I think I cannot add anything significant to this issue. Pfister (1990) summarizes: *Mit [dieser Arbeit] stößt er ein neues Tor auf und leitet die Loslösung von der Zahlentheorie und die Verselbständigung der Theorie der quadratischen Formen ein. Die damals revolutionierenden, heute aber wohlbekannten und in jedem guten Kurs über Lineare Algebra gelehrten Inhalte sind ...*

However, it should perhaps be mentioned that Witt's paper had almost no immediate effect: In the following 25 years, only very few papers on the algebraic theory of quadratic forms appeared and even fewer were inspired directly by Witt's paper. One may say: the gate was open but nobody walked through it. (An exception is Arf's (1941) treatment of the characteristic 2 case along the lines suggested by Witt.)

The real breakthrough in the algebraic theory of quadratic forms came with Pfister's papers (1965, 1965a, 1966).

I heard (or rather read) the name Pfister for the first time on 22nd November 1963 when I received a postcard from H. Lenz telling me: *Unterdessen hat Herr Dr. A. Pfister, Math. Institut Göttingen, in den dort angeschnittenen Fragen ganz wesentliche Fortschritte erzielt.* In a letter dated 4th April 1999, Pfister informed me in some detail about the discovery of his structure theorems. The story begins with a course by H. Lenz on *Geometrische Algebra* in the summer semester 1962 at the University of Munich. It contained material from Artin's book (1957) and Dieudonné's book (1955) on classical groups, but mentioned also Hurwitz' work on the composition of forms, H. Kneser's remarks (1934) on the level of fields, some inequalities relating level and number of square classes, the u-invariant and a related result of M. Kneser. Pfister writes: *Das war gewissermaßen meine Grundlage.* (In particular, Kneser's paper (1934) must be considered as one of the germs of the algebraic theory of quadratic forms. It formulated a really significant (and accessible) problem and contributed to its solution.) A little later, Lenz published a short paper (1963) expanding on this earlier work. It contains the first "structure theorem" for the Witt group: *Satz 1. Die Wittsche Gruppe eines nicht formalreellen Körpers K ist eine 2-Gruppe* ... and it finishes with an important open problem: *Es ist ja immer noch offen, ob die Fälle $s > 4$ überhaupt vorkommen; und falls ja, ob dann die Anzahl q der Quadratklassen endlich sein kann.* (The second half of this problem is still open.)

The story continues in Göttingen, where Pfister became M. Kneser's assistant in May 1963. The decisive event was a colloquium talk by Cassels on 11th July 1963, where Cassels proved that $1 + X_1^2 + \ldots + X_n^2$ is not a sum of n squares in $\mathbb{R}(X_1, \ldots, X_n)$. Pfister reported in (1984): *I heard this talk and immediately realized that the result should be connected with the unsolved problem* [of the possible levels]. One day later he had proved, using Cassels' theorem, that H. Kneser's construction (1934) gives a field of level 8; three days later the case of level 16 was solved. In August Lenz and Pfister met and discussed various problems, in particular Lenz formulated the problem of the possible Pythagoras numbers. Pfister became more and more convinced that a general solution of the level problem should be related to the fact that the elements represented as sums of 2^n squares form a group. He was able to prove this decisive fact on 17th September and submitted his first paper (1965) a few weeks later.

The exchange of ideas between Lenz and Pfister continued during the following year. In February 1964 Lenz asked whether, in the composition formula

$$\sum_{i=1}^{n} z_i^2 = \left(\sum_{i=1}^{n} x_i^2\right)\left(\sum_{i=1}^{n} y_i^2\right), \ n = 2^m,$$

the z_i can be taken as linear functions in the y_j, a fact proved immediately by Pfister. In March Pfister proved that the values of "Pfister" forms form a group; in May he submitted the corresponding paper (1965a); Lenz conjectured that the Witt group has only 2-torsion, a fact proved by Pfister the same day; a few days later Lenz conjectured that the Witt ring does not contain zero divisors of odd dimension, etc. Also, during these months, Pfister's results became known to the public through colloquia, and talks in Oberwolfach by himself, Cassels and Lenz. In a letter dated 1st December 1964 Pfister informed me about the essential results of his Inventiones paper (1966) and mentioned also: *Ungelöstes Problem: $\bigcap_{n=1}^{\infty} M^n = 0$ für jeden Körper K ?* This "Hauptsatz" was proved a few years later by Arason and Pfister (1971).

Regarding the cooperation with Lenz, Pfister writes (4.4.99): *Lenz arbeitete alle meine Briefe gründlich durch, fand eigene Beweisvarianten und massenhaft neue Fragen. Er investierte enorm viel Interesse, Zeit und Arbeit. Sein Einfluß auf mich war daher in diesem guten Jahr der stärkste und wesentlichste.* – I think that this cooperation is a fine example of the fact, that also – and perhaps quite often – not so well known mathematicians contribute to the progress of our science.

It is well known that several authors – in particular Witt himself, who suddenly reappeared on the stage – simplified and extended Pfister's results. It is however my impression that all these contributions are only marginal compared to Pfister's achievements. Of course, Witt's approach, using "round" forms – which I shall mention again later – is simpler, very elegant and clever. But Pfister's theory of multiplicative forms gives more and also deeper results. One should note, for example, that only Pfister's "strong" methods give a proof of the Arason-Pfister Hauptsatz.

Because it gives me the opportunity to mention an other important name, I want to report a little anecdote concerning the exchange of ideas between Pfister and Witt: It is well-known that one of the few examples where the classification is possible are the C_2-fields of Tsen (1936). In fact, Tsen's work is the basis for all later work on quadratic forms over algebraic function fields. By definition, a field K has the C_i-property if every homogeneous polynomial $f \in K[X_1, X_2, \dots, X_n]$ of degree $d > 0$ with $n > d^i$ has a root $(x_1, x_2, \dots, x_n) \neq 0 \in K^n$. In particular, the C_n-property of $\mathbb{C}(X_1, \dots, X_n)$ is essential in Pfister's work on Hilbert's 17th problem. But it seems that by the sixties Tsen had been completely forgotten; instead references were made to Lang and Nagata. So, when also Pfister attributed the C_n-property to Lang, Witt replied: *Das hat doch schon mein Freund Chungtze Tsen bewiesen!* – This may have saved Tsen's name from falling into oblivion.[3]

Certainly Pfister's first papers on the level, on multiplicative forms and on the structure of the Witt ring contain his best known results. The algebraic theory of quadratic forms rests on this foundation. It seems to me that after this breakthrough Pfister's main interests shifted from structure theorems more in the direction of the study of concrete "individual" forms. (Of course, it is an essential point, that the structure theorems resulted from the study of the properties of individual forms, in particular Pfister forms.) Anyway, his lecture notes (Pfister (1995)) show his main interests very clearly: the u-invariant and related invariants, Pythagoras numbers, systems of quadratic forms, etc. They contain a wealth of information which is difficult to find elsewhere. One should particularly mention Pfister's work on systems of quadratic forms and his interest in "strong" properties of forms over function fields (e.g. estimates for the "size" of isotropic vectors). Today it is clear that these problems are much more difficult than the more formal theory.

4. Algebraic topology

The flourishing of a theory depends on interesting examples and applications. For the algebraic theory of quadratic forms, algebraic topology has played an important rôle, which – in my opinion – has perhaps been underestimated a little bit so far. Already in the twenties and thirties quadratic forms appeared several times as algebraic invariants of topological objects. I want to mention and to discuss briefly the following (related) examples:

1) the Kronecker-Poincaré intersection form,
2) the quadratic form of a knot,
3) the linking form (Verschlingungsform).

If M is a compact oriented connected $4k$-dimensional manifold, one has the cup product symmetric bilinear form

$$\cup : H^{2k}(M, \mathbb{Z}) \times H^{2k}(M, \mathbb{Z}) \longrightarrow H^{4k}(M, \mathbb{Z}) = \mathbb{Z}.$$

H. Weyl (1924) – and independently but in less explicit form Veblen (1923) – proved that this form is unimodular. Weyl already expected this form to be an interesting topological invariant of the manifold: *Por tanto, con m par, la clase a que pertenece la forma caracteristica de grado m (en particular su indice de inercia) constituye una nueva peculiaridad de las superficies 2m-dimensionales respecto al Análisis-Situs.*

The quadratic form of a knot was first defined by the little known mathematician Lebrecht Goeritz (1933), who apparently worked out suggestions and ideas of K. Reidemeister. Already the title of his paper "Knoten und quadratische Formen" indicates a close connection. For every knot an integral quadratic form is defined. The class of this form is determined by the isotopy class of the knot up to integral equivalence and adding or cancelling direct summands $\langle \pm 1 \rangle$. In particular the Minkowski symbols c_p are well defined knot invariants.

The linking form is perhaps a less well known invariant. It is defined on a finite abelian group taking values in $\mathbb{Q}/\mathbb{Z}$. This finite abelian group is the torsion subgroup of $H_k(M)$ of a $(2k+1)$-dimensional oriented manifold M. If k is odd this form is symmetric, if k is even it is skew-symmetric. It is defined and discussed in some detail by de Rham (1931) who also gives references to earlier work by Brouwer and Lebesgue. If M is the twofold covering of a knot complement, Seifert (1936) points out a close connection to the quadratic form of the knot. This form leads naturally to a purely algebraic study of forms on finite abelian groups. The first steps of such an investigation are already contained in de Rham's paper. He considers the case of skew-symmetric forms and gives a complete description. However he fails to notice the orthogonal decomposition in primary components, using elementary divisors instead, and he does not discuss the symmetric case. Unaware of de Rham's paper, Kneser and Puppe (1953) investigate the situation more carefully, and they proved essentially (and up to terminology) a rather fundamental algebraic result in the theory of quadratic forms, namely the existence of the exact sequence

$$0 \to W(\mathbb{Z}) \to W(\mathbb{Q}) \to WT \to 0 \qquad (*)$$

where WT denotes the Witt group of symmetric bilinear forms on finite abelian groups. It is interesting to note what R.H. Fox writes about this paper in his report for the Mathematical Reviews (1954 I 100): *The proof depends on a little-known theorem of A. Meyer ... and an equally recondite theorem of M. Eichler ... It seems to the reviewer that this theorem is of considerable importance, not only to knot theory but to the theory of quadratic forms itself; can it be that there is no simple direct proof of it?* (I must confess that I never looked at this paper before preparing this talk and that I was quite surprised when I saw that this result $(*)$ had essentially been proved so early.)

However, I am not the only one who had not read the Kneser-Puppe paper. In 1963 C.T.C. Wall once again discussed quadratic forms on finite groups mentioning neither de Rham nor Kneser-Puppe. Like the earlier papers he does not work out the 2-part completely. It seems that the details of quadratic forms on 2-groups were written down explicitly only around 1970 when also the relations to Gaussian sums and quadratic reciprocity were pointed out by several authors. A complete discussion is provided by Durfee (1977).

I will later discuss another instance where a question from knot theory led to very significant developments in the theory of quadratic forms.

I come now to the intersection form of a $4k$-manifold. It is well known that this has been in the center of mathematical research for more than 50 years. The first thing to mention is perhaps Hirzebruch's (1956) index theorem (Hauptsatz 8.2.2) answering the question posed (implicitly) by Weyl and which is one of Hirzebruch's great achievements. Since the index of inertia depends only on the real intersection form, the theory of quadratic forms does not yet enter significantly at this point. However, it was recognized very early that the intersection form has the "right" functorial properties. For example it is hyperbolic (over any coefficient field) if M bounds a $(n+1)$-manifold. Thus it gives an invariant from the cobordism ring to the Witt ring. A much deeper connection between quadratic forms and algebraic topology is established in Milnor's paper (1958). Though it is to some extent a report on known results it had a significant influence on the further development on both theories. Milnor discusses the problem to which extent an oriented simply connected 4-manifold is determined by its quadratic form. This continues earlier work of Whitehead (1949). Milnor proves that two such manifolds have the same homotopy type if and only if their quadratic forms are isometric. So, for a complete classification of the homotopy types, one needs a classification of integral unimodular forms. With the assistance of O.T. O'Meara he summarizes what was known about this problem, thus introducing odd and even forms (type I and II), mentioning E_8 and Gaussian sums and referring to earlier work of A. Meyer, H.

Braun, B.W. Jones and Eichler. He also asks which quadratic forms occur as intersection forms, a question which leads directly to the marvellous work of Donaldson and Freedman. An other closely related topic discussed by topologists during these years is the so called van der Blij invariant *mod* 8 of an integral quadratic form (van der Blij (1959)).

A few years later, Milnor (1961) discusses a more delicate question originating from a concrete problem in topology (surgery): When is the intersection form isotropic? As an application of the Hasse-Minkowski theorem, he proves that this is the case if and only if it is indefinite (remember that the determinant is ± 1). At this point a really significant connection between topology and algebra is established. (See also the remarks on the work of Freedman below.) Certainly, Milnor became interested in the theory of quadratic forms through these examples. This had far-reaching consequences for the further development of the theory of quadratic forms.

Looking at papers in algebraic topology from this time, one quite often finds some remarks on quadratic forms. Serre (1962) – at the request of H. Cartan – took up the task of giving a systematic treatment. His talks in the Cartan seminar of 26th February and 5th March 1962 are remarkable examples of clear and perspicuous presentation.

He begins with the words: *Il s'agit de résultats arithmétiques qui sont utiles en topologie différentielle.* He defines several new concepts (Grothendieck rings, λ-ring structure) and hints at the existence of entire new theories: *La catégorie S peut se définir pour un anneau commutatif quelconque (et même en fait pour un schéma de base quelconque, non nécessairement affine); un élément de S est l'analogue algébrique d'un fibré vectoriel ayant pour groupe structural le groupe orthogonal.*

In Serre's talk we not only find new concepts but also new results or at least new proofs. In particular, he gives a complete classification of indefinite unimodular lattices, clearly a fundamental result. One may argue that, in principle, this had already been obtained: Eichler's (1912 – 1992) result (1952, Kap III, §15) on indefinite lattices is a much more general and deeper result. (For $n > 2$ every spinor genus contains only one class.) However, Eichler fails to mention the explicit application to unimodular $\mathbb{Z}$-lattices, and also O'Meara (1963) misses this point quite narrowly in the last paragraph of his book. Also, Serre's proof – using only the Hasse-Minkowski theorem in Meyer's version and Kneser's method of neighbors – is much easier than the application of Eichler's theory.

Though this is quite clear from the discussion of Milnor's algebraic work in Bass (1993), I would like to add one further remark concerning J. Milnor: he arrived at the theory of quadratic forms from (at least) two different directions. One has been sketched already; its starting point is the intersection form and the classification of 4-dimensional manifolds. The other one begins with Whitehead's work on simple homotopy equivalence of finite CW-complexes. This leads to the Whitehead group and Whitehead torsion which is related to h- and s-cobordism. The calculation of the Whitehead group is closely related to the functor K_1 and the congruence subgroup problem. This in turn leads to central extensions and the functor K_2, to Steinberg symbols, norm residue symbols, reciprocity laws, etc. The definition of K_2 is perhaps Milnor's best known achievement in algebra (Milnor 1971a). When it appeared it certainly came as a great surprise to the mathematical community. Bass' (1968) monumental monograph on K_0 and K_1 had already been published, but it was not at all clear how to define K_2 with the right properties. As far as quadratic forms are concerned, the essential point seems to be that Milnor realized that for fields the functors K_2 (via Matsumoto's theorem) and Witt ring (and perhaps Galois cohomology) have very similar descriptions by generators and relations. This led to important results (e.g. on the Witt group of a rational function field $K(X)$) and eventually to the Milnor conjecture.

The paper (1970) which contains the Milnor conjecture is certainly Milnor's most important contribution to the theory of quadratic forms. But it should not be overlooked that also (1969) on the classification of isometries had some significant influence on the further development. The paper originated also in topology, namely in some problems about knot cobordism discussed by Levine, but its content is entirely algebraic. First of all, it contains a very perspicuous presentation of the classification problem for isometries. As mentioned in 1.4, this problem had, in principle, been solved before. But Milnor's treatment opened the way to a much more general theory. His approach is a model for similar classification problems (self-adjoint transformations, sesquilinear forms, pairs of quadratic forms, in general spaces with additional structures). Together with the relevant chapters in Bass' book (1968) it led directly to the introduction of additive and abelian categories with duality, a convenient setting for many aspects of quadratic form theory. Secondly the paper contains what may seem to be a purely technical result, namely that over p-adic fields the transfer of quadratic forms preserves the Hasse symbol. Today it is well known that this theorem has many applications. When I saw it the first time, I realized immediately that this allowed to reduce the proof of Hilbert's quadratic reciprocity law for arbitrary algebraic number fields to the field $\mathbb{Q}$. During my stay in Princeton, I mentioned this to A. Weil and he informed me that already his paper (1964) contained Milnor's result and what is now known as Weil's reciprocity law.

But now we have digressed perhaps too far from algebraic topology! Lack of time, space, energy and information – the basic concepts of our world – prevent me from discussing the relations between quadratic forms and topology in more detail, and I want to close this section with two remarks. Firstly, since the sixties also non-simply-connected oriented $2k$-dimensional manifolds were considered. In this case the fundamental group acts as a group of isometries of the intersection form. This leads to the study of ε-symmetric and ε-quadratic forms over group rings. More than anybody else, C.T.C.

Wall initiated a systematic investigation of this situation, from the topological side as well as from the algebraic. Surgery obstruction groups and Witt groups of various kinds were defined, their functorial properties studied systematically, exact localization sequences, Mayer-Vietoris sequences and periodicity results were proved and more or less explicit calculations were performed. This is a field of active research to the present day.

Secondly, I want to mention the marvellous work of M. H. Freedman and S. Donaldson. Compact simply connected 4-manifolds M are characterized up to homeomorphism by the intersection form ω (and the Kirby-Siebenmann invariant $\sigma \in \mathbb{Z}/2\mathbb{Z}$, indicating whether M stably has a differentiable structure or not). Every unimodular quadratic form ω arises in this way. The invariants ω, σ are independent except for the relation prescribed by Rochlin's theorem: if M is stably differentiable, then the signature is a multiple of 16. Moreover, if a differentiable 4-manifold has a positive definite intersection form, then this is actually the unit form. (For all this, see: Proceedings of the International Congress of Mathematicians 1986.)

5. Unpublished work

Digressing even more from a systematic treatment of my topic, I now want to report on some unpublished work which nevertheless had significant influence on the development of the algebraic theory of quadratic forms.

The first is a talk by M. Kneser (1962) in Paris. It seems his private notes of this talk did not circulate widely, but certainly Kneser's results were prototypical for later work by a considerable number of authors (Bak-Scharlau, Fröhlich, Knebusch, Knebusch-Rosenberg-Ware, Wall, Scharlau). I received the manuscript in July 1967; the (unpublished) second part of my Queen's University course on quadratic forms in 1968/69 was partially based on it. Kneser's notes contain in particular: the definition of various Witt-Grothendieck groups over arbitrary rings; a brief discussion of the Clifford algebra; the description (over fields) of the discriminant and the Witt invariant via non-abelian Galois cohomology; a discussion of the first terms of the filtration of the Witt-Grothendieck ring; a discussion of special ground fields, in particular of local and global number fields. He then discusses Witt and Witt-Grothendieck groups $\widehat{W}$ of rings of algebraic integers considering unimodular and also non-degenerate ($det \neq 0$) forms. In fact his results carry over easily to an arbitrary Dedekind domain R. As an application of the strong approximation theorem he proves the exactness of the sequence

$$0 \longrightarrow C/C^2 \longrightarrow \widehat{W}(R) \longrightarrow \widehat{W}(K)$$

(where C denotes the ideal class group). In retrospect one sees a rather close connection to the Kneser-Puppe paper (1953) and gets the impression that already at that time Kneser knew all basic results concerning Witt groups of Dedekind domains and their quotient fields as exposed e.g. in Scharlau (1985), Chap. 5 and 6.

Next I want to mention some work of G. Harder which was not published by Harder himself but is contained (at least partially) in Knebusch's Habilitationsschrift (1970). Harder had worked on a much more general and difficult problem, namely semisimple groups over complete curves. Specializing one of his results to the orthogonal group and the projective line (Harder (1968), 3.5) he obtained the result that every quadratic form over the polynomial ring $K[X]$ is extended from K. Obviously this led him to think about quadratic forms over the rational function field. One should note perhaps at this point that the analogous problem in algebraic K-theory, namely the structure of projective modules over $R[X]$ had already received a lot of attention at this time; in fact it was one of the central problems of algebraic K-theory (see Weibel (1999) for more details).

Knebusch told me in a letter of 17th December 1969 that during the summer semester 1968 Harder had established the existence of the reciprocity (or sum) formula for the second residue forms. That a result like this should hold was clear from the analogous result for the Brauer group. Details and generalizations of Harder's ideas were worked out a little later by Knebusch (1970), Scharlau (1972), and Geyer, Harder, Knebusch, Scharlau (1970). In some colloquium talks Harder gave simple proofs of his result which used only elementary lattice theory and the Riemann-Roch theorem for rational function fields (a more or less trivial result).

At about the same time, J. Tate worked on a very similar and closely related problem, namely the computation of K_2 of $\mathbb{Q}$ and of a rational function field $K(X)$. In the course of this computation he essentially rediscovered Gauss' first proof of the quadratic reciprocity law. Tate talked about these things several times and thus inspired Milnor's analogous computation of $W(K(X))$, which is more elementary than Harder's approach. Milnor (1970) also published Tate's results. As mentioned before, all this work of Kneser, Harder, Tate, and Milnor is closely related, and has a strong number theoretic flavour.

I now come to a different topic, namely Witt's highly influential colloquium talks in 1967 and 1968. The first was given in Cologne on 10th November 1967; it is now published in Witt's Collected Papers (1998). It is well known that in this talk Witt gave simplified proofs of Pfister's structure theorems. He achieved this by the introduction of round forms, and he obviously enjoyed to talk about "round quadratic forms". (He also liked to mention that he had learned the Grothendieck construction from his Chinese nanny. And when he passed from the Grothendieck ring to the Witt ring he said: and now we introduce the relation $1 + -1 = 0$.) As one may guess, I had attended the Cologne lecture and was quite impressed. Witt spoke however for almost two hours and the audience got a little nervous, in particular when he mentioned very close to the end that he had a lot more to say about lattices, their Θ-series and modular forms. I want to mention one further little detail: at some point he mentioned pythagorean fields. It was the first time that I heard of such a thing but one year later I noticed that the pythagorean closure of a field may in some sense replace the family of all real closures. I think in this way pythagorean (and related) fields became an interesting special topic in the algebraic theory of quadratic forms. – Witt's results - though unpublished – became quickly known in the community, in particular through Lorenz' (1970) and my own lecture notes.

The last unpublished paper which I would like to mention is Ax (1967) which was submitted for publication on 11th January 1967 but withdrawn after Pfister (1967) had completely solved the problem posed there. It is concerned with a quantitative version of Hilbert's 17th problem: every positive definite function in $\mathbb{R}(X_1, \dots, X_n)$ is a sum of 2^n squares. Ax proved this for $n \leq 3$ and contributed substantially to the general case. In particular he had the idea to apply Tsen's theorem. Pfister (1995 p. 89) writes: *I am very grateful to J. Ax for sending me a preprint of his manuscript .... This was one of the most lucky occurrences in my life. After a careful study of the manuscript I realized that the reduction to C_n-fields should be retained, but that the use of cohomology should be replaced by multiplicative forms.* - Of course, this is exactly the principle of the Milnor conjecture: replace cohomology by quadratic forms, cup products by Pfister forms. (Unfortunately, Ax's important paper is not mentioned in any of the standard books on quadratic forms.)

Perhaps one may conclude this section with the remark that the course of history cannot entirely be reconstructed from published documents but other sources have to be considered also in order to get a more complete (and accurate) picture. (Concerning the history, one may mention the work of Voevodsky and his collaborators and quite a few of Rost's influential preprints.)

6. Some reminiscences 1962 – 1970

I began my university studies in Bonn in the summer semester 1959. During the summer term 1962 I attended a little course (one hour weekly) by F. Hirzebruch on "Quadratische Formen". It covered roughly the first four paragraphs of Hirzebruch's lecture notes (Hirzebruch, Neumann, Koh (1971)). During the summer vacation I undertook a two-months backpack trip to the Near East resulting in one of my first ornithological publications, Scharlau (1963) (on the birds of the Egyptian oasis El-Dachla). The following winter semester I attended two seminars of Hirzebruch, one on "Charakteristische Klassen", the other the "Oberseminar". I gave talks on "Stiefel-Whitney-Klassen" and the "Hurwitz-Radon-Eckmann Theorem" on the number of independent linear vector fields on spheres. - Algebraic topology was the big thing in those days! – Some time later, I asked Hirzebruch for a subject for a "Diplomarbeit" and I suggested myself the field of quadratic forms. This was a little bit unusual since almost all of Hirzebruch's students worked in topology. Nevertheless, he agreed and suggested that I should start reading Delzant's note (1962). This lead me to Galois cohomology and Serre's Corps locaux (1962) that had just appeared.

From the first moment on it was obvious that it should be possible to develop a theory of quadratic forms similar to – but much easier than - the theory of vector bundles and K-theory. I made for myself lists of analogies between the two theories, and talked a lot about this with Karl-Heinz Mayer, the assistant supervising my progress. By the summer semester 1964 I had finished my Diplomarbeit "Quadratische Formen und Stiefel-Whitney-Klassen". Among other things, I worked out the effect of the Bockstein homomorphism on the Stiefel-Whitney classes and showed that the non-abelian Galois cohomology of

$$1 \to SO \to O \to \{\pm 1\} \to 1; \quad 1 \to \{\pm 1\} \to Pin \to O \to 1$$

leads to the first two Stiefel-Whitney classes. Both results were suggested by topology; I was not aware of Springer's paper (1959) containing the same result.

My Diplomarbeit was never published, not even submitted to the "Prüfungsamt". I hesitated to take the examination because then I might have been drafted for military service. In the meantime S. Lang had invited me to spend one year at Columbia University. While preparing my stay there, I continued my work and computed the Witt group of a power series field $k((t))$ – again being unaware of Springer's result (1955) – and discovered as an application fields where quadratic forms are not classified by Stiefel-Whitney classes. (This is of course a very trivial fact. But in March 1970, at the inauguration of "New Fine Hall" in Princeton, J. Milnor, who spoke on the results of his paper (1970), mentioned it and wrote my name in huge letters across the board, one of the prouder moments of my mathematical life.) I talked to J. Tits, who had become a professor in Bonn, about these results. He communicated them to Serre, who, of course, noticed immediately the generalization to local fields.

At Columbia I attended a very substantial course on "Quadratic Forms" by H. Bass. It started with algebraic number theory and treated quadratic forms in a quite general and modern fashion. Later Bass extended this material to his Tata notes (1967). Bass' course was also attended by T.-Y. Lam and A. Bak.

Back in Bonn, I started to work on my thesis on "Quadratische Formen und Galois-Cohomologie" which I finished by November 1966. It appeared in somewhat revised form in Scharlau (1967) and contained also some results on the Brauer group of henselian fields. After the Ph.D.-examination I wrote my paper on the Brauer group of algebraic function fields. I think, I must have tried to compute also the Witt group of a rational function field but I was not successful. The academic year 1968/69 I spent with my friends J. Neukirch (1937 - 1997) and W.-D. Geyer and their families at Queen's University. We had a really good time; I enjoyed very much giving a course on quadratic forms and learned a lot in Geyer's course on algebraic function fields. One of the things I tried to work out was a cohomological proof of Springer's result on odd degree extensions. I did not succeed, but in the process I discovered the method of transfer in the first days of October 1968. The point was to take not the trace but a suitable different linear map. I realized immediately that this would prove to be a useful construction.

The following year I spent at the Institute in Princeton. My main interest were structure theorems for the Witt ring and applications of the method of transfer on the one hand and quadratic reciprocity laws for number and function fields on the other. I profited very much from the work and talks of Milnor, Bass, Tate, Harder and Knebusch. As mentioned before, A. Weil informed me about his results on the transfer and "Weil reciprocity". He also suggested that the characteristic 2 case always should be included as far as possible. It seems the quadratic forms community has not followed this advice. (Knus et al. (1998) is an exception). M. Knebusch and I exchanged several letters resulting a little later in some common publications on the questions mentioned above. – Looking back at that time it seems to me that it was quite easy to discover and prove theorems – theorems, perhaps not very deep and difficult, but certainly basic and aesthetically satisfying. Life was like in paradise, one had not to work very hard but just to pick up the fruits.

7. An outlook

It is probably not possible to predict the future course of a mathematical theory. But one may assume that already visible trends will continue for some time. The algebraic theory of quadratic forms is an offspring of abstract algebra, in particular of linear and geometric algebra and of field theory. We have seen that also problems originating in topology played an important rôle. However, for about twenty years algebraic geometry has had the strongest impact on the algebraic theory of quadratic forms. In some sense this is true also for the preceding period. Significant work of Cassels, Ax, Pfister and others made essential use of polynomial rings etc., that is of algebraic geometry in a broad sense. The same is true, of course, for Knebusch's concept of generic splitting fields. I remember that I talked around 1964 to M. Kneser about possible research problems, and that he mentioned quadratic forms over function fields, adding that it would be necessary to study algebraic geometry for this purpose. In this context it may be interesting to see how M. Knebusch remembers these years (letter dated 1st March 1999): *Ganz im Gegenteil glaubte ich ursprünglich, daß eine direkte Theorie der quadratischen Formen über Körpern einfach zu schwierig ist. Man müsste erst mehr über quadratische Formen über vollständigen Schemata wissen, um bei Körpern weiter zu kommen. Ein Körper erlaubt zu wenig Geometrie. Es gibt dort sozusagen zu viele quadratische Formen, während es über den vollständigen Schemata nur wenige Vektorbündel mit quadratischer Form gibt, aber gute Möglichkeiten für eine geometrische Betrachtung. - Auch heute glaube ich, daß diese Philosophie nicht ganz verkehrt ist.*

In fact, these remarks seem to describe very precisely the present situation: the Milnor conjecture is a statement about quadratic forms over fields, but Voevodsky's proof requires a lot of difficult and to some extent highly technical algebraic geometry. The future will show how much of this is really necessary. But in the meantime we can expect that algebraic geometry will continue to exert a strong influence on quadratic form theory. Many of the most important unsolved problems – odd u-invariants[4], the behaviour of generic splitting towers, the validity of Hasse principles, the computation of u-invariants and Pythagoras numbers, Witt groups of function fields, curves and higher-dimensional varieties, relations to Chow groups, K-groups, and cohomology – all this and much more seems to depend on further progress in algebraic geometry. However, it is interesting to observe that the constructions by Voevodsky (and others, like Bloch, Friedlander, Morel) involve many techniques from algebraic topology. One of the main obstacles to progress seems to be the fact that the Witt group functor does not have good properties e.g. one does not know analogues of the exact sequence of cohomology or K-theory for Witt groups. Perhaps also this problem will be resolved in the context of algebraic geometry.

Footnotes

1) In the introduction of his paper (1937) Witt makes the following interesting remark: Da aus der neueren Theorie der Algebren bekannt ist, daß eine Algebra (a, b) im wesentlichen dasselbe ist, wie ein System von Normrestsymbolen, hat *Artin* in einer Vorlesung die quadratischen Formen von vornherein auf dieser Grundlage behandelt. Er ordnete jeder *Form* $f = \sum_1^n a_i x_i^2$ die *Algebra* $S(f) = \prod_{i \leq k}(a_i, a_k)$ zu. Die Invarianz dieser Algebra bei allen Transformationen war allerdings nicht ganz leicht einzusehen, ... This leads to the question how Artin could in fact prove the invariance of $S(f)$; usually one applies cancellation and Witt's "Satz 7". This remark seems also to indicate that what today is usually called Hasse algebra should perhaps be named after Artin. And finally it is fascinating to observe, that already in this first paper the crucial word "Normrestsymbol" occurs. This is certainly a "Leitmotiv" of the whole theory, leading directly to the Milnor and the Bloch-Kato conjecture.

2) It seems that Durfee's paper is not very well known. The standard books do not refer to it. But Durfee had essentially proved the same results as later Springer.

3) The history of the C_n-property is strange: It was discovered by Tsen and Lang, and certainly Lang was not aware of Tsen's priority. Nevertheless, their work was perhaps not "independent" because both were students of E. Artin. D. Leep pointed out to me that the situation is even more confusing: almost simultaneously with Lang's work, also Carlitz (1951) proved the C_n-property of function fields over finite fields. Necessarily his methods and results are very close to Tsen's and Lang's. F. Lorenz (Münster) has collected some biographical information on Tsen (unpublished). More information on Tsen and his work is contained in Sh. Ding, M. Kang, E. Tan (1999).

4) In July 1999 Izhboldin proved the existence of fields of u-invariant 9.

References

A.A. Albert (1939): Structure of algebras. Amer. Math. Soc. Coll. Publ. **24**, Providence: Amer. Math. Soc.

J.Kr. Arason (1975): Cohomologische Invarianten quadratischer Formen. J. Algebra **36**, 448–491

– (1980): Der Wittring projektiver Räume. Math. Ann. **253**, 205–212

J.Kr. Arason, A. Pfister (1971): Beweis des Krullschen Durchschnittsatzes für den Wittring. Invent. Math. **12**, 173–176

– (1982): Quadratische Formen über affinen Algebren und ein algebraischer Beweis des Satzes von Borsuk–Ulam. J. reine angew. Math. **331**, 181–184

C. Arf (1941): Untersuchungen über quadratische Formen in Körpern der Charakteristik 2 (Teil I). J. reine angew. Math. **183**, 148–167

E. Artin (1957): Geometric Algebra. London, New York: Interscience

E. Artin, O. Schreier (1927): Algebraische Konstruktion reeller Körper. Abh. Math. Sem. Hamburg **5**, 85–99

J. Ax (1967): On ternary definite rational functions (unpublished)

R. Baeza (1978): Quadratic forms over semilocal rings. Lecture Notes in Mathematics **655**, Berlin, Heidelberg, New York: Springer

A. Bak (1969): On modules with quadratic forms. Lecture Notes in Mathematics **108**, 55–66, Berlin, Heidelberg, New York: Springer

– (1981): K–theory of forms. Ann. Math. Studies **98**. Princeton

H. Bass (1967): Lectures on topics in algebraic K–theory. Bombay: Tata Institute.

– (1968): Algebraic K–theory. New York and Amsterdam: Benjamin.

– (1993): John Milnor the Algebraist, p. 45–84 in: Topological Methods in Modern Mathematics. Houston: Publish or Perish.

E. Bayer–Fluckiger, C. Kearton, S.M.J. Wilson (1989): Hermitian forms in additive categories: finiteness results. J. Algebra **123**, 336–350

E. Bayer–Fluckiger, R. Parimala (1995): Galois cohomology of the classical groups over fields of cohomological dimension ≤ 2. Invent. Math. **122**, 195–229

– (1998): Classical groups and the Hasse principle. Ann. of Math. **147**, 651–693

E. Becker (1978): Hereditarily–pythagorean fields and orderings of higher level. IMPA Monografias de matemática No. **29**: Rio de Janeiro.

E. Becker, L. Bröcker (1978): On the description of the reduced Wittring. J. Algebra **52**, 328–346

A.A. Belskii (1968): Cohomological Witt rings (in Russian). Math. USSR–Izvestija **2**, 1101–1115

F. van der Blij (1959): An invariant of quadratic forms mod 8. Indag. Math. **21**, 291–293

J. Bochnak, M. Coste, M.–F. Roy (1987): Géométrie algébrique réelle. Ergeb. Math. **12**, Berlin, Heidelberg, New York: Springer

L. Bröcker (1974): Zur Theorie der quadratischen Formen über formal reellen Körpern. Math. Ann. **210**, 233–256

– (1977): Über die Anzahl der Anordnungen eines kommutativen Körpers. Arch. Math. **29**, 458–464

– (1991): On basic semialgebraic sets. Expo. Math. **9**, 289–334

L. Carlitz (1951): Some applications of a theorem of Chevalley. Duke Math. J. **18**, 811–819

J.W.S. Cassels (1964): On the representation of rational functions as sums of squares. Acta Arith. **9**, 79–82

J.W.S. Cassels, W.J. Ellison, A. Pfister (1971): On sums of squares and on elliptic curves over function fields. J. Number Theory **3**, 125–149

J.–L. Colliot–Thélène (1992): L'arithmétique des variétés rationnelles. Ann. Fac. des Sciences Toulouse **1**, 295–336

J.–L. Colliot–Thélène, M. Coste, L. Mahé, M.–F. Roy (Eds.) (1982): Géométrie algébrique réelle et formes quadratiques. Lecture Notes in Mathematics **959**, Berlin, Heidelberg, New York: Springer

J.–L. Colliot–Thélène, U. Jannsen (1991): Sommes de carrés dans les corps de fonctions. C.R. Acad. Sci. Paris, Sér. I. **312**, 759–762

M. Coste, L. Mahé, M.–F. Roy (Eds.) (1991): Real algebraic geometry. Lecture Notes in Mathematics **1524**, Berlin, Heidelberg, New York: Springer

Z.D. Dai, T.Y. Lam (1984): Levels in algebra and topology. Comment. Math. Helv. **59**, 376–424

Z.D. Dai, T.Y. Lam, C.K. Peng (1980): Levels in algebra and topology. Bull. Amer. Math. Soc. **3**, 845–848

A. Delzant (1962): Définition des classes de Stiefel–Whitney d'un module quadratique sur un corps de caractéristique différente de 2. C.R. Acad. Sci. Paris **255**, 1366–1368

L.E. Dickson (1899): Determination of the structure of all linear homogenous groups in a Galois field which are defined by a quadratic invariant. Amer. J. Math. **21**, 193–256. Mathematical Papers, Vol. I, 1–64

– (1906): On quadratic, Hermitian and bilinear forms. Trans. Amer. Math. Soc. **7**, 279–292. Mathematical Papers, Vol. VI

– (1907): On quadratic forms in a general field. Bull. Amer. Math. Soc. **14**, 108–115. Mathematical Papers, Vol. IV, 512–519

J. Dieudonné (1955): La géométrie des groupes classiques, Berlin, Heidelberg, New York: Springer

Sh. Ding, M. Kang, E.–Tj. Tan (1999): Chiungtze C. Tsen (1898–1940) and Tsen's Theorems. Rocky Mountain J. Math. **29**, 1237–1269

A. Dress (1971): A connection between Burnside– and Wittrings. Unpublished preprint, Bielefeld

D.W. Dubois (1969): A nullstellensatz for ordered fields. Ark. Math. **8**, 111–114

A. Durfee (1977): Bilinear and quadratic forms on torsion modules. Advances in Math. **25**, 133–164

W. Durfee (1948): Quadratic forms over fields with a valuation. Bull. Amer. Math. Soc. **54**, 338–351

M. Eichler (1952): Quadratische Formen und orthogonale Gruppen, Berlin, Heidelberg, New York: Springer

R. Elman, T.Y. Lam (1973a): Quadratic forms and the u–invariant I. Math. Z. **131**, 283–304

– (1973b): Quadratic forms and the u–invariant II. Invent. Math. **21**, 125–137

– (1974): Classification theorems for quadratic forms over fields. Comm. Math. Helv. **49**, 373–381

– (1976): Quadratic forms under algebraic extensions. Math. Ann. **219**, 21–42

A. Fröhlich, A.M.McEvett (1969): Forms over rings with involution. J. Algebra **12**, 79–104

W.D. Geyer, G. Harder, M. Knebusch, W. Scharlau (1970): Ein Residuensatz für symmetrische Bilinearformen. Invent. Math. **11**, 319–328

L. Goeritz (1933): Knoten und quadratische Formen. Math. Z. **36**, 647–654

D.K. Harrison (1970): Witt rings. Lecture Notes, Univ. of Kentucky, Lexington

H. Hasse (1923a): Über die Darstellbarkeit von Zahlen durch quadratische Formen im Körper der rationalen Zahlen. J. reine angew. Math. **152**, 129–148

– (1923b): Über die Äquivalenz quadratischer Formen im Körper der rationalen Zahlen. J. reine angew. Math. **152**, 205–224

– (1924a): Darstellbarkeit von Zahlen durch quadratische Formen in einem beliebigen algebraischen Zahlkörper. J. reine angew. Math. **153**, 113–130

– (1924b): Äquivalenz quadratischer Formen in einem beliebigen algebraischen Zahlkörper. J. reine angew. Math. **153**, 158–162

F. Hirzebruch (1956): Neue topologische Methoden in der algebraischen Geometrie. Ergeb. Math. **9**, Berlin, Heidelberg, New York: Springer

F. Hirzebruch, W.D. Neumann, S. Koh (1971): Differentiable manifolds and quadratic forms. Lecture Notes in Pure and Applied Math. **4**, New York: Marcel Dekker.

D. Hoffmann (1999): Pythagoras numbers of fields. J. Amer. Math. Soc. **12** (3), 839–848

D. Hoffmann, J. van Geel (1998): Zeros and norm groups of quadratic forms over function fields in one variable over a local non–dyadic field. J. Ramanujan Math. Soc. **13** (2), 85–110

J. S. Hsia, R. Johnson (1974): On the representation in sums of squares for definite functions in one variable over an algebraic number field. Amer. J. Math. **96**, 448–453

U. Jannsen (1989/90): Principe de Hasse cohomologique. Séminaire de théorie des nombres de Paris. Unpublished

B.W. Jones (1950): The arithmetic theory of quadratic forms. New York: Wiley.

I. Kaplansky (1953): Quadratic forms. J. Math. Soc. Japan **6**, 200–207

K. Kato (1982): Symmetric bilinear forms, quadratic forms and Milnor K–theory in characteristic two. Invent. Math. **66**, 493–510

– (1986): A Hasse principle for two dimensional global fields. J. reine angew. Math **366**, 142–181

M. Knebusch (1970): Grothendieck– und Wittringe von nichtausgearteten symmetrischen Bilinearformen. S.–Ber. Heidelberg. Akad. Wiss. Math.–naturw. Kl. 1969/70 3. Abh.

– (1971): Ein Satz über die Werte von quadratischen Formen über Körpern. Invent. Math. **12**, 300–303

– (1976): Generic splitting of quadratic forms I. Proc. London Math. Soc. **33**, 65–93

– (1977): Generic splitting of quadratic forms II. Proc. London Math. Soc. **34**, 1–31

– (1977a): Symmetric bilinear forms over algebraic varieties. Conference on quadratic forms 1976. Ed. G. Orzech, Queen's papers in pure and applied math. No. **46**, Kingston, Ontario

– (1982): An algebraic proof of the Borsuk–Ulam theorem for polynomial mappings. Proc. Amer. Math. Soc. **84**, 29–32

– (1984): An invitation to real spectra. Canad. Math. Soc. Conf. Proc. **4**, 51–105, 337–338, Providence: Amer. Math. Soc.

M. Knebusch, A. Rosenberg, R. Ware (1972): Structure of Wittrings and quotients of abelian group rings. Amer. J. Math. **94**, 119–155

M. Knebusch, W. Scharlau (1971): Über das Verhalten der Witt–Gruppe bei galoisschen Körpererweiterungen. Math. Ann. **193**, 189–196

– (1971a): Quadratische Formen und quadratische Reziprozitätsgesetze über algebraischen Zahlkörpern. Math. Z. **121**, 346–368

M. Knebusch, C. Scheiderer (1989): Einführung in die reelle Algebra. Braunschweig/Wiesbaden: Vieweg

H. Kneser (1934): Verschwindende Quadratsummen in Körpern. Jber. Dt. Math. Ver. **44**, 143–146

M. Kneser (1962): Colloquium talk (unpublished)

– (1969): Lectures on Galois cohomology of classical groups. Lecture Notes in Mathematics **47**. Tata Institute of Fundamental Research, Bombay

M. Kneser, D. Puppe (1953): Quadratische Formen und Verschlingungsinvarianten von Knoten. Math. Z. **58**, 376–385

M.–A. Knus (1991): Quadratic and hermitian forms over rings. Grundlehren der mathematischen Wissenschaften **294**, Berlin, Heidelberg, New York: Springer

M.–A. Knus, A. Merkurjev, M. Rost, J.–P. Tignol (1998): The Book of Involutions. Colloquium Publications **44**, Providence: Amer. Math. Soc.

M.–A. Knus, M. Ojanguren, R. Parimala (1982): Positive definite quadratic bundles over the plane. Comment. Math. Helv. **57**, 400–411

J.–L. Krivine (1964): Anneaux préordonnés. J. Anal. Math. **12**, 307–326

L. Kronecker (1890): Algebraische Reduktion der Schaaren bilinearer Formen.

– (1890, 1891): Algebraische Reduktion der Schaaren quadratischer Formen. In Leopold Kronecker's Werke Bd. 3,2. Berlin, Leipzig: Teubner (1931), New York: Chelsea (1968)

T.Y. Lam (1983): Orderings, valuations and quadratic forms. Conference Board of the Mathematical Sciences, Providence: Amer. Math. Soc.

W. Landherr (1938): Liesche Ringe vom Typus A über einem algebraischen Zahlkörper und hermitesche Formen über einem Schiefkörper. Abh. Math. Sem. Hamburg **12**, 200–241

S. Lang (1952): On quasi algebraic closure. Annals of Math. **55**, 317–390

D.B. Leep (1984): Systems of quadratic forms. J. reine angew. Math. **350**, 109–116

H. Lenz (1963): Einige Ungleichungen aus der Algebra der quadratischen Formen. Arch. Math. **14**, 373–382

D. Lewis (1989): New proofs of the structure theorems for Witt rings. Expo. Math. **7**, 83–88

F. Lorenz (1970): Quadratische Formen über Körpern. Lecture Notes in Mathematics **130**, Berlin, Heidelberg, New York: Springer

F. Lorenz, J. Leicht (1970): Die Primideale des Wittschen Ringes. Invent. Math. **10**, 82–88

M. Marshall (1980): Abstract Witt rings. Queen's papers in pure and applied mathematics **57**, Kingston

A. Merkurjev (1981): On the norm residue symbol of degree 2. Dokladi Akad. Nauk SSSR **261**, 542–547, English translation: Soviet Math. Doklady **23**, 546–551

– (1992): Simple algebras and quadratic forms. Math. USSR Isvestija **38**, 215–221

J. Milnor (1958): On simply connected 4–manifolds. Symp. Topologia Algebraica, Mexico, 122–128

– (1961): A procedure for killing homotopy groups of differentiable manifolds. In Differential Geometry. Proc. Symp. in Pure Math. Vol. III, Providence: Amer. Math. Soc.

– (1969): On isometries of inner product spaces. Invent. Math. **8**, 83–97

– (1970): Algebraic K–theory and quadratic forms. Invent. Math. **9**, 318–344

– (1971): Symmetric inner product spaces in characteristic two. In Prospects in Mathematics, Ann. of Math. Studies **70**, 59–75, Princeton

– (1971a): Introduction to algebraic K–theory. Annals of Math. Studies **72**. Princeton

J. Milnor, D. Husemoller (1973): Symmetric bilinear forms. Erg. der Math. u. ihrer Grenzgebiete **73**, Berlin, Heidelberg, New York: Springer

H. Minkowski (1890): Über die Bedingungen, unter welchen zwei quadratische Formen mit rationalen Koeffizienten ineinander rational transformiert werden können. J. reine angew. Math. **106**, 5–26. Gesammelte Abhandlungen, Erster Band, 219–239

O.T. O'Meara (1963): Introduction to quadratic forms, Grundlehren der mathematischen Wissenschaften **117**, Berlin, Heidelberg, New York: Springer

R. Parimala, R. Sujatha (1996): Hasse principle for Witt groups of function fields with special reference to elliptic curves. Duke Math. Journal **85**, 555–582

R. Parimala, V. Suresh (1998): Isotropy of quadratic forms over function fields of p-adic curves. Publ. Math. I.H.E.S. **88**, 129–150

A. Pfister (1965): Darstellung von –1 als Summe von Quadraten in einem Körper. J. London Math. Soc. **40**, 159–165

– (1965a): Multiplikative quadratische Formen. Arch. Math. **16**, 363–370

– (1966): Quadratische Formen in beliebigen Körpern. Invent. Math. **1**, 116–132

– (1967): Zur Darstellung definiter Funktionen als Summe von Quadraten. Invent. Math. **4**, 229–237

– (1984): Some remarks on the historical development of the algebraic theory of quadratic forms. Canad. Math. Soc. Conf. Proc. **4**, 1–16, Providence: Amer. Math. Soc.

– (1990): Quadratische Formen. In: Ein Jahrhundert Mathematik (Ed. G. Fischer et al.), 657–671, Braunschweig/Wiesbaden: Vieweg

– (1995): Quadratic forms with applications to algebraic geometry and topology. Cambridge

– (2000): On the Milnor Conjectures: History, Influence, Applications. Jahresber. Deutsche Math.–Ver. **102**, 15–41

A. Pfister, S. Stolz (1987): On the level of projective spaces. Comment. Math. Helv. **62**, 286–291

Y. Pourchet (1971): Sur la représentation en somme de carrés des polynômes à une indéterminée sur un corps de nombres algébriques. Acta Arith. **19**, 89–104

A. Prestel (1984): Lectures on formally real fields. Lecture Notes in Mathematics **1093**, Berlin, Heidelberg, New York: Springer

H.G. Quebbemann, R. Scharlau, W. Scharlau, M. Schulte (1976): Quadratische Formen in additiven Kategorien. Bull. Soc. Math. France Mémoire **48**, 93–101

H.G. Quebbemann, W. Scharlau, M. Schulte (1979): Quadratic and hermitian forms in additive and abelian categories. J. Algebra **59**, 264–289

D. Quillen (1973): Higher algebraic K-theory: I. In H. Bass (ed.), Higher K-theories. Lecture Notes in Mathematics **341**, 85–147, Berlin, Heidelberg, New York: Springer

G. de Rham (1931): Sur l'Analysis situs des variétés à n dimensions. Journ. de Math. pures et appliquées **10**, 115–201

C. Riehm (1970): The corestriction of algebraic structures. Invent. Math. **11**, 73–98

J.–J. Risler (1976): Le théorème des zéros en géométries algébrique et analytique réelle. Bull. Soc. Math. France **104**, 113–127

P. Roquette (1966): Splitting of algebras by function fields of one variable. Nagoya Math. J. **27**, 625–642

A. Rosenberg, R. Ware (1970): The zero–dimensional galois cohomology of Witt rings. Invent. Math. **11**, 65–72

C.–H. Sah (1972): Symmetric bilinear forms and quadratic forms. J. Algebra **20**, 144–160

D. Saltman (1997): Division Algebras over p–adic curves. J. Ramanujan Math. Soc. **12** (1), 25–47. (cf. review Zentralblatt für Math. 902.16021)

W. Scharlau (1963): Ornithologische Beobachtungen in der ägyptischen Oase El–Dachla. Orn. Mitt. **15**, 246–248

– (1964): The cohomological theory of quadratic forms. Preprint Columbia Univ.

– (1967): Quadratische Formen und Galois–Cohomologie. Invent. Math. **4**, 238–264

– (1969): Quadratic forms. Queen's papers in pure and applied mathematics **22**, Kingston

– (1969a): Zur Pfisterschen Theorie der quadratischen Formen. Invent. Math. **6**, 327–328

– (1970): Induction theorems and the structure of the Witt group. Invent. Math. **11**, 37–44

– (1972): Quadratic reciprocity laws. J. Number Theory **4**, 78–97

– (1975): Zur Existenz von Involutionen auf einfachen Algebren. Math. Z. **145**, 29–32

– (1985): Quadratic and hermitian forms. Grundlehren der mathematischen Wissenschaften **270**, Berlin, Heidelberg, New York: Springer

C. Scheiderer (1996): Hasse principles and approximation theorems. Invent. Math. **125**, 307–365

H. Seifert (1936): Die Verschlingungsinvarianten der zyklischen Knotenüberlagerungen. Hamb. Math. Abh. **11**, 84–101

J.–P. Serre (1962): Formes bilinéaires symétriques entierès de discriminant $\pm$ 1. Sem. H. Cartan 14e année, 1961/62

– (1962): Corps locaux. Paris: Hermann

– (1964): Cohomologie galoisienne. Lecture Notes in Mathematics **5**, Berlin, Heidelberg, New York: Springer

– (1970): Cours d'arithmétique, Paris: Hermann. English translation (1978): A course in arithmetic. Graduate Texts in Mathematics **7**, Berlin, Heidelberg, New York: Springer

T.A. Springer (1952): Sur les formes quadratique d'indice zéro. C.R. Acad. Sci. Paris **234**, 1517–1519

– (1955): Quadratic forms over a field with a discrete valuation. Indag. Math. **17**, 352–362

– (1959): On the equivalence of quadratic forms. Indag. Math. **21**, 241–253

G. Stengle (1974): A nullstellensatz and a positivstellensatz in semialgebraic geometry. Math. Ann. **207**, 87–97

S. Stolz (1989): The level of real projective spaces. Comment. Math. Helv. **64**, 661–674

R. Sujatha (1995): On the Hasse principle for Witt groups. Math. Ann. **302**, 215–224

R. Swan (1962): Vector bundles and projective modules. Trans. Amer. Math. Soc. **105**, 264–277

R. Swan (1985): K-Theory of quadric hypersurfaces. Ann. of Maths. **112**, 113–153

Ch. Tsen (1936): Zur Stufentheorie der Quasialgebraisch-Abgeschlossenheit kommutativer Körper. J. Chinese Math. Soc. **1**, 81–92

O. Veblen (1923): The intersection numbers. Trans. Amer. Math. Soc. **25**, 540–550

G.E. Wall (1963): On the conjugacy classes in unitary, symplectic and orthogonal groups. J. Austral. Math. Soc. **3**, 1–62

C.T.C. Wall (1963): Quadratic forms on finite groups and related topics. Topology **2**, 281–298

– (1970): On the axiomatic foundations of the theory of hermitian forms. Proc. Cambridge Phil. Soc. **67**, 243–250

– (1973): Foundations of algebraic L–theory. In H. Bass (ed.), Hermitian K-Theory and Geometric Applications. Lecture Notes in Mathematics **343**, 266–300, Berlin, Heidelberg, New York: Springer

C. Weibel (1999): The development of algebraic K–theory before 1980. Algebra, K-theory, groups, and education (New York, 1997), 211–238, Providence: Amer. Math. Soc.

A. Weil (1964): Sur certains groupes d'opérateurs unitaires. Acta Math. **111**, 143–211

H. Weyl (1924): Análisis situs combinatorio (continuación). Rev. Mat. Hisp.–Americana **6**, 33–41. Gesammelte Abh. II, 416–432, Berlin, Heidelberg, New York: Springer, 1968

J.H.C. Whitehead (1949): On simply connected 4–dimensional polyhedra. Comment. Math. Helv. **22**, 48–92

E. Witt (1937): Theorie der quadratischen Formen in beliebigen Körpern. J. reine angew. Math. **176**, 31–44

– (1998): Collected papers. Gesammelte Abhandlungen. Berlin, Heidelberg, New York: Springer

Fachbereich Mathematik und Informatik
Westfälische Wilhelms-Universität Münster
Einsteinstr. 62
48149 Münster
Germany

E-mail address: scharla@math.uni-muenster.de

Contemporary Mathematics
Volume **272**, 2000

Local fundamental classes derived from higher K-groups: III

Victor P. Snaith

ABSTRACT. To a Galois extension of local fields is associated a canonical fundamental class constructed from algebraic K-groups in dimensions $2r, 2r+1$ for any $r \geq 0$. The case $r = 0$ is a familiar part of classical local class field theory and the higher dimensional analogues satisfy all the same naturality properties. In positive characteristic we study the connection between the K_2/K_3 fundamental class and Lichtenbaum's motivic complex of weight two. Also the fundamental classes are used in the construction of higher-dimensional Chinburg invariants, $\Omega_r(E/F, 2)$, associated to a Galois extension of global fields. We explain how to calculate $\Omega_1(E/F, 2)$ in the case of tamely ramified Galois extension of function fields.

1. Introduction

At the conference I gave a talk entitled: "Does the weight-two motivic complex have an Euler characteristic?" and in this paper I shall describe the background to this question – the existence of local fundamental classes associated to higher K-groups – and ramifications in a number of directions, including the connection with the motivic complex of weight two.

Let us begin with the local fundamental classes. Suppose that G is a finite group and that

$$\underline{\underline{E}}: \quad A \longrightarrow B \longrightarrow C \longrightarrow D$$

is a 2-extension of $\mathbf{Z}[G]$-modules in which B and C are cohomologically trivial. Such a sequence defines an element of $Ext^2_{\mathbf{Z}[G]}(D, A)$.

There are two natural operations associated to a subgroup, $J \subseteq G$. The first - passage to subgroups - is merely to consider the modules as $\mathbf{Z}[J]$-modules. The second - passage to quotient groups - is more complicated and applies to the case of a normal subgroup, $J \lhd G$. In this case, let A^J, A_J denote the J-invariants and J-coinvariants of A, respectively ([**19**] p.3). We have a commutative diagram of $\mathbf{Z}[G/J]$-modules in which the rows and columns are exact and N_J denotes the norm, $N(x) = \sum_{g \in J} gx$,

2000 *Mathematics Subject Classification.* Primary 11R37; Secondary 19F05.

$$\begin{array}{ccccccccc} & & & & B_J & \longrightarrow & C_J & \longrightarrow & D_J & \longrightarrow & 0 \\ & & & & \cong\Big\downarrow N_J & & \cong\Big\downarrow N_J & & & & \\ 0 & \longrightarrow & A^J & \longrightarrow & B^J & \longrightarrow & C^J & & & & \end{array}$$

resulting in the associated J-invariant/coinvariant 2-extension

$$A^J \longrightarrow B^J \longrightarrow C^J \longrightarrow D_J$$

in which the $\mathbf{Z}[G/J]$-modules, B^J and C^J, are cohomologically trivial.

These operations are relevant to the naturality properties of the fundamental classes constructed in ([**15**] [**16**] [**17**]).

THEOREM 1.1.
Let L/K be a Galois extension of local fields with group, $G(L/K)$. Then for each $r \geq 0$ there exists a canonical 2-extension of $\mathbf{Z}[G(L/K)]$-modules of the form

$$0 \longrightarrow K_{2r+1}(L) \longrightarrow A_r(L) \longrightarrow B_r(L) \longrightarrow K_{2r}(L) \longrightarrow 0$$

satisfying the following conditions:

(*i*) *The $\mathbf{Z}[G(L/K)]$-modules $A_r(L)$ and $B_r(L)$ are cohomologically trivial.*

(*ii*) *If $G(L/E) \subseteq G(L/K)$ then the canonical 2-extension associated with L/E is canonically isomorphic to the canonical 2-extension for L/K, considered as a 2-extension of $\mathbf{Z}[G(L/E)]$-modules.*

(*iii*) *If $G(L/E) \lhd G(L/K)$ then the canonical 2-extension associated with E/K is canonically isomorphic to*

$$K_{2r+1}(L)^{G(L/E)} \longrightarrow (A_r(L))^{G(L/E)} \longrightarrow (B_r(L))^{G(L/E)} \longrightarrow K_{2r}(L)_{G(L/E)}.$$

Actually, in [**15**] and [**16**] we showed how to construct the canonical, natural 2-extension (depending up to quasi-isomorphism upon a choice of a root of unity, ξ_t) of Theorem 1.1 for finite Galois extensions of p-adic local fields. Furthermore, when $r \geq 2$ in this case we needed to assume that the Lichtenbaum-Quillen Conjecture ([**16**] p.326) was true for the mod p algebraic K-theory of p-adic local fields. When [**16**] was written this was known for 2-adic fields by [**22**]. For p-adic fields when p is odd the conjecture was proved recently [**8**]. When L/K is a Galois extension of local fields of characteristic p the Lichtenbaum-Quillen Conjecture is now known to be true by [**7**], which shows that the K-theory of L has no p-torsion, combined with the results of [**20**].

These advances make it possible to construct the local fundamental classes associated to the higher K-groups of local fields without any assumptions.

When $r = 0$ the result is well-known from local class field theory and the theory of the Brauer group. The construction in higher dimensions is accomplished by imitating the construction of the classical local fundamental class given in ([**14**] p.202; see also [**18**] p.303, [**19**] p.9). Since [**15**] and [**16**] only dealt with p-adic local fields I shall devote §2 to a sketch of the construction in the characteristic p case. I shall do this by giving a complete treatment of the construction of the K_2/K_3

case in §2.1-§2.18 and in §2.19 I shall explain the simple modifications necessary for the K_{2r}/K_{2r+1} case. In §3 we concentrate on the case of a tamely ramified Galois extension in characteristic p. From the construction of §2 we deduce a simpler representative (Theorem 3.3) of the fundamental 2-extension in the tame case. We also compute, in Remark 3.6, the Ext^2, in all cases, in which the local fundamental 2-extensions live. We would like to compare our K_2/K_3 local fundamental class with the motivic complex $\Gamma(2, L)$ of ([**9**] [**10**]). As a first approximation to this, in Theorem 4.2, we sketch how to prove that the K_2/K_3 local fundamental class and $\Gamma(2, L)$ induce the same cup-product isomorphism in Tate cohomology, at least if we invert 2. It is here that the homological results of Remark 3.6 are used to reduce to the tame case. The last two sections explain how one uses the K_{2r}/K_{2r+1} local fundamental classes to associate a Chinburg invariant, $\Omega_r(E/F, 2)$, to each Galois extension of global fields E/F in characteristic $p > 0$. The Chinburg invariant takes its values in the class-group of the integral group-ring of the Galois group, as explained in §5.1. This means that it may be described in terms of projective modules or in terms of idélic-valued functions on the representation ring, via the Hom-description of §6.2. We sketch how to evaluate $\Omega_1(E/F, 2)$ in the tame case in terms of modules (Theorem 5.6) and of the Hom-description (Theorem 6.5).

2. Constructing the local fundamental classes in characteristic p

2.1. Let L/K be a Galois extension of local fields in characteristic $p > 0$ with group, $G(L/K)$. Let K_0 denote the maximal unramified extension of K, $L_0 = K_0L$ and $d = [K_0 \cap L : K]$. Let $\overline{L}$ denote the residue field of L so that $\overline{L}$ is a finite field of characteristic p. Since $K_{2r}(L)$ and $K_{2r+1}(L)$ have no p-torsion [**7**] the results of [**20**] imply that the tame symbol

$$\delta : K_{2r}(L) \longrightarrow K_{2r-1}(\overline{L})$$

and the map induced by the inclusion of the field of constants

$$K_{2r+1}(\overline{L}) \longrightarrow K_{2r+1}(L)$$

each has uniquely divisible kernel and cokernel. Since $G(L/K)$ is finite we have canonical isomorphisms of the form

$$\begin{aligned} Ext^2_{\mathbf{Z}[G(L/K)]}(K_{2r}(L), K_{2r+1}(L)) &\cong Ext^2_{\mathbf{Z}[G(L/K)]}(TorsK_{2r}(L), K_{2r+1}(L)) \\ &\cong Ext^2_{\mathbf{Z}[G(L/K)]}(K_{2r-1}(\overline{L}), K_{2r+1}(\overline{L})) \end{aligned}$$

where $TorsM$ denotes the torsion subgroup of M. Incidentally, for one-dimensional local fields L the indecomposable K-group $K_j^{ind}(L)$ [**11**] differs from $K_j(L)$ only by uniquely divisible groups when $j \geq 3$ by [**2**].

For the moment let us concentrate on the construction of the $K_2 - K_3$ fundamental class. The preceding isomorphisms of Ext-groups will be very useful since, in the $K_2 - K_3$ case, they reduce the construction of the 2-extension of Theorem 1.1 to Theorem 2.2 below, which will be explained in the remainder of this section.

In §2.19 I will explain how to modify the following construction to the $K_{2r} - K_{2r+1}$ case when $r \geq 2$. However in §3.2 and Theorem 3.3 I will give a much simpler construction of an equivalent 2-extension in the tamely ramified case. The main advantages in using the following modification of the construction of [**15**] and [**16**] are that it works even in the wildly ramified case and using it the proof of the naturality property of Theorem 2.2 (iii) is identical to that of ([**15**] §4). Possibly one might attempt to verify the naturality properties by using Remark 3.6, which establishes an isomorphism of Ext^2's in the wild and the tame cases, to reduce the verification to the tame case, to reduce the verification to the case of the tame 2-extension of Theorem 3.3.

THEOREM 2.2.

Let L/K be a Galois extension of (one-dimensional) local fields with group, $G(L/K)$, and char$(L) = p$. Then there exists a canonical 2-extension of $\mathbf{Z}[G(L/K)]$*-modules of the form*

$$K_3(L) \longrightarrow \widetilde{\oplus}_{i=1}^{d} K_3(L_0) \stackrel{1-\hat{F}}{\longrightarrow} \widetilde{\mathcal{W}} \longrightarrow Tors K_2(L)$$

satisfying the following conditions:

(i) The $\mathbf{Z}[G(L/K)]$*-modules $\widetilde{\oplus}_{i=1}^{d} K_3(L_0)$ and $\widetilde{\mathcal{W}}$ are cohomologically trivial.*

(ii) If $G(L/E) \subseteq G(L/K)$ then the canonical 2-extension associated with L/E is quasi-isomorphic to the canonical 2-extension for L/K, considered as a 2-extension of $\mathbf{Z}[G(L/E)]$*-modules.*

(iii) If $G(L/E) \triangleleft G(L/K)$ then the canonical 2-extension associated with E/K is quasi-isomorphic to

$$K_3(L)^{G(L/E)} \longrightarrow (\widetilde{\oplus}_{i=1}^{d} K_3(L_0))^{G(L/E)} \stackrel{1-\hat{F}}{\longrightarrow} (\widetilde{\mathcal{W}})^{G(L/E)} \longrightarrow Tors K_2(L)_{G(L/E)}$$

where M_G denotes the G-coinvariants of M.

2.3. Assume for the moment that L/K is totally ramified, so that $K_0 \cap L = K$ and $d = 1$. We have isomorphisms, $G(K_0/K) \cong G(L_0/L) \cong \hat{\mathbf{Z}}$, the adic integers, and we may define $L(s) = LK(s) \subset L_0$ to be the fixed field of $s\hat{\mathbf{Z}}$ for each positive integer, s. Let $\kappa(s)$ denote the kernel of the norm, $N_{L(s)/L} : K_3(L(s)) \longrightarrow K_3(L)$. From [**11**], bearing in mind that $K_3^{ind}(L)$ and $K_3(L)$ differ only by a uniquely divisible group, we have a canonical exact sequence of the form

$$\begin{aligned} 0 \longrightarrow H^1(G(L(s)/L); K_3(L(s))) \longrightarrow K_2(L) \\ \longrightarrow K_2(L(s))^{G(L(s)/L)} \longrightarrow H^1(G(L(s)/L); K_3(L(s))) \longrightarrow 0. \end{aligned}$$

Let $F \in G(L(s)/L)$ denote the lifting of the Frobenius under the isomorphism, $G(L_0/L) \xrightarrow{\cong} G(\overline{L}_0/\overline{L})$, where $\overline{L}_0 = \overline{\mathbf{F}}_p$, the algebraic closure of $\mathbf{F}_p$, and $\overline{L}$ respectively denote the residue fields of L_0 and L. Hence we have an exact sequence

$$0 \longrightarrow K_3(L) \longrightarrow K_3(L(s)) \stackrel{1-F}{\longrightarrow} \kappa(s) \longrightarrow Ker(K_2(L) \longrightarrow K_2(L(s))^{G(L(s)/L)}) \longrightarrow 0$$

since $K_3(L(s))^{G(L(s)/L)} \cong K_3(L)$. Taking the direct limit over s we obtain an exact sequence of $\mathbf{Z}[G(L/K)]$-modules of the form

$$0 \longrightarrow K_3(L) \longrightarrow K_3(L_0) \stackrel{1-F}{\longrightarrow} \mathcal{U} \longrightarrow \bigcup_s Ker(K_2(L) \longrightarrow K_2(L(s))^{G(L(s)/L)}) \longrightarrow 0$$

where $\mathcal{U} = \bigcup_s \kappa(s)$. By a result of Tate ([**21**], c.f. [**15**] Lemma 2.1; see also [**12**]) $K_2(L) \cong \mu(L) \oplus D_L$ where D_L is uniquely divisible and $\mu(L) \cong \mu(\overline{L})$ is the cyclic group of order prime to p given by the groups of roots of unity, $\mu(L)$. As in ([**15**] Proposition 2.8)

$$\bigcup_s Ker(K_2(L) \longrightarrow K_2(L(s))^{G(L(s)/L)}) = {}_{p^\infty}K_2(L),$$

the p-primary part of $K_2(L)$, which is zero.

The torsion in $K_3(L_0)$ is isomorphic to $(\mathbf{Q}/\mathbf{Z})(2)[1/p]$, which also lies in $\mathcal{U}$, both being isomorphic to $K_3(\overline{L}_0)$. Furthermore, if $\overline{L} = \overline{K} = \mathbf{F}_q$, then $L(s) \cong \mathbf{F}_{q^s}((X))$ ([**5**] p.9) and the cokernel of the injective homomorphism induced by the inclusion of the field of constants, $K_3(\mathbf{F}_{q^s}) \longrightarrow K_3(L(s))$, is uniquely divisible ([**11**] I(1.4), [**13**]). Therefore each of the inclusions

$$K_3(\overline{\mathbf{F}}_p) \cong (\mathbf{Q}/\mathbf{Z})(2)[1/p] \longrightarrow K_3(L_0)$$

and

$$K_3(\overline{\mathbf{F}}_p) \longrightarrow \mathcal{U}$$

induces isomorphisms on Tate cohomology groups, $\hat{H}^i(G(L/K); -)$, for all i (c.f. [**15**] Corollary 2.12).

2.4. Now let us consider the general case of a finite Galois extension of (one-dimensional) local fields, L/K, in characteristic p with $W' = L \cap K_0$ and $d = [W' : K]$. Hence W'/K is the maximal unramified subextension of L/K. Since L/W' is totally ramified we have, from §2.3, a short exact sequence of $\mathbf{Z}[G(L/W')]$-modules

$$0 \longrightarrow K_3(L) \longrightarrow K_3(L_0) \xrightarrow{1-F} \mathcal{U} \longrightarrow 0,$$

which we shall modify to construct the canonical 2-extension of Theorem 2.2 following the method of ([**15**] §3) which was in turn an imitation of the fundamental class of local class field theory, as described in ([**18**] §7.1; [**19**] §1.2).

Notice that $G(L/W') = G_0(L/K)$ and $G_1(L/W') = G_1(L/K)$, by ([**14**] p.62), so that

$$r = [G(L/W') : G_1(L/W')] = [G_0(L/K) : G_1(L/K)] .$$

If $G(L/E) = G_1(L/K)$ then

$$G(L/K)/G(L/E) \cong G(E/K) = \{a, g \mid a^r = 1, g^d = a^c, gag^{-1} = a^v\}$$

where $< a >= G(E/W')$ and $v = |\overline{K}|$.

Now let us consider the Galois groups which will be involved in the construction. We have a map, given by the restriction of g to W' and denoted by $(g|W')$, inducing a homomorphism

$$h_1 : G(K_0/K) \times G(L/K) \cong \hat{\mathbf{Z}} \times G(L/K) \longrightarrow G(W'/K)$$

defined by $h_1(F^i, z) = (F^i|W')\cdot(z|W')^{-1}$. Here, as in §2.3, F denotes the Frobenius automorphism. To a pair, $(F^i, z) \in Ker(h_1)$, we may associate the Galois automorphism of $L_0 = LK_0$ which is equal to F^i on K_0 and to z on L. This induces an isomorphism

$$Ker(h_1) \xrightarrow{\cong} G(L_0/K).$$

If $d = [W' : K]$ define $F_0 \in G(L_0/K)$ by $(F_0|K_0) = F^d$ and $(F_0|L) = 1$, the identity map.

As explained in ([**18**] pp. 303-304; [**19**] pp. 9-10), there is an isomorphism of K-algebras

$$\lambda : L \otimes_K K_0 \longrightarrow \oplus_{i=1}^{d} L_0$$

given by the formula $\lambda(\alpha \otimes \beta) = (F^{d-1}(\beta)\alpha, F^{d-2}(\beta)\alpha, \ldots, \beta\alpha)$ and fitting into a commutative diagram of the form

$$\begin{array}{ccc} L \otimes_K K_0 & \xrightarrow{(1 \otimes F)} & L \otimes_K K_0 \\ \downarrow \lambda & & \lambda \downarrow \\ \oplus_{i=1}^d L_0 & \xrightarrow{\hat{F}} & \oplus_{i=1}^d L_0 \end{array}$$

where $\hat{F}(x_1, \ldots, x_d) = (F_0(x_d), x_1, \ldots, x_{d-1})$. Since K-theory is additive there is an induced isomorphism

$$\lambda : K_s(L \otimes_K K_0) \longrightarrow \oplus_{i=1}^d K_s(L_0)$$

for each $s \geq 0$. This isomorphism induces a $G(L/K)$-action on the right-hand group, induced by the Galois action on L. Since the Frobenius acts on the other factor, K_0, in the tensor product we may define a $\mathbf{Z}[G(L/K)]$-homomorphism

$$\hat{F} : \oplus_{i=1}^d K_s(L_0) \longrightarrow \oplus_{i=1}^d K_s(L_0)$$

by the same formula as before. Therefore the induced map

$$\hat{F} : \oplus_{i=1}^d K_3^{ind}(L_0) \longrightarrow \oplus_{i=1}^d K_3^{ind}(L_0)$$

is also a $\mathbf{Z}[G(L/K)]$-homomorphism.

LEMMA 2.5. ([**15**] *Lemma 3.2*)
Suppose that $g \in G(L/K)$ satisfies $(g|W') = (F^j|W')$ for some $0 \leq j \leq d-1$. Then the action of g on $(u_1, u_2, \ldots, u_d) \in \oplus_{i=1}^d K_3^{ind}(L_0)$ is given by

$$g(u_1, u_2, \ldots, u_d) \quad = \quad \hat{F}^{-j}((F^j, g)(u_1), (F^j, g)(u_2), \ldots, (F^j, g)(u_d)).$$

DEFINITION 2.6. Let $\mathcal{U} \subset K_3^{ind}(L_0)$ be as in §2.3. Define a subgroup, $\mathcal{W} \subset \oplus_{i=1}^d K_3^{ind}(L_0)$ by

$$\mathcal{W} \quad = \quad \{(u_1, u_2, \ldots, u_d) \mid \sum_{i=1}^d u_i \in \mathcal{U}\}.$$

Hence $\mathcal{W} = \mathcal{U}$ when $d = 1$.

LEMMA 2.7. ([**15**] *Lemma 3.4*)
(i) The subgroup, $\mathcal{W} \subset \oplus_{i=1}^d K_3^{ind}(L_0)$, is a $\mathbf{Z}[G(L/K)]$-submodule and
(ii) the image of $1 - \hat{F} : \oplus_{i=1}^d K_3^{ind}(L_0) \longrightarrow \oplus_{i=1}^d K_3^{ind}(L_0)$ lies in $\mathcal{W}$.

PROPOSITION 2.8. (*c.f.* [**15**] *Proposition 3.5*)
In the notation of §2.4, there is a natural exact sequence of $\mathbf{Z}[G(L/K)]$-modules of the form

$$0 \longrightarrow K_3^{ind}(L) \xrightarrow{\Delta} \oplus_{i=1}^d K_3^{ind}(L_0) \xrightarrow{1-\hat{F}} \mathcal{W} \longrightarrow 0$$

where Δ denotes the diagonal map.

THEOREM 2.9.
With the notation of §2.4, each of the inclusions

$$\oplus_{i=1}^d (\mathbf{Q}/\mathbf{Z})(2)[1/p] \cong \oplus_{i=1}^d K_3(\overline{\mathbf{F}}_p) \subset \mathcal{W} \subset \oplus_{i=1}^d K_3(L_0)$$

induces an isomorphism on Tate cohomology groups, $\hat{H}^i(G(L/K);-)$, for all i. Furthermore

$$\hat{H}^i(G(L/K);\oplus_{i=1}^d K_3(L_0)) \cong \begin{cases} \mathbf{Z}/r & \text{if } i \text{ is odd,} \\ 0 & \text{if } i \text{ is even} \end{cases}$$

where $r = [G_0(L/K) : G_1(L/K)]$.

Proof

The statement concerning cohomology isomorphisms follows from the discussion of §2.3.

Consider the spectral sequence for ordinary group cohomology

$$\begin{aligned} E_2^{s,t} &= H^s(G(W'/K); H^t(G(L/W'); \oplus_{i=1}^d K_3(L_0))) \\ &\Longrightarrow H^{s+t}(G(L/K); \oplus_{i=1}^d K_3(L_0)). \end{aligned}$$

As in §2.4, let $r = [G(L/W) : G_1(L/W')] = [G_0(L/K) : G_1(L/K)]$. By Lemma 2.5, $G(L/W')$ acts diagonally, component-by-component on $\oplus_{i=1}^d K_3^{ind}(L_0)$ and trivially on the subgroup, $\oplus_{i=1}^d (\mathbf{Q}/\mathbf{Z})(2)[1/p]$. Also, since $G_1(L/K) = G(L/E)$ is a p-group,

$$\begin{aligned} H^t(G(L/W'); K_3(L_0)) &\cong H^t(G(L/W'); (\mathbf{Q}/\mathbf{Z})(2)[1/p]) \\ &\cong H^t(G(E/W'); (\mathbf{Q}/\mathbf{Z})(2)[1/p]) \end{aligned}$$

which is isomorphic to $\mathbf{Z}/r = {}_r(\mathbf{Q}/\mathbf{Z})(2)[1/p])$, the r-torsion subgroup, if $0 < t$ is odd and is zero if $0 < t$ is even, since $G(E/W') \cong G_0(L/W')/G_1(L/W') \cong \mathbf{Z}/r$.

Note that $L_0 = W_0'$, although in the rest of this proof we shall refer to it as W_0' since it helps in keeping track of the Galois actions which are to follow.

Hence,

$$H^t(G(L/W'); \oplus_{i=1}^d K_3(L_0)) \cong \begin{cases} \oplus_{i=1}^d \mathbf{Z}/r & \text{if } t \text{ is odd,} \\ 0 & \text{if } 0 < t \text{ is even,} \\ \oplus_{i=1}^d K_3(W_0') & \text{if } t = 0. \end{cases}$$

If $g \in G(W'/K) \cong \mathbf{Z}/d$ is a generator then the cohomology, $H^*(G(W'/K); M)$, is computed from the cochain complex

$$\ldots \longrightarrow M \xrightarrow{1-g} M \xrightarrow{N} M \xrightarrow{1-g} M \longrightarrow \ldots$$

where $N = \sum_{j=0}^{d-1} g^j$. The generator, g, may be lifted to

$$(F, F) \in G(L_0/K) \subset G(K_0/K) \times G(L/K)$$

in the notation of §2.4. By Lemma 2.5, the action on $\oplus_{i=1}^d \mathbf{Z}/r$ is given by

$$\begin{aligned} g(a_1, \ldots, a_d) &= (F(a_2), F(a_3), \ldots, F(a_d), F_0^{-1}(F(a_1))) \\ &= (F(a_2), F(a_3), \ldots, F(a_d), F(a_1)) \\ &= (qa_2, qa_3, \ldots, qa_d, qa_1) \end{aligned}$$

since F acts on $\mathbf{Z}/r = {}_r(\mathbf{Q}/\mathbf{Z})(2)[1/p]$ by "multiplication" by $q = |\overline{K}|^2$, $F_0 = F^d$ and r divides $q^d - 1$.

Therefore $Ker(1-g)$ consists of d-tuples, $(a_1, \ldots, a_d)$, such that $a_i = q^{d-i} a_d$ which comprise a copy of $\mathbf{Z}/r$ generated by $(q^{d-1}, q^{d-2}, \ldots, q, 1)$. On the other hand

$$\begin{aligned} N(a_1,\dots,a_d) &= (a_1,a_2,\dots,a_{d-1},a_d) \\ &\quad +(qa_2,qa_3,\dots,qa_d,qa_1) \\ &\quad +(q^2a_3,q^2a_4,\dots,q^2a_1,q^2a_2) \\ &\quad \vdots \\ &= xq^{1-d}(q^{d-1},q^{d-2},\dots,q,1) \end{aligned}$$

where $x = \sum_{i=1}^{d} q^{i-1}a_i \in \mathbf{Z}/r$. Therefore $H^{2s}(G(W'/K);\oplus_{i=1}^{d}\mathbf{Z}/r)$ vanishes if $s > 0$ and is isomorphic to $\mathbf{Z}/r$ if $s = 0$.

An element, $(a_1,\dots,a_d)$, lies in $Ker(N)$ if and only if $x = \sum_{i=1}^{d} q^{i-1}a_i \equiv 0$ (modulo r). However, for such elements, we may solve the equation

$$(a_1,\dots,a_d) = (1-g)(b_1,\dots,b_d) = (b_1-qb_2,b_2-qb_3,\dots,b_d-qb_1)$$

by choosing b_1 and b_d to satisfy $b_d - qb_1 = a_d$ and then setting

$$b_i = q^{d-i+1}b_1 + \sum_{j=i}^{d} q^{j-1}a_j$$

for $2 \le i \le d-1$. This is consistent because, in $\mathbf{Z}/r$

$$a_1 + qb_2 = (a_1 + qa_2 + q^2a_3 + \ldots + q^{d-1}a_d) + q^d b_1 = b_1 \ .$$

Hence $H^{2s+1}(G(W'/K);\oplus_{i=1}^{d}\mathbf{Z}/r)$ vanishes for all $s \ge 0$.

Now let us consider $H^s(G(W'/K);\oplus_{i=1}^{d}K_3(W_0'))$. Since $F_0 = F^d$ acts trivially on W_0' the action on $(u_1,\dots,u_d) \in \oplus_{i=1}^{d}K_3(W_0')$ is given by

$$g(u_1,\dots,u_d) = (F(u_2),F(u_3),\dots,F(u_d),F(u_1)) \ .$$

Therefore $(u_1,\dots,u_d) \in Ker(1-g)$ if and only if $u_i = F^{d-i}(u_d)$ for each $1 \le i \le d$. On the other hand

$$\begin{aligned} N(v_1,\dots,v_d) &= (v_1,v_2,\dots,v_{d-1},v_d) \\ &\quad +(F(v_2),F(v_3),\dots,F(v_d),F(v_1)) \\ &\quad +(F^2(v_3),F^2(v_4),\dots,F^2(v_1),F^2(v_2)) \\ &\quad \vdots \\ &= (w,F^{-1}(w),F^{-2}(w),\dots,F^{1-d}(w)) \end{aligned}$$

where $w = \sum_{j=1}^{d} F^{j-1}(v_j)$. Setting $v_1 = v_2 = \ldots = v_{d-1} = 0$ and $v_d = u_d$ shows that $Ker(1-g) = Im(N)$ and $H^{2s}(G(W'/K);\oplus_{i=1}^{d}K_3(W_0'))$ vanishes for $s > 0$ while $H^0(G(W'/K);\oplus_{i=1}^{d}K_3(W_0')) \cong K_3(W_0')$.

The element, $(u_1,\dots,u_d)$, lies in $Ker(N)$ if and only if $0 = \sum_{j=1}^{d} F^{j-1}(u_j) \in K_3(W_0')$ while $(1-g)(v_1,\dots,v_d) = (v_1-F(v_2),v_2-F(v_3),\dots,v_d-F(v_1))$. In this case the equation $(1-g)(v_1,\dots,v_d) = (u_1,\dots,u_d)$ may be solved by choosing v_1 and v_d to satisfy $v_d - F(v_1) = u_d$ and then setting $v_i = F^{d-i+1}(v_1) + \sum_{j=i}^{d} F^{j-i}(u_j)$ for $2 \le i \le d-1$. This is consistent because

$$u_1 + F(v_2) = u_1 + \sum_{j=2}^{d} F^{j-2+1}(u_j) + F^d(v_1) = v_1 \ .$$

Hence $H^{2s+1}(G(W'/K);\oplus_{i=1}^{d}K_3(W_0'))$ vanishes for all $s \ge 0$.

Therefore $E_2^{s,t} = 0$ if $0 < s+t$ is even so that $\hat{H}^{2s}(G(L/K); \oplus_{i=1}^d K_3(L_0))$ vanishes when $s > 0$, as required. When $s+t = 2u+1$ the only contributions in the spectral sequence come from $E_2^{0,2u+1} \cong \mathbf{Z}/r$ and so $\hat{H}^{2u+1}(G(L/K); \oplus_{i=1}^d K_3(L_0)) \cong \mathbf{Z}/r$, as required. Finally, the only contribution to $H^0(G(L/K); \oplus_{i=1}^d K_3(L_0))$ comes from $E_2^{0,0} \cong K_3(W_0')$, from which is it easy to calculate the Tate cohomology in dimensions less than or equal to zero. □

2.10. Suppose that L/K is a totally ramified Galois extension of local fields in characteristic p with group, $G(L/K)$ and set $r = [G(L/K) : G_1(L/K)]$. Choose a primitive r-th root of unity, ξ_r. If X is any $\mathbf{Z}[G(L/K)]$-module containing ${}_r(\mathbf{Q}/\mathbf{Z})(2)[1/p]$, depending on the choice of ξ_r, we shall define a new module $X\tilde{\oplus}\mathbf{Z}$, and an extension of the form

$$0 \longrightarrow X \longrightarrow X\tilde{\oplus}\mathbf{Z} \longrightarrow \mathbf{Z} \longrightarrow 0.$$

For example, this construction may be applied with X equal to any module in the chain

$$K_3(\overline{\mathbf{F}}_p) \cong (\mathbf{Q}/\mathbf{Z})(2)[1/p] \subset \mathcal{U} \subset K_3(L_0).$$

Note that the action of $G(L/K)$ on $(\mathbf{Q}/\mathbf{Z})(2)[1/p]$ is trivial, since L/K is totally ramified, while the Frobenius, F of §2.4, acts like "multiplication" by t^2 where $t = |\overline{L}|$.

As an abelian group $X\tilde{\oplus}\mathbf{Z}$ is just the direct sum, $X \oplus \mathbf{Z}$. As a $\mathbf{Z}[G(L/K)]$-module the embedding of X is given by sending $x \in X$ to $(x, 0)$. If $G_1(L/K) = G(L/E)$, as in Theorem 2.9 (proof), the action on $(0,1) \in X\tilde{\oplus}\mathbf{Z}$ will factor through $G(L/K)/G(L/E) \cong G(E/K)$. Furthermore, if $G(E/K) \cong \mathbf{Z}/r$ with generator, g, then $g(0,1) = (\xi_r, 1)$. This action is well-defined since $g^i(0,1) = (i\xi_r, 1)$, in additive notation.

Similarly we may define a $\mathbf{Z}[G(L/K)]$-module, $X\tilde{\oplus}\mathbf{Z}[1/p]$, and an extension of the form

$$0 \longrightarrow X \longrightarrow X\tilde{\oplus}\mathbf{Z}[1/p] \longrightarrow \mathbf{Z}[1/p] \longrightarrow 0$$

by setting

$$X\tilde{\oplus}\mathbf{Z}[1/p] = \lim_{\rightarrow}(X\tilde{\oplus}\mathbf{Z} \stackrel{(1,t)}{\longrightarrow} X\tilde{\oplus}\mathbf{Z} \stackrel{(1,t)}{\longrightarrow} X\tilde{\oplus}\mathbf{Z} \stackrel{(1,t)}{\longrightarrow} \ldots),$$

the direct limit of iterations of the map sending (z, m) to (z, mt) where $t = |\overline{L}|$. This is a $\mathbf{Z}[G(L/K)]$-module, since $t \equiv 1$ (*modulo* r).

PROPOSITION 2.11. ([**15**] *Proposition 2.8*)
In the notation of §2.10 let X be any of the $\mathbf{Z}[G(L/K)]$-modules in the chain

$$Tors(K_3(L_0))[1/p] \cong (\mathbf{Q}/\mathbf{Z})(2)[1/p] \subset \mathcal{U} \subset K_3(L_0).$$

Then $X\tilde{\oplus}\mathbf{Z}[1/p]$ is a cohomologically trivial $\mathbf{Z}[G(L/K)]$-module.

2.12. In the situation of §2.4 we shall define a $\mathbf{Z}[G(L/K)]$-module

$$\tilde{\oplus}_{i=1}^d K_3(L_0).$$

As an abelian group this module is simply the direct sum

$$(\oplus_{i=1}^d K_3(L_0)) \oplus (\oplus_{i=1}^d \mathbf{Z}[1/p]).$$

The summand, $\oplus_{i=1}^d K_3(L_0)$, has the $\mathbf{Z}[G(L/K)]$-module structure described in the proof of Theorem 2.9. Thus $g \in G(L/K)$ lifts to $(F, F) \in G(L_0/K)$ and acts like

$$g(a_1, \ldots , a_d) \ = \ (F(a_2), F(a_3), \ldots , F(a_d), F_0^{-1}(F(a_1))).$$

Any element, $h \in G(L/W')$, acts component-by-component as $h(a_1, \ldots , a_d) = (h(a_1), \ldots , h(a_d))$.

Consider the $\mathbf{Z}[G(L/K)]$-submodule, $\oplus_{i=1}^d(\mathbf{Q}/\mathbf{Z})(2)[1/p] \subset \oplus_{i=1}^d K_3(L_0)$. Since L/W' is totally ramified, $G(L/W')$ acts trivially on this submodule. Hence $G(L/K)$ acts through the quotient, $G(W'/K) =< g >$, with

$$g(a_1, \ldots , a_d) \ = \ (F(a_2), F(a_3), \ldots , F(a_d), F_0^{-1}F(a_1)) \ = \ (qa_2, qa_3, \ldots , qa_d, q^{1-d}a_1)$$

where $q = v^2$ and $v = |\overline{K}|$. An arbitrary element of the $\mathbf{Z}[G(L/K)]$-module

$$Inf_{G(W'/K)}^{G(L/K)} Ind_{\{1\}}^{G(W'/K)}((\mathbf{Q}/\mathbf{Z})(2)[1/p])$$

has the form $z = 1 \otimes b_d + g \otimes b_{d-1} + \ldots + g^{d-1} \otimes b_1$ and

$$g(z) \ = \ 1 \otimes b_1 + g \otimes b_d + \ldots + g^{d-1} \otimes b_2 \ .$$

Hence, if we write $z = (b_1, \ldots , b_d)$ then $g(z) = (b_2, b_3, \ldots , b_d, b_1)$.

Define an isomorphism

$$\phi : \oplus_{i=1}^d(\mathbf{Q}/\mathbf{Z})(2)[1/p] \xrightarrow{\cong} Inf_{G(W'/K)}^{G(L/K)} Ind_{\{1\}}^{G(W'/K)}((\mathbf{Q}/\mathbf{Z})(2)[1/p])$$

by

$$\phi(a_1, \ldots , a_d) \ = \ (q^{1-d}a_1, q^{2-d}a_2, \ldots , q^{-1}a_{d-1}, a_d) \ .$$

This is an isomorphism of $\mathbf{Z}[G(L/K)]$-modules, since

$$\begin{aligned} \phi(g(a_1, \ldots , a_d)) \ &= \ \phi(qa_2, qa_3, \ldots , qa_d, q^{1-d}a_1) \\ &= \ (q^{2-d}a_2, q^{3-d}a_3, \ldots , a_d, q^{1-d}a_1) \end{aligned}$$

and

$$\begin{aligned} g(\phi(a_1, \ldots , a_d)) \ &= \ g(q^{1-d}a_1, q^{2-d}a_2, \ldots , q^{-1}a_{d-1}, a_d) \\ &= \ (q^{2-d}a_2, q^{3-d}a_3, \ldots , a_d, q^{1-d}a_1). \end{aligned}$$

We have an isomorphism of $\mathbf{Z}[G(L/K)]$-modules

$$Inf_{G(W'/K)}^{G(L/K)} Ind_{\{1\}}^{G(W'/K)}((\mathbf{Q}/\mathbf{Z})(2)[1/p]) \cong Ind_{G(L/W')}^{G(L/K)}((\mathbf{Q}/\mathbf{Z})(2)[1/p])$$

where $G(L/W')$ acts trivially on $(\mathbf{Q}/\mathbf{Z})(2)[1/p]$. Hence we may form the following push-out diagram, which defines the module $\widetilde{\oplus}_{i=1}^d K_3(L_0)$.

$$\begin{array}{ccc} \oplus_{i=1}^d(\mathbf{Q}/\mathbf{Z})(2)[1/p] \cong Ind_{G(L/W')}^{G(L/K)}((\mathbf{Q}/\mathbf{Z})(2)[1/p]) & \longrightarrow & \oplus_{i=1}^d K_3(L_0) \\ \downarrow & & \downarrow \\ Ind_{G(L/W')}^{G(L/K)}((\mathbf{Q}/\mathbf{Z})(2)[1/p]\widetilde{\oplus}\mathbf{Z}[1/p]) & \longrightarrow & \widetilde{\oplus}_{i=1}^d K_3(L_0) \end{array}$$

PROPOSITION 2.13. ([**15**] *Proposition 3.10*)
The $\mathbf{Z}[G(L/K)]$*-module,* $\widetilde{\oplus}_{i=1}^{d} K_3(L_0)$*, of §2.12 is cohomologically trivial.*

2.14. Now consider the $\mathbf{Z}[G(L/K)]$-submodule, $\mathcal{W} \subset \oplus_{i=1}^{d} K_3^{ind}(L_0)$, of Definition 2.6. Since $(\mathbf{Q}/\mathbf{Z})(2)[1/p] \subset \mathcal{U}$ we have $\oplus_{i=1}^{d}(\mathbf{Q}/\mathbf{Z})(2)[1/p] \subset \mathcal{W}$. Also $G(L/W')$ acts component-by-component on $\mathcal{W}$ so that $\mathcal{W} \cong (\oplus_{i=1}^{d-1} K_3^{ind}(L_0)) \oplus \mathcal{U}$ as a $\mathbf{Z}[G(L/W')]$-module. Therefore, by Theorem 2.9, the inclusion induces a cohomology isomorphism of the following form

$$\hat{H}^t(G(L/W'); \mathcal{W}) \xrightarrow{\cong} \hat{H}^t(G(L/W'); \oplus_{i=1}^{d} K_3^{ind}(L_0)).$$

Define $\widetilde{\mathcal{W}}$ by the pushout diagram

$$\begin{array}{ccc} \oplus_{i=1}^{d}(\mathbf{Q}/\mathbf{Z})(2)[1/p] \cong Ind_{G(L/W')}^{G(L/K)}((\mathbf{Q}/\mathbf{Z})(2)[1/p]) & \longrightarrow & \mathcal{W} \\ \downarrow & & \downarrow \\ Ind_{G(L/W')}^{G(L/K)}((\mathbf{Q}/\mathbf{Z})(2)[1/p]\widetilde{\oplus}\mathbf{Z}[1/p]) & \longrightarrow & \widetilde{\mathcal{W}} \end{array}$$

so that we immediately obtain the following result from Proposition 2.13.

PROPOSITION 2.15. ([**15**] *Proposition 3.12*)
The $\mathbf{Z}[G(L/K)]$*-module,* $\widetilde{\mathcal{W}}$*, of §2.14 is cohomologically trivial.*

2.16. On $(\mathbf{Q}/\mathbf{Z})(2)[1/p]$ F_0 acts by multiplication by q^d where $q = v^2$, $v = |\overline{K}|$. Hence, if $a_i \in (\mathbf{Q}/\mathbf{Z})(2)[1/p]$, then

$$(1 - \hat{F})(a_1, \ldots, a_d) \;=\; (a_1 - q^d a_d, a_2 - a_1, \ldots, a_d - a_{d-1}).$$

Now consider the isomorphism of §2.12

$$\phi : \oplus_{i=1}^{d}(\mathbf{Q}/\mathbf{Z})(2)[1/p] \xrightarrow{\cong} Ind_{G(L/W')}^{G(L/K)}((\mathbf{Q}/\mathbf{Z})(2)[1/p])$$

where, on the right, $G(L/W')$ acts trivially on $(\mathbf{Q}/\mathbf{Z})(2)[1/p]$ and

$$\phi(a_1, \ldots, a_d) \;=\; (q^{1-d}a_1, q^{2-d}a_2, \ldots, q^{-1}a_{d-1}, a_d) = \sum_{i=0}^{d-1} g^i \otimes q^{-i} a_{d-i}.$$

Hence $\hat{F}$ translates to $\phi\hat{F}\phi^{-1}$, which is given by

$$\begin{aligned} & \phi(\hat{F}(\phi^{-1}(\sum_{i=0}^{d-1} g^i \otimes b_{d-i}))) \\ = \ & \phi(\hat{F}(\phi^{-1}(b_1, \ldots, b_d))) \\ = \ & \phi(\hat{F}(q^{d-1}b_1, q^{d-2}b_2, \ldots, qb_{d-1}, b_d)) \\ = \ & \phi(q^d b_d, q^{d-1}b_1, q^{d-2}b_2, \ldots, qb_{d-1})) \\ = \ & (qb_d, qb_1, \ldots, qb_{d-1}) \\ = \ & q(1 \otimes b_{d-1} + g \otimes b_{d-2} + \ldots + g^{d-1} \otimes b_d) \\ = \ & qg^{d-1}(\sum_{i=0}^{d-1} g^i \otimes b_{d-i}) \\ = \ & qg^{-1}(\sum_{i=0}^{d-1} g^i \otimes b_{d-i}). \end{aligned}$$

Define

$$\hat{F} : Ind_{G(L/W')}^{G(L/K)}((\mathbf{Q}/\mathbf{Z})(2)[1/p]\widetilde{\oplus}\mathbf{Z}[1/p]) \longrightarrow Ind_{G(L/W')}^{G(L/K)}((\mathbf{Q}/\mathbf{Z})(2)[1/p]\widetilde{\oplus}\mathbf{Z}[1/p])$$

by the formula ($b \in (\mathbf{Q}/\mathbf{Z})(2)[1/p]$, $m \in \mathbf{Z}[1/p]$))

$$\hat{F}(g^i \otimes (b, m)) = g^{i-1} \otimes (qb, vm).$$

If $u \in \mathbf{Z}$ satisfies $uv \equiv 1$ (modulo r) then

$$g^{-i}ag^i(0,1) = a^{u^i}(0,1) = (u^i\xi_r, 1) \in (\mathbf{Q}/\mathbf{Z})(2)[1/p]\widetilde{\oplus}\mathbf{Z}[1/p]) \ ,$$

writing the first coordinate additively, as usual. Here a and g are as in §2.4. Then $\hat{F}$ is a $\mathbf{Z}[G(L/K)]$-module homomorphism since

$$\begin{aligned} \hat{F}(a(g^i \otimes (0,1))) &= \hat{F}(g^i \otimes a^{u^i}(0,1)) \\ &= \hat{F}(g^i \otimes (u^i\xi_r, 1)) \\ &= g^{i-1} \otimes (qu^i\xi_r, v) \end{aligned}$$

while

$$\begin{aligned} a(\hat{F}(g^i \otimes (0,1))) &= a(g^{i-1} \otimes (0,v)) \\ &= g^{i-1} \otimes a^{u^i}(0,v)) \\ &= g^{i-1} \otimes (vu^{i-1}\xi_r, v) \end{aligned}$$

and these are equal since $vu^{i-1} \equiv quu^{i-1} \equiv qu^i$ (modulo r). Also

$$\hat{F}(g(g^i \otimes (0,1))) = \hat{F}(g^{i+1} \otimes (0,1)) = g^i \otimes (0,v)$$

while

$$g(\hat{F}(g^i \otimes (0,1))) = g(g^{i-1} \otimes (0,v)) = g^i \otimes (0,v).$$

THEOREM 2.17. *(c.f. [**15**] Theorem 3.14)*
In the notation of §2.4 and §2.16, there is a 2-extension of $\mathbf{Z}[G(L/K)]$-modules of the form

$$K_3(L) \xrightarrow{\Delta} \widetilde{\oplus}_{i=1}^{d} K_3(L_0) \xrightarrow{1-\hat{F}} \widetilde{\mathcal{W}} \longrightarrow \overline{L}^* \cong TorsK_2(L).$$

Proof

We have merely to evaluate the kernel and cokernel of the map

$$1-\hat{F} : \widetilde{\oplus}_{i=1}^{d} K_3^{ind}(L_0) \longrightarrow \widetilde{\mathcal{W}}$$

where $\hat{F}$ was defined in §2.16. However, as abelian groups there are isomorphism

$$\widetilde{\oplus}_{i=1}^{d} K_3^{ind}(L_0) \cong (\oplus_{i=1}^{d} K_3^{ind}(L_0)) \oplus (\oplus_{i=1}^{d} \mathbf{Z}[1/p])$$

and

$$\widetilde{\mathcal{W}} \cong \mathcal{W} \oplus (\oplus_{i=1}^{d} \mathbf{Z}[1/p]).$$

Furthermore the resulting map on the quotients of these direct sums by the submodules given by the first summands has the form

$$\lambda : \oplus_{i=1}^{d} \mathbf{Z}[1/p] \longrightarrow \oplus_{i=1}^{d} \mathbf{Z}[1/p]$$

and is given by

$$\lambda(m_1, \ldots, m_d) = (m_1 - vm_d, \ldots, m_d - vm_{d-1}).$$

If $(m_1, \ldots, m_d) \in Ker(\lambda)$ then $m_1 = vm_d, m_2 = vm_1, \ldots$ and $m_d(1-v^d) = 0$ so that λ is injective. Also we can solve the equation $\lambda(m_1, \ldots, m_d) = (x_1, \ldots, x_d)$ by choosing $x_d, x_1, \ldots, x_{d-2}$ successively but the equations are consistent if and only if

$$m_d(1-v^d) = v^{d-1}x_1 + v^{d-2}x_2 + \ldots + vx_{d-1} + x_d.$$

Therefore, if π_2 is defined to depend only on the coordinates in the second summand and to be given by

$$\pi_2(x_1, x_2, \ldots, x_d) = v^{d-1}x_1 + v^{d-2}x_2 + \ldots + vx_{d-1} + x_d$$

then $Ker(\pi_2) = Im(\lambda)$, which proves the exactness of the sequence.

It remains to verify that $G(L/K)$ acts on $\mathbf{Z}/(v^d - 1) \cong \overline{L}^*$ via the quotient to $G(W'/K)$ whose generator acts by "multiplication" by $v = |\overline{K}|$. Certainly, $G(L/E)$ acts trivially on $\widetilde{\oplus}_{i=1}^{d} K_3^{ind}(L_0)$ and $\widetilde{\mathcal{W}}$. Also the action by $a \in G(E/W')$ is trivial on the second coordinates in the direct sums. Finally the generator, $g \in G(W'/K)$, acts via

$$\begin{aligned} g(x_1, \ldots, x_d) &= g(\textstyle\sum_{i=0}^{d-1} g^i \otimes x_{d-i}) \\ &= \textstyle\sum_{i=0}^{d-1} g^{i+1} \otimes x_{d-i} \\ &= (x_2, x_3, \ldots, x_d, x_1) \end{aligned}$$

and $x_1 + vx_d + v^2x_{d-1} + \ldots + v^{d-1}x_2 \equiv v(v^{d-1}x_1 + v^{d-2}x_2 + \ldots + vx_{d-1} + x_d)$ (modulo $v^d - 1$) so that g acts on $\mathbf{Z}/(v^d - 1)$ by multiplication by v, as required.□

2.18. *The proof of Theorem 2.2*
The verification of the naturality properties for the 2-extension of Theorem 2.17 is identical to that of the p-adic case in ([**15**] §4). □

2.19. *Modifications when $r \geq 2$*

In order to construct the local fundamental class of Theorem 1.1 when $r \geq 2$ one replaces

$$\hat{F} : Ind_{G(L/W')}^{G(L/K)}((\mathbf{Q}/\mathbf{Z})(2)[1/p]\tilde{\oplus}\mathbf{Z}[1/p]) \longrightarrow Ind_{G(L/W')}^{G(L/K)}((\mathbf{Q}/\mathbf{Z})(2)[1/p]\tilde{\oplus}\mathbf{Z}[1/p])$$

in 2.16 by the homomorphism

$$\hat{F} : Ind_{G(L/W')}^{G(L/K)}((\mathbf{Q}/\mathbf{Z})(r+1)[1/p]\tilde{\oplus}\mathbf{Z}[1/p]) \longrightarrow Ind_{G(L/W')}^{G(L/K)}((\mathbf{Q}/\mathbf{Z})(r+1)[1/p]\tilde{\oplus}\mathbf{Z}[1/p])$$

given by the formula ($b \in (\mathbf{Q}/\mathbf{Z})(r+1)[1/p]$, $m \in \mathbf{Z}[1/p]$)

$$\hat{F}(g^i \otimes (b,m)) = g^{i-1} \otimes (v^{r+1}b, v^r m).$$

Thereafter one follows the construction of §2.3 - §2.18 after the manner of [**16**].

3. The local fundamental classes in the tame case in characteristic p

3.1. Suppose we are in the tame situation. That is, L/K is a tamely ramified Galois extension of local fields with Galois group $G(L/K)$ of the following form:

$$G(L/K) = < a, g \mid g^d = a^c,\ a^r = 1,\ gag^{-1} = a^v >$$

where $v = |\overline{K}|$, the order of the residue field of K. Here, if W'/K is the maximal unramified subextension then $G(L/W') =< a >$ and the image of g in $G(\overline{L}/\overline{K})$ is the Frobenius automorphism. Note that, as in ([**3**] p.369), in general we may arrange that *c is a divisor of r*. However, when $char(K) = p > 0$ we may arrange that $c = r$ since $K \cong \mathbf{F}_v((X))$, the field of fractions of $\mathbf{F}_v[[X]]$, and L is a Kummer extension of $L^{<a>} = \mathbf{F}_{v^d}((X))$. Hence $G(L/K)$ is equal to the semi-direct product $< g > \propto < a >$ of the cyclic group of order d, $< g >$, acting on the cyclic group of order r, $< a >$, by $gag^{-1} = a^v$.

For each positive integer, m, set $L(m) = \mathbf{F}_{v^{dm}}L$ as in §2.3 so that $\overline{L(m)} = \mathbf{F}_{v^{dm}}$ and

$$G(L(m)/K) = < a, g \mid g^{dm} = 1 = a^r,\ gag^{-1} = a^v >$$

and there is an extension of the form

$$G(L(m)/L) \longrightarrow G(L(m)/K) \xrightarrow{\pi_m} G(L/K)$$

in which $\pi_m(a) = a$ and $\pi_m(g) = g$. The kernel of π_m, $G(L(m)/L)$, is isomorphic to $G(\mathbf{F}_{v^{dm}}/\mathbf{F}_{v^d})$ which is cyclic of order m generated by g^d. If L_0 is the maximal unramified extension of L, as in §2.1, then L_0/K is equal to the limit of the extensions, $L(m)/K$.

As explained in §2.1, for $r \geq 1$ there are homomorphisms of the form

$$\delta : K_{2r}(L(m)) \longrightarrow K_{2r-1}(\overline{L(m)}) = K_{2r-1}(\mathbf{F}_{v^{dm}}) \cong \mathbf{F}^*_{v^{dmr}}$$

and

$$\mathbf{F}^*_{v^{dm(r+1)}} \cong K_{2r+1}(\mathbf{F}_{v^{dm}}) = K_{2r+1}(\overline{L(m)}) \longrightarrow K_{2r+1}(L(m))$$

each having uniquely divisible kernel and cokernel.

Hence we have canonical cohomology isomorphisms for $r \geq 1$

$$H^i(G(L(m)/K); K_{2r}(L(m))) \cong H^i(G(L(m)/K); \mathbf{F}^*_{v^{dmr}}),$$
$$H^i(G(L(m)/K); K_{2r+1}(L(m))) \cong H^i(G(L(m)/K); \mathbf{F}^*_{v^{dm(r+1)}})$$

for all $i > 0$ and

$$Ext^2_{\mathbf{Z}[G(L(m)/K)]}(K_{2r}(L(m)), K_{2r+1}(L(m))) \cong Ext^2_{\mathbf{Z}[G(L(m)/K)]}(\mathbf{F}^*_{v^{dmr}}, \mathbf{F}^*_{v^{dm(r+1)}}).$$

3.2. *The economical 2-extension*

Let $\mu_\infty[1/p]$ denote the group of roots of unity of order prime to p. This group is $\mathbf{Q}/\mathbf{Z}[1/p]$ written multiplicatively. Let $\mu_\infty[1/p]\tilde{\oplus}\mathbf{Z}$ denote the $\mathbf{Z}[< a >]$-module given by the direct sum of $\mu_\infty[1/p]$ with the integers where a acts trivially on $\mu_\infty[1/p]$ but satisfies $a(1,1) = (\xi_r, 1)$ for some primitive r-th root of unity, ξ_r. Note that the first coordinate of $(1,1)$ is the trivial element of $\mu_\infty[1/p]$ while the second coordinate is the integer, 1. Hence we have the induced $\mathbf{Z}[G(L(m)/K)]$-module, $Ind_{<a>}^{G(L(m)/K)}(\mu_\infty[1/p]\tilde{\oplus}\mathbf{Z})$. This is a submodule of the module appearing in the bottom left corner of the diagrams of §2.12 and §2.14 when $r = 1$, the only difference being that $\mathbf{Z}[1/p]$ has been replaced by the integers, $\mathbf{Z}$.

THEOREM 3.3.

Let $L(m)/K$ be as in §3.1. Then

(i) $Ind_{<a>}^{G(L(m)/K)}(\mu_\infty[1/p]\tilde{\oplus}\mathbf{Z})$ is a cohomologically trivial $\mathbf{Z}[G(L(m)/K)]$-module and

(ii) there is a 2-extension of $\mathbf{Z}[G(L(m)/K)]$-modules of the form

$$\mathbf{F}^*_{v^{dm(r+1)}} \longrightarrow Ind_{<a>}^{G(L(m)/K)}(\mu_\infty[1/p]\tilde{\oplus}\mathbf{Z}) \stackrel{\hat{F}-1}{\longrightarrow} Ind_{<a>}^{G(L(m)/K)}(\mu_\infty[1/p]\tilde{\oplus}\mathbf{Z}) \longrightarrow \mathbf{F}^*_{v^{dmr}}$$

where

$$\hat{F}(g^i \otimes (\eta, m)) = g^{i-1} \otimes (\eta^{v^{r+1}}, mv^r)$$

whose class in $Ext^2_{\mathbf{Z}[G(L(m)/K)]}(\mathbf{F}^*_{v^{dm}}, \mathbf{F}^*_{v^{2dm}})$ corresponds under the isomorphism of §3.1 to the K_{2r}/K_{2r+1} local fundamental class of Theorem 1.1.

Proof

We shall consider only the case when $r = 1$, since for $r \geq 2$ one modifies the argument in the manner described in §2.19.

Part (i) follows from Proposition 2.11. As mentioned in §3.2, the module $Ind_{<a>}^{G(L(m)/K)}(\mu_\infty[1/p]\tilde{\oplus}\mathbf{Z})$ is a submodule of both $\tilde{\oplus}_{i=1}^d K_3(L_0)$ and $\widetilde{\mathcal{W}}$. Furthermore the homomorphism

$$\tilde{\oplus}_{i=1}^d K_3(L_0) \stackrel{1-\hat{F}}{\longrightarrow} \widetilde{\mathcal{W}}$$

restricts to the homomorphism $1 - \hat{F}$ where $\hat{F}$ is defined by the formula of Theorem 3.3(ii). Hence we obtain a homomorphism of 2-extensions of $\mathbf{Z}[G(L(m)/K)]$-modules. It is straightforward to verify that this homomorphism induces the canonical maps of §3.1 between the modules at the two ends. □

In the next section we shall compare the 2-extension of Theorem 3.3 when $r = 1$ with the motivic-cohomology complex $\Gamma(2, L)$ of ([**9**] [**10**]). For that purpose we shall need the following two computational results.

PROPOSITION 3.4.

The edge homomorphism in the Serre spectral sequence of the extension in §3.1 yields an isomorphism of the form

$$H^*(G(L(m)/K); \mathbf{F}^*_{v^{dm}}) \stackrel{\cong}{\longrightarrow} H^*(G(L/K); \mathbf{F}^*_{v^d})$$

for all positive integers, m.

There is a similar isomorphism of the form

$$H^*(G(L(m)/K); K_3(\mathbf{F}_{\upsilon^{dm}})) \xrightarrow{\cong} H^*(G(L/K); K_3(\mathbf{F}_{\upsilon^d})).$$

Proof

For the first part the cohomology spectral sequence has the form

$$E_2^{s,t} = H^s(G(L/K); H^t(G(L(m)/L); \mathbf{F}^*_{\upsilon^{dm}})) \Longrightarrow H^{s+t}(G(L(m)/K); \mathbf{F}^*_{\upsilon^{dm}}).$$

We have $E_2^{s,t} = 0$ for all $t > 0$. This is because $L(m)/L$ is unramified and so $G(L(m)/L) \cong G(\mathbf{F}_{\upsilon^{dm}}/\mathbf{F}_{\upsilon^d})$, which implies that $H^t(G(L(m)/L); \mathbf{F}^*_{\upsilon^{dm}}) = 0$ for all $t > 0$ and $H^0(G(L(m)/L); \mathbf{F}^*_{\upsilon^{dm}}) = \mathbf{F}^*_{\upsilon^d}$. This proves the first part and replacing $\mathbf{F}_{\upsilon^{dm}}$ by $\mathbf{F}_{\upsilon^{2dm}} \cong K_3(\mathbf{F}_{\upsilon^{dm}})$ yields the second part. □

COROLLARY 3.5.

The natural map yields an isomorphism

$$H^t(G(L(m)/K); \mathbf{F}^*_{\upsilon^{dm}}) \xrightarrow{\cong} H^t(G(L(ms)/K); \mathbf{F}^*_{\upsilon^{dms}})$$

for all $t \geq 0; m, s \geq 1$.

In the limit we obtain an isomorphism

$$H^t(G(L/K); \mathbf{F}^*_{\upsilon^d}) \xrightarrow{\cong} \lim_{\substack{\rightarrow \\ m}} H^t(G(L(m)/K); \mathbf{F}^*_{\upsilon^{dm}})$$

for all $t \geq 0$.

REMARK 3.6. In the circumstances of §3.1, using Theorem 3.3, one can show that there is an isomorphism of the form

$$Ext^2_{\mathbf{Z}[G(L/K)]}(\mathbf{F}^*_{\upsilon^d}, \mathbf{F}^*_{\upsilon^{2d}}) \cong \mathbf{Z}/r \oplus \mathbf{Z}/r$$

where r is the order of the inertia group, $G_0(L/K) =< a >$.

In fact this isomorphism holds even if L/K is not tamely ramified. Let L/K be any Galois extension of local fields in characteristic $p > 0$ with Galois group, $G(L/K)$. Let $G_1(L/K) \subset G(L/K)$ denote the first wild ramification group ([**14**] p.62), which is a finite p-group. If M is the fixed field of $G_1(L/K) = G(L/M)$ then M/K is the maximal tamely ramified subextension.

Consider the spectral sequence

$$E_2^{s,t}(L/K) = H^s(G(L/K); Ext^t_{\mathbf{Z}}(K_2(L), K_3(L))) \Longrightarrow Ext^{s+t}_{\mathbf{Z}[G(L/K)]}(K_2(L), K_3(L))$$

and the corresponding one for M/K. Since $G(L/M) \triangleleft G(L/K)$ is a p-group and multiplication by p is an isomorphism on each of $K_2(L), K_3(L), K_2(M), K_3(M)$ it is not difficult to show that the natural map induces an isomorphism, $E_2^{s,t}(M/K) \xrightarrow{\cong} E_2^{s,t}(L/K)$. Therefore it gives an isomorphism between corresponding Ext^2's so that

$$\mathbf{Z}/r \oplus \mathbf{Z}/r \cong Ext^2_{\mathbf{Z}[G(M/K)]}(K_2(M), K_3(M)) \cong Ext^2_{\mathbf{Z}[G(L/K)]}(K_2(L), K_3(L))$$

as claimed.

4. The motivic complex $\Gamma(2, L)$

4.1. In [**9**] (see also [**10**]) Lichtenbaum constructs the motivic complex, $\Gamma(2, L)$, for a field L. It is a natural 2-extension of the form

$$0 \longrightarrow K_3^{ind}(L) \longrightarrow C_{2,1}(L) \stackrel{\phi_{2,L}}{\longrightarrow} C_{2,2}(L) \longrightarrow K_2(L) \longrightarrow 0.$$

The motivic-cohomology complex for L is given by

$$0 \longrightarrow C_{2,1}(L) \stackrel{\phi_{2,L}}{\longrightarrow} C_{2,2}(L) \longrightarrow 0$$

and is denoted by $\Gamma(2, L)$.

The related motivic cohomology complex $\Gamma(1, L)$ is given by a short exact sequence of the form ([**9**] Proposition 2.4)

$$0 \longrightarrow C_{1,1}(L) \longrightarrow C_{1,2}(L) \stackrel{\phi}{\longrightarrow} K_1(L) \cong L^* \longrightarrow 0$$

where

$$C_{1,2}(L) \;=\; \mathbf{Z}[\mathbf{P}^1_L - \{0, 1, \infty\}],$$

the free abelian group on the projective line minus three points. The homomorphism, ϕ, may be taken to be

$$\phi(\sum_i n_i[a_i, b_i]) = \prod_i\,(1 - (a_i/b_i))^{n_i} \in L^*.$$

There is a pairing which gives rise to an important commutative diagram of the form

$$\begin{array}{ccccccc}
Tor(L^*, L^*) & \longrightarrow & L^* \otimes C_{1,1}(L) & \longrightarrow & L^* \otimes C_{1,2}(L) & \stackrel{1\otimes\phi}{\longrightarrow} & L^* \otimes L^* \\
\downarrow \rho_L & & \downarrow & & \cong \downarrow \mu_L & & \downarrow \phi_L \\
K_3(L) & \longrightarrow & C_{2,1}(L) & \longrightarrow & C_{2,2}(L) & \longrightarrow & K_2(L)
\end{array}$$

In this diagram ϕ_L is the Steinberg symbol map and, if L is a local field of characteristic $p > 0$, ρ_L becomes an isomorphism upon inverting 2.

When L/K is a Galois extension of local fields in characteristic $p > 0$ the 2-extension associated to $\Gamma(2, L)$ defines a class in

$$\begin{aligned}[\Gamma(2, L)] \in Ext^2_{\mathbf{Z}[G(L/K)]}(K_2(L), K_3^{ind}(L)) &\cong Ext^2_{\mathbf{Z}[G(L/K)]}(K_2(L), K_3(L)) \\ &\cong Ext^2_{\mathbf{Z}[G(L/K)]}(\mathbf{F}^*_{v^d}, \mathbf{F}^*_{v^{2d}})\end{aligned}$$

where $\overline{K} = \mathbf{F}_v$ and $\overline{L} = \mathbf{F}_{v^d}$. This class defines cup-product homomorphisms in Tate cohomology of the form

$$([\Gamma(2, L)] \bigcup -) : \hat{H}^i(G(L/K); K_2(L)) \longrightarrow \hat{H}^{i+2}(G(L/K); K_3^{ind}(L)).$$

On the other hand cup-product with the K_2/K_3 local fundamental class of Theorem 1.1 and Theorem 2.2 (equivalently Theorem 3.3) yields isomorphisms in Tate cohomology

$$\hat{H}^i(G(L/K); \mathbf{F}^*_{v^d}) \stackrel{\cong}{\longrightarrow} \hat{H}^{i+2}(G(L/K); \mathbf{F}^*_{v^{2d}})$$

for all i.

THEOREM 4.2.

Let L/K be any Galois extension of local fields in characteristic p. Then, upon inverting 2, the cup-product of §2.2 is an isomorphism for all i

$$([\Gamma(2,L)]\bigcup -) : \hat{H}^i(G(L/K); K_2(L)\otimes \mathbf{Z}[1/2]) \longrightarrow \hat{H}^{i+2}(G(L/K); K_3^{ind}(L)\otimes \mathbf{Z}[1/2])$$

which may be identified with the isomorphism given by the cup-product with the K_2/K_3 local fundamental class for L/K of Theorem 3.3

$$\hat{H}^i(G(L/K); K_2(L)\otimes \mathbf{Z}[1/2])) \longrightarrow \hat{H}^i(G(L/K); \mathbf{F}^*_{v^d}\otimes \mathbf{Z}[1/2]).$$

Proof

I shall only sketch the steps of the proof, since it is rather computational and because I intend elsewhere to prove the stronger result that the 2-extensions are equivalent.

The first step is to use the observations of Remark 3.6 to reduce to the tamely ramified case. The second step is to make an equivalent extension to $[\Gamma(2,L)]$, after inverting 2, out of the upper row of the diagram of §4.1 because it is easier to produce a map of 2-extensions into the upper row. Since we are permitting ourselves to invert 2 we lose nothing by this simplification as $\rho\otimes \mathbf{Z}[1/2]$ is an isomorphism, as mentioned in §4.1. Even with these simplifications it does not seem possible to map the 2-extension of Theorem 3.3 for L/K into the modified upper row of the diagram. On the other hand it is not difficult to find such a map for the 2-extension associated to the infinite Galois extension, L_0/K. However, such comparison of 2-extensions is sufficient by virtue of the cohomology isomorphisms of Proposition 3.4 and Corollary 3.5. □

5. Euler characteristics and Chinburg invariants

5.1. Suppose, in §1, that A and D are finitely generated $\mathbf{Z}[G]$-modules. Then the 2-extension defines a class, $[\underline{\underline{E}}] \in Ext^2_{\mathbf{Z}[G]}(D,A)$ which may be represented by a (possibly different) 2-extension in which B and C are finitely generated and cohomologically trivial. In this case B (resp. C) has a finitely generated, projective $\mathbf{Z}[G]$-resolution of the form $P_{1,B}\longrightarrow P_{0,B}\longrightarrow B$ (resp. $P_{1,C}\longrightarrow P_{0,C}\longrightarrow C$). The *Euler characteristic* of $[\underline{\underline{E}}]$ is defined to be the element

$$\chi_{[\underline{\underline{E}}]} = \sum_{i=0}^{1}(-1)^i([P_{i,B}] - [P_{i,C}]) \in K_0(\mathbf{Z}[G]).$$

The Euler characteristic depends only on the isomorphism class – which is weaker than the 2-extension equivalence class – of $[\underline{\underline{E}}]$ and is defined if and only if the cup-product

$$([\underline{\underline{E}}]\bigcup -) : \hat{H}^i(G; D) \longrightarrow \hat{H}^{i+2}(G; A)$$

is an isomorphism on Tate cohomology for all i.

In particular, suppose that we are in the situation of §2.1. That is, L/K is a Galois extension of local fields in characteristic $p > 0$ with group, $G(L/K)$. Suppose that the residue fields satisfy $\overline{L} = \mathbf{F}_{v^d}$ and $\overline{K} = \mathbf{F}_v$. As explained in §3.1 in the tame case, $K_{2r}(L)$ and $K_{2r+1}(L)$ differ from $\mathbf{F}^*_{v^{dr}}$ and $\mathbf{F}^*_{v^{d(r+1)}}$, respectively, only by uniquely divisible groups. Hence we have an isomorphism of the form

$$Ext^2_{\mathbf{Z}[G(L/K)]}(K_{2r}(L), K_{2r+1}(L)) \cong Ext^2_{\mathbf{Z}[G(L/K)]}(\mathbf{F}^*_{v^{dr}}, \mathbf{F}^*_{v^{d(r+1)}}).$$

Therefore the K_{2r}/K_{2r+1} local fundamental class of Theorem 1.1, constructed in §2 in characteristic $p > 0$, has an associated Euler characteristic denoted by

$$\Omega_r(L/K, 2) \in Cl(\mathbf{Z}[G(L/K)]) \ = \ Ker(rank : K_0(\mathbf{Z}[G]) \longrightarrow \mathbf{Z}) \ .$$

This Euler characteristic lies in the reduced K_0-group, given by the kernel of the rank homomorphism, because $\mathbf{F}^*_{v^{d(r+1)}}$ and $\mathbf{F}^*_{v^{dr}}$ are both finite (i.e. of rank zero). We shall call $\Omega_r(L/K, 2)$ a *local Chinburg invariant.*

When L/K is a tamely ramified Galois extension of local fields of characteristic $p > 0$, as in §3.1, we may use the particularly simple form of the local fundamental class given by Theorem 3.3 for computations of $\Omega_r(L/K, 2)$. For this reason I shall restrict to the tamely ramified case for the rest of this section.

The notation $\Omega_r(L/K, 2)$ is intended to suggest the local Chinburg invariants of ([**18**] Ch VII). The main role of the $\Omega_r(L/K, 2)$'s is to construct the higher K-theory analogues of the *second Chinburg invariant*, $\Omega_r(E/F, 2)$, of a Galois extension of global fields in characteristic $p > 0$. Let $\mathcal{O}_E$ denote the ring of integers of E.

For each prime $P \lhd \mathcal{O}_F$ choose a prime $Q \lhd \mathcal{O}_E$ above P and let E_Q/F_P denote the Galois extension of local fields given by the completions of F and E at P and Q, respectively. The inclusion of the decomposition group $G(E_Q/F_P) \subseteq G(E/F)$ induces a homomorphism of class-groups

$$Ind_{G(E_Q/F_P)}^{G(E/F)} : \mathcal{CL}(\mathbf{Z}[G(E_Q/F_P)]) \longrightarrow \mathcal{CL}(\mathbf{Z}[G(E/F)]).$$

The K_{2r}/K_{2r+1} second Chinburg invariant of E/F is defined by

$$\Omega_r(E/F, 2) = \sum_{P \lhd \mathcal{O}_F,\ prime} Ind_{G(E_Q/F_P)}^{G(E/F)}(\Omega_r(G(E_Q/F_P), 2)) \in \mathcal{CL}(\mathbf{Z}[G(E/F)]).$$

It is easy to see that the sum which defines $\Omega_r(E/F, 2)$ is finite. All but finitely many extensions, E_Q/F_P, are unramified. However, in the unramified case the Galois modules, $\mathbf{F}^*_{v^{d(r+1)}}$ and $\mathbf{F}^*_{v^{dr}}$, are cohomologically trivial in Theorem 3.3. In fact each module has trivial class in the class-group ([**18**] Ch VII) so that

$$\Omega_r(E_Q/F_P, 2) = [\mathbf{F}^*_{v^{d(r+1)}}] - [\mathbf{F}^*_{v^{dr}}] = 0 \in \mathcal{CL}(\mathbf{Z}[G(E_Q/F_P)]).$$

REMARK 5.2. It is worthwhile comparing the definition of $\Omega_r(E/F, 2)$ given in §5.1 for global fields in characteristic $p > 0$ with the construction in characteristic zero given in [**15**] for $r = 1$ and in [**16**] in general. The problem in characteristic zero is that $K_{2r+1}(E_Q)$ is not a finitely generated $\mathbf{Z}[G(E_Q/F_P)]$-module (even modulo uniquely divisible modules). Hence one proceeds, by means of the syntomic regulator map, to find a free submodule of $K_{2r+1}(E_Q)$ with a finitely generated quotient to play the role of $K_{2r+1}(E_Q)$. The intrusion of a regulator in the construction promises to make explicit computations very difficult.

QUESTION 5.3. Theorem 4.2 implies that the 2-extension of $[G(L/K)]$-modules associated to $\Gamma(2, L)$ has an Euler characteristic

$$\chi_{[\Gamma(2,L)]} \in \mathcal{CL}(\mathbf{Z}[1/2][G(L/K)]).$$

This raises the question: Do the Euler characteristics associated to Theorem 4.2 coincide? That is, do we have the equation

$$\chi_{[\Gamma(2,L)]} = \Omega_1(L/K, 2) \in \mathcal{CL}(\mathbf{Z}[1/2][G(L/K)]) \ ?$$

I imagine that this relation is true and hope to return to its proof in a subsequent paper.

An analogous question is discussed in the number field situation in §5 of [**1**].

5.4. *The K_2/K_3 second Chinburg invariant in the tame case*

As in §3.1, let L/K be a tamely ramified Galois extension of local fields with $char(K) = p > 0$ and Galois group

$$G(L/K) \;= \;< a, g \mid g^d = a^r = 1,\; gag^{-1} = a^v >$$

where $v = |\overline{K}|$, the order of the residue field of K.

PROPOSITION 5.5.

Let L/K be as in §5.4. Then there is a finitely generated projective $\mathbf{Z}[G(L/K)]$-module, $M(L/K)$, defined by the following pushout diagram:

$$\begin{array}{ccc} \mathbf{Z}[G(L/K)](a-1) & \longrightarrow & \mathbf{Z}[G(L/K)] \\ \big\downarrow (-v^{d-1}g) & & \big\downarrow \Psi \\ \mathbf{Z}[G(L/K)](a-1) & \longrightarrow & M(L/K) \end{array}$$

Proof

It is not difficult to see that the Tate cohomology of $\mathbf{Z}[G(L/K)](a-1)$ is given by

$$\hat{H}^i(G(L/K); \mathbf{Z}[G(L/K)](a-1)) \cong \begin{cases} \mathbf{Z}/r & \text{if } i \text{ is odd,} \\ 0 & \text{if } i \text{ is even,} \end{cases}$$

since $\mathbf{Z}[G(L/K)]/\mathbf{Z}[G(L/K)](a-1) \cong Ind_{<a>}^{G(L/K)}(\mathbf{Z})$. Since $HCF(v,r) = 1$ the left-hand vertical homomorphism induces an isomorphism on $\hat{H}^*(G(L/K); -)$, which ensures that $M(L/K)$ is cohomologically trivial and therefore projective, being finitely generated and torsion-free. □

THEOREM 5.6.

Let L/K be a tamely ramified Galois extension of local fields in characteristic $p > 0$, as in §3.1. Then the local K_2/K_3 Chinburg invariant of §5.1 is given by

$$\Omega_1(L/K, 2) = [M(L/K)] - [\mathbf{Z}[G(L/K)]] \in Cl(\mathbf{Z}[G(L/K)]).$$

Proof

I shall only sketch the proof, which is rather laborious homological algebra.

Firstly one constructs a commutative diagram of $\mathbf{Z}[G(L/K)]$-modules of the form

$$\begin{array}{ccccccc} \hat{X}(2) & \longrightarrow & P_1 & \xrightarrow{\delta} & P_0 & \longrightarrow & \mathbf{F}_{v^d}^* \\ \big\downarrow g_4 & & \big\downarrow g_3 & & \big\downarrow g_2 & & \big\downarrow 1 \\ \mathbf{F}_{v^{2d}}^* & \longrightarrow & \widetilde{Ind}(2) & \xrightarrow{\hat{F}-1} & \widetilde{Ind}(2) & \longrightarrow & \mathbf{F}_{v^d}^* \end{array}$$

where $\hat{X}(2) = ker(\delta)$ and $\widetilde{Ind}(2) = Ind_{<a>}^{G(L/K)}(\mu_\infty[1/p]\tilde{\oplus}\mathbf{Z})$. The lower 2-extension in the diagram is that of Theorem 3.3 when $m = 1$. The upper 2-extension is made from the start of a projective resolution of $\mathbf{F}_{v^d}^*$. Secondly one ensures that g_4 is surjective. Then it is straightforward (cf. [**3**] pp. 369/370) to show that

$$\Omega_1(L/K, 2) = [Ker(g_4)] - rank(Ker(g_4)) \cdot [\mathbf{Z}[G(L/K)]] \in Cl(\mathbf{Z}[G(L/K)]).$$

A suitable choice for the upper 2-extension is given by modifying the tame resolution of ([**3**] Lemma 6.3 p.370). □

6. The Hom-description of $\Omega_1(E/F, 2)$

It is often useful and illuminating to describe arithmetic invariants lying in the class-group, $CL(\mathbf{Z}[G(E/F)])$, in terms of idèlic-valued functions on the representation ring $R(G(E/F)$ by means of the Hom-description. In this section I shall do this for $\Omega_1(E/F, 2)$ in the case when E/F is a tamely ramified Galois extension of global fields in characteristic $p > 0$.

DEFINITION 6.1. *Adèles and Idèles*

A more extensive reference for the material of this section is ([**4**]II,p.334 et seq).

Let N be an number field. The *adèle ring* of N is defined to be the ring given by the *restricted product*

$$J(N) = \prod_{P \text{ prime}}{}' N_P,$$

where $\prod'$ signifies that we take those elements of the topological ring, $\prod_P N_P$, for whom almost all entries lie in the ring of integers, $\mathcal{O}_{N_P}$. When P is an Archimedean prime we adopt the convention that $\mathcal{O}_{N_P} = N_P$. The group of *idèles*, $J^*(N)$, is the group of units in $J(N)$

$$J^*(N) = \{(x_P) \in J(N) \mid x_P \neq 0 \text{ and almost everywhere } \quad x_P \in \mathcal{O}^*_{N_P}\}$$

where, as usual, $\mathcal{O}^*_{N_P}$ denotes the multiplicative group of units in $\mathcal{O}_{N_P}$. The *unit idèles* is the subgroup in which every entry is a unit

$$U(\mathcal{O}_N) = \prod_{P \text{ prime}} \mathcal{O}^*_{N_P}.$$

Now let G be a finite group. We may extend the adèles and idèles to the group-rings, $\mathcal{O}_N[G]$ and $N[G]$. Define

$$\begin{aligned} J(N[G]) &= \prod_{P \text{ prime}}{}' N_P[G], \\ J^*(N[G]) &= \{(\alpha_P) \in J(N[G]) \mid \alpha_P \in \mathcal{O}_{N_P}[G]^* \text{ for almost} \\ &\qquad\qquad \text{all } P \text{ and } \quad \alpha_P \in N_P[G]^* \text{otherwise } \} \\ U(\mathcal{O}_N[G]) &= \prod_{P \text{ prime}} \mathcal{O}_{N_P}[G]^*. \end{aligned}$$

6.2. Now suppose that M/N is a finite Galois extension of number fields with Galois group, $G(M/N)$. In this case $G(M/N)$ acts on the set of primes of M and hence acts upon the groups $J^*(M), U(\mathcal{O}_M), J^*(M[G])$ and $U(\mathcal{O}_M[G])$. Therefore the absolute Galois group, Ω_N, also acts continuously on these groups via projection to $G(M/N)$.

Suppose now that M is large enough to contain all $|G|$-th roots of unity, so that M is a splitting field for G. In this case the ring of finite dimensional M-representations of G, $R_M(G) \cong R(G)$ is also a $\mathbf{Z}[G(M/N)]$-module. Therefore we may consider the group of $G(M/N)$-equivariant maps

$$Hom_{G(M/N)}(R(G), J^*(M)) \cong Hom_{\Omega_N}(R(G), J^*(M)).$$

Let V be a projective $\mathbf{Z}[G]$-module of rank one. This means that $V \otimes \mathbf{Z}_p$ is a free $\mathbf{Z}_p[G]$-module on one generator, x_p, for each prime p and that $V \otimes \mathbf{Q}$ is a free $\mathbf{Q}[G]$-module on one generator, x_0. Since $\mathbf{Q}[G]$ and $\mathbf{Z}_p[G]$ are subrings of $\mathbf{Q}_p[G]$ we may compare these bases for the $\mathbf{Q}_p[G]$-module, $\mathbf{Z}_p \otimes \mathbf{Q}_p$. This means that there is a unit

$$\lambda_p \in \mathbf{Q}_p[G]^*$$

which is defined by

$$\lambda_p \cdot x_0 = x_p \in V \otimes \mathbf{Q}_p.$$

In fact, λ_p will almost always lie in $\mathbf{Z}_p[G]^*$ so that we obtain an idèle

$$(\lambda_p) \in J^*(\mathbf{Q}[G]).$$

More generally, if V is a projective module of rank n this comparison of bases will yield an invertible $n \times n$-matrix with adèlic entries

$$(\lambda_p) \in GL_n(J(\mathbf{Q}[G])).$$

Now suppose that T is a representation

$$T : G \longrightarrow GL_m(M).$$

We may apply T to each $\lambda_p \in GL_n(\mathbf{Q}_p[G])$ to obtain an invertible matrix

$$T(\lambda_p) \in M_{mn}(\mathbf{Q}_p \otimes_{\mathbf{Q}} M)$$

where $M_n(A)$ denotes the $n \times n$ matrices with entries in A. There is a ring isomorphism of the form

$$\mathbf{Q}_p \otimes_{\mathbf{Q}} M \cong \prod_{\substack{Q|p \\ Q \text{ prime of } M}} M_Q.$$

Therefore we obtain an element

$$det(T(\lambda_p)) \in \prod_{\substack{Q|p \\ Q \text{ prime of } M}} M_Q^*.$$

Suppose that $M/\mathbf{Q}$ is Galois. Since $(\lambda_p) \in GL_n(J(\mathbf{Q}[G]))$ we obtain an $\Omega_{\mathbf{Q}}$-equivariant map, given by $det(T(\lambda_p))$ at the primes of M which divide p,

$$Det((\lambda_p)) \in Hom_{\Omega_{\mathbf{Q}}}(R(G), J^*(M)).$$

Now let us consider the dependence upon the choices of the basis elements, x_0 and x_p. The discussion is similar in the case of modules of rank greater than one. If we replace x_p by another generator, x'_p, these choices will be related by an equation

$$x'_p = u_p x_p$$

for some $u_p \in \mathbf{Z}_p[G]^*$ so that we obtain a unit idèle

$$u = (u_P) \in U(\mathbf{Z}[G])$$

and $Det((\lambda_p))$ will be altered by multiplication by

$$Det(u) \in Det(U(\mathbf{Z}[G])) \subset Hom_{\Omega_{\mathbf{Q}}}(R(G), J^*(M)).$$

Also there is a diagonal embedding of M^* into $J^*(M)$ which induces an inclusion

$$Hom_{\Omega_{\mathbf{Q}}}(R(G), M^*) \subset Hom_{\Omega_{\mathbf{Q}}}(R(G), J^*(M)).$$

By a similar argument, changing x_0 to x_0' will change $Det((\lambda_p))$ by a function which lies in the subgroup, $Hom_{\Omega_{\mathbf{Q}}}(R(G), M^*)$. Therefore we have associated to each finitely generated projective $\mathbf{Z}[G]$-module, V, a well-defined element

$$Det[V] \in \frac{Hom_{\Omega_{\mathbf{Q}}}(R(G), J^*(M))}{Hom_{\Omega_{\mathbf{Q}}}(R(G), M^*) \cdot Det(U(\mathbf{Z}[G]))}.$$

The following connection between the Det-construction and the class-group is called the 'Hom-description' and is originally due to Fröhlich.

THEOREM 6.3. *([**4**]II,p.334; [**6**])*
With the notation introduced above there is an isomorphism

$$Det : \mathcal{CL}(\mathbf{Z}[G]) \xrightarrow{\cong} \frac{Hom_{\Omega_{\mathbf{Q}}}(R(G), J^*(M))}{Hom_{\Omega_{\mathbf{Q}}}(R(G), M^*) \cdot Det(U(\mathbf{Z}[G]))}$$

which sends the class of a projective module, V, to the class of $Det((\lambda_p))$ of 6.2.

6.4. *The Hom-description representative $\Omega_1(L/K, 2)$*
I shall now calculate the representative for $[M(L/K)] \in \mathcal{CL}(\mathbf{Z}[G(L/K)])$, which occurred in Theorem 5.6, in the Hom-description of the class-group.
Let $\mathcal{N}$ denote the element

$$(1 + v^2 g^{d-1} + \ldots + v^{2d-2} g) = \mathcal{N} \in \mathbf{Z}[G(L/K)]$$

and set

$$\alpha_0 = \mathcal{N} v(1 + a + \ldots + a^{r-1}) - v^{d-1} g \in \mathbf{Z}[G(L/K)].$$

Since v is a unit in the rationals and in $\mathbf{Z}_p$ for all primes, p, except for the residue characteristic we may take $\Psi(\alpha_0 \otimes 1)$ as a basis element for $M(L/K) \otimes \Lambda$ when $\Lambda = \mathbf{Q}$ or $\mathbf{Z}_p$ when v is not a power of p. This means that the Hom-description representative is trivial for all primes except the residue characteristic.
The same is true even for the residue characteristic if $d = 1$.
If p is the residue characteristic and $d \geq 2$ we need a basis element for $M(L/K) \otimes \mathbf{Z}_p$.
Choose $\beta_p \in M(L/K) \otimes \mathbf{Z}_p$ to be the image of

$$(a - 1, 1) \in \mathbf{Z}_p[G(L/K)](a-1) \oplus \mathbf{Z}_p[G(L/K)],$$

where the coordinates refer to the upper right and lower left corners of the pushout diagram of Theorem 5.6 so that $(0, 1) = \alpha_0$. Then the image of β_p under the quotient map

$$M(L/K) \otimes \mathbf{Z}_p \longrightarrow \frac{M(L/K) \otimes \mathbf{Z}_p}{\mathbf{Z}_p[G(L/K)](a-1)} \cong \mathbf{Z}_p[g]/(g^d - 1)$$

is equal to the generator, 1. On the other hand, in the bottom left corner of the pushout,

$$\begin{aligned}
(a-1)^{v^d - 1} \beta_p &= (a-1)^{v^d} + (a-1)^{v^d - 2}(a-1)\alpha_0 \\
&= (a-1)^{v^d} + (a-1)^{v^d - 1}(-v^{d-1} g) \\
&\equiv a^{v^d} - 1 \ (modulo\ p) \\
&\equiv a - 1 \ (modulo\ p).
\end{aligned}$$

Therefore $(a-1)^{v^d-1}\beta_p$ generates $\mathbf{Z}_p[G(L/K)](a-1)$ in the bottom left corner and so

$$M(L/K)\otimes \mathbf{Z}_p \ = \ \mathbf{Z}_p[G(L/K)]\beta_p.$$

When tensored with $\mathbf{Q}_p$, β_p is equal in the top right $\mathbf{Q}_p[G(L/K)]$ to $1+(a-1)g^{-1}v^{1-d}$ while the rational basis is 1 in the top right so that the Hom description representative is equal to

$$Det(1+(a-1)g^{-1}v^{1-d}) \in Hom_{\Omega_{\mathbf{Q}_p}}(R(G(L/K),\mathcal{O}_S^*)$$

for any Galois extension, $S/\mathbf{Q}_p$, containing all $[L:K]$-th roots of unity.

Therefore we have established the following result:

THEOREM 6.5.

Suppose that E/F is a tamely ramified Galois extension of global fields in characteristic $p>0$. Then there exists a representative of the K_2/K_3 Chinburg invariant of §5.1

$$\Omega_1(E/F,2) \ = \ \sum_{P\lhd\mathcal{O}_F,\ prime} Ind_{G(E_Q/F_P)}^{G(E/F)}(\Omega_1(G(E_Q/F_P),2)) \in \mathcal{CL}(\mathbf{Z}[G(E/F)])$$

in the Hom-description of the class-group which is trivial except at places above p. At places above p the representative is given by sending $\chi\in R(G(E/F))$ to

$$\prod_P Det(1+(a_P-1)g_P^{-1}v_P^{1-d_P})(Res_{G(E_Q/F_P)}^{G(E/F)}(\chi))$$

where the product is taken over those primes of $P\lhd\mathcal{O}_F$ for which

$$G(E_Q/F_P) \ = <a_P, g_P \mid g_P^{d_P}=a^{r_P}=1,\ g_Pa_Pg_P^{-1}=a_P^{v_P}>$$

with $v_P=|\mathcal{O}_F/P|$ and $d\geq 2$ in the notation of §5.4.

Proof

This follows from the definition of $\Omega_1(E/F,2)$ together with the fact that the homomorphism

$$Ind_{G(E_Q/F_P)}^{G(E/F)} : \mathcal{CL}(\mathbf{Z}[G(E_Q/F_P)]) \longrightarrow \mathcal{CL}(\mathbf{Z}[G(E/F)])$$

corresponds in the Hom-description to the map induced by restriction of representations to the decomposition group. □

References

[1] D. Burns and M. Flach: On Galois structure invariants associated to Tate motives; Amer. J. Math. 120 (1998) 1343–1397.

[2] R.A. Bogomolov: Two theorems on the divisibility and torsion in the Milnor K-groups; Mat. Sb. 130 (1986) 3, 404–412; English transl. Math. USSR-Sb. 58 (1987) 2, 407–416.

[3] T. Chinburg: Exact sequences and Galois module structure; Annals Math. 121 (1985) 351–376.

[4] C.W. Curtis and I. Reiner: *Methods of Representation Theory*, vols. I (1981) & II (1987) Wiley.

[5] I.B. Fesenko and S.V. Vostokov: *Local fields and their extensions - a constructive approach*; Trans. of Math. Monographs vol. 121, A.M.S. (1991).

[6] A. Fröhlich: *Galois Module Structure of Algebraic Integers*, Ergeb. Math. 3 Folge-Band #1 (1983) Springer-Verlag.

[7] T. Geisser and M. Levine: The K-theory of fields in characteristic p; Inventiones Math. 139 (2000) 459–493.

[8] L. Hesselholt and I. Madsen: private communication (May 1999).

[9] S. Lichtenbaum: The construction of weight two arithmetic cohomology; Inventiones Math. 88 (1987) 183–215.
[10] S. Lichtenbaum: New results on weight two arithmetic cohomology; Grothendieck Festschrift v.3, Birkhäuser Progress in Math. #88 (1990) 35–55.
[11] B. Kahn: Descente Galoisienne et K_2 des corps de nombres; K-theory 7 (1) 55–100 (1993).
[12] A.S. Merkurjev: On the torsion of K_2 of local fields; Annals of Maths. (2) 118 (1983) 375–381.
[13] A.S. Murkurjev and A.A. Suslin: The K_3 group of a field; Izv. Nauk. SSSR 54 (1990) 339–356 (Eng. trans. Math. USSR 36 (1990) 541–565).
[14] J-P. Serre: *Local Fields* ; Grad. Texts in Math. # 67 (1979) Springer-Verlag.
[15] V.P. Snaith: Local fundamental classes derived from higher dimensional K-groups; Proc. Great Lakes K-theory Conf., Fields Institute Communications Series #16 (A.M.S. Publications) 285–324 (1997).
[16] V.P. Snaith: Local fundamental classes derived from higher dimensional K-groups II; Proc. Great Lakes K-theory Conf., Fields Institute Communications Series #16 (A.M.S. Publications) 325–344 (1997).
[17] V.P. Snaith: Local fundamental classes derived from higher K-groups in characteristic p; McMaster preprint #2 (1996/7).
[18] V.P. Snaith: *Explicit Brauer Induction (with applications to algebra and number theory)*; Cambridge Studies in Advanced Math. #40 (1994) Cambridge University Press.
[19] V.P. Snaith: *Galois Module Structure*; Fields Institute Monographs # 2 A.M.S. (1994).
[20] A.A. Suslin: On the K-theory of local fields; J.P.A.Alg. (2-3) 34 (1984) 301–318.
[21] J. Tate: On the torsion in K_2 of fields; *Algebraic Number Theory* Proc. International Symposium, Kyoto (1976) 243–261, published by the Japanese Society for the Promotion of Science.
[22] V. Voevodsky: The Milnor conjecture; Max-Planck Inst. preprint series or http://www.math.uiuc.edu/K-theory (170) 1996.

Faculty of Mathematical Studies
University of Southampton
Southampton
Hants. SO17 1BJ, England

E-mail address: vps@maths.soton.ac.uk

Contemporary Mathematics
Volume **272**, 2000

Hilbert's Theorem on Positive Ternary Quartics

Richard G. Swan

ABSTRACT. We give a detailed exposition of Hilbert's proof that a non-negative real quartic form in 3 variables is a sum of 3 squares.

In [H] Hilbert proved the following theorem.

THEOREM. *Let $F(x, y, z)$ be a homogeneous form of degree 4 over the real numbers such that $F(x, y, z) \geq 0$ for all real (x, y, z). Then $F = \varphi^2 + \psi^2 + \chi^2$ where φ, ψ, χ are real quadratic forms.*

Hilbert's proof is very condensed and assumes the reader is familiar with ideas then current. Consequently, the proof is almost incomprehensible to a modern reader. T. Y. Lam remarked that it would be a good idea to give a new exposition of this proof in modern terminology and with more of the details filled in. This is the aim of the present paper. The argument follows roughly that indicated by Hilbert but I have not attempted to follow his method exactly.

In [CL], Choi and Lam gave a short elementary proof of the fact that F in the theorem is a sum of squares of quadratic forms. Their method, however, does not show that only 3 quadratic forms are required.

This paper was originally intended for publication in the proceedings of the Santa Barbara conference [JR] but I withdrew it after Lam informed me that an exposition of this theorem had been published by Rajwade [R]. Since there is still considerable interest in Hilbert's theorem and my version of the proof differs in many details from that of Rajwade, I am publishing it here in the hope that it also may prove useful to those working on this subject. I would like to thank B. Reznick for pointing out that there may still be some value in publishing this version. I would also like to thank A. Ranicki for agreeing to publish it in the present proceedings.

I will begin with some preliminary results needed for the proof.

1. Counting constants. If F is a ternary quartic form over $\mathbb{C}$, let $[F] \in \mathbb{P}^{14}(\mathbb{C})$ be the point whose homogeneous coordinates are the coefficients of F, and let $\{F = 0\}$ be the curve in $\mathbb{P}^2(\mathbb{C})$ defined by $F = 0$.

2000 *Mathematics Subject Classification.* Primary 11E10, Secondary 11E25, 11E76.
Key words and phrases. positive ternary quartic form.
Partly supported by the NSF

LEMMA 1.1. *$\{[F] \in \mathbb{P}^{14}(\mathbb{C}) \mid \{F=0\}$ is not smooth$\}$ is a closed subset of $\mathbb{P}^{14}$ of dimension ≤ 13 (actually $= 13$).*

PROOF. In $\mathbb{P}^{14} \times \mathbb{P}^2$ let

$$\begin{aligned} X &= \{([F],p) \mid \{F=0\} \text{ is singular at } p\} \\ &= \{([F],p) \mid F(p)=0, \frac{\partial F}{\partial x}(p) = \frac{\partial F}{\partial y}(p) = \frac{\partial F}{\partial z}(p) = 0\}\,. \end{aligned}$$

By Euler's theorem there are only 3 independent conditions on F so the fibers of $\mathrm{pr}_2 : X \to \mathbb{P}^2$ are isomorphic to $\mathbb{P}^{11}$. In fact we can choose coordinates so that $p = (0,0,1)$ and the fiber consists of all $[F]$ where the coefficients of xz^3, yz^3, z^4 are zero. By the dimension theorem [**Ha, Ch. 2, Ex. 3.22**], $\dim X = 11 + 2 = 13$ and the required set is $\mathrm{pr}_1 X$. It is closed since $\mathbb{P}^2$ is proper.

REMARK. Since, except for a set of $[F]$ of dimension 7, $\{F=0\}$ has only a finite number of singular points, $\mathrm{pr}_1 : X \to \mathrm{pr}_1(X)$ has finite general fibers and the dimension theorem shows that $\dim\, \mathrm{pr}_1(X) = \dim X = 13$.

LEMMA 1.2. *Let $Y = \{[F] \mid \{F=0\}$ has at least 2 distinct singular points$\}$. Then $\dim \overline{Y} \le 12$.*

PROOF. Let Δ be the diagonal of $\mathbb{P}^2 \times \mathbb{P}^2$ and let $X \subset \mathbb{P}^{14} \times (\mathbb{P}^2 \times \mathbb{P}^2 - \Delta)$ be the set of $([F],p,q)$ such that $\{F=0\}$ is singular at p and at q. In other words,

$$F(p) = \frac{\partial F}{\partial x}(p) = \frac{\partial F}{\partial y}(p) = \frac{\partial F}{\partial z}(p) = 0$$

and

$$F(q) = \frac{\partial F}{\partial x}(q) = \frac{\partial F}{\partial y}(q) = \frac{\partial F}{\partial z}(q) = 0\,.$$

Then $Y = \mathrm{pr}_1(X)$. Choose coordinates so $p = (0,0,1)$ and $q = (0,1,0)$. The fiber of $\mathrm{pr}_2 : X \to \mathbb{P}^2 \times \mathbb{P}^2 - \Delta$ over (p,q) consists of all $[F]$ with the coefficients of $xz^3, yz^3, z^4, xy^3, y^4, zy^3$ all zero so the fiber is $\mathbb{P}^8$. By the dimension theorem, $\dim X = 8 + 4 = 12$ and $X \to \overline{Y}$ is dominant so $\dim \overline{Y} \le 12$. The general fibers are again finite so $\dim \overline{Y} = 12$.

Presumably similar arguments give Hilbert's assertions about the linear system $\alpha F + \alpha' F' + \alpha'' F''$ but we will not need this here.

2. Regular sequence. Let $S = k[x_0, \dots, x_n]$ be a polynomial ring over a field k and let $f_1, \dots, f_m$ be forms of positive degree in S. Let $V(f_1, \dots, f_m) \subset \mathbb{P}^n = \mathrm{Proj}\, S$ be defined by $f_1 = 0, \dots, f_m = 0$ so that $V(f_1, \dots, f_m) = \mathrm{Proj}\, A$ with $A = S/(f_1, \dots, f_m)$.

LEMMA 2.1. *If $m \le n+1$ then $f_1, \dots, f_m$ is a regular sequence on S if and only if $\dim V(f_1, \dots, f_m) \le n - m$. (This means $V(f_1, \dots, f_m) = \emptyset$ if $m = n+1$).*

PROOF. Let $\mathfrak{m} = S^+ = \oplus_{i>0} S_i$. If M is a finitely generated graded S-module then $M_\mathfrak{m} = 0$ if and only if $M = 0$. Therefore $f_1, \dots, f_m$ is regular on S if and only if it is regular on $S_\mathfrak{m}$. Now

$$\dim V(f_1, \dots, f_m) = \dim\, \mathrm{Proj}\, A = \dim A - 1 = \dim A_\mathfrak{m} - 1\,.$$

The last equality follows from the fact that all minimal primes P of A are homogeneous and so lie in A^+. Since all maximal ideals in an affine domain have the same height, $\dim A/P = \dim A_{\mathfrak{m}}/P_{\mathfrak{m}}$. Now $A_{\mathfrak{m}} = S_{\mathfrak{m}}/(f_1, \ldots, f_m)$ so we have to show that $f_1, \ldots, f_m$ is regular on $S_{\mathfrak{m}}$ if and only if $\dim S_{\mathfrak{m}}/(f_1, \ldots, f_m) \leq \dim S_{\mathfrak{m}} - m$ (and so $= \dim S_{\mathfrak{m}} - m$). This is clear since $S_{\mathfrak{m}}$ is Cohen-Macaulay.

COROLLARY 2.2. *Let φ, ψ, χ be forms of degree > 0 in $S = k[x, y, z]$. Then φ, ψ, χ is a regular sequence on S if and only if φ, ψ, χ have no common zero in $\mathbb{P}^2(\overline{k})$.*

Here $\overline{k}$ is the algebraic closure of k. By Hilbert's Nullstellensatz φ, ψ, χ have no common zero in $\mathbb{P}^2(\overline{k})$ if and only if $V(\varphi, \psi, \chi) = \emptyset$.

REMARK. We only need the "if" statement here. This special case can also be deduced from Max Noether's fundamental theorem so we cannot conclude that Hilbert knew about Cohen-Macaulay rings. If φ, ψ, χ have no common zero in $\mathbb{P}^2$ then φ and ψ have no common factor since such a factor would have a common zero with χ. Suppose that $\varphi\varphi' + \psi\psi' + \chi\chi' = 0$. If p lies on $\{\varphi = 0\} \cap \{\psi = 0\}$, choose coordinates so that $z \neq 0$ at p and let $\varphi_* = \varphi(x, y, 1)$, etc. in the local ring $\mathcal{O}_{\mathbb{P}^2,p}$.Then χ_* is a unit in $\mathcal{O}_{\mathbb{P}^2,p}$ so $\chi'_* \in (\varphi_*, \psi_*)$. By Noether's theorem [**F, Ch.5, §5**], we have $\chi' \in (\varphi, \psi)$ in S as required.

The relevance of Noether's theorem for Hilbert's proof was already observed in [CL].

3. Tangent map. Let $f : \mathbb{P}^m(\mathbb{R}) \rightsquigarrow \mathbb{P}^n(\mathbb{R})$ be a rational map given by $x \mapsto (f_0(x), \ldots, f_n(x))$ where the f_i are forms of degree r. Let z be a point of $\mathbb{P}^m(\mathbb{R})$ with $f_i(z) \neq 0$ for some i so that f is defined at z. Then f induces a map of tangent spaces $T_z(\mathbb{P}^m) \to T_{f(z)}(\mathbb{P}^n)$. Consider the affine cones $\mathbb{R}^{m+1} - \{0\} \to \mathbb{P}^m$ and $\mathbb{R}^{n+1} - \{0\} \to \mathbb{P}^n$. Let $f : \mathbb{R}^{m+1} \to \mathbb{R}^{n+1}$ be given by $f(x) = (f_0(x), \ldots, f_n(x))$.

LEMMA 3.1. *Let x be a point of $\mathbb{R}^{m+1} - \{0\}$ over z. Then*

$$\ker[T_z(\mathbb{P}^m) \to T_{f(z)}(\mathbb{P}^n)] \approx \ker[T_x(\mathbb{R}^{m+1}) \to T_{f(x)}(\mathbb{R}^{n+1})] .$$

PROOF. Let $T_x^{\mathrm{vert}}(\mathbb{R}^{m+1})$ consist of tangent vectors to the fiber of $\mathbb{R}^{m+1} - \{0\} \to \mathbb{P}^m$. We get

$$\begin{array}{ccccccccc} 0 & \longrightarrow & T_x^{\mathrm{vert}}(\mathbb{R}^{m+1}) & \longrightarrow & T_x(\mathbb{R}^{m+1}) & \longrightarrow & T_z(\mathbb{P}^m) & \longrightarrow & 0 \\ & & \downarrow & & \downarrow & & \downarrow & & \\ 0 & \longrightarrow & T_{f(x)}^{\mathrm{vert}}(\mathbb{R}^{n+1}) & \longrightarrow & T_{f(x)}(\mathbb{R}^{n+1}) & \longrightarrow & T_{f(z)}(\mathbb{P}^n) & \longrightarrow & 0 . \end{array}$$

Identify $T_x(\mathbb{R}^{m+1})$ with $\mathbb{R}^{m+1}$ so $\xi \in \mathbb{R}^{m+1}$ corresponds to the tangent at x to the curve $\epsilon \mapsto x + \epsilon\xi$. The vertical tangents in $T_x(\mathbb{R}^{m+1})$ correspond to multiples of x ($\xi = \lambda x$, $\lambda \in \mathbb{R}$). The map $T_x(\mathbb{R}^{m+1}) \to T_{f(x)}(\mathbb{R}^{n+1})$ sends ξ to the tangent at $f(x)$ to the curve

$$\epsilon \mapsto f(x + \epsilon\xi) = f(x) + \epsilon(\sum_j \frac{\partial f_i}{\partial x_j}(x)\xi_j) + O(\epsilon^2)$$

and so is given, as expected, by the Jacobian $\frac{\partial f_i}{\partial x_j}$. The map $T_x^{\mathrm{vert}}(\mathbb{R}^{m+1}) \to T_{f(x)}^{\mathrm{vert}}(\mathbb{R}^{n+1})$ sends $\xi = \lambda x$ to the tangent vector at $f(x)$ to the curve

$$\epsilon \mapsto f(x+\epsilon\xi) = (1+\epsilon\lambda)^r f(x) = (1+r\epsilon\lambda + O(\epsilon^2))f(x)$$

and so sends λx to $r\lambda f(x)$ which is non-trivial since $f(x) \neq 0$ by assumption (and $r \neq 0$ in $\mathbb{R}$). Therefore, $T_x^{\mathrm{vert}}(\mathbb{R}^{m+1}) \to T_{f(x)}^{\mathrm{vert}}(\mathbb{R}^{n+1})$ is an isomorphism and the snake lemma shows that the kernels in Lemma 3.1 are isomorphic.

The same result also holds over $\mathbb{C}$.

4. Topological dimension. The following must be well-known.

LEMMA 4.1. *Let $X \subset \mathbb{R}^n$ be a real algebraic set,*

$$I = \{f \in \mathbb{R}[x_1, \dots, x_n] \mid f|X = 0\} ,$$

and $A = \mathbb{R}[x_1, \dots, x_n]/I$. Then

$$\dim_{\mathrm{top}} X = \text{Krull dim } A .$$

PROOF. If X has irreducible components X_i then $\dim_{\mathrm{top}} X = \sup \dim_{\mathrm{top}} X_i$ [**HW, Th. III 2**]. Therefore we can assume X irreducible so that A is a domain. If Krull dim $A = 0$ then X is finite and we are done. If $s \in A$, the coordinate ring of $Z = X \cap \{s = 0\}$ is a quotient of $A/(s)$ and so has lower Krull dimension. Therefore $\dim_{\mathrm{top}} Z <$ Krull dim A by induction on the Krull dimension and it will suffice to show that $X_s = X - Z$ has $\dim_{\mathrm{top}} X_s =$ Krull dim A. The coordinate ring of $X_s = X \cap \{s \neq 0\}$ is A_s which can be seen by embedding X_s in $\mathbb{R}^n \times \mathbb{R}$ by $x \mapsto (x, s(x)^{-1})$ so that X_s is defined by the equations of X and $ys(x) = 1$. If B is a finitely generated $\mathbb{R}$-algebra and has the same quotient field K as A, we can find $A_s = B_t$. Write $K = \mathbb{R}(x_1, \dots, x_d, y)$ where the x_i are algebraically independent, $d =$ Krull dim A, and y satisfies

$$f(x,y) = y^m + f_1(x)y^{m-1} + \cdots + f_m(x) = 0$$

with $\frac{\partial f}{\partial y}(x,y) \neq 0$ in K. Replace A by a localization A_s of $B = (\mathbb{R}[x_1, \dots, x_d, y]/f)_{\frac{\partial f}{\partial y}}$. Now X_s is an open set of the set defined by $f(x,y) = 0$ and the implicit function theorem shows that $X_s \to \mathbb{R}^d$ by $(x,y) \mapsto x$ is a local homeomorphism.

COROLLARY 4.2. *If $Y \subset \mathbb{C}^n$ is an algebraic set then* $\dim_{\mathrm{top}}(Y \cap \mathbb{R}^n) \leq \dim_{\mathrm{alg}}(Y)$. *If $Y \subset \mathbb{P}^n(\mathbb{C})$ is algebraic,* $\dim_{\mathrm{top}}(Y \cap \mathbb{P}^n(\mathbb{R})) \leq \dim_{\mathrm{alg}}(Y)$.

PROOF. Apply Lemma 4.1 to $X = Y \cap \mathbb{R}^n$. The Zariski closure of X in $\mathbb{C}^n$ is $\operatorname{Spec} \mathbb{C} \otimes_{\mathbb{R}} A$ and lies in Y so $\dim Y \geq \dim \mathbb{C} \otimes_{\mathbb{R}} A = \dim A \geq \dim_{\mathrm{top}}(X)$. The second assertion follows from the first.

5. Proof of Hilbert's theorem. Let F stand for a ternary quartic form. Write $F > 0$ if $F(x,y,z) > 0$ for all real $(x,y,z) \neq (0,0,0)$, and similarly for $F \geq 0$. Let $U = \{[F] \in \mathbb{P}^{14}(\mathbb{R}) \mid F > 0\}$. The topological closure $\overline{U}$ is $\{[F] \in \mathbb{P}^{14}(\mathbb{R}) \mid F \geq 0\}$ since we can approximate F by $F + \epsilon(x^4+y^4+z^4)$ in U. Note that U is connected since we can join $[F]$ and $[G]$ in U by the curve $t \mapsto [tF + (1-t)G]$, $0 \leq t \leq 1$. Let $V = \{[F] \in U \mid \{F = 0\}$ is smooth as a curve over $\mathbb{C}\}$.

LEMMA 5.1. *V is open, dense in U, and connected.*

PROOF. By Lemma 1.1, $W = \{[F] \in \mathbb{P}^{14}(\mathbb{R}) \mid \{F = 0\} \text{ is smooth}\}$ is open. Since $V = W \cap U$, V is also open. Let Y be the set of Lemma 1.2. We claim $V = U - Y$. Clearly $V \subset U - Y$. If $[F] \in U - Y$ then $F > 0$ so $\{F = 0\}$ has no real points. Since F is real, the singular points of $\{F = 0\}$ occur in complex conjugate pairs but $[F] \notin Y$ so $\{F = 0\}$ has at most one singular point. Therefore it is smooth. Let $T = Y \cap \mathbb{P}^{14}(\mathbb{R})$. Then $T \subset \overline{Y} \cap \mathbb{P}^{14}(\mathbb{R})$ which has topological dimension ≤ 12 by Lemma 1.2 and Corollary 4.2, so $\dim T \leq 12$. Since $V = U - T$, $U - \overline{V} \subset T$ but $U - \overline{V}$ is a 14 dimensional manifold (if not empty) and $\dim T \leq 12$. It follows that $U \subset \overline{V}$. Since $\dim T \leq \dim U - 2$, $V = U - T$ is connected by [HW, Th. IV 4 Cor. 1].

Now define $\pi : \mathbb{P}^{17}(\mathbb{R}) \to \mathbb{P}^{14}(\mathbb{R})$ by $[\varphi, \psi, \chi] \mapsto [F]$ where φ, ψ, χ are quadratic forms, $[\varphi, \psi, \chi]$ is the point whose homogeneous coordinates are the coefficients of φ, ψ, χ, and $F = \varphi^2 + \psi^2 + \chi^2$. If $[\varphi, \psi, \chi] \in \pi^{-1}(V)$ then φ, ψ, χ have no common zero in $\mathbb{P}^2(\mathbb{C})$ since at such a point p we would have

$$F(p) = 0 \ , \ \frac{\partial F}{\partial x}(p) = 2\varphi\frac{\partial \varphi}{\partial x} + \cdots = 0 = \frac{\partial F}{\partial y} = \frac{\partial F}{\partial z}$$

and $\{F = 0\}$ would be singular. Consider the map of tangent spaces $T_{[\varphi,\psi,\chi]}(\mathbb{P}^{17}(\mathbb{R})) \to T_{[F]}(\mathbb{P}^{14}(\mathbb{R}))$. By Lemma 3.1, its kernel is the same as that of $T_{\{\varphi,\psi,\chi\}}(\mathbb{R}^{18}) \to T_{\{F\}}(\mathbb{R}^{15})$ where $\{\varphi, \psi, \chi\}$ is the point whose coordinates are the coefficients of φ, ψ, χ, and similarly for $\{F\}$. To find this kernel, consider the curve

$$\epsilon \mapsto \{\varphi, \psi, \chi\} + \epsilon\{\varphi', \psi', \chi'\}$$

which maps to

$$\epsilon \mapsto (\varphi + \epsilon\varphi')^2 + (\psi + \epsilon\psi')^2 + (\chi + \epsilon\chi')^2 = F + 2\epsilon(\varphi\varphi' + \psi\psi' + \chi\chi') + O(\epsilon^2) \ .$$

The map of tangent spaces is $\{\varphi', \psi', \chi'\} \mapsto 2\{\varphi\varphi' + \psi\psi' + \chi\chi'\}$ and the kernel is $\big\{\{\varphi', \psi', \chi'\} \mid \varphi\varphi' + \psi\psi' + \chi\chi' = 0\big\}$. By Corollary 2.2 with $k = \mathbb{R}$, this is generated by the "Koszul elements" $\{\psi, -\varphi, 0\}$, $\{\chi, 0, -\varphi\}$, $\{0, \chi, -\psi\}$ so

$$\{\varphi', \psi', \chi'\} = \alpha\{\psi, -\varphi, 0\} + \beta\{\chi, 0, -\varphi\} + \gamma\{0, \chi, -\psi\} \ .$$

All are quadratic so α, β, γ are constant (i.e. lie in $\mathbb{R}$) and the kernel has dimension ≤ 3. This forces $T_{[\varphi,\psi,\chi]}(\mathbb{P}^{17}(\mathbb{R})) \to T_{[F]}(\mathbb{P}^{14}(\mathbb{R}))$ to be onto since these are vector spaces of dimension 17 and 14. The implicit function theorem now shows that in the neighborhood of $[\varphi, \psi, \chi]$, π is diffeomorphic to a projection $(x_1, \ldots, x_{17}) \mapsto (x_1, \ldots, x_{14})$. In particular, $\pi^{-1}(V) \to V$ is an open map so its image is an open set of V. On the other hand, $\pi(\pi^{-1}(V)) = \pi(\mathbb{P}^{17}(\mathbb{R})) \cap V$ which is closed in V since $\pi(\mathbb{P}^{17}(\mathbb{R}))$ is compact. Since $\pi(\pi^{-1}(V))$ is clearly non-empty and V is connected, $\pi(\pi^{-1}(V)) = V$ so that $V \subset \pi(\mathbb{P}^{17}(\mathbb{R})) \subset \overline{U}$. Since $\pi(\mathbb{P}^{17}(\mathbb{R}))$ is closed and $\overline{V} = \overline{U}$, $\pi(\mathbb{P}^{17}(\mathbb{R})) = \overline{U}$ as required.

REMARK. This argument doesn't work over $\mathbb{C}$ since $\pi : \mathbb{P}^{17}(\mathbb{C}) \to \mathbb{P}^{14}(\mathbb{C})$ is not defined everywhere because $\varphi^2 + \psi^2 + \chi^2$ can be zero so it is not clear whether the image is closed.

REFERENCES

[CL] M. D. Choi and T. Y. Lam, Extremal positive semidefinite forms, Math. Ann. 231 (1977), 1–18

[F] W. Fulton, Algebraic Curves, Benjamin, New York, 1969

[Ha] R. Hartshorne, Algebraic Geometry, Springer-Verlag, New York, 1977

[H] D. Hilbert, Über die Darstellung definiter Formen als Summe von Formenquadraten, Math. Ann. 32 (1888), 342–350

[HW] W. Hurewicz and H. Wallman, Dimension Theory, Princeton, 1948

[JR] B. Jacob and A. Rosenberg, eds., K-Theory and Algebraic Geometry: Connections with Quadratic Forms and Division Algebras, Proc. Symp. Pure Math. 58, Amer. Math. Soc. 1995

[R] A. R. Rajwade, Squares, London Math. Society Lecture Notes Series 171, Cambridge Univ. Press 1993

Department of Mathematics
The University of Chicago
5734 University Avenue
Chicago, IL 60637
E-mail address: swan@math.uchicago.edu

Contemporary Mathematics
Volume **272**, 2000

Quadratic forms and normal surface singularities

C.T.C. Wall

ABSTRACT. A rapid survey of the current understanding of normal surface singularities (over $\mathbb{C}$). After introducing the main examples, we recall the topological classification, and proceed to properties of the intersection quadratic form on the lattice of cycles on a resolution. Next the main analytic invariants are defined, and the known results on their relation to topological invariants (especially for rational and elliptic singularities) are presented. Quadratic forms also play a major part in the discussion of smoothings of singularities. We conclude with a brief mention of the Durfee conjecture and of the Neumann-Wahl conjecture.

The title announced for my talk[1] was 'Quadratic forms in singularity theory'. But that is too wide a topic for a lecture. There are important applications of quadratic forms to the study of deformations and hence, for example, to enumeration of possible configurations of singularities on quartic surfaces. A good survey of such applications is given by Nikulin [**29**]. I have chosen the above topic partly because it remains an area of current activity, with challenging problems.

This is intended as a survey article for non-specialists. No new results are claimed, and I will attempt to explain things as we go.

1. Terminology and examples

Roughly speaking, a normal surface singularity is an isolated singular point of a complex surface (i.e. 2 complex dimensions). We may regard (a piece of) the surface as embedded in affine space $\mathbb{C}^n$, with the singular point at O: then the surface X is locally defined by equations $f_i = 0$ where each f_i is holomorphic in some neighbourhood of the origin. More precisely, one may choose a neighbourhood U of O in $\mathbb{C}^n$ such that all f_i are holomorphic on U; X is the locus of their simultaneous zeros in U, and X is of dimension 2, with O as its only singular point.

We are only interested in a neighbourhood of O in X; we use the standard notation (X, O) to denote the germ of X at O: a normal surface singularity

[1]Expanded version of talk given at conference on quadratic forms, Dublin, July 1999.

consists of such a germ. Later we will need to be more precise about the neighbourhood U. The local ring $\mathcal{O}_{X,O}$ is defined as the quotient of the ring $\mathcal{O}_n$ of germs at O of holomorphic functions on $\mathbb{C}^n$ (or equivalently, the ring of power series in $(z_1, \dots, z_n)$ with non-zero radius of convergence) by the ideal generated by the f_i. This ring determines the singularity up to the natural equivalence relation, local analytic isomorphism. A convenient reference for foundational material on normal surface singularities is [**15**].

The word *normal* refers to the fact that we require the local ring $\mathcal{O}_{X,O}$ to be normal — i.e. integrally closed in its quotient field. For the case of surface singularities, this implies that O is an isolated singularity of X, and that X has only one branch at O. The converse holds if [**15**, 3.1] (X, O) is an IHS or more generally if it is an ICIS; the example of the image of the map $(\mathbb{C}^2, O) \to (\mathbb{C}^4, O)$ given by $(x, y) \mapsto (x^2, x^3, y, xy)$ shows that some further condition is needed. In general, an isolated surface singularity is normal if and only if it is Cohen-Macaulay.

We illustrate these definitions with some important examples.

Let $f(x, y, z) = 0$ be an algebraic equation in 3 complex variables having an isolated singularity at the origin. This is said to define an *isolated hypersurface singularity*, or IHS[2] for short. Examples of particular importance are the ADE singularities:

$$\begin{array}{lll} A_n & x^{n+1} + y^2 + z^2 = 0 & (n \geq 1) \\ D_n & x^{n-1} + xy^2 + z^2 = 0 & (n \geq 4) \\ E_6 & x^4 + y^3 + z^2 = 0, & \\ E_7 & x^3y + y^3 + z^2 = 0, & \\ E_8 & x^5 + y^3 + z^2 = 0. & \end{array}$$

The next most complicated examples may be defined by equations of the form

$$x^p + y^q + z^r + \lambda xyz = 0,$$

where $p, q, r \geq 2$ are integers. For the *parabolic* singularities we have $\frac{1}{p} + \frac{1}{q} + \frac{1}{r} = 1$ (and in each case certain values of λ are excluded). If $\frac{1}{p} + \frac{1}{q} + \frac{1}{r} < 1$ we say we have a *cusp* singularity: here $\lambda \neq 0$ and we may change variables to reduce to $\lambda = 1$.

There exist extensive further classifications (see e.g. [**1**], [**46**]) of IHS's, mostly included in the subclass of examples of the form $g(x, y) + z^n = 0$.

A surface defined in $\mathbb{C}^{n+2}$ by just n equations is said to be a *complete intersection*. If in addition we have an isolated singularity (usually taken to be at the origin) we speak of an *isolated complete intersection singularity*, or ICIS for short. In many ways these are similar to IHS's. The most important examples are surfaces given by 2 equations in $\mathbb{C}^4$.

If a finite group G acts linearly on $\mathbb{C}^2$, with the action free on the complement of the origin, one may define a quotient space $X = \mathbb{C}^2/G$, which will have an isolated singularity at the image of the origin: these are

[2]Warning: some authors use IHS to mean integral homology sphere.

called *quotient singularities*. For example, we write $(X_{n,r}, x)$ for the quotient of the unit ball by the action of the cyclic group $\mathbb{Z}_n$ of order n given by $T.(z_1, z_2) = (\epsilon_n z_1, \epsilon_n^r z_2)$, where $\epsilon_n := e^{2\pi i/n}$ (we will retain this notation for other examples) and r is prime to n.

Although this definition does not exhibit X as a subset of affine space, an embedding may be constructed using a set of generators of the ring of invariants $\mathbb{C}[z_1, z_2]^G$. For example, the action of $\mathbb{Z}_3$ defining $X_{3,1}$ has generating invariants $w_1 = z_1^3, w_2 = z_1^2 z_2, w_3 = z_1 z_2^2, w_4 = z_2^3$, and $X_{3,1}$ can be identified with the subset of $\mathbb{C}^4$ defined by $w_1 w_3 = w_2^2$, $w_1 w_4 = w_2 w_3$, $w_2 w_4 = w_3^2$.

One may always suppose $G \subset U_2$; if further $G \subset SU_2$ then, as was shown by Klein, we have an ADE singularity, as shown in the following table, where Q_{4a} denotes the quaternion group of order $4a$ and T^*, O^*, I^* the binary polyhedral groups.

Singularity	$A_n = X_{n+1,-1}$	D_n	E_6	E_7	E_8
Group	$\mathbb{Z}_{n+1}$	Q_{4n-4}	T^*	O^*	I^*

Otherwise we have a singularity which is not an ICIS: the example given above required 3 defining equations in $\mathbb{C}^4$.

One may combine the above constructions by taking the quotient of an IHS or ICIS by a finite group acting on it in a manner which is free outside the singular point, giving a *hyperquotient* singularity.

In a certain sense, quotients of spheres can be replaced by quotients of the hyperbolic plane $\mathfrak{H}$. Let a, b, c be natural numbers with $a^{-1}+b^{-1}+c^{-1} < 1$. Consider a triangle in $\mathfrak{H}$ with interior angles $\pi/a, \pi/b$ and π/c. Let G be the group of isometries of $\mathfrak{H}$ generated by the reflections in the edges of this triangle, and G^+ the subgroup of index 2 of orientation-preserving isometries. There is an induced action of G^+ on the tangent line bundle $T\mathfrak{H}$. Form the quotient space $T\mathfrak{H}/G^+$, and collapse the zero section $\mathfrak{H}/G^+$ to a point. The quotient can be proved to have an induced structure as complex space, and is a normal surface singularity. These are called triangle singularities. It is known that there are just 14 triples $a \leq b \leq c$ for which the singularity is an IHS, and a further 8 for which we have an ICIS. For example, the triple $(2, 3, 7)$ corresponds to the singularity $x^2 + y^3 + z^7 = 0$.

Cusp singularities in general are so called because they arise naturally as 'cusps' in the compactification of Hilbert modular surfaces [**11**]. We briefly recall the definition. Let K be a real quadratic field; denote conjugation over $\mathbb{Q}$ by a prime. Let M be an additive subgroup of K which is free abelian of rank 2; we may suppose M generated by 1 and ω with $\omega > 1 > \omega' > 0$. The group G of elements $\gamma \in K$ with $\gamma M = M$ and $\gamma, \gamma' > 0$ is infinite cyclic: let G_j be the subgroup of (finite) index j. The semi-direct product $M.G_j$ acts freely and properly on $\mathfrak{H} \times \mathfrak{H}$ by $(\gamma, m)(z_1, z_2) = (\gamma z_1 + m, \gamma' z_2 + m')$: denote the quotient by $X'(M, G_j)$. Then define $X(M, G_j)$ by compactifying at one point, corresponding to $(i\infty, i\infty)$.

In [**11**], Hirzebruch gives a detailed analysis of these singularities and their resolutions. There is a purely periodic negative continued fraction expansion

$$\omega = b_1 - \frac{1}{b_2-}\frac{1}{b_3-}\cdots\frac{1}{b_k-}\frac{1}{b_1-}\cdots,$$

with each $b_r \geq 2$, where we take k as j times the minimal period. We may distinguish the b_r equal to 2 from those exceeding 2 and then write the sequence $b_1, \ldots, b_k$ (up to cyclic reordering) as

$$\mathbf{b} = 2, \ldots, 2, k_1 + 2, 2, \ldots, 2, k_2 + 2, \ldots\ldots k_g + 2,$$

where there are $k_1^* - 1$ numbers 2 in the first group, $k_2^* - 1$ in the second, and so on. We have (up to cyclic reordering) the continued fraction expansion

$$\omega - 1 = k_1 + \frac{1}{k_1^*+}\frac{1}{k_2+}\frac{1}{k_2^*+}\cdots.$$

Shifting this by 1 to interchange the roles of k and k^* yields another cyclic sequence

$$\mathbf{b}^* = k_1^* + 2, 2, \ldots, 2, k_2^* + 2, \ldots, \ldots, 2, \ldots, 2,$$

where there are $k_1 - 1$ numbers 2 in the first group, and so on till $k_g - 1$ numbers 2 in the last. Then $\Delta = \sum(b_r - 2) = \sum k_i$ is the length of the sequence $\mathbf{b}^*$ and $\Delta^* = \sum(b_r^* - 2) = \sum k_i^*$ is the length of the sequence $\mathbf{b}$.

Cusp singularities are classified by sets $\mathbf{b}$ of integers ≥ 2, defined up to cyclic reordering, or equivalently by the dual set $\mathbf{b}^*$. The cusp singularity is an IHS if $\Delta^* \leq 3$ and an ICIS if (and only if) $\Delta^* \leq 4$.

Another important class are the *quasihomogeneous* singularities: here we require all the equations defining X to be homogeneous (of various degrees) with respect to assigning z_i to have weight a_i for each i. We may then choose a common multiple A of the a_i and define an action of the multiplicative group $\mathbb{C}^\times$ by $t.(z_1, \ldots, z_n) = (t^{A/a_1}z_1, \ldots, t^{A/a_n}z_n)$. This group action, with O as fixed point, is the essential feature of this class of singularities. Thus an IHS is quasihomogeneous if the equation f is equivariant under an action of the multiplicative group $\mathbb{C}^\times$ of the form $f(t^a x, t^b y, t^c z) \cong t^d f(x, y, z)$.

Of the above examples, the ADE, parabolic, quotient and triangle singularities are quasihomogeneous; the cusp singularities are not.

The above examples are the easiest to describe, and hence in some sense the simplest. We shall see that even for them there are unsolved problems.

2. Topological classification

Somewhat surprisingly, the most accessible part of the structure of a normal surface singularity is its topology. Let us begin by choosing the neighbourhood U of O in $\mathbb{C}^n$ to be a closed neighbourhood, with boundary ∂U, such that $X_0 = X \cap U$ is homeomorphic to the cone on its boundary $\partial X_0 = X \cap \partial U$. Routine arguments show that we may choose U to be given by $\sum_1^n |z_i|^2 \leq \epsilon$ for any small enough ϵ. This is not always a convenient choice: if X is quasihomogeneous as just described, then we may take U as $\sum_1^n |z_i|^{2A/a_i} \leq \epsilon$ for small enough ϵ. This gives a neighbourhood X_0 invariant by the action of $\{t \in \mathbb{C} \mid |t| \leq 1\}$, with ∂X_0 invariant by the circle group S^1.

From now on, when we write (X, x) for a normal surface singularity, we suppose X itself to be a good neighbourhood of x in this sense.

It is known that one may construct, by a sequence of blowings-up and normalisations, a *resolution*, viz. a proper map

$$\pi : (\tilde{X}, E) \to (X, x)$$

with the following properties:

$\tilde{X}$ is a non-singular complex surface (with boundary);

E has dimension 1, and is a union of irreducible curves E_i;

π induces an isomorphism of $\tilde{X} - E$ onto $X - \{x\}$, and $\pi(E) = x$.

We have a *good resolution* if also

each component E_i of E is a smooth embedded curve;

the components E_i have *normal crossings*: at each point of E lying on more than one E_i there are local coordinates (z_1, z_2) (in $\tilde{X}$) such that the components are given respectively by $z_1 = 0$ and $z_2 = 0$; in particular, there are only 2 components through such a point.

This result has a somewhat involved history. In Zariski's 1935 book he examines critically each of 4 claimed proofs of resolution of singularities of surfaces and finds them all wanting; a satisfactory argument was given in 1936 by R.J. Walker. A convenient proof for present purposes may be found in [**15**, Chap. 2].

The resolution procedure is explicit enough to be used for effective calculations, and has been implemented in computer algorithms. The result is almost unique, but it remains possible to blow up a point of E, giving a further resolution. The effect of blowing up a point on a smooth surface is to produce one exceptional curve, of genus 0, with self-intersection -1: such a curve is traditionally called an *exceptional curve of the first kind.* A resolution is said to be *minimal* if it includes no exceptional curve of the first kind; a good resolution is minimal if it includes no exceptional curve of the first kind which meets at most 2 other components of E. There exist an essentially unique minimal resolution and an essentially unique minimal good resolution [**15**, 5.12]. Mostly we prefer to deal with an arbitrary good resolution, and take care when it matters that the invariants we define are unchanged by this process.

The resolution provides the following numerical data:

for each component E_i, the genus $g_i = g(E_i)$;

the self-intersection numbers $b_{i,i} = E_i^2$ of the cycles E_i in $\tilde{X}$;

the intersection numbers $b_{i,j} = E_i.E_j = \#(E_i \cap E_j)$.

The numerical data are usually conveniently presented in the form of a weighted graph Γ. This has one vertex v_i for each component E_i of the exceptional set, which is weighted by the genus (in brackets: $[g_i]$) and negative self-intersection $a_i = -E_i^2$: the default values $g_i = 0$, $a_i = 2$ are omitted where they occur. Two vertices v_i, v_j are joined by $E_i.E_j$ edges.

THEOREM 2.1. *A set of integers $\{g_i : 1 \leq i \leq n\}$ and $\{b_{i,j} : 1 \leq i, j \leq n\}$ can be realised as arising from an isolated surface singularity if and only if*

$g_i \geq 0$ for each i, $b_{j,i} = b_{i,j} \geq 0$ for each $i \neq j$, and the quadratic form $\sum_{i,j=1}^{n} b_{i,j}x_ix_j$ is negative definite.

The numerical data completely determine the topology of $\tilde{X}$ and X.

If the singularity is normal, then Γ is connected.

PROOF. The fact that the quadratic form defined by intersection numbers is negative definite was proved by Mumford [**23**] in a paper which marks the beginning of modern research in this area. A proof is also given in [**15**, 4.4].

The self-intersections E_i^2 determine the normal bundles of the curves E_i, which are embedded and of given genera g_i, and hence the topology of the neighbourhood is determined. The neighbourhoods are plumbed together as indicated by the intersection numbers $E_i.E_j$. Moreover, since we chose a good neighbourhood X of x, $\tilde{X}$ is itself a smooth regular neighbourhood [**10**] of E, so may be described topologically as such a plumbing.

Conversely, we can use plumbing to construct a neighbourhood $\tilde{X}$ of a set E with the given invariants. Now by a theorem of Grauert [**8**] (a proof is also given in [**15**, 4.9]), since the quadratic form is negative definite, there exists a map $\pi : (\tilde{X}, E) \to (X, x)$ defining an isolated surface singularity: this proves sufficiency. If the singularity is normal, there is just one branch at x: this is equivalent to connectedness of E, and hence to that of Γ. □

In particular, the numerical invariants determine the topology of the *link* $L = \partial X \cong \partial\tilde{X}$, which has a natural orientation as boundary of the complex (hence oriented) manifold $\tilde{X}$. (To determine the analytic type one would need also the analytic types of the E_i, of their normal bundles, and the precise positions of the intersection points.) There is also a strong converse result, due to Neumann [**26**].

THEOREM 2.2. *(i) The topology of the oriented link L uniquely determines the numerical data.*
(ii) Both L and $-L$ are singularity links if and only if either (a) L is a lens space or (b) L is a T^2 bundle over a circle whose monodromy matrix has trace ≥ 3.
(iii) If L is not as in (a)or (b), then $\pi_1(L)$ determines L uniquely.
(iv) In all cases, L is an irreducible 3-manifold; unless $\pi_1(L)$ is cyclic or a triangle group, L is Haken.

As to (a), the following are equivalent:

$\pi_1(L)$ is cyclic,

L is a lens space,

(X, x) is a cyclic quotient singularity.

In fact, $\partial X_{n,r}$ is the lens space $L(n, r)$, which is orientation-preserving homeomorphic to $L(n, s)$ if and only if $s \equiv r$ or $rs \equiv 1 \pmod{n}$ and is orientation-reversing homeomorphic to $L(n, -r)$.

The exceptional case (b) corresponds to cusp singularities, where the monodromy matrix may be taken as

$$A(\mathbf{b}) = \begin{pmatrix} b_1 & 1 \\ -1 & 0 \end{pmatrix} \cdots \begin{pmatrix} b_n & 1 \\ -1 & 0 \end{pmatrix},$$

and $-L$ is the link of the cusp singularity defined by the dual sequence $\mathbf{b}^*$.

For all the simpler examples, the graph Γ is a tree. For the cyclic quotient singularity $X_{n,r}$, Γ is a *bamboo* or *chain*, so $E_i.E_j = 1$ if $j = i+1$ and 0 if $i < j$ otherwise; all g_i vanish, but the $a_i = -E_i^2$ are determined as the coefficients in the expansion of $\frac{r}{n}$ as a negative continued fraction $\frac{1}{a_1-}\ \frac{1}{a_2-}\ \cdots \frac{1}{a_k}$.

For a cusp singularity, Γ consists of rational curves E_i arranged in a cycle with the negative self-intersection numbers being $(b_1, \ldots, b_k)$ (this needs interpretation if $k = 1$).

For a parabolic singularity E has just one component, which has genus 1. In general, a singularity for which E consists of a single elliptic curve is called *simple elliptic* [**33**].

For any quasihomogeneous singularity, we have an action of S^1 on the link L, and L/S^1 has the structure of orbifold, with genus g, say, and some cone points. The corresponding resolution graph is a star, with central vertex having genus g, and one arm for each cone point corresponding to the resolution of a cyclic quotient singularity. The details are explained in full in [**26**].

For an ADE singularity, Γ is the Dynkin graph associated to the Lie group of the same name (with each $g_i = 0, a_i = -2$). For a triangle singularity there are 3 cone points; the central vertex has $g_0 = 0, a_0 = 1$; and each arm has length 1.

Each 2-dimensional orbifold has a geometric structure in the sense of Thurston, whose type is determined by the sign of the orbifold Euler characteristic. Correspondingly it was shown by Neumann [**27**] that each quasihomogeneous singularity link L has a geometric structure, whose type is S^3 if $\chi(L/S^1) > 0$, *Nil* if $\chi(L/S^1) = 0$, and $\widetilde{SL_2(\mathbb{R})}$ if $\chi(L/S^1) < 0$.

For a cusp singularity, the link has a geometric structure of type *Sol*.

Among the more naive invariants of the topology of any normal surface singularity are the Betti numbers (with rational coefficients) $\beta_1(\Gamma)$ of the graph and $\beta_1(L)$ of the link. Elementary homology calculations show that

$$\beta_1(L) = \beta_1(\tilde{X}) = \beta_1(E) = \beta_1(\Gamma) + 2\sum g_i,$$

while of course $\chi(\Gamma) = 1 - \beta_1(\Gamma) = \#I - \sum_{i<j} b_{i,j}$.

3. The algebra of cycles

Some interesting topological invariants of (X, x) can be obtained from a close study of the quadratic form of Theorem 2.1. In this section we will give more details than in the rest of this article, since some of the results may be new to specialists of pure quadratic form theory. Write H for $H_2(\tilde{X}; \mathbb{Z})$ and $H_{\mathbb{Q}}$ for $H_2(\tilde{X}; \mathbb{Q})$. Then the quadratic form is just that given on H

by intersection numbers on $\tilde{X}$. We make the standing hypothesis that the form is negative definite, with $b_{i,j} \geq 0$ for $i \neq j$; we usually (but not always) suppose also Γ connected.

The group H has a preferred basis consisting of the classes of the cycles E_i (which we also denote by E_i). This defines a partial order on H: set $\sum a_i E_i \geq 0$ if $a_i \geq 0$ for each i.

We say that $D \in H$ is *nef* if $D.E_i \geq 0$ for each i. The negatives of the nef classes form a semigroup $\mathcal{E} \subset H$ (it is the intersection of H with a convex cone), and we write $\mathcal{E}^+$ for $\mathcal{E}$ with $\{0\}$ removed.

LEMMA 3.1. *(i)If $D \in \mathcal{E}$, then $D \geq 0$.*
(ii) If Γ is connected, and $D \in \mathcal{E}^+$, then $D \geq \sum E_i$; in particular, $D > 0$.
(iii) If Γ is connected, and $\mathcal{W} \subseteq \mathcal{E}^+$, then $\inf \mathcal{W} \in \mathcal{E}^+$.

PROOF. (i) Set $D = \sum a_i E_i = D_+ - D_-$, where $D_+ := \sup\{D, 0\} = \sum_{a_i>0} a_i E_i$, so also $D_- := \sum_{a_i<0}(-a_i)E_i \geq 0$. Since $-D$ is nef and $D_- \geq 0$, $D.D_- \leq 0$. Since $E_i.E_j \geq 0$ for $i \neq j$ and D_+, D_- have no components in common, $D_+.D_- \geq 0$. Thus $D_-.D_- = D_+.D_- - D.D_- \geq 0$. Since the form is negative definite, it follows that $D_- = 0$.

(ii) Now $D \geq 0$. If any $a_i = 0$, $D.E_i \geq 0$ while as $-D$ is nef, $D.E_i \leq 0$, so $D.E_i = 0$. Thus for each j such that v_j is adjacent to v_i in Γ, we must have $a_j = 0$. By connectivity of Γ, it follows that $D = 0$, contrary to assumption.

(iii) Set $W := \inf \mathcal{W} = \sum a_i E_i$, and choose $D^i \in \mathcal{W}$ having a_i as coefficient of E_i. Then $D^i - W \geq 0$ and has zero coefficient of E_i, so $0 \leq E_i.(D^i - W) \leq -E_i.W$ since $D^i \in \mathcal{E}^+$. Thus $-W$ is nef. By (ii) we cannot have $W = 0$. □

The element $Z_{num} := \inf \mathcal{E}^+$ was introduced by Artin [**3**], who called it the *fundamental cycle*. It follows from the lemma that $Z_{num} \geq \sum E_i$. Now if $Z_{num} \geq D \geq 0$ and $D.E_i > 0$ we have $Z_{num} \geq D + E_i$. For if $Z_{num} - D = \sum m_i E_i \geq 0$, since $(Z_{num} - D).E_i < 0$ we must have $m_i > 0$. We may thus begin with $D = \sum E_i$ and whenever $D.E_i > 0$, add E_i: the process must terminate with Z_{num}.

Our first important invariant of the singularity is the integer $\Delta = -Z_{num}^2$.

We next consider the canonical class $K \in H^2(\tilde{X}; \mathbb{Z})$. For any divisor D on any smooth algebraic surface $\tilde{X}$, we have the formula $\chi(\mathcal{O}_D) = -\frac{1}{2}(D^2 + \langle K, D\rangle)$. If D is the class of a reduced curve C, then $\chi(C) = 2\chi(\mathcal{O}_C) + \mu(C)$, where $\chi(C)$ is the usual Euler characteristic and $\mu(C)$ a measure of the singularities of C. In particular as E_i is non-singular, we obtain

$$2 - 2g_i = \chi(E_i) = 2\chi(\mathcal{O}_{E_i}) = -E_i^2 - \langle K, E_i\rangle.$$

Thus in the presence of the quadratic form, the knowledge of the $\langle K, E_i\rangle$ is equivalent to a knowledge of the genera g_i. Since $E_i^2 < 0$ we see that if $\langle K, E_i\rangle < 0$ then $g_i = 0, E_i^2 = -1$ and we have an exceptional curve of the first kind. If $\langle K, E_i\rangle = 0$ then $g_i = 0$ and $E_i^2 = -2$. In particular, if we have a minimal resolution, $\langle K, E_i\rangle \geq 0$ for each i.

Since the quadratic form is nondegenerate, there is a unique $K' \in H_{\mathbb{Q}}$ such that, for each i, $K'.E_i = \langle K, E_i \rangle = 2g_i - 2 - E_i^2$. We may now define further numerical invariants $K'.Z_{num} = \langle K, Z_{num} \rangle$ (equivalently we may consider $\chi(\mathcal{O}_{Z_{num}}) = -\frac{1}{2}(Z_{num}^2 + K'.Z_{num}))$ and K'^2. Unlike the others the latter changes, being decreased by 1, if we blow up an additional point of E. Thus we must either insist that we have a minimal (perhaps good) resolution or add the number of components of E, which is $\#I = \beta_2(E) = \beta_2(\tilde{X})$, thus giving $K'^2 + \beta_2(\tilde{X})$.

There is an alternative 'fundamental cycle' defined as follows (cf. eg. [**14**]). Set $K' = \sum a_i E_i$ and define $Z_k := \sum [-a_i] E_i$, where the square bracket denotes integer part. The relation of this to Z_{num} is not clear to the writer.

An invariant of a rather different kind is obtained from the Zariski decomposition. The following is an adaptation by Sakai [**34**] and Wahl [**45**] of Zariski's original result, and presents a rather strange feature of quadratic form theory.

THEOREM 3.2. *For each $D \in H_{\mathbb{Q}}$ there is a unique decomposition $D = P + N$ such that*
P is nef, so for each i, $p_i = P.E_i \geq 0$;
$N \geq 0$, so $N = \sum q_i E_i$ with each $q_i \geq 0$;
for each i, $p_i q_i = 0$.

PROOF. The proof will use induction: observe that our standing hypothesis is inherited by the restriction of the form to the sublattice generated by any collection of the E_i (we are not assuming Γ connected).

We begin with existence. Set $D_1 := D$, $I_1 := \{i \mid D_1.E_i \leq 0\}$, and let N_1 be the unique linear combination of $\{E_i \mid i \in I_1\}$ such that $N_1.E_i = D_1.E_i$ for each $i \in I_1$: this exists since the restriction of a negative definite quadratic form to any subspace is nondegenerate.

Applying (i) of Lemma 3.1 to this subspace, we find $N_1 \geq 0$. Now set $D_2 := D_1 - N_1$, $I_2 := \{i \mid D_2.E_i \leq 0\}$, and repeat the process.

More formally, suppose inductively that we have constructed D_k and $N_k \geq 0$ by this process. Set $D_{k+1} := D_k - N_k$, $I_{k+1} := \{i \mid D_{k+1}.E_i \leq 0\}$, and let N_{k+1} be the unique linear combination of $\{E_i \mid i \in I_{k+1}\}$ such that $N_{k+1}.E_i = D_{k+1}.E_i$ for each $i \in I_{k+1}$: then by Lemma 3.1, $N_{k+1} \geq 0$.

Then by construction if $i \in I_k$, $D_{k+1}.E_i = 0$ so $i \in I_{k+1}$, so $I_{k+1} \supseteq I_k$. Since there are only finitely many subsets, this sequence must become constant. But if $I_{k+1} = I_k$, then D_{k+1} is nef. It then suffices to take $P = D_{k+1}$, $N = \sum_1^k N_r$.

As to uniqueness, suppose we have two decompositions $D = P + N = P' + N'$; write $N = N_c + N_d$, $N' = N'_c + N'_d$, where N_c, N'_c contain the components E_i appearing in both N and N' and N_d, N'_d are linear combinations of the other components. Thus for E_i in N, $E_i.N'_d \geq 0$. Hence $N'_c.E_i \leq N'.E_i$, and since $D - N'$ is nef, this $\leq D.E_i$, which equals $N.E_i$ since E_i is orthogonal to P. Now $N'_c - N$ involves only components of N, and we have seen that its product with each is non-negative. Thus by (i) of Lemma 3.1 again, $N'_c - N \geq 0$, so $N' - N \geq N'_c - N \geq 0$. By symmetry, $N - N' \geq 0$. Hence $N = N'$. □

To illustrate the character of this result, consider the quadratic form with matrix $\begin{pmatrix} -2 & 1 \\ 1 & -2 \end{pmatrix}$. The decomposition of $xE_1 + yE_2$ has

$$\begin{array}{ll} N = xE_1 + yE_2 & \text{if } x \geq 0, y \geq 0, \\ N = (y - \frac{1}{2}x)E_2 & \text{if } x \leq 0, -x + 2y \geq 0, \\ N = (x - \frac{1}{2}y)E_1 & \text{if } 2x - y \geq 0, y \leq 0, \text{ and} \\ N = 0 & \text{if } -x + 2y, 2x - y \leq 0. \end{array}$$

We now apply the decomposition theorem to write $K' + \sum E_i$ in the form $P + N$. It is then shown by Wahl [**45**] that P^2 is a useful invariant. In particular, he shows that $P^2 \leq 0$, and depends only on $\pi_1(L)$.

4. Some analytic invariants

A surprising — and challenging — feature of the theory of normal surface singularities is that their analytic properties are only weakly related to the relatively straightforward topological classification. In this section we define some of the most important analytic invariants which we would like to understand.

The simplest of these is the *embedding dimension* $ed\,(X, x)$: the least number n such that there exists a singularity $(Y, O) \subset (\mathbb{C}^n, O)$ isomorphic to (X, x). The components of any map from (X, x) to $(\mathbb{C}^n, O)$ must be functions on X, i.e. elements of $\mathcal{O}_{X,x}$, vanishing at x; and thus belonging to the maximal ideal $\mathfrak{m}$ of $\mathcal{O}_{X,x}$. It is not hard to show that a collection of n elements of $\mathfrak{m}$ defines an embedding of (X, x) in $(\mathbb{C}^n, O)$ if and only if it spans $\mathfrak{m}$ modulo $\mathfrak{m}^2$. Thus we have $ed\,(X, x) = \dim(\mathfrak{m}/\mathfrak{m}^2)$. We may regard the embedding dimension as the first in the sequence of invariants $\dim(\mathfrak{m}^k/\mathfrak{m}^{k+1})$, which may be collected as the coefficients in a power series, the *Hilbert-Samuel function* of (X, x). This determines in turn the *multiplicity mult* (X, x): the local intersection number of (X, x) at O with a generic $(n-2)$-dimensional linear subspace of $\mathbb{C}^n$. In fact we have

$$2\,mult\,(X, x) = \lim_{k \to \infty} k^{-2} \dim(\mathfrak{m}^k/\mathfrak{m}^{k+1}).$$

Next write ω_X for the sheaf of regular holomorphic 2-forms on X. On the smooth part $(X - \{x\})$ of X, we can identify this with the determinant line bundle of the complex cotangent bundle. The singularity (X, x) is said to be *Gorenstein* if this bundle is trivial over the punctured neighbourhood $X - \{x\}$ of x. This turns out to be an extremely important condition. It holds for all IHS, ICIS, cusp and triangle singularities, but not for quotient singularities other than the ADE singularities. For the IHS $f(x, y, z) = 0$, for example, elementary calculations show that

$$\omega := \frac{dx \wedge dy}{\partial f/\partial z} = \frac{dy \wedge dz}{\partial f/\partial x} = \frac{dz \wedge dx}{\partial f/\partial y}$$

spans ω_X at any point $\neq O$, since at least one of the partial derivatives is non-zero. If (X, x) is Gorenstein, then $K' \in H$ is integral, and hence $Z_k = -K'$.

There are several equivalent definitions for the *genus* $p_g(X, x)$ of the singularity. We may set $p_g(X, x) = \dim H^1(\tilde{X}; \mathcal{O}_{\tilde{X}}) = \dim (R^1\pi_*\mathcal{O}_{\tilde{X}})_x$, or regard $p_g(X, x) = \dim (\Gamma(\tilde{X} - E, \omega_{\tilde{X}})/\Gamma(\tilde{X}, \omega_{\tilde{X}}))$ as the dimension of the space of obstructions to extending a holomorphic 2-form defined on $\tilde{X} - E$ to one defined on the whole of $\tilde{X}$ (note that numerator and denominator in this definition are each separately infinite-dimensional).

The genus can also be regarded as the first in a sequence of invariants, the *plurigenera*. There are several variations of the definition (see e.g. [**22**]). Consider

$$\dim \Big(\Gamma(\tilde{X} - E, \mathcal{O}(mK_{\tilde{X}})) \Big/ \Gamma(\tilde{X}, \mathcal{O}(mK_{\tilde{X}} + rE))\Big)$$

Taking $r = m - 1$ we obtain the L^2-plurigenus $P^l_m(X, x)$ (see particularly [**47**]), and with $r = m$ the log-plurigenus $P^g_m(X, x)$ (it is natural to take $r = 0$, but this does not give good results). Following (as we have done with the other invariants) the standard development for algebraic varieties, we extract invariants that measure the rate at which P_m tends to infinity. For $b = l$ or g we define the *Kodaira dimension* $\kappa^b(X, x)$ as $\lim\sup_{m\to\infty} \log P^b_m / \log m$, where $\log 0$ is interpreted as $-\infty$. Thus $\kappa^b(X, x)$ is $-\infty$ if all P^b_m vanish, and is 0 if not, but they are bounded. For normal surface singularities, $\kappa^l(X, x)$ takes only the values $\{-\infty, 0, 2\}$ and $\kappa^g(X, x)$ the values $\{-\infty, 2\}$.

One may also measure the growth of P^b_m by the invariant

$$C^b := \lim\sup P^b_m/m^2 .$$

Thus $C^b = 0$ unless $\kappa^b = 2$. It was shown by Morales [**21**] that $C^g = -\frac{1}{2}K^2$ (for the minimal resolution), and by Wahl [**45**] that $C^l = -\frac{1}{2}P^2$.

Finally, the definitions of classes of singularities important for the classification theory of algebraic varieties specialise to surfaces as follows (see e.g. [**14**, §2.3]). For a normal surface singularity (X, x), $K' = \sum e_i E_i$ — or sometimes $\min e_i$ — is called the *discrepancy*. Now (X, x) is said to be

terminal if each $e_i > 0$; *canonical* if each $e_i \geq 0$;
log terminal if each $e_i > -1$; and *log canonical* if each $e_i \geq -1$.

5. Characterisations of classes of singularities

5.1. Rational singularities. The singularity (X, x) is said to be *rational* if $p_g(X, x) = 0$. This condition was defined by du Val [**40**] and developed by Artin [**2**] and [**3**]. Artin showed that it was equivalent to the condition $\chi(\mathcal{O}_{Z_{num}}) = 1$ (we always have $\chi(\mathcal{O}_{Z_{num}}) \leq 1$). Artin also showed that within this class, the embedding dimension and multiplicity could be computed topologically: $ed\,(X,x) = 1 - Z^2_{num}$ and $mult\,(X,x) = -Z^2_{num}$.

A singularity is rational and Gorenstein if and only if it is an ADE singularity; these are also the only canonical singularities (and a terminal surface singularity is, in fact, non-singular). This class of singularities has a long history: algebraic geometers often call them du Val singularities after [**40**]; they may also be characterised as rational singularities of multiplicity 2, and so are sometimes called rational double points. Singularity theorists call them simple singularities after [**1**]. Further equivalent conditions are $K^2 = 0$ and $\kappa^g(X,x) = -\infty$.

We have $\kappa^l(X,x) = -\infty$ if and only if (X,x) is a quotient singularity. All such singularities are rational. They were first studied by Brieskorn [**4**]. Equivalent conditions are that $\pi_1(\partial X)$ is finite, and that X is quasihomogeneous, with ∂X having geometric structure of S^3 type: an enumeration is easily derived from this. A singularity is log terminal if and only if it is a quotient singularity [**14**, 4.18].

Since rational singularities may be characterised topologically they may in principle be enumerated by combinatorial arguments. In each case, Γ is a tree, and all components E_i are rational curves. Enumerations were given for rational singularities of multiplicity 3 by Artin [**2**], multiplicity 4 by Stevens [**38**] and multiplicity 5 by Tosun [**39**] (those of multiplicity 2 are the ADE singularities).

The sandwiched singularities defined by Spivakovsky [**36**] as singularities which can be mapped onto a smooth surface are also rational singularities of some importance.

5.2. Elliptic singularities. Following Wagreich [**41**] many authors use the term 'elliptic' to refer to singularities with $\chi(\mathcal{O}_{Z_{num}}) = 0$. However we will reserve the term for the singularities with genus $p_g(X,x) = 1$: $p_g(X,x) = 1$ implies $\chi(\mathcal{O}_{Z_{num}}) = 0$, but not conversely. The condition $p_g(X,x) = 1$ cannot be characterised topologically.

Elliptic Gorenstein singularities were studied by Laufer [**16**] and Reid [**32**], who (each) characterised them topologically by the condition

$$\chi(\mathcal{O}_{Z_{num}}) = 0, \text{ and } 0 < D < Z_{num} \text{ implies } \chi(\mathcal{O}_D) > 0.$$

In view of this latter condition, they are often called 'minimally elliptic singularities'. Both authors showed that the embedding dimension and multiplicity are topological invariants for this class, and established Theorem 5.2 for this case.

A different, but also important class is given by

PROPOSITION 5.1. *The following conditions on a normal surface singularity are equivalent, and imply that $p_g(X,x) \leq 1$: (X,x) is log canonical; $\pi_1(L)$ is solvable or finite; (X,x) is either smooth, simple elliptic, or a cusp singularity or a finite quotient of one of these; $\kappa^l(X,x) \leq 0$; $P^2 = 0$.*

Singularities with solvable or finite fundamental group were studied by Wagreich [**42**], who enumerated the graphs Γ. One may deduce directly that in each case some finite cover is either smooth, simple elliptic, or a cusp singularity. The equivalence of this to log canonical was proved by Kawamata [**13**] (see also [**14**, 4.1]), who again enumerated the graphs. If (X, x) is a finite quotient of a simple elliptic singularity, ∂X is geometric of type Nil; a classification may be obtained directly via listing orbifolds with $\chi = 0$. The only possible type of covering giving a cusp singularity is a double covering; these too are easily listed (reversal of order in **b** must give the same sequence up to cyclic rearrangement). Equivalence of these conditions to $\kappa^l(X, x) \leq 0$ is given by Ishii [**12**].

5.3. Other singularities. Although there exists an extensive literature, and many classifications (particularly for singularities with $\chi(\mathcal{O}_{Z_{num}}) = 0$), very few general results are known beyond the above classes.

If $\kappa^l(X, x) > 0$, then $\kappa^l(X, x) = 2$; $\kappa^g(X, x)$ takes only the values $-\infty$ and 2; so neither of these invariants yields further information.

A Gorenstein surface with $p_g(X, x) = 2$ satisfies $\chi(\mathcal{O}_{Z_{num}}) = 0$. However, this class is not topologically defined.

The following recent result is the best currently known on the questions we have mentioned.

THEOREM 5.2. [**24**] *Suppose (X, x) a Gorenstein surface singularity such that $\chi(\mathcal{O}_{Z_{num}}) = 0$ and $\beta_1(\tilde{X}) = 0$. Then*
(i) there is a topological way to calculate $p_g(X, x)$ (it is the length of the 'elliptic sequence' of divisors);
(ii) $ed\,(X, x) = \max(3, -Z^2_{num})$;
(iii) If $Z^2_{num} \leq -3$, then $\dim(\mathfrak{m}^k/\mathfrak{m}^{k+1}) = -kZ^2_{num}$ for all $k \geq 1$, (there are also formulae for $-Z^2_{num} = 1, 2$) and hence
(iv) if $Z^2_{num} \leq -2$, then mult $(X, x) = -Z^2_{num}$.
(v) (X, x) is an ICIS if and only if $-Z^2_{num} \leq 4$.

6. Smoothings

If $f(x, y, z) = 0$ defines an IHS, then nearby surfaces $f(x, y, z) = \eta$ are smooth (at least, in a neighbourhood of O). If we intersect with the ball $B_\epsilon : |x|^2 + |y|^2 + |z|^2 \leq \epsilon$, where $\eta << \epsilon << 1$), and set $D^*_\eta := \{t | 0 < |t| \leq \eta\}$, then according to Milnor [**20**] f induces a fibration $B_\epsilon \cap f^{-1} D^*_\eta \to D^*_\eta$. The fibres of this are called *Milnor fibres*; we write M for a typical one. Milnor also showed that M is simply-connected. For the restriction to the boundary ∂B_ϵ, there is no need to exclude the value $f = 0$, and we find that ∂M can be identified with L.

We thus have two 4 (real) dimensional manifolds with boundary L: $\tilde{X}$ and M. We will see that usually they are very different, and will seek to compare them.

A slight variation of the argument also goes through for the case of an ICIS. Here, as in the IHS case, M is simply-connected.

More generally, for any normal surface singularity (X_0, x), where we suppose (as usual) X_0 chosen as a good neighbourhood, a deformation consists of a flat, proper map $\pi : \mathcal{M} \to S$, which we may suppose topologically trivial on the boundary, of complex spaces, such that $O \in S$ and $\pi^{-1}(O) \cong X_0$. It can be shown that there always exists a deformation which is universal in an appropriate sense. The base space S of this universal deformation is in general reducible (see Example 1 below). A component S_i of S is called a *smoothing component* if, for generic $v \in S_i$, $M_v := \pi^{-1}(v)$ is smooth, and such a fibre is called a *smoothing* of (X_0, x). The topology of M_v does not depend on the choice of $v \in S_i$ (provided v avoids a suitable discriminant variety), but may well depend on the choice of component S_i (see Example 1 below). We shall see that there may be no smoothing components, but Artin showed that if (X_0, x) is rational, there is one; indeed in this ca e, all components are smoothing components. If (X_0, x) is an ICIS, then S is smooth, of dimension τ, say.

Example 1 (Pinkham [**30**]). Let $X = X_{4,1}$: X is a cone on a rational normal quartic curve, and is parametrised by $(x^4, x^3y, x^2y^2, xy^3, y^4)$.

We may write the equations defining X as the conditions that the matrix $\begin{pmatrix} z_0 & z_1 & z_2 & z_3 \\ z_1 & z_2 & z_3 & z_4 \end{pmatrix}$ has rank ≤ 1. Introduce parameters t_1, t_2, t_3 and replace the lower row by $(z_1 + t_1, z_2 + t_2, z_3 + t_3, z_4)$. This defines a family $\{M_t\}$ with base space S_1 having three coordinates (t_1, t_2, t_3).

Secondly, we write the equations using a 3×3 matrix, which we deform to

$$\begin{pmatrix} z_0 & z_1 & z_2 \\ z_1 & z_2 + u_2 & z_3 \\ z_2 & z_3 & z_4 \end{pmatrix}.$$

Here we have one parameter u_2 giving the coordinate on the base space S_2. Again writing the conditions that the rank of the matrix is 1 defines a deformation $\{M_u\}$ of X, which is smooth for generic values of the parameter.

The deformation space S is the union of the 3-parameter space S_1 and the 1-parameter space S_2, meeting in a single point. Each of S_1 and S_2 is a smoothing component. For S_1, the Betti number of the fibre $\beta_2(M) = 1$; for S_2, $\beta_2(M) = 0$.

We have the following interesting characterisation.

THEOREM 6.1. *(Wahl* [**44**]*) For (X, x) an ICIS, $\beta_2(M) \geq \tau$, with equality if and only if (X, x) is quasihomogeneous.*

Suppose M the fibre of a smoothing of (X, x); we may, as above, suppose $\partial M \cong L$. Consider the quadratic form of intersection numbers on $H_2(M)$. This is no longer in general negative definite (it is if (X, x) is rational). Diagonalise the form over $\mathbb{R}$ on $H_2(M; \mathbb{R})$, and write μ_-, μ_0 and μ_+ for the respective numbers of negative, zero and positive terms, so that $\mu_- + \mu_0 + \mu_+ = \mu = \dim H_2(M; \mathbb{R})$. We concentrate on these invariants. Write also σ for the signature $\mu_+ - \mu_-$.

Elementary homology calculation shows that $\mu_0 = \beta_1(L) - \beta_1(M)$. But in [**9**] we find the nontrivial result:

THEOREM 6.2. *For M a smoothing of a normal surface singularity, $\beta_1(M) = 0$.*

Thus $\mu_0 = \beta_1(L)$ is determined by the topology of the singularity. Also, $\chi(M) = 1 + \beta_2(M) = 1 + \mu$. Now we have

THEOREM 6.3. *The genus is given by $\mu_0 + \mu_+ = 2p_g$.*

This was established by Durfee [**5**] in the case of IHS and, following some partial results, proved for arbitrary smoothings of normal surface singularities by Steenbrink [**37**].

Thus the quadratic form on $H_2(M)$ is negative definite if and only if the singularity is rational (we recall that the form on $H_2(\tilde{X})$ is always negative definite). Moreover, (X, x) (though not just the topology of (X, x)) determines μ_+ as well as μ_0. It follows from Example 1 that it does not always determine μ_-. However there are several essentially equivalent formulae for μ.

First, for the case of IHS, Durfee [**5**] established the formula (1), and Laufer [**17**] proved (2).

$$3\sigma + 2\mu + 2\beta_1(L) + K^2 + \beta_2(\tilde{X}) = 0. \tag{1}$$

$$12p_g - \mu - \beta_1(L) + K^2 + \beta_2(\tilde{X}) = 0. \tag{2}$$

Subtracting (2) from (1) and dividing by 3 gives $\sigma + \mu + \beta_1(L) = 4p_g$. Using $\beta_1(L) = \mu_0$ and $\sigma = \mu_+ - \mu_-$, this reduces to the assertion of Theorem 6.3.

Then Looijenga and Wahl proved in [**19**] that (3) holds, and reduces to (4), for arbitrary smoothings of normal surface singularities.

$$(K_M^2 - 2\chi(M)) - \sigma(M) = (K_{\tilde{X}}^2 - 2\chi(\tilde{X})) - 3\sigma(\tilde{X}). \tag{3}$$

$$c_1^2(M) + \chi(M) = c_1^2(\tilde{X}) + \chi(\tilde{X}) + 12p_g. \tag{4}$$

Substituting $\chi(M) = 1 + \mu$, $c_1(\tilde{X}) = K$, and $\chi(\tilde{X}) = 1 - \beta_1(\tilde{X}) + \beta_2(\tilde{X})$ with $\beta_1(\tilde{X}) = \beta_1(L)$, we see that (4) reduces to

$$c_1^2(M) + \mu = K^2 - \beta_1(L) + \beta_2(\tilde{X}) + 12p_g, \tag{5}$$

which coincides with (2) apart from the term $c_1^2(M)$. Now we have

LEMMA 6.4. *(Seade [**35**]) If (X, x) is Gorenstein, then M admits a nowhere 0 holomorphic 2-form, so $c_1(M) = 0$.*

Hence (2) holds for all smoothings of normal Gorenstein singularities (X, x) — in fact, this result is due to Steenbrink [**37**] — in particular, (X, x) determines μ and hence also μ_- in these cases.

For any normal Gorenstein singularity we may use these formulae to calculate a value of μ_-. If this value turns out to be negative, there cannot exist a smoothing. This conclusion is already non-trivial — for example, it implies that if a simple-elliptic singularity with $Z^2_{num} = -D$ is smoothable, we must have $D \leq 9$, which is best possible. For smoothability of a cusp singularity it yields the necessary condition (in the above notation) $\Delta \leq \Delta^* + 9$.

Example 2 (Laufer) The IHS $x^2 + y^7 + z^{14} = 0$ and $x^3 + y^4 + z^{12} = 0$ have the same topological type, with Γ consisting of only one vertex, with $g = 3$ and $E^2 = -1$ (so $\mu_0 = \beta_1(L) = 6$). But the invariants $(\mu_-, \mu_0, \mu_+, p_g)$ take the values $(60, 6, 12, 9)$ in the first case and (50,6,10,8) in the second. Moreover, the singularities have different multiplicities.

Now let M be a simply-connected 4-manifold with boundary L such that $\beta_1(L) = 0$. The quadratic form of intersection numbers on $H_2(M;\mathbb{Z})$ is non-degenerate, so we can identify the dual lattice in $H_2(M;\mathbb{Q})$ with $H^2(M;\mathbb{Z})$, and the quotient $H^2(M;\mathbb{Z})/H_2(M;\mathbb{Z})$ with $H_1(L;\mathbb{Z})$. The bilinear pairing of $H_2(M;\mathbb{Q})$ with itself to $\mathbb{Q}$ (the *discriminant bilinear form*) thus induces a bilinear pairing of $H_1(L;\mathbb{Z})$ with itself to $\mathbb{Q}/\mathbb{Z}$: the *linking pairing*. For a general 4-manifold we obtain a non-degenerate quadratic form on the quotient $\overline{H_2}(M;\mathbb{Z})$ of $H_2(M;\mathbb{Z})$ by the torsion subgroup and the radical of the intersection form, and this induces a linking pairing on the torsion subgroup $H_1(L;\mathbb{Z})_t$ of $H_1(L;\mathbb{Z})$.

If (X, x) is a normal Gorenstein singularity, and M a smoothing, we may apply the above considerations to both M and $\tilde{X}$, each having boundary L. The discriminant bilinear forms of the intersection forms on $\overline{H_2}(M;\mathbb{Z})$ and $\overline{H_2}(\tilde{X};\mathbb{Z})$ must both coincide with the linking pairing on $H_1(L;\mathbb{Z})_t$. Thus the singularity (X, x) determines the discriminant bilinear form of the intersection form on $\overline{H_2}(M;\mathbb{Z})$ as well as μ_-, μ_0 and μ_+. Thus a necessary condition for the existence of a smoothing of (X, x) is the existence of a quadratic form over $\mathbb{Z}$, with the given values of μ_-, μ_0, μ_+ and the given discriminant bilinear form: this is already quite a strong condition.

This condition was enhanced by Looijenga and Wahl [**19**], with a very careful presentation of the details. If M is simply-connected and $\beta_1(L) = 0$ and moreover the intersection pairing on $H_2(M;\mathbb{Z})$ is even, the linking pairing on $H_1(L;\mathbb{Z})$ can be enhanced to a quadratic map to $\mathbb{Q}/2\mathbb{Z}$ (the *discriminant quadratic form* of the quadratic form on $H_2(M;\mathbb{Z})$). A closer analysis shows that such an enhancement can be defined whenever M is almost-complex, and that the discriminant quadratic form for L is determined (independently of M) by the almost complex structure on $\tau_L \oplus \mathbb{R}$, provided only that c_1 gives a torsion class in $H_1(L)$.

Matters are complicated by the possibility that $H_1(M;\mathbb{Z})$ contains non-trivial torsion. However, the possibilities for the torsion are bounded by the torsion in $H_1(L;\mathbb{Z})$. Eventually, it is shown in [**19**] that if we set $A := H_1(L;\mathbb{Z})$ and write $q : A_t \to \mathbb{Q}/\mathbb{Z}$ for the form that has been constructed, a smoothing component induces an (isomorphism class of) quintuple (V, Q, I, i, s) where (V, Q) is an ordinary lattice; s an orientation on a maximal positive semidefinite subspace; I is an isotropic subspace of A_t; and

$i : V^*/V \to A/I$ an injective homomorphism with finite cokernel inducing an isomorphism $\overline{V^*/V} \to I^\perp/I$ of quadratic groups.

Work of Nikulin [**29**] gives (in most cases) necessary and sufficient conditions for the existence of a quadratic form with assigned signature and discriminant form. Using this, Looijenga and Wahl essentially determine the set $\mathcal{S}$ of isomorphism classes of the relevant quintuples for the problems, which triangle singularities, or which cusp singularities are smoothable?

These methods do not address the converse problem, of finding smoothings, for which global methods seem more effective. For triangle singularities work of Looijenga [**18**] using the structure of the moduli space for K3 surfaces shows in many cases that the set of smoothing components maps bijectively to $\mathcal{S}$. The situation for cusp singularities is similar but less clear; the relevant references here are [**6**] and [**7**].

7. Two conjectures

As a result of a number of calculations, Durfee [**5**] was led to the conjecture that for all IHS, and probably for all ICIS (where there is still a canonical smoothing) $\sigma \leq 0$ and to the sharper conjecture that $p_g \leq \mu/6$ (which is equivalent to $\sigma \leq -\frac{1}{3}\mu - \mu_0$). Numerous partial results are known. Stronger inequalities are available if $\chi(\mathcal{O}_{Z_{num}}) \geq 0$, or if we have an IHS with equation of the form $g(x, y) + z^2 = 0$: see the review in [**25**]. The most general result to date is due to Némethi [**25**]: the conjecture holds for all IHS with equation of the form $g(x, y) + z^n = 0$ provided that g is irreducible. The verification depends on direct calculations of both sides.

The following example shows that the inequality does not hold for all smoothings of normal surface singularities.

Example 3 (Wahl [**43**]) Consider the IHS (Y, O) defined by $z_1z_2^3 + z_2z_3^3 + z_3z_1^3 = 0$. The formula $T.(z_1, z_2, z_3) = (\epsilon_7 z_1, \epsilon_7^2 z_2, \epsilon_7^4 z_3)$ defines an action of a group G of order 7; the quotient $(X, x) = (Y, O)/G$ is a triangle singularity corresponding to the triple $(6, 6, 6)$. There is a smoothing obtained by factoring out G from an equivariant smoothing of (Y, O). It has $(\mu_-, \mu_0, \mu_+) = (1, 0, 2)$, so $\sigma = 1 > 0$.

Again arising out of experimental verifications, it was conjectured by Neumann and Wahl [**28**] that if (X, x) is an ICIS such that $H_1(L; \mathbb{Z}) = 0$, so L is a homology sphere over $\mathbb{Z}$, the Casson invariant $\lambda(L)$ is equal to $\frac{1}{8}\sigma$. They verified the conjecture if (X, x) is weighted homogeneous, or an IHS with equation of the form $g(x, y) + z^n = 0$, and in a few other cases.

Perhaps the conjecture holds for any smoothable Gorenstein singularity; perhaps it can be extended to the case when L is a rational homology sphere (i.e. $\beta_1(L) = 0$); but the evidence for any such extension of the conjecture is very weak.

The cases that have been verified again depend on direct evaluations of both sides (using, e.g. Dedekind sums ...).

References

[1] V.I. Arnold, Critical points of smooth functions and their normal forms, Russian Math Surveys **30v** (1975) 1–75. Reprinted in *Singularity Theory*, London Math. Soc. Lecture Notes **53** (1981) 132–206.

[2] M. Artin, Some numerical criteria for contractibility of curves on algebraic surfaces, Amer. J. Math. **84** (1962) 485–496.

[3] M. Artin On isolated rational singularities of surfaces, Amer. J. Math. **88** (1966) 129–136.

[4] E. Brieskorn, Rationale Singularitäten komplexer Flächen, Invent. Math. **4** (1968) 336–358.

[5] A. Durfee, The signature of smoothings of complex surface singularities, Math. Ann. **232** (1978) 85–98.

[6] R. Friedman and R. Miranda, Smoothing cusp singularities of small length, Math. Ann. **263** (1983) 185–212.

[7] R. Friedman and H. Pinkham, Smoothings of cusp singularities via triangle singularities, Comp. Math. **53** (1984) 303–316.

[8] H. Grauert, Über Modifikationen und exzeptionelle analytischen Menge, Math. Ann. **146** (1962) 331–368.

[9] G.-M. Greuel and J.H.M. Steenbrink, On the topology of smoothable singularities, pp 535–545 in *Singularities* (ed. P. Orlik), Proc. Symp. in Pure Math. **40**, part I, Amer. Math. Soc., 1983.

[10] M.W. Hirsch, Smooth regular neighbourhoods, Ann. of Math. **76** (1962) 524–530.

[11] F. Hirzebruch, Hilbert modular surfaces, l'Enseignement Math. **19** (1973) 183–281.

[12] S. Ishii, Two dimensional singularities with bounded plurigenera δ_m are $\mathbb{Q}$-Gorenstein singularities, *Singularities* (ed. R. Randell), Contemp. Math. **90** (1989) 135–145.

[13] Y. Kawamata, Crepant blowing-up of 3-dimensional canonical singularities and its application to degenerations of surfaces, Ann. of Math. **127** (1988) 93–164.

[14] J. Kollár and S. Mori, *Birational geometry of algebraic varieties*, Cambridge University Press, 1998.

[15] H. Laufer, *Normal two-dimensional singularities*, Ann. of Math. Study **71**, Princeton Univ. Press, 1971.

[16] H. Laufer, On minimally elliptic singularities, Amer. Jour. Math. **99** (1977) 1257–1295.

[17] H. Laufer, On μ for surface singularities, pp 45–49 in Proc. Symp in Pure Math. **30**, Amer. Math. Soc., 1977.

[18] E.J.N. Looijenga, The smoothing components of a triangle singularity II, Math. Ann. **269** (1984) 357–387.

[19] E.J.N. Looijenga and J. Wahl, Quadratic functions and smoothing surface singularities, Topology **25** (1986) 261–291.

[20] J.W. Milnor, *Singular points of complex hypersurfaces*, Ann. of Math. Study **91**, Princeton University Press, 1968.

[21] M. Morales, Calcul de quelques invariants des singularitès de surfaces normales, Monographes de l'Enseignement Math. **31** (1983) 191–203.

[22] M. Morales, Resolution of quasi-homogeneous singularities and plurigenera, Compositio Math. **64** (1987) 311–327.

[23] D. Mumford, The topology of normal singularities of an algebraic surface and a criterion for simplicity, Publ. Math. IHES **9** (1961) 5–22.

[24] A. Némethi, 'Weakly' elliptic Gorenstein singularities of surfaces, Invent. Math. **137** (1999) 145–167.

[25] A. Némethi, The signature of $f(x, y) + z^N$, pp 131-150 in *Singularity Theory* (eds. J.W. Bruce and D. Mond) London Math. Soc. Lecture Notes **263** (1999).

[26] W.D. Neumann, A calculus for plumbing applied to the topology of complex surface singularities and degenerating complex curves, Trans. Amer. Math. Soc. **268** (1981) 299–344.

[27] W.D. Neumann, Geometry of quasihomogeneous surface singularities, pp 245–258 in *Singularities* (ed. P. Orlik), Proc. Symp. in Pure Math. **40**, part II, Amer. Math. Soc., 1983.

[28] W.D. Neumann and J. Wahl, Casson invariant of links of singularities, Comment. Math. Helvet. **65** (1990) 58–78.

[29] V.V. Nikulin, Integral symmetric bilinear forms and some of their applications, Math. USSR. Izv. **14i** (1980) 103–167.

[30] H. Pinkham, Deformations of algebraic varieties with G_m-action, Astérisque **20** (1974) 1–131.

[31] H. Pinkham, Deformations of normal singularities with $\mathbb{C}^*$-action, Math. Ann. **232** (1978) 65–84.

[32] M.A. Reid, *Elliptic Gorenstein singularities*, unpublished manuscript, 1975.

[33] K. Saito, Einfach-elliptische Singularitäten, Invent. Math, **14** (1971) 123–142.

[34] F. Sakai, Anticanonical models of rational surfaces, Math. Ann. **269** (1984) 389–410.

[35] J. Seade, A cobordism invariant for surface singularities, pp 479–484 in *Singularities* (ed. P. Orlik), Proc. Symp. in Pure Math. **40**, part II, Amer. Math. Soc., 1983.

[36] M. Spivakovsky, Sandwich singularities and desingularisation of surfaces by normalised Nash transformations, Ann. of Math. **131** (1990) 411–491.

[37] J.H.M. Steenbrink, Mixed Hodge structure associated with isolated singularities, pp 513–536 in *Singularities* (ed. P. Orlik), Proc. Symp. in Pure Math. **40**, part II, Amer. Math. Soc., 1983.

[38] J. Stevens, Partial resolutions of rational quadruple points, Int. J. Math. **2** (1991) 205–221.

[39] M. Tosun, Singularités rationelles des surfaces complexes, Ph.D. thesis, Université de Provence, 1998.

[40] P. du Val, On isolated singularities which do not affect the conditions of adjunction I, Proc. Camb. Phil. Soc. **30** (1934) 453–465.

[41] P. Wagreich, Elliptic singularities of surfaces, Amer. Jour. Math. **92** (1970) 419–454.

[42] P. Wagreich, Singularities of complex surfaces with solvable local fundamental group, Topology **11** (1972) 51–72.

[43] J. Wahl, Smoothings of normal surface singularities, Topology **20** (1981) 219–246.

[44] J. Wahl, A characterisation of quasi-homogeneous Gorenstein surface singularities, Comp. Math. **55** (1985) 269–288.

[45] J. Wahl, A characteristic number for links of surface singularities, Jour. Amer. Math. Soc. **3** (1990) 625–637.

[46] C.T.C. Wall, Notes on the classification of singularities, Proc. London Math. Soc. **48** (1984) 461–513.

[47] K. Watanabe, On plurigenera of normal isolated singularities I, Math. Ann. **250** (1980) 65–94.

DEPT. OF MATHEMATICAL SCIENCES
LIVERPOOL UNIVERSITY
LIVERPOOL L69 3BX, UK

E-mail address: ctcw@liverpool.ac.uk

Selected Titles in This Series

(Continued from the front of this publication)

248 **Naihuan Jing and Kailash C. Misra, Editors,** Recent developments in quantum affine algebras and related topics, 1999

247 **Lawrence Wasson Baggett and David Royal Larson, Editors,** The functional and harmonic analysis of wavelets and frames, 1999

246 **Marcy Barge and Krystyna Kuperberg, Editors,** Geometry and topology in dynamics, 1999

245 **Michael D. Fried, Editor,** Applications of curves over finite fields, 1999

244 **Leovigildo Alonso Tarrío, Ana Jeremías López, and Joseph Lipman,** Studies in duality on noetherian formal schemes and non-noetherian ordinary schemes, 1999

243 **Tsit Yuan Lam and Andy R. Magid, Editors,** Algebra, K-theory, groups, and education, 1999

242 **Bernhelm Booss-Bavnbek and Krzysztof Wojciechowski, Editors,** Geometric aspects of partial differential equations, 1999

241 **Piotr Pragacz, Michał Szurek, and Jarosław Wiśniewski, Editors,** Algebraic geometry: Hirzebruch 70, 1999

240 **Angel Carocca, Víctor González-Aguilera, and Rubí E. Rodríguez, Editors,** Complex geometry of groups, 1999

239 **Jean-Pierre Meyer, Jack Morava, and W. Stephen Wilson, Editors,** Homotopy invariant algebraic structures, 1999

238 **Gui-Qiang Chen and Emmanuele DiBenedetto, Editors,** Nonlinear partial differential equations, 1999

237 **Thomas Branson, Editor,** Spectral problems in geometry and arithmetic, 1999

236 **Bruce C. Berndt and Fritz Gesztesy, Editors,** Continued fractions: From analytic number theory to constructive approximation, 1999

235 **Walter A. Carnielli and Itala M. L. D'Ottaviano, Editors,** Advances in contemporary logic and computer science, 1999

234 **Theodore P. Hill and Christian Houdré, Editors,** Advances in stochastic inequalities, 1999

233 **Hanna Nencka, Editor,** Low dimensional topology, 1999

232 **Krzysztof Jarosz, Editor,** Function spaces, 1999

231 **Michael Farber, Wolfgang Lück, and Shmuel Weinberger, Editors,** Tel Aviv topology conference: Rothenberg Festschrift, 1999

230 **Ezra Getzler and Mikhail Kapranov, Editors,** Higher category theory, 1998

229 **Edward L. Green and Birge Huisgen-Zimmermann, Editors,** Trends in the representation theory of finite dimensional algebras, 1998

228 **Liming Ge, Huaxin Lin, Zhong-Jin Ruan, Dianzhou Zhang, and Shuang Zhang, Editors,** Operator algebras and operator theory, 1999

227 **John McCleary, Editor,** Higher homotopy structures in topology and mathematical physics, 1999

226 **Luis A. Caffarelli and Mario Milman, Editors,** Monge Ampère equation: Applications to geometry and optimization, 1999

225 **Ronald C. Mullin and Gary L. Mullen, Editors,** Finite fields: Theory, applications, and algorithms, 1999

224 **Sang Geun Hahn, Hyo Chul Myung, and Efim Zelmanov, Editors,** Recent progress in algebra, 1999

223 **Bernard Chazelle, Jacob E. Goodman, and Richard Pollack, Editors,** Advances in discrete and computational geometry, 1999